Nutrients and Cell Signaling

OXIDATIVE STRESS AND DISEASE

Series Editors

LESTER PACKER, PH.D.
ENRIQUE CADENAS, M.D., PH.D.

University of Southern California School of Pharmacy
Los Angeles, California

Nutrients and Cell Signaling

edited by

Janos Zempleni
Krishnamurti Dakshinamurti

Taylor & Francis
Taylor & Francis Group

Boca Raton   London   New York   Singapore

A CRC title, part of the Taylor & Francis imprint, a member of the
Taylor & Francis Group, the academic division of T&F Informa plc.

Published in 2005 by
CRC Press
Taylor & Francis Group
6000 Broken Sound Parkway NW, Suite 300
Boca Raton, FL 33487-2742

© 2005 by Taylor & Francis Group, LLC
CRC Press is an imprint of Taylor & Francis Group

No claim to original U.S. Government works
Printed in the United States of America on acid-free paper
10 9 8 7 6 5 4 3 2 1

International Standard Book Number-10: 0-8247-2248-5 (Hardcover)
International Standard Book Number-13: 978-0-8247-2248-7 (Hardcover)
Library of Congress Card Number 2004063455

Library of Congress Cataloging-in-Publication Data

Nutrients and cell signaling / [edited] by Krishnamurti Dakshinamurti and Janos Zempleni.
   p. ; cm. -- (Oxidative stress and disease ; 10)
   Includes bibliographical references and index.
   ISBN 0-8247-2248-5 (alk. paper)
   1. Cellular signal transduction. 2. Nutrient interactions. I. Dakshinamurti, Krishnamurti, 1928- II. Zempleni, Janos. III. Series.
   [DNLM: 1. Cell Communication--physiology. 2. Amino Acids--physiology. 3. Carbohydrates--physiology. 4. Lipids--physiology. 5. Trace Elements--physiology. 6. Vitamins-physiology. W1 OX626 v.10 2005]
   QP517.C45N88 2005
   612.3--dc22   2004063455

Taylor & Francis Group
is the Academic Division of T&F Informa plc.

Visit the Taylor & Francis Web site at
http://www.taylorandfrancis.com

and the CRC Press Web site at
http://www.crcpress.com

Series Introduction

Oxygen is a dangerous friend. Overwhelming evidence indicates that oxidative stress can lead to cell and tissue injury. However, the same free radicals that are generated during oxidative stress are produced during normal metabolism and thus are involved in both human health and disease, with the following characteristics:.

Free radicals are molecules with an odd number of electrons. The odd, or unpaired, electron is highly reactive as it seeks to pair with another free electron.
Free radicals are generated during oxidative metabolism and energy production in the body.
Free radicals are involved in:
    Enzyme-catalyzed reactions
    Electron transport in mitochondria
    Signal transduction and gene expression
    Activation of nuclear transcription factors
    Oxidative damage to molecules, cells, and tissues
    Antimicrobial action of neutrophils and macrophages
    Aging and disease

Normal metabolism is dependent on oxygen, a free radical. Through evolution, oxygen was chosen as the terminal electron acceptor for respiration. The two unpaired electrons of oxygen spin in the same direction; thus, oxygen is a biradical, but is not a very dangerous free radical. Other oxygen-derived free radical species, such as superoxide or hydroxyl radicals, formed during metabolism or by ionizing radiation, are stronger oxidants and are therefore more dangerous.

In addition to research on the biological effects of these reactive oxygen species, research on reactive nitrogen species (ROS) has been gathering momentum. NO, or nitrogen monoxide (nitric oxide), is a free radical generated by NO synthase (NOS). This enzyme modulates physiological responses such as vasodilation or signaling in the brain. However, during inflammation, synthesis of NOS (iNOS) is induced. This iNOS can result in the overproduction of NO, causing damage. More worrisome, however, is the fact that excess NO can react with superoxide to produce the very toxic product peroxynitrite. Oxidation of lipids, proteins, and DNA can result, thereby increasing the likelihood of tissue injury.

Both reactive oxygen and nitrogen species are involved in normal cell regulation in which oxidants and redox status are important in signal transduction. Oxidative stress is increasingly seen as a major upstream component in the signaling cascade involved in inflammatory responses, stimulating adhesion molecule and chemoattractant production. Hydrogen peroxide, which breaks down to produce hydroxyl radicals, can also activate NF-kB, a transcription factor involved in stimulating inflammatory responses. Excess production of these reactive species is toxic, exerting cytostatic effects, causing membrane damage, and activating pathways of cell death (apoptosis or necrosis).

Virtually all diseases thus far examined involve free radicals. In most cases, free radicals are secondary to the disease process, but in some instances free radicals are causal. Thus, there is a delicate balance between oxidants and antioxidants in health and disease. Their proper balance is essential for ensuring healthy aging.

The term *oxidative stress* indicates that the antioxidant status of cells and tissues is altered by exposure to oxidants. The redox status is thus dependent on the degree to which a cell's components are in the oxidized state. In general, the reducing environment inside cells helps to prevent oxidative damage. In this reducing environment, disulfide bonds (S–S) do not spontaneously form because sulfhydryl groups kept in the reduced state (SH) prevent protein misfolding or aggregation. This reducing environment is maintained by oxidative metabolism and by the action of antioxidant enzymes and substances, such as glutathione, thioredoxin, vitamins E and C, and enzymes such as superoxide dismutase (SOD), catalase, and the selenium-dependent glutathione and thioredoxin hydroperoxidases, which serve to remove reactive oxygen species.

Changes in the redox status and depletion of antioxidants occur during oxidative stress. The thiol redox status is a useful index of oxidative stress mainly because metabolism and NADPH-dependent enzymes maintain cell glutathione (GSH) almost completely in its reduced state. Oxidized glutathione (glutathione disulfide, GSSG) accumulates under conditions of oxidant exposure, and this changes the ratio of oxidized to reduced glutathione; an increased ratio indicates oxidative stress. Many tissues contain large amounts of glutathione, 2–4 mM in erythrocytes or neural tissues and up to 8 mM in hepatic tissues. Reactive oxygen and nitrogen species can directly react with glutathione to lower the levels of this substance, the cell's primary preventative antioxidant.

Current hypotheses favor the idea that lowering oxidative stress can have a clinical benefit. Free radicals can be overproduced or the natural antioxidant system defenses weakened, first resulting in oxidative stress, and then leading to oxidative injury and disease. Examples of this process include heart disease and cancer. Oxidation of human low-density lipoproteins is considered the first step in the progression and eventual development of atherosclerosis, leading to cardiovascular disease. Oxidative DNA damage initiates carcinogenesis.

Compelling support for the involvement of free radicals in disease development comes from epidemiological studies showing that an enhanced antioxidant status is associated with reduced risk of several diseases. The association of vitamin E and prevention of cardiovascular disease is a notable example. Elevated antioxidant status is also associated with decreased incidence of cataracts and cancer, and some recent reports have suggested an inverse correlation between antioxidant status and occurrence of rheumatoid arthritis and diabetes mellitus. Indeed, the number of indications in which antioxidants may be useful in the prevention or the treatment of disease is increasing.

Oxidative stress, rather than being the primary cause of disease, is more often a secondary complication in many disorders. Oxidative stress diseases include inflammatory bowel diseases, retinal ischemia, cardiovascular disease and restenosis, AIDS, Acute Respiratory Distress Syndrome, and neurodegenerative diseases such as stroke, Parkinson's disease, and Alzheimer's disease. Such indications may prove amenable to antioxidant treatment because there is a clear involvement of oxidative injury in these disorders.

In this series of books, the importance of oxidative stress in diseases associated with organ systems of the body is highlighted by exploring the scientific evidence and the medical applications of this knowledge. The series also highlights the major natural antioxidant enzymes and antioxidant substances, such as vitamins E, A, and C, flavonoids, polyphenols, carotenoids, lipoic acid, and other nutrients present in foods and beverages.

Oxidative stress is an underlying factor in health and disease. More and more evidence indicates that a proper balance between oxidants and antioxidants is involved in maintaining health and longevity and that altering this balance in favor of oxidants may result in pathological responses causing functional disorders and disease. This series is intended for researchers in the basic biomedical sciences and clinicians. The potential for healthy aging and disease prevention necessitates gaining further knowledge about how oxidants and antioxidants affect biological systems.

Lester Packer
Enrique Cadenas

Preface

Living organisms constantly receive and interpret signals from the environment. In unicellular organisms the inputs are mainly from nutrient availability and the presence of noxious stimuli. In multicellular organisms, the cumulative action of signals from other cells in their proximity as well as cells at a distance are required in addition to inputs about nutrient availability, for the normal functioning of the cell, which includes cell growth, differentiation, and maintenance of homeostasis. In addition to external signals, cells derive internal signals based on their energy and metabolic status. Thus, cell signaling is at the core of most biological processes. The etiology of many disease processes has been traced to aberrant cell signaling. In the post-genomic era, the understanding of how these complex biomolecular net works function as dynamic systems in controlling cell function has become a vibrant area of research.

Endocrine, paracrine, and autocrine signals have specific sites of action. Contact-dependent signaling requires cells to be in direct membrane-to-membrane contact for signals to be transferred from one cell to another. Signal molecules may trigger immediate changes in the metabolism of the cell, changes in gene expression, or immediate changes in electrical charge across the plasma membrane. Signaling involves a receptor that recognizes the signal with resultant effect at the plasma membrane or the transduction of the signal information to the cell interior and subcellular systems that respond to the signal received. The receptors are proteins with specific binding sites for the signaling molecule, the ligand. Membrane soluble ligands include the steroid hormones, which bind to cytoplasmic receptors. Other signaling molecules — peptides, proteins, gases, ions, and nucleotides — bind to integral membrane proteins. Receptors are of several types and include gated ion channels and enzyme-linked receptors that activate protein kinases. Ligand binding to receptors also results in production of second messengers, cAMP, cGMP, IP3, Ca^{2+}, DAG, and NO. Activation involves protein phosphorylation cascades, a series of amplification steps resulting in a rapid response.

The focus of this book is the delineation of the role of nutrients in cell signaling. Nutrients comprise amino acids, lipids, carbohydrates, vitamins, and minerals. That nutrients regulate the hormonal status and thus affect metabolic processes is well known. However, the direct roles of nutrients in regulating signal cascades, transcription, redox status, and intracellular calcium levels are now being recognized. Metabolic regulatory processes in turn send signals regulating nutrient intake. We are beginning to understand many aspects of these interactions. Signaling pathways used by cells to regulate transcriptional/translational response to nutrient signals are evolutionarily conserved. For example, the TOR (target of rapamycin) pathway coordinates nutrient availability with cell growth and proliferation. In higher organisms, TOR function is also linked to hormone-dependent mitogenic signaling.

As editors, we sought to emphasize broad concepts rather than isolated facts whenever possible. We have assembled an array of leading experts in nutrient-dependent cell signaling to cover as broad an area as possible. This text provides select examples from the following areas of nutrient-dependent cell signaling: nuclear receptors; transcription factors and signaling cascades; amino acids, lipids and glycation; insulin release, signaling and insulin resistance; calcium-dependent signaling; and feeding and nutrient homeostasis. We hope that the chapters presented here will expand the horizons of researchers, officials, administrators, and graduate students interested in nutrition and cell signaling.

We express our gratitude to our colleagues who have created time in their busy schedules to contribute chapters to this text. Also, we thank our families, who bore with us during the preparation of this book.

Janos Zempleni
Krishnamurti Dakshinamurti

Contributors

Bernard Beck
Centre de Recherches UHP - EA 3453 Systèmes Neuromodulateurs des Comportements Ingestifs
Université Henri Poincaré
Nancy, France

Luc Bertrand
Hormone and Metabolic Research Unit
University of Louvain Medical School
Louvain-a-Neuve, Belgium
and
Institute of Cellular Pathology
Louvain, Belgium

Edward F. C. Blommaart
Department of Medical Biochemistry
Academic Medical Center
University of Amsterdam
The Netherlands

Roger F. Butterworth
Neuroscience Research Unit
Hôpital Saint-Luc
University of Montreal
Montreal, Canada

Gabriela Camporeale
Department of Nutrition and Health Sciences
University of Nebraska
Lincoln, Nebraska, USA

Helen Chan
Neuroscience Research Unit
Hôpital Saint-Luc
University of Montreal
Montreal, Canada

Jie Chen
Department of Cell and Structural Biology
University of Illinois
Urbana, Illinois, USA

Patrice Codogno
Glycobiology and Cell Signaling Unit
INSERM U504
Paris, France

Michael P. Czubryt
Division of Stroke and Vascular Disease
National Centre for Agri-Food Research in Medicine
St. Boniface Hospital Research Centre
Faculty of Medicine
University of Manitoba
Winnipeg, Manitoba, Canada

Krishnamurti Dakshinamurti
Department of Biochemistry and Medical Genetics
University of Manitoba
Winnipeg, Manitoba, Canada

Shyamala Dakshinamurti
Department of Pediatrics and Child Health
University of Manitoba
Winnipeg, Manitoba, Canada

Peter F. Dubbelhuis
Department of Medical Biochemistry
Academic Medical Center
University of Amsterdam
The Netherlands

Ebru Erbay
Department of Cell and Structural Biology
University of Illinois
Urbana, Illinois, USA

Randolph S. Faustino
Division of Stroke and Vascular Disease
National Centre for Agri-Food Research in Medicine
St. Boniface Hospital Research Centre
Faculty of Medicine
University of Manitoba
Winnipeg, Manitoba, Canada

Nick E. Flynn
Department of Chemistry and Biochemistry
Angelo State University
San Angelo, Texas, USA

Nicole T. Gavel
Division of Stroke and Vascular Disease
National Centre for Agri-Food Research in Medicine
St. Boniface Hospital Research Centre
Faculty of Medicine
University of Manitoba
Winnipeg, Manitoba, Canada

Jonathan Geiger
Department of Pharmacology, Physiology, and Therapeutics
University of North Dakota Medical School
Fargo, North Dakota, USA

Annapaula Giuletti
Laboratorium voor Experimentele Geneeskunde en Endocrinologie
Katholieke Universiteit Leuven
Leuven, Belgium

Vadim N. Gladyshev
Department of Biochemistry
University of Nebraska
Lincoln, Nebraska, USA

Conny Gysemans
Laboratorium voor Experimentele Geneeskunde en Endocrinologie
Katholieke Universiteit Leuven
Leuven, Belgium

Jason C. G. Halford
School of Psychology
University of Liverpool
Liverpool, United Kingdom

John A. Hanover
Laboratory of Cell Biochemistry and Biology
NIDDK at the National Institutes of Health
Bethesda, Maryland, USA

Tony E. Haynes
Faculty of Nutrition
Texas A&M University
College Station, Texas, USA

Raymond J. Hohl
Department of Pharmacology
University of Iowa
Iowa City, Iowa, USA

Sarah A. Holstein
Department of Pharmacology
University of Iowa
Iowa City, Iowa, USA

Louis Hue
Hormone and Metabolic Research Unit
University of Louvain Medical School
Louvain-a-Neuve, Belgium
and
Institute of Cellular Pathology
Louvain, Belgium

Miriam N. Jacobs
School of Biomedical and Molecular
Sciences
University of Surrey
Guildford, United Kingdom

Timothy J. Kieffer
Departments of Physiology and Surgery
Laboratory of Molecular and Cellular Medicine
University of British Columbia
Vancouver, British Columbia, Canada

Jae Eun Kim
Department of Cell and Structural Biology
University of Illinois
Urbana, Illinois, USA

Mitsuhisa Komatsu
Department of Aging Medicine and Geriatrics
Shinshu University Graduate School of Medicine
Matsumoto, Japan

Nagarama Kothapalli
Department of Biochemistry
University of Nebraska
Lincoln, Nebraska, USA

Melanie N. Landry
Division of Stroke and Vascular Disease
National Centre for Agri-Food Research in Medicine
St. Boniface Hospital Research Centre
Faculty of Medicine
University of Manitoba
Winnipeg, Manitoba, Canada

Xiaoxia Li
Department of Immunology
The Cleveland Clinic Foundation
Cleveland, Ohio, USA

Dona Love
Laboratory of Cell Biochemistry and Biology
National Institutes of Health
Bethesda, Maryland, USA

Christopher J. Lynch
Department of Cellular and Molecular Physiology
Pennsylvania State University College of Medicine
Hershey, Pennsylvania, USA

Chantal Mathieu
Laboratorium voor Experimentele Geneeskunde en Endocrinologie
Katholieke Universiteit Leuven
Leuven, Belgium

Alfred J. Meijer
Department of Medical Biochemistry
Academic Medical Center
University of Amsterdam
Amsterdam, The Netherlands

Cynthia J. Meininger
Department of Medical Physiology and Cardiovascular Research Institute
The Texas A&M University Health Science Center
College Station, Texas, USA

Lut Overbergh
Laboratorium voor Experimentele Geneeskunde en Endocrinologie
Katholieke Universiteit Leuven
Leuven, Belgium

Wulf Paschen
Laboratory of Molecular Neurobiology
Max Planck Institute for Neurological Research
Cologne, Germany

Grant N. Pierce
Division of Stroke and Vascular Disease
National Centre for Agri-Food Research in Medicine
St. Boniface Hospital Research Centre
Faculty of Medicine
University of Manitoba
Winnipeg, Manitoba, Canada

Pu Qin
Department of Microbiology and Immunology
Temple University School of Medicine
Philadelphia, Pennsylvania, USA

William D. Rees
The Rowett Research Institute, Aberdeen, Scotland
United Kingdom

Gautam Sarath
USDA-ARS
University of Nebraska
Lincoln, Nebraska, USA

Wenjuan Shi
Faculty of Nutrition
Texas A&M University
College Station, Texas, USA

Dianne R. Soprano
Department of Biochemistry
Temple University School of Medicine
Philadelphia, Pennsylvania, USA

Kenneth J. Soprano
Department of Microbiology and Immunology
Temple University School of Medicine
Philadelphia, Pennsylvania, USA

Dan Su
Department of Biochemistry
University of Nebraska
Lincoln, Nebraska, USA

Evelyne van Etten
Laboratorium voor Experimentele Geneeskunde en Endocrinologie
Katholieke Universiteit Leuven
Leuven, Belgium

Thomas C. Vary
Department of Cellular and Molecular Physiology
Pennsylvania State University College of Medicine
Hershey, Pennsylvania, USA

Annemieke Verstuyf
Laboratorium voor Experimentele Geneeskunde en Endocrinologie
Katholieke Universiteit Leuven
Leuven, Belgium

JoEllen Welsh
Department of Biological Sciences
University of Notre Dame
South Bend, Indiana, USA

Rhonda D. Wideman
Departments of Physiology and Surgery
Laboratory of Molecular and Cellular Medicine
University of British Columbia
Vancouver, Canada

Guoyao Wu
Department of Animal Science and Faculty of Nutrition
Texas A&M University,
College Station, Texas, USA

Janos Zempleni
Department of Nutrition and Health Sciences
University of Nebraska
Lincoln, Nebraska, USA

Contents

I

Nuclear Receptors, Transcription Factors, and Signaling Cascades

1

NFκB-Dependent Signaling Pathways

XIAOXIA LI

INTRODUCTION

The family of NFκB transcription factors includes a collection of proteins, conserved from drosophila to humans and related through a highly conserved DNA-binding and dimerization region, the Rel homology (RH) domain.[1] However, the NFκB family can be divided into two groups, based on differences in their structures, functions, and modes of synthesis.[2,3] Members of one group (p105, p100, and drosophila Rel) have long C-terminal domains that contain multiple copies of ankyrin repeats, which act to inhibit these molecules. Members of this group give rise to active, shorter proteins that contain the Rel homology domain (p50 from p105, p52 from p100) either by limited proteolysis[4-8] or arrested translation.[9,10] Members of this group do not function as transcription activators, except when they form dimers with members of the second group, which includes p65 (RelA), Rel (c-Rel), RelB, and the drosophila Rel proteins dorsal and Dif.[11] These proteins are not synthesized as precursors, and in addition to the N-terminal Rel homology domain, they possess one or more C-terminal transcriptional activation domains. Members of both groups of NFκB proteins can form homo- or heterodimers. NFκB was the original name for the p50-p65 heterodimer.

In unstimulated cells, NFκB-family proteins exist as hetero- or homodimers that are sequestered in the cytoplasm by virtue of their association with a member of the IκB family of inhibitory proteins. Extracellular signals can lead to the activation of the IκB kinase (IKK), which in turn phosphorylates two specific serine residuals on IκB proteins (S32 and S36).[4,12] Phospho-IκB is then recognized by the β-TrCP-containing SCF ubiquitin ligase complex, leading to its ubiquitination and degradation by the proteasome. The destruction of IκB unmasks the nuclear localization signal of NFκB, leading to its nuclear translocation and binding to the promoters of target genes.[4,13] Recently, more and more evidence suggests that the phosphorylation and degradation of IκB and the consequent liberation of NFκB are not sufficient to activate NFkB-dependent transcription. A second level of regulation of NFκB activity relies on phosphorylation of members of the second group NFκB proteins (p65/RelA, RelB, and c-Rel), resulting in the activation of transcriptional activity of NFκB.[14-21] This chapter will review the mechanisms of NFκB activation and then outline the functions of activated NFκB in the immune system, stress responses, and cell survival and development.

MOLECULAR MECHANISMS OF NFκB ACTIVATION

Activation of the IKK

Many of the extracellular signals that lead to the activation of NFκB converge on a high-molecular-weight oligomeric protein, the serine-specific IKK.[22-27] IKK is composed of at least three subunits, the catalytic subunits IKKα and β, and the regulatory subunit, IKKγ. IKKα and β have very similar primary structures (52% identity), with protein kinase domains near their N-termini and leucine zipper (LZ) and helix–loop–helix (HLH) motifs near their C-termini. IKKγ lacks a kinase domain but contains long stretches of coiled-coil sequence, which function in protein–protein binding, including a C-terminal LZ motif. The IKKα and IKKβ subunits preferentially form heterodimers, and both can directly phosphorylate the critical S32 and S36 residues of IκBα. IKKγ serves a structural and regulatory role, and is thought to mediate interactions with upstream activators of IKK in response to cellular activation signals. However, the identity of the immediately upstream IKK activators remains to be identified.

Activation of the IKK complex involves the phosphorylation of two serine residues located in the activation loop within the kinase domain of IKKα (S176 and S180) or IKKβ (S177 and S181). IKKβ-null mouse embryonic fibroblasts (MEFs) have a greatly reduced ability to degrade IκBα or activate NFκB in response to various stimuli including TNFα and LPS.[28–30] IKKγ-null cells fail to degrade IκB or activate NFκB in response to tumor necrosis factor (TNF) and lipopolysaccharide (LPS), indicating that IKKγ is required for these processes.[27,31–34]

However, IKKα-null cells retain nearly normal IκB degradation and NFκB DNA-binding activity in response to stimulation.[35–37] Instead, IKKα leads to the phosphorylation and processing of p100, resulting in the formation of an active RelB-p52 heterodimer.[38–40] NFκB-inducing kinase (NIK) was shown to play a role in activating IKKα, leading to p100 processing.[41–47]

Modulation of NFκB Transcriptional Activity

In addition to the signal-induced liberation of NFκB from IκB and the consequent nuclear localization of NFκB, the transcriptional activity of NFκB is also regulated in response to stimulation.[14] The NFκB p65 protein has a site for phosphorylation by protein kinase A on serine 276, and phosphorylation of this residue is required for efficient binding to the transcriptional activator protein CBP. The catalytic subunit of protein kinase A was shown to be bound to inactive NFκB complexes and, upon IκB degradation, to phosphorylate p65, resulting in a conformational change of p65 and consequent interaction with CBP.[15,16] Moreover, TNFα treatment of cells results in phosphorylation of serine 529 in the transactivation domain of p65, resulting in activation of its transcriptional activity. Recently, casein kinase II (CK II) was implicated in the TNFα-dependent phosphorylation of serine 529.[17,18] IκBα protects p65 from phosphorylation by constitutively active CKII, but signal-dependent degradation of IκB exposes the p65 phosphorylation site to CKII activity. Thus, once released from IκB, at least two kinases, PKA and CKII, phosphorylate p65 at different serine residues to increase its transcriptional activity.

On the other hand, PI3K and Akt have been shown to be required for interleukin-1 (IL-1) and TNFα-induced NFκB activity. IL-1 and TNF induce the activation of PI3K and Akt, which leads to the phosphorylation and activation of p65.[19] Akt has also been implicated in Ras-induced NFκB activation through the activation of IKKβ and phosphorylation of p65 at serines 529 and 536.[20,21] Furthermore, IKK complexes from IKKα-null and IKKβ-null MEFs are both deficient in PI3K-mediated phosphorylation of the transactivation domain of the p65 subunit of NFκB in response to IL-1 and TNF. These results indicate that both IKKα and IKKβ are activated by the IL-1 and TNF-induced PI3K/Akt, leading to the phosphorylation and activation of the p65 NFκB subunit.

FUNCTION OF NFκB IN INNATE IMMUNITY

The IL-1/Toll Receptor Superfamily

The IL-1/Toll receptors play essential roles in inflammation and innate immunity. The defining feature of a member of the superfamily is a Toll-like domain on the cytoplasmic side of the receptor (Figure 1.1).[48,49] The first members of this superfamily are in the IL-1R family. These receptors contain three Ig domains in their extracellular regions. The second group of IL-1/Toll receptors includes the recently identified pattern recognition receptors, the Toll-like receptors (TLRs, nine members so far), which consist of two major domains characterized by extracellular leucine-rich repeats (LRRs) and an intracellular Toll-like domain.[49–53] TLR4 has been genetically

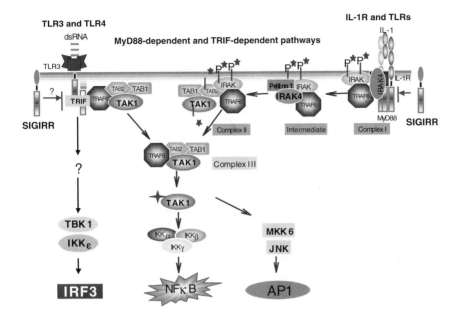

Figure 1.1 Model for IL-1R/TLR-mediated signaling.

identified as a signaling molecule essential for the responses to LPS.[54] Mice with targeted disruption of the TLR4 gene are LPS unresponsive. Unlike TLR4, TLR2 responds to mycobacteria, yeast, and Gram-positive bacteria.[55–58] TLR9 has been shown to recognize bacterial DNA,[53] whereas TLR5 mediates the induction of the immune response by bacterial flagellins.[59] Recent studies show that TLR3 recognizes double-stranded RNA (dsRNA), and that activation of this receptor induces the activation of NFκB and the production of type 1 interferons.[60] The last group includes only one receptor so far, single Ig IL-1R-related molecule (SIGIRR), which has only a single extracellular Ig domain.[61]

IL-1R-Mediated NFκB Activation

The Toll-IL-1R superfamily enables biological activities mainly by activating the transcription of various genes in different target cells. Great effort has been devoted to understanding the signaling pathways mediated by this receptor superfamily. Because their cytoplasmic domains are similar, these receptors may employ related yet distinct signaling components and downstream pathways. Because IL-1R was the first discovered receptor in this superfamily, the IL-1-mediated signaling pathway serves as a "prototype" for other family members.

Genetic and biochemical studies revealed that IL-1R mediates a very complex pathway, involving a cascade of kinases organized by multiple adapter molecules into signaling complexes, leading to activation of the transcription factors NFκB, ATF, and AP-1.[62-64] Li et al. postulated the following model of the IL-1 pathway (Figure 1.1), which was based on recent studies.[65-67] Upon IL-1 stimulation, adapter molecules (including MyD88,[68] TIRP,[69] and Tollip[70]) are first recruited to the IL-1 receptor, followed by the recruitment of two serine–threonine kinases, IRAK4[71,72] and IRAK,[73,74] and the adapter TRAF6,[75] resulting in the formation of the receptor complex (Complex 1). During the formation of Complex I, IRAK4 is activated, leading to the hyperphosphorylation of IRAK, which creates an interface for its interaction with adapter Pellino 1.[65]

The formation of Pellino 1-IRAK4-IRAK-TRAF6 causes conformational changes in the receptor complex (Complex I), releasing these signaling molecules from the receptor. The released components then interact with the membrane-bound preassociated TAK1-TAB1-TAB2,[66] resulting in the formation of Complex II (IRAK-TRAF6-TAK1-TAB1-TAB2). TAK1 and TAB2 are phosphorylated in the membrane-bound Complex II, triggering the dissociation and translocation of TRAF6-TAK1-TAB1-TAB2 (Complex III) from the membrane to the cytosol.

The translocated Complex III interacts with additional factors in the cytosol, leading to TAK1 activation. The activation of TAK1 eventually results in the activation of IκB kinase (IKK), in turn leading to the phosphorylation and degradation of IκB proteins, and liberation of NFκB to activate transcription in the nucleus.[23-26] Activated TAK1 has also been implicated in the IL-1-induced activation of MKK6 and JNK,[41] leading to the activation of other transcription factors including ATF and AP1, thereby also activating gene transcription.

Chen and colleagues, using an *in vitro* system to study TRAF6-mediated IKK activation,[46,47] showed that this activation requires nonclassical ubiquitination catalyzed by the ubiquitination proteins Ubc13 and Uev1A, which plays a regulatory role and do not lead to proteasome-mediated degradation. TRAF6 functions as part of a unique E3 complex with Ubc13 and Uev1A, and TRAF6 itself is the target of ubiquitination. Furthermore, it was shown that the TAK1 complex (TAK1-TAB1-TAB2) is activated by association with ubiquitinated TRAF6. Once activated, TAK1 can directly phosphorylate IKKβ and MKK6, leading to the activation of both the JNK and NFκB signaling pathways.

TLR3-Mediated NFκB Activation

Several Toll-IL-1 receptors also utilize variations of the above common signaling pathway. For example, TLR3 and TLR4 use MyD88-independent pathways to activate the transcription factors NFκB and IRF3.[76,56] IRF3 is part of the IRF family of transcription factors, which play important roles in host defenses against pathogens, immunomodulation, and growth control. Recently, the IKK-related kinases IKKε and TANK-binding kinase 1 have been implicated in the phosphorylation and activation of IRF3.[77,78] The TIR domain-containing adapter inducing IFN-β (TRIF) was recently identified as an adapter for TLR3 and TLR4.[79,80] TRIF-deficient mice are defective in both TLR3- and TLR4-mediated activation of IRF3.[79] Whereas TRIF-deficient mice showed complete loss of TLR3-induced NFκB activation, TLR4-mediated NFκB activation was only abolished completely in mice deficient in both MyD88 and TRIF.[79]

We have also shown previously that dsRNA-triggered, TLR3-mediated NFκB activation is independent of MyD88, IRAK4, and IRAK.[81] Instead, TRAF6, TAK1, and TAB2 are recruited to TLR3 upon PolyI:C stimulation. TRAF6-TAK1-TAB2 are then translocated to the cytosol, where TAK1 is phosphorylated and activated, leading to the activation of IKK and NFκB.[81] We recently found that adapter molecule TRIF recruits TRAF6-TAK1-TAB2 to TLR3 through its TRAF6-binding site, which is required for NFκB but not IRF3 activation (unpublished data, Jiang and Li). Therefore, dsRNA-induced TLR3/TRIF-mediated NFκB and IRF3 activation diverge at TRIF. Based on our work and the published studies of others,[77-81] we have proposed a model for TLR3-mediated signaling (Figure 1.1). Upon PolyI:C stimulation, TRIF is recruited to TLR3 through TIR-TIR domain interactions. TRIF then recruits TRAF6 to TLR3 through its TRAF6 binding site, followed by recruitment of TAK1 and TAB2 via their interaction with TRAF6. TRAF6-TAK1-TAB2 then dissociate from the receptor and translocate to the cytosol, where TAK1 is activated. Activated TAK1 leads to the activation of IKK and NFκB. Independent of TRAF6-TAK1-TAB2, through interaction with unknown intermediate signaling components, TRIF leads to the activation of TBK1/IKKε, resulting in IRF3 activation.

Negative Regulation of Toll-IL-1R-Mediated Signaling

Although the positive regulation of NFκB, AP1, and IRF3 through the Toll-IL-1R superfamily has been studied extensively, very little is known about how these pathways are negatively regulated. Recently, an IRAK family member, IRAK-M, was implicated in the negative regulation of TLR signaling.[82] The IRAK family consists of two active kinases, IRAK and IRAK4, and two inactive kinases, IRAK2 and IRAKM. IRAKM expression is induced upon TLR stimulation and negatively regulates TLR signaling. IRAKM-deficient cells exhibit increased cytokine production upon Toll-IL-1R stimulation and bacterial challenge, and IRAKM-deficient mice show increased inflammatory responses to bacterial infection. At the molecular level, IRAKM was shown to prevent the dissociation of IRAK and IRAK4 from MyD88 and the formation of IRAK-TRAF6 complexes. The detailed molecular mechanism for how IRAKM negatively regulates Toll-IL-1R-mediated signaling is still unclear. Besides IRAKM, another protein, MyD88s, has also been demonstrated to act as a negative regulator of Toll-IL-1 signaling. MyD88s is an alternatively spliced form of MyD88 that blocks the recruitment of IRAK-4.[83,84]

Through database searching, we identified a novel Toll domain-containing receptor. During characterization of this protein, it was independently reported as SIGIRR or the single immunoglobulin IL-1 receptor-related molecule.[61] SIGIRR represents a unique subgroup of the Toll-IL-1R superfamily because its extracellular domain consists of a single Ig domain. Although most members of this superfamily have been shown to be capable of activating NFκB, AP1, or IRF3 constitutively or after structural modification, this activation has not been observed with SIGIRR. Further, SIGIRR has been shown not to bind IL-1 or enhance IL-1 signaling, but to inhibit IL-1-mediated signaling. In order to understand the biological functions of SIGIRR, we constructed an SIGIRR-deficient mouse.[85] Increased expression of inflammatory genes was observed in the lungs and colons of SIGIRR-deficient mice, compared with control wild-type mice, upon IL-1 stimulation. SIGIRR-deficient cells exhibited increased activation of NFκB and JNK upon IL-1, LPS, and CpG DNA stimulation. Endotoxin tolerance, a protective mechanism against endotoxic shock, was dramatically reduced in SIGIRR-deficient mice. Thus, SIGIRR probably plays critical roles in regulating Toll-IL-1R signaling and innate immune homeostasis. The fact that SIGIRR forms a complex with the IL-1 receptor, IRAK, and TRAF6 upon IL-1 stimulation suggests that SIGIRR functions through its interaction with the Toll-IL-1 receptor complex.

EFFECT OF NFκB ACTIVATION IN STRESS RESPONSES

Many activators of NFκB are not bacterial or viral pathogens. NFκB activity, for instance, is induced in response to various physiological stresses such as ischemia or reperfusion, liver regeneration, and hemorrhagic shock. Physical stress in the form of irradiation, as well as oxidative stress to cells, also induces NFκB. In addition, NFκB is activated by both environmental stresses (such as heavy metals) and various chemotherapeutic agents. Therefore, NFκB is activated by and induces responses to many forms of cellular stress and can, therefore, more generally be thought of as a central mediator of stress responses.[86]

NFκB Activation by Oxidative Stress

In respiring cells, a small amount of the consumed oxygen is reduced to highly reactive chemical entities that are collectively called reactive oxygen species (ROS) or reactive oxidative intermediates (ROIs). A state of moderately increased levels of intracellular ROS is referred to as oxidative stress.[87] Cells respond to these adverse conditions by modulating their antioxidant levels, inducing new gene expression, and modifying proteins. Direct evidence that the level of ROS may regulate NFκB was provided by frank exposure of cells to H_2O_2.[88,89] The available data indicate that H_2O_2-induced NFκB activation is highly cell-type dependent and, therefore, that H_2O_2 is unlikely to be a general mediator of NFκB activation.[90–93] Adding H_2O_2 to HeLa cells induces the appearance of a slowly-migrating form of IκBα in SDS-polyacrylamide gels, which is rapidly degraded unless cells are treated with a proteasome inhibitor.[94] Addition of antioxidants (PDTC or NAC) or overexpression of peroxidases blocked the IκBα phosphorylation and degradation induced by TNF, phorbol myristate acetate (PMA), and LPS.[91,95] Therefore, IκB phosphorylation and degradation might be a step that is responsive to oxidative stress.[89] However, the IKK activity induced by TNF was not affected by these antioxidants. One possibility is that antioxidants affect the recognition of IκBα by the IKK complex, slowing down the phosphorylation of IκB upon TNF stimulation. Alternatively, antioxidants may directly inhibit the ubiquitination or degradation of IκBα.

Stress-Induced NFκB Activation through IKK-Independent Pathways

Short-wavelength UV (UV-C) light activates NFκB in certain cell types concomitantly with IκB degradation. Pretreatment of cells with proteasome inhibitors blocked IκBα degradation and NFκB activation induced by UV, indicating that IκB degradation is required. However, neither IKK activation nor the phosphorylation of IκBα on Ser 32 and Ser 36 was observed after UV-C.[96,97] Furthermore, even the IκBα mutant that contains alanines at position 32 and 36 was still susceptible to UV-C-induced degradation. Similar to UV-C, treatment of cells with amino acid analogs also activates NFκB through proteosome-mediated IκB degradation without apparent phosphorylation at Ser 32 or Ser 36.[94] It is likely that exposure to amino acid analogs activates a stress response similar to the one triggered by UV-C. Another pathway leading to NFκB activation was activated when cells were treated with tyrosine phosphatase inhibitors (e.g., pervanadate) or upon reoxygenation of hypoxic cells.[98,99] In these cases, NFκB was activated through the tyrosine phosphorylation of IκBα without degradation. The phosphorylation site was identified as Tyr-42, a site that is present only in IκBα and not IKKβ. The tyrosine phosphorylation of IκBα led to its dissociation from NFκB.[98]

ROLE OF NFκB IN CELL SURVIVAL

NFκB Activation by the TNF Receptor Superfamily

The TNF receptor superfamily is another growing group of NFκB-activating receptors (Table 1.1),[100] which plays a critical role in regulating cell survival. These receptors have a characteristic cysteine-rich pseudorepeat in their extracellular regions. The corresponding ligands isolated so far also share significant sequence homology and belong to the TNF family. The cytoplasmic domains of TNF receptor superfamily members are relatively short and contain no known catalytic motifs. They also lack significant sequence homology among themselves, except for the death domains in TNFR1, DR4, and DR5. Therefore, unlike the IL-1/Toll receptor superfamily, the TNF receptor superfamily uses quite distinct downstream signaling components, such as different TRAF proteins. The formation of homo- or heterodimers (or both) among the six members of the TRAF family generates useful complexity.

Table 1.1 Members of the TNF Receptor Superfamily and Their Ligands

Receptor	Standard Name	Other Names	Ligand
NGFR	TNFRSF16	p75	NGF
Troy	TNFRSF19	Taj	—
EDAR	—	—	EDA1
XEDAR	—	EDA-A2R	EDA2
CD40	TNFRSF5	p50, Bp50	CD40L
DcR3	TNFRSF6B	—	—
FAS	TNFRSF6	CD95, APO-1, APT1	FASL
OX40	TNFRSF4	CD134, ACT35, TXGP1L	OX40L
AITR	TNFRSF18	GITR	AITRL
CD30	TNFRSF8	Ki-1, D1S166E	CD30L
HveA	TNFRSF14	HVEM, ATAR, TR2, LIGHTR	—
4-1BB	TNFRSF9	CD137, ILA	4-1BBL
TNFR2	TNFRSF1B	CD120b, p75, TNFBR, TNFR80, TNF-R-11	TNF
DR3	TNFRSF12	TRAMP, WSL-1, LARD, WSL-LR, DDR3, TR3, APO-3	DR3L
CD27	TNFRSF7	Tp55, S152	CD27L
TNFR1	TNFRSF1A	CD120a p55-R, TNFAR, TNFR60, TNF-R-1	TNF
LTβR	TNFRSF3	TNFR2-RP, TNFCR, TNF-R-1	LTα, LTβ, Light
RANK	TNFRSF11A	TRANCE-R	RANKL
TACI	—	CAML interactor	APR1L, BLYS
BCMA	TNFRSF17	BCM	APR1L, BLYS
DR6	—	TR7	—
OPG	TNFRSF11B	OCIF, TR1, osteoprotegerin	—
DR4	TNFRSF10A	Apo2, TRAILR-1	TRAIL
DR5	TNFRSF10B	KILLER, TRICK2A, TRAIL-R2, TRICKB	TRAIL
DcR1	TNFRSF10C	TRAILR3, LIT, TRID	—
DcR2	TNFRSF10D	TRUNDD, TRAILR4	—

NFκB as a Survival Factor

The CD40 (CD40L/CD154)-CD40 ligand and BAFF-BAFF receptor (BAFFR) cytokine systems, members of the TNF superfamily, play critical roles in the homeostatic regulation of B cell functions.[101–107] CD40 and BAFFR ligation activate the NFκB family of transcription factors, which are critical for the regulation of B cell survival. Upon stimulation with TNF receptor family ligands (including CD40L and BAFF) and other inflammatory mediators, the IκB proteins are phosphorylated by IKK.[22–26] IKKβ-mediated phosphorylation of IκB represents the canonical NFκB activation pathway. Conditional deletion of *IKKβ* results in the rapid loss of B cells,[108,109] indicating that the canonical NFκB activation pathway mediated by IKKβ is probably required for the general differentiation and homeostasis of B cells. CD40 and BAFFR are also able to induce a noncanonical NFκB2 processing pathway; that is, IKKα leads to the phosphorylation and processing of p100, resulting the formation of an active RelB-p52 heterodimer.[38–40] NIK was shown to play a role in activating IKKα, leading to p100 processing,[41–47] The fact that mature B cell numbers are reduced in mice lacking NIK or p52, and in irradiated mice reconstituted with IKKα-deficient lymphocytes, suggests the important role of this noncanonical NFκB activation pathway in CD40- and BAFFR-mediated B cell survival.

A Protective Role for NFκB in Signal-Induced Cell Death

Some of the TNF ligands that activate NFκB also lead to apoptosis, although NFκB itself induces responses that suppress apoptosis. One example is the TNFR-1-mediated signaling pathway. While the N-terminal half of TRADD interacts with TRAF2 to activate NFκB, TRADD also interacts with FADD through its C-terminal death domain.[110–114] The death-effector domain at the N-terminal end of FADD/MORT1 then interacts with a related motif in the prodomain of caspase 8 (also called FLICE, MACH, or Mch5), thereby leading to the activation of apoptotic protease cascades.[110] Several members of the death domain-containing TNF receptor superfamily (including TNFR-1, DR4/Trail-R1/Apo-2, and DR5/Trail-R2) can induce apoptosis.[115,116] Although death domains are absent in other members of the TNFR family, some of these receptors, such as TNFR2, CD30, and LTβR, are nevertheless capable of inducing cell death under certain circumstances.[117,118] Recently, it has been shown that the activation of TLR2 by bacterial lipoprotein (BLP) can also lead to apoptosis, presumably through the adaptor molecule MyD88.[119,120] Activation of the putative intracellular LPS receptors, Nod1 and Nod2, has also been shown to lead to apoptosis.

It is quite intriguing that many signals that can initiate apoptosis also activate NFκB, which suppresses apoptosis. This seemingly contradictory phenomenon has been elucidated for the TNF pathway. NFκB induces a group of gene products, including TRAF1 and TRAF2 and c-IAP1 and c-IAP2, that function cooperatively at the earliest checkpoint to suppress TNFα-mediated apoptosis.[121] Recently, it has been shown that NFκB activation upregulates the caspase 8 inhibitor FLIP, resulting in increased resistance to Fas ligand (FasL) or TNF-mediated apoptosis.[122]

NFκB plays an important role in preventing apoptosis in embryonic liver and during liver regeneration. Both RelA-null and IKKβ-null mice are embryonic lethal due to liver degeneration and hepatocyte apoptosis.[28,30] The apoptosis of embryonic liver cells in RelA-null and IKKβ-null mice appears to be due to their increased sensitivity to circulating TNFα, because there is no apoptosis of embryonic livers in mice doubly deficient for either RelA and TNFα, or IKKβ and TNF-R1.[123,28]

NFκB Activation upon BCR and TCR Engagement

B lymphocytes lacking the adaptor protein B cell linker (BLNK) do not proliferate in response to B cell antigen receptor (BCR) engagement.[124] Examination of the various BCR-activated signaling pathways in mouse BLNK-null B cells reveals intact activation of Akt and mitogen-activated protein kinases but impaired activation of NFκB. Recently, phospholipase C (PLC)-γ2 and Bruton's tyrosine kinase (Btk) have also been demonstrated to be essential for NFκB activation upon BCR engagement.[125,126] Because BLNK interacts with Btk and PLC-γ2, BLNK might be involved in mediating the formation of a Btk-PLC-γ2 signaling axis that regulates NFκB activation. The NFκB activation defect may be sufficient to explain similar defects in BCR-induced B cell proliferation in BLNK-null,[127–129] Btk-null,[130,131] and PLC-γ2-null mice.[132]

Bcl10, a CARD-containing protein identified from the t(1;14)(p22;q32) breakpoint in MALT lymphomas, has also been implicated in B cell and T cell proliferation through antigen receptor-mediated NFκB activation.[133–135] Bcl10-null mice are severely immunodeficient and bcl10-null lymphocytes are defective in antigen-receptor- or PMA-induced proliferation. Tyrosine phosphorylation, MAPK and AP-1 activation, and Ca^{2+}-dependent signaling are normal in mutant lymphocytes, but antigen receptor-induced NFκB activation is absent. Thus, Bcl10 functions as a positive regulator of lymphocyte proliferation, which specifically connects antigen receptor signaling in B and T cells to NFκB activation.

Activation of NFκB and pre-T cell receptor (pre-TCR) expression are tightly correlated during thymocyte development.[136,137] Inhibition of NFκB in isolated thymocytes results in spontaneous apoptosis of cells expressing pre-TCR, whereas inhibition of NFκB in transgenic mice through the expression of a mutated, suppressor form of IκBα leads to a loss of β-selected thymocytes. In contrast, the forced activation of NFκB through the expression of a dominant active IκB kinase allows differentiation to proceed to the CD4+, CD8+ stage in a Rag1-null mouse that cannot assemble pre-TCR. Therefore, signals emanating from pre-TCR are mediated at least in part by NFκB, which provides a selective survival signal for developing thymocytes with productive β chain rearrangements.

Constitutive Activation of NFκB in Human Cancers

NFκB proteins also participate in cellular growth control and neoplasia. The viral oncoprotein v-Rel was the first member of this family to be identified and is still the only one that is acutely transforming *in vitro* and *in vivo*.[138] Several oncogenic viruses, such as human T cell leukemia virus type 1 and Epstein-Barr virus, activate NFκB as part of the transformation process. Moreover, there is growing evidence implicating all vertebrate NFκB factors in human cancer.[139] Chromosomal aberrations involving the human c-rel, relA, NFκB1 (encoding p105/p50), and NFκB2 (encoding p100/p52) genes are found in many hematopoietic and solid tumors. Constitutively high levels of nuclear NFκB activity have also been described in many cancer cell types, as a result of the constitutive activation of upstream signaling kinases or mutations inactivating IκBs. Constitutive IKK activity was observed in Hodgkin's disease and childhood acute lymphoblastic leukemia.

In addition to its antiapoptotic role, as discussed above, NFκB also induces cell proliferation and cell cycle progression by regulating the expression of target genes, including *c-myc* and cyclinD1.[139,140] In certain systems the upregulation of NFκB is associated with advanced stages of oncogenesis, supporting a role in tumor progression. In this respect, NFκB activates the expression of genes that are important for invasion and metastasis, including those encoding angiogenic factors such as VEGF, proteolytic enzymes such as matrix metalloproteinases, urokinase plasminogen activators (uPAs), and cell adhesion molecules such as ICAM-1. The uPA, significantly increased in most breast cancer cell lines that contain constitutively active NFκB, is required for intravasation and is associated with poor prognosis,[141] supporting a role for NFκB in metastasis.

FUTURE DIRECTIONS

As discussed in this chapter, NFκB proteins play an important role in innate immunity, stress responses, and cell proliferation and survival. The identification of IκB kinase marks the turning point for elucidating the molecular mechanisms of NFκB activation. Signal-induced activation of IKKβ leads to the phosphorylation and degradation of IκB, liberating NFκB from the IκB inhibitory proteins. In addition to the canonical NFκB activation pathway, IKKα leads to the phosphorylation and processing of p100, resulting in the formation of an active RelB-p52 heterodimer. The identity of the kinase that activates IκB kinase remains controversial, although several candidates have been proposed, including MEKK1, MEKK3, and TAK1. NIK was shown to play a role in activating IKKα, leading to p100 processing. Further biochemical and genetic studies are required to unify the identity of the IκB kinase in the field, although it is also quite possible that more than one kinase is capable of activating IκB kinase. Although many stimuli lead to activation of NFκB, including cytokines, bacterial products, viruses, and environmental insults, the receptor proximal signaling components of the pathways activated by most of these stimuli need to be better defined. The further understanding of the NFκB-dependent signaling pathways will lay a solid foundation for the development of small-molecule drugs for controlling inflammation and treatment of cancers.

References

1. JA Hoffmann, FC Kafatos, CA Janeway, RA Ezekowitz. Phylogenetic perspectives in innate immunity. *Science* 284: 1313–1318, 1999.

2. PA Baeuerle, T Henkel. Function and activation of NF-kappa B in the immune system. *Annu Rev Immunol* 12: 141–179, 1994.

3. U Siebenlist, G Franzoso, K Brown. Structure, regulation, and function of NF-kappa B. *Annu Rev Cell Biol* 10: 405–455, 1994.

4. M Karin, Y Ben Neriah. Phosphorylation meets ubiquitination: the control of NF-[kappa]B activity. *Annu Rev Immunol* 18: 621–663, 2000.

5. C Lee, MP Schwartz, S Prakash, M Iwakura, A Matouschek. ATP-dependent proteases degrade their substrates by processively unraveling them from the degradation signal. *Mol Cell* 7: 627–637, 2001.

6. L Lin, S Ghosh. A glycine-rich region in NF-kappaB p105 functions as a processing signal for the generation of the p50 subunit. *Mol Cell Biol* 16: 2248–2254, 1996.

7. A Orian, AL Schwartz, A Israel, S Whiteside, C Kahana, A Ciechanover. Structural motifs involved in ubiquitin-mediated processing of the NF-kappaB precursor p105: roles of the glycine-rich region and a downstream ubiquitination domain. *Mol Cell Biol* 19: 3664–3673, 1999.

8. ZJ Chen, T Maniatis. Role of the ubiquitin-proteasome pathway in NF-kappaB activation. *Cell* 91: 303–322, 1998.

9. L Lin, GN DeMartino, WC Greene. Cotranslational biogenesis of NF-kappaB p50 by the 26S proteasome. *Cell* 92: 819–828, 1998.

10. L Lin, GN DeMartino, WC Greene. Cotranslational dimerization of the Rel homology domain of NF-kappaB1 generates p50-p105 heterodimers and is required for effective p50 production. *EMBO J* 19: 4712–4722, 2000.

11. PA Baeuerle, D Baltimore. NF-kappa B: ten years after. *Cell* 87: 13–20, 1996.

12. S Ghosh, MJ May, EB Kopp. NF-kappa B and Rel proteins: evolutionarily conserved mediators of immune responses. *Annu Rev Immunol* 16: 225–260, 1998.

13. MA Read, JE Brownell, TB Gladysheva, M Hottelet, LA Parent, MB Coggins, JW Pierce, VN Podust, RS Luo, V Chau, VJ Palombella. Nedd8 modification of cul-1 activates SCF(beta(TrCP))-dependent ubiquitination of IkappaBalpha. *Mol Cell Biol* 20: 2326–2333, 2000.

14. N Silverman, T Maniatis. NF-kappaB signaling pathways in mammalian and insect innate immunity. *Genes Dev* 15: 2321–2342, 2001.

15. H Zhong, H SuYang, H Erdjument-Bromage, P Tempst, S Ghosh. The transcriptional activity of NF-kappaB is regulated by the IkappaB-associated PKAc subunit through a cyclic AMP-independent mechanism. *Cell* 89: 413–424, 1997.

16. H Zhong, RE Voll, S Ghosh. Phosphorylation of NF-kappa B p65 by PKA stimulates transcriptional activity by promoting a novel bivalent interaction with the coactivator CBP/p300. *Mol Cell* 1: 661–671, 1998.

17. D Wang, AS Baldwin, Jr. Activation of nuclear factor-kappaB-dependent transcription by tumor necrosis factor-alpha is mediated through phosphorylation of RelA/p65 on serine 529. *J Biol Chem* 273: 29411–29416, 1998.

18. D Wang, SD Westerheide, JL Hanson, AS Baldwin, Jr. Tumor necrosis factor alpha-induced phosphorylation of RelA/p65 on Ser529 is controlled by casein kinase II. *J Biol Chem* 275: 32592–32597, 2000.

19. N Sizemore, S Leung, GR Stark. Activation of phosphatidylinositol 3-kinase in response to interleukin-1 leads to phosphorylation and activation of the NF-kappaB p65/RelA subunit. *Mol Cell Biol* 19: 4798–4805, 1999.

20. LV Madrid, CY Wang, DC Guttridge, AJ Schottelius, AS Baldwin, Jr., MW Mayo. Akt suppresses apoptosis by stimulating the transactivation potential of the RelA/p65 subunit of NF-kappaB. *Mol Cell Biol* 20: 1626–1638, 2000.

21. LV Madrid, MW Mayo, JY Reuther, AS Baldwin, Jr. Akt stimulates the transactivation potential of the RelA/p65 subunit of NF-kappa B through utilization of the Ikappa B kinase and activation of the mitogen-activated protein kinase p38. *J Biol Chem* 276: 18934–18940, 2001.

22. JA DiDonato, M Hayakawa, DM Rothwarf, E Zandi, M Karin. A cytokine-responsive IkappaB kinase that activates the transcription factor NF-kappaB. *Nature* 388: 548–554, 1997.

23. F Mercurio, H Zhu, BW Murray, A Shevchenko, BL Bennett, J Li, DB Young, M Barbosa, M Mann, A Manning, A Rao. IKK-1 and IKK-2: cytokine-activated IkappaB kinases essential for NF-kappaB activation. *Science* 278: 860–866, 1997.

24. CH Regnier, HY Song, X Gao, DV Goeddel, Z Cao, M Rothe. Identification and characterization of an IkappaB kinase. *Cell* 90: 373–383, 1997.

25. JD Woronicz, X Gao, Z Cao, M Rothe, DV Goeddel. IkappaB kinase-beta: NF-kappaB activation and complex formation with IkappaB kinase-alpha and NIK. *Science* 278: 866–869, 1997.

26. E Zandi, DM Rothwarf, M Delhase, M Hayakawa, M Karin. The IkappaB kinase complex (IKK) contains two kinase subunits, IKKalpha and IKKbeta, necessary for IkappaB phosphorylation and NF-kappaB activation. *Cell* 91: 243–252, 1997.

27. S Yamaoka, G Courtois, C Bessia, ST Whiteside, R Weil, F Agou, HE Kirk, RJ Kay, A Israel. Complementation cloning of NEMO, a component of the IkappaB kinase complex essential for NF-kappaB activation. *Cell* 93: 1231–1240, 1998.

28. Q Li, D Van Antwerp, F Mercurio, KF Lee, IM Verma. Severe liver degeneration in mice lacking the IkappaB kinase 2 gene. *Science* 284: 321–325, 1999.

29. ZW Li, W Chu, Y Hu, M Delhase, T Deerinck, M Ellisman, R Johnson, M Karin. The IKKbeta subunit of IkappaB kinase (IKK) is essential for nuclear factor kappaB activation and prevention of apoptosis. *J Exp Med* 189: 1839–1845, 1999.

30. M Tanaka, ME Fuentes, K Yamaguchi, MH Durnin, SA Dalrymple, KL Hardy, DV Goeddel. Embryonic lethality, liver degeneration, and impaired NF-kappa B activation in IKK-beta-deficient mice. *Immunity* 10: 421–429, 1999.

31. D Rudolph, WC Yeh, A Wakeham, B Rudolph, D Nallainathan, J Potter, AJ Elia, TW Mak. Severe liver degeneration and lack of NF-kappaB activation in NEMO/IKKgamma-deficient mice. *Genes Dev* 14: 854–862, 2000.

32. A Smahi, G Courtois, P Vabres, S Yamaoka, S Heuertz, A Munnich, A Israel, NS Heiss, SM Klauck, P Kioschis, S Wiemann, A Poustka, T Esposito, T Bardaro, F Gianfrancesco, A Ciccodicola, M D'Urso, H Woffendin, T Jakins, D Donnai, H Stewart, SJ Kenwrick, S Aradhya, T Yamagata, M Levy, RA Lewis, DL Nelson. Genomic rearrangement in NEMO impairs NF-kappaB activation and is a cause of incontinentia pigmenti. The International Incontinentia Pigmenti (IP) Consortium. *Nature* 405: 466–472, 2000.

33. C Makris, VL Godfrey, G Krahn-Senftleben, T Takahashi, JL Roberts, T Schwarz, L Feng, RS Johnson, M Karin. Female mice heterozygous for IKK gamma/NEMO deficiencies develop a dermatopathy similar to the human X-linked disorder incontinentia pigmenti. *Mol Cell* 5: 969–979, 2000.

34. M Schmidt-Supprian, W Bloch, G Courtois, K Addicks, A Israel, K Rajewsky, M Pasparakis. NEMO/IKK gamma-deficient mice model incontinentia pigmenti. *Mol Cell* 5: 981–992, 2000.

35. Y Hu, V Baud, M Delhase, P Zhang, T Deerinck, M Ellisman, R Johnson, M Karin. Abnormal morphogenesis but intact IKK activation in mice lacking the IKKalpha subunit of IkappaB kinase. *Science* 284: 316–320, 1999.

36. Q Li, Q Lu, JY Hwang, D Buscher, KF Lee, JC Izpisua-Belmonte, IM Verma. IKK1-deficient mice exhibit abnormal development of skin and skeleton. *Genes Dev* 13: 1322–1328, 1999.

37. K Takeda, O Takeuchi, T Tsujimura, S Itami, O Adachi, T Kawai, H Sanjo, K Yoshikawa, N Terada, S Akira. Limb and skin abnormalities in mice lacking IKKalpha. *Science* 284: 313–316, 1999.

38. S Ghosh, M Karin. Missing pieces in the NF-kappaB puzzle. *Cell* 109 Suppl: S81–S96, 2002.

39. Coope H.J. CD40 regulates the processing of NF-kB2 p100 to p52. *EMBO J* 21: 5375–5385, 2002.

40. Kayagaki N. BAFF/BLyS receptor 3 binds the B cell survival factor BAFF ligand through a discrete surface loop and promotes processing of NF-kB2. *Immunity* 10: 515–524, 2003.

41. J Ninomiya-Tsuji, K Kishimoto, A Hiyama, J Inoue, Z Cao, K Matsumoto. The kinase TAK1 can activate the NIK-I kappaB as well as the MAP kinase cascade in the IL-1 signalling pathway. *Nature* 398: 252–256, 1999.

42. T Irie, T Muta, K Takeshige. TAK1 mediates an activation signal from toll-like receptor(s) to nuclear factor-kappaB in lipopolysaccharide-stimulated macrophages. *FEBS Lett* 467: 160–164, 2000.

43. G Takaesu, S Kishida, A Hiyama, K Yamaguchi, H Shibuya, K Irie, J Ninomiya-Tsuji, K Matsumoto. TAB2, a novel adaptor protein, mediates activation of TAK1 MAPKKK by linking TAK1 to TRAF6 in the IL-1 signal transduction pathway. *Mol Cell* 5: 649–658, 2000.

44. G Takaesu, J Ninomiya-Tsuji, S Kishida, X Li, GR Stark, K Matsumoto. Interleukin-1 (IL-1) receptor-associated kinase leads to activation of TAK1 by inducing TAB2 translocation in the IL-1 signaling pathway. *Mol Cell Biol* 21: 2475–2484, 2001.

45. Y Qian, M Commane, J Ninomiya-Tsuji, K Matsumoto, X Li. IRAK-mediated translocation of TRAF6 and TAB2 in the interleukin-1-induced activation of NFkappa B. *J Biol Chem* 276: 41661–41667, 2001.

46. L Deng, C Wang, E Spencer, L Yang, A Braun, J You, C Slaughter, C Pickart, ZJ Chen. Activation of the IkappaB kinase complex by TRAF6 requires a dimeric ubiquitin-conjugating enzyme complex and a unique polyubiquitin chain. *Cell* 103: 351–361, 2000.

47. C Wang, L Deng, M Hong, GR Akkaraju, J Inoue, ZJ Chen. TAK1 is a ubiquitin-dependent kinase of MKK and IKK. *Nature* 412: 346–351, 2001.

48. A Bowie, LA O'Neill. The interleukin-1 receptor/Toll-like receptor superfamily: signal generators for proinflammatory interleukins and microbial products. *J Leukoc Biol* 67: 508–514, 2000.

49. FL Rock, G Hardiman, JC Timans, RA Kastelein, JF Bazan. A family of human receptors structurally related to Drosophila Toll. *Proc Natl Acad Sci USA* 95: 588–593, 1998.

50. R Medzhitov, P Preston-Hurlburt, CA Janeway, Jr. A human homologue of the Drosophila Toll protein signals activation of adaptive immunity. *Nature* 388: 394–397, 1997.

51. O Takeuchi, T Kawai, H Sanjo, NG Copeland, DJ Gilbert, NA Jenkins, K Takeda, S Akira. TLR6: A novel member of an expanding toll-like receptor family. *Gene* 231: 59–65, 1999.

52. TH Chuang, RJ Ulevitch. Cloning and characterization of a subfamily of human toll-like receptors: hTLR7, hTLR8 and hTLR9. *Eur Cytokine Netw* 11: 372–378, 2000.

53. H Hemmi, O Takeuchi, T Kawai, T Kaisho, S Sato, H Sanjo, M Matsumoto, K Hoshino, H Wagner, K Takeda, S Akira. A Toll-like receptor recognizes bacterial DNA. *Nature* 408: 740–745, 2000.

54. A Poltorak, X He, I Smirnova, MY Liu, CV Huffel, X Du, D Birdwell, E Alejos, M Silva, C Galanos, M Freudenberg, P Ricciardi-Castagnoli, B Layton, B Beutler. Defective LPS signaling in C3H/HeJ and C57BL/10ScCr mice: mutations in Tlr4 gene. *Science* 282: 2085–2088, 1998.

55. O Takeuchi, K Hoshino, T Kawai, H Sanjo, H Takada, T Ogawa, K Takeda, S Akira. Differential roles of TLR2 and TLR4 in recognition of Gram-negative and Gram-positive bacterial cell wall components. *Immunity* 11: 443–451, 1999.

56. O Takeuchi, A Kaufmann, K Grote, T Kawai, K Hoshino, M Morr, PF Muhlradt, S Akira. Cutting edge: preferentially the R-stereoisomer of the mycoplasmal lipopeptide macrophage-activating lipopeptide-2 activates immune cells through a toll-like receptor 2- and MyD88-dependent signaling pathway. *J Immunol* 164: 554–557, 2000.

57. DM Underhill, A Ozinsky, AM Hajjar, A Stevens, CB Wilson, M Bassetti, A Aderem. The Toll-like receptor 2 is recruited to macrophage phagosomes and discriminates between pathogens. *Nature* 401: 811–815, 1999.

58. DM Underhill, A Ozinsky, KD Smith, A Aderem. Toll-like receptor-2 mediates mycobacteria-induced proinflammatory signaling in macrophages. *Proc Natl Acad Sci USA* 96: 14459–14463, 1999.

59. G Sebastiani, G Leveque, L Lariviere, L Laroche, E Skamene, P Gros, D Malo. Cloning and characterization of the murine toll-like receptor 5 (Tlr5) gene: sequence and mRNA expression studies in Salmonella-susceptible MOLF/Ei mice. *Genomics* 64: 230–240, 2000.

60. L Alexopoulou, AC Holt, R Medzhitov, RA Flavell. Recognition of double-stranded RNA and activation of NF-kappaB by Toll-like receptor 3. *Nature* 413: 732–738, 2001.

61. E Thomassen, BR Renshaw, JE Sims. Identification and characterization of SIGIRR, a molecule representing a novel subtype of the IL-1R superfamily. *Cytokine* 11: 389–399, 1999.

62. PJ Barnes, M Karin. Nuclear factor-kappaB: a pivotal transcription factor in chronic inflammatory diseases. *N Engl J Med* 336: 1066–1071, 1997.

63. LA O'Neill. Towards an understanding of the signal transduction pathways for interleukin 1. *Biochim Biophys Acta* 1266: 31–44, 1995.

64. LA O'Neill. Molecular mechanisms underlying the actions of the proinflammatory cytokine interleukin 1. Royal Irish Academy Medal Lecture. *Biochem Soc Trans* 25: 295–302, 1997.

65. Z Jiang, HJ Johnson, H Nie, J Qin, TA Bird, X Li. Pellino 1 is required for IL-1-mediated signaling through its interaction with IRAK4-IRAK-TRAF6. *J Biol Chem*, 2002.

66. Z Jiang, J Ninomiya-Tsuji, Y Qian, K Matsumoto, X Li. Interleukin-1 (IL-1) receptor-associated kinase-dependent IL-1-induced signaling complexes phosphorylate TAK1 and TAB2 at the plasma membrane and activate TAK1 in the cytosol. *Mol Cell Biol* 22: 7158–7167, 2002.

67. S Janssens, R Beyaert. Functional diversity and regulation of different interleukin-1 receptor-associated kinase (IRAK) family members. *Mol Cell* 11: 293–302, 2003.

68. H Wesche, WJ Henzel, W Shillinglaw, S Li, Z Cao. MyD88: an adapter that recruits IRAK to the IL-1 receptor complex. *Immunity* 7: 837–847, 1997.

69. LH Bin, LG Xu, HB Shu. TIRP: a novel TIR domain-containing adapter protein involved in Toll/interleukin-1 receptor signaling. *J Biol Chem,* 2003.

70. K Burns, J Clatworthy, L Martin, F Martinon, C Plumpton, B Maschera, A Lewis, K Ray, J Tschopp, F Volpe. Tollip, a new component of the IL-1RI pathway, links IRAK to the IL-1 receptor. *Nat Cell Biol* 2: 346–351, 2000.

71. S Li, A Strelow, EJ Fontana, H Wesche. IRAK-4: a novel member of the IRAK family with the properties of an IRAK-kinase. *Proc Natl Acad Sci USA* 99: 5567–5572, 2002.

72. N Suzuki, S Suzuki, GS Duncan, DG Millar, T Wada, C Mirtsos, H Takada, A Wakeham, A Itie, S Li, JM Penninger, H Wesche, PS Ohashi, TW Mak, WC Yeh. Severe impairment of interleukin-1 and Toll-like receptor signalling in mice lacking IRAK-4. *Nature* 416: 750–756, 2002.

73. Z Cao, WJ Henzel, X Gao. IRAK: a kinase associated with the interleukin-1 receptor. *Science* 271: 1128–1131, 1996.

74. X Li, M Commane, C Burns, K Vithalani, Z Cao, GR Stark. Mutant cells that do not respond to interleukin-1 (IL-1) reveal a novel role for IL-1 receptor-associated kinase. *Mol Cell Biol* 19: 4643–4652, 1999.

75. Z Cao, J Xiong, M Takeuchi, T Kurama, DV Goeddel. TRAF6 is a signal transducer for interleukin-1. *Nature* 383: 443–446, 1996.

76. T Kawai, O Adachi, T Ogawa, K Takeda, S Akira. Unresponsiveness of MyD88-deficient mice to endotoxin. *Immunity* 11: 115–122, 1999.

77. KA Fitzgerald, SM McWhirter, KL Faia, DC Rowe, E Latz, DT Golenbock, AJ Coyle, S Liao, and T Maniatis. IKKe and TBK1 are essential components of the IRF3 signaling pathway. *Nature Immunology* 4: 491–496, 2003.

78. S Sharma, BR tenOever, N Grandvaux, G Zhou, R Lin, J Hiscott. Triggering the interferon antiviral responses through an IKK-related pathway. *Science* 300: 1148–1151, 2003.

79. M Yamamoto, S Sato, H Hemmi, K Hoshino, T Kaisho, H Sanjo, O Takeuchi, M Sugiyama, M Okabe, K Takeda, S Akira. Role of adapter TRIF in the MyD88-independent Toll-like receptor signaling pathway. *Science* 301: 640–643, 2003.

80. K Hoebe, X Du, P Georgel, E Janssen, K Tabeta, SO Kim, J Goode, P Lin, N Mann, S Mudd, K Crozat, S Sovath, J Han, and B Beutler. Identification of LPS2, a key transducer of MyD88-independent TIR signaling. *Nature* 424: 743–748, 2003.

81. Z Jiang, M Zamanian-Daryoush, H Nie, AM Silva, BR Williams, X Li. PolyI:C-induced Toll-like receptor 3 (TLR3)-mediated activation of NFkappa B and MAP kinase is through an interleukin-1 receptor-associated kinase (IRAK)-independent pathway employing the signaling components TLR3-TRAF6-TAK1-TAB2-PKR. *J Biol Chem* 278: 16713–16719, 2003.

82. K Kobayashi, LD Hernandez, JE Galan, CA Janeway, Jr., R Medzhitov, RA Flavell. IRAK-M is a negative regulator of Toll-like receptor signaling. *Cell* 110: 191–202, 2002.

83. K Burns, S Janssens, B Brissoni, N Olivos, R Beyaert, J Tschopp. Inhibition of interleukin 1 receptor/Toll-like receptor signaling through the alternatively spliced, short form of MyD88 is due to its failure to recruit IRAK-4. *J Exp Med* 197: 263–268, 2003.

84. S Janssens, K Burns, J Tschopp, R Beyaert. Regulation of interleukin-1- and lipopolysaccharide-induced NF-kappaB activation by alternative splicing of MyD88. *Curr Biol* 12: 467–471, 2002.

85. D Wald, J Qin, Z Zhao, Y Qian, M Naramura, L Tian, J Towne, JE Sims, GR Stark, X Li. SIGIRR, a negative regulator of Toll-like receptor-interleukin 1 receptor signaling. *Nat Immunol* 4: 920–927, 2003.

86. HL Pahl. Activators and target genes of Rel/NF-kappaB transcription factors. *Oncogene* 18: 6853–6866, 1999.

87. BS Berlett, ER Stadtman. Protein oxidation in aging, disease, and oxidative stress. *J Biol Chem* 272: 20313–20316, 1997.

88. CK Sen, L Packer. Antioxidant and redox regulation of gene transcription. *FASEB J* 10: 709–720, 1996.

89. N Li, M Karin. Is NF-kappaB the sensor of oxidative stress? *FASEB J* 13: 1137–1143, 1999.

90. R Schreck, P Rieber, PA Baeuerle. Reactive oxygen intermediates as apparently widely used messengers in the activation of the NF-kappa B transcription factor and HIV-1. *EMBO J* 10: 2247–2258, 1991.

91. SK Manna, HJ Zhang, T Yan, LW Oberley, BB Aggarwal. Overexpression of manganese superoxide dismutase suppresses tumor necrosis factor-induced apoptosis and activation of nuclear transcription factor-kappaB and activated protein-1. *J Biol Chem* 273: 13245–13254, 1998.

92. M Meyer, R Schreck, PA Baeuerle. H2O2 and antioxidants have opposite effects on activation of NF-kappa B and AP-1 in intact cells: AP-1 as secondary antioxidant-responsive factor. *EMBO J* 12: 2005–2015, 1993.

93. MT Anderson, FJ Staal, C Gitler, LA Herzenberg. Separation of oxidant-initiated and redox-regulated steps in the NF-kappa B signal transduction pathway. *Proc Natl Acad Sci USA* 91: 11527–11531, 1994.

94. C Kretz-Remy, EE Bates, AP Arrigo. Amino acid analogs activate NF-kappaB through redox-dependent IkappaB- alpha degradation by the proteasome without apparent IkappaB-alpha phosphorylation. Consequence on HIV-1 long terminal repeat activation. *J Biol Chem* 273: 3180–3191, 1998.

95. C Kretz-Remy, P Mehlen, ME Mirault, AP Arrigo. Inhibition of I kappa B-alpha phosphorylation and degradation and subsequent NF-kappa B activation by glutathione peroxidase overexpression. *J Cell Biol* 133: 1083–1093, 1996.

96. N Li, M Karin. Ionizing radiation and short wavelength UV activate NF-kappaB through two distinct mechanisms. *Proc Natl Acad Sci USA* 95: 13012–13017, 1998.

97. K Bender, M Gottlicher, S Whiteside, HJ Rahmsdorf, P Herrlich. Sequential DNA damage-independent and -dependent activation of NF- kappaB by UV. *EMBO J* 17: 5170–5181, 1998.

98. V Imbert, RA Rupec, A Livolsi, HL Pahl, EB Traenckner, C Mueller-Dieckmann, D Farahifar, B Rossi, P Auberger, PA Baeuerle, JF Peyron. Tyrosine phosphorylation of I kappa B-alpha activates NF-kappa B without proteolytic degradation of I kappa B-alpha. *Cell* 86: 787–798, 1996.

99. S Singh, BG Darnay, BB Aggarwal. Site-specific tyrosine phosphorylation of IkappaBalpha negatively regulates its inducible phosphorylation and degradation. *J Biol Chem* 271: 31049–31054, 1996.

100. RM Locksley, N Killeen, MJ Lenardo. The TNF and TNF receptor superfamilies: integrating mammalian biology. *Cell* 104: 487–501, 2001.

101. Schonbeck U and Libby P. The CD40/CD154 receptor/ligand dyad. CMLS *Cell Mol Life Sci* 58: 4–43, 2002.

102. C van Kooten, J Banchereau. Functional role of CD40 and its ligand. *Int Arch Allergy Immunol* 113: 393–399, 1997.

103. GA Bishop, BS Hostager. Signaling by CD40 and its mimics in B cell activation. *Immunol Res* 24: 97–109, 2001.

104. XF Lei, Y Ohkawara, MR Stampfli, C Mastruzzo, RA Marr, D Snider, Z Xing, M Jordana. Disruption of antigen-induced inflammatory responses in CD40 ligand knockout mice. *J Clin Invest* 101: 1342–1353, 1998.

105. JS Thompson, SA Bixler, F Qian, K Vora, ML Scott, TG Cachero, C Hession, P Schneider, ID Sizing, C Mullen, K Strauch, M Zafari, CD Benjamin, J Tschopp, JL Browning, C Ambrose. BAFF-R, a newly identified TNF receptor that specifically interacts with BAFF. *Science* 293: 2108–2111, 2001.

106. F Mackay, JL Browning. BAFF: a fundamental survival factor for B cells. *Nat Rev Immunol* 2: 465–475, 2002.

107. F Mackay, P Schneider, P Rennert, J Browning. BAFF AND APRIL: a tutorial on B cell survival. *Annu Rev Immunol* 21: 231–264, 2003.

108. ZW Li, SA Omori, T Labuda, M Karin, RC Rickert. IKK beta is required for peripheral B cell survival and proliferation. *J Immunol* 170: 4630–4637, 2003.

109. M Pasparakis, M Schmidt-Supprian, K Rajewsky. IkappaB kinase signaling is essential for maintenance of mature B cells. *J Exp Med* 196: 743–752, 2002.

110. MP Boldin, TM Goncharov, YV Goltsev, D Wallach. Involvement of MACH, a novel MORT1/FADD-interacting protease, in Fas/APO-1- and TNF receptor-induced cell death. *Cell* 85: 803–815, 1996.

111. H Hsu, J Xiong, DV Goeddel. The TNF receptor 1-associated protein TRADD signals cell death and NF-kappa B activation. *Cell* 81: 495–504, 1995.

112. H Hsu, HB Shu, MG Pan, DV Goeddel. TRADD-TRAF2 and TRADD-FADD interactions define two distinct TNF receptor 1 signal transduction pathways. *Cell* 84: 299–308, 1996.

113. AM Chinnaiyan, K O'Rourke, M Tewari, VM Dixit. FADD, a novel death domain-containing protein, interacts with the death domain of Fas and initiates apoptosis. *Cell* 81: 505–512, 1995.

114. AM Chinnaiyan, CG Tepper, MF Seldin, K O'Rourke, FC Kischkel, S Hellbardt, PH Krammer, ME Peter, VM Dixit. FADD/MORT1 is a common mediator of CD95 (Fas/APO-1) and tumor necrosis factor receptor-induced apoptosis. *J Biol Chem* 271: 4961–4965, 1996.

115. LA Tartaglia, TM Ayres, GH Wong, DV Goeddel. A novel domain within the 55 kDa TNF receptor signals cell death. *Cell* 74: 845–853, 1993.

116. N Itoh, S Nagata. A novel protein domain required for apoptosis. Mutational analysis of human Fas antigen. *J Biol Chem* 268: 10932–10937, 1993.

117. SY Lee, SY Lee, G Kandala, ML Liou, HC Liou, Y Choi. CD30/TNF receptor-associated factor interaction: NF-kappa B activation and binding specificity. *Proc Natl Acad Sci USA* 93: 9699–9703, 1996.

118. WR Force, AA Glass, CA Benedict, TC Cheung, J Lama, cf. Ware. Discrete signaling regions in the lymphotoxin-beta receptor for tumor necrosis factor receptor-associated factor binding, subcellular localization, and activation of cell death and NF-kappaB pathways. *J Biol Chem* 275: 11121–11129, 2000.

119. AO Aliprantis, RB Yang, MR Mark, S Suggett, B Devaux, JD Radolf, GR Klimpel, P Godowski, A Zychlinsky. Cell activation and apoptosis by bacterial lipoproteins through toll-like receptor-2. *Science* 285: 736–739, 1999.

120. O Adachi, T Kawai, K Takeda, M Matsumoto, H Tsutsui, M Sakagami, K Nakanishi, S Akira. Targeted disruption of the MyD88 gene results in loss of IL-1- and IL-18-mediated function. *Immunity* 9: 143–150, 1998.

121. CY Wang, MW Mayo, AS Baldwin, Jr. TNF- and cancer therapy-induced apoptosis: potentiation by inhibition of NF-kappaB. *Science* 274: 784–787, 1996.

122. O Micheau, S Lens, O Gaide, K Alevizopoulos, J Tschopp. NF-kappaB signals induce the expression of c-FLIP. *Mol Cell Biol* 21: 5299–5305, 2001.

123. TS Doi, MW Marino, T Takahashi, T Yoshida, T Sakakura, LJ Old, Y Obata. Absence of tumor necrosis factor rescues RelA-deficient mice from embryonic lethality. *Proc Natl Acad Sci USA* 96: 2994–2999, 1999.

124. JE Tan, SC Wong, SK Gan, S Xu, KP Lam. The adaptor protein BLNK is required for B cell antigen receptor-induced activation of nuclear factor-kappa B and cell cycle entry and survival of B lymphocytes. *J Biol Chem* 276: 20055–20063, 2001.

125. JB Petro, SM Rahman, DW Ballard, WN Khan. Bruton's tyrosine kinase is required for activation of IkappaB kinase and nuclear factor kappaB in response to B cell receptor engagement. *J Exp Med* 191: 1745–1754, 2000.

126. JB Petro, WN Khan. Phospholipase C-gamma 2 couples Bruton's tyrosine kinase to the NF- kappaB signaling pathway in B lymphocytes. *J Biol Chem* 276: 1715–1719, 2001.

127. H Jumaa, B Wollscheid, M Mitterer, J Wienands, M Reth, PJ Nielsen. Abnormal development and function of B lymphocytes in mice deficient for the signaling adaptor protein SLP-65. *Immunity* 11: 547–554, 1999.

128. R Pappu, AM Cheng, B Li, Q Gong, C Chiu, N Griffin, M White, BP Sleckman, AC Chan. Requirement for B cell linker protein (BLNK) in B cell development. *Science* 286: 1949–1954, 1999.

129. K Hayashi, R Nittono, N Okamoto, S Tsuji, Y Hara, R Goitsuka, D Kitamura. The B cell-restricted adaptor BASH is required for normal development and antigen receptor-mediated activation of B cells. *Proc Natl Acad Sci USA* 97: 2755–2760, 2000.

130. WN Khan, FW Alt, RM Gerstein, BA Malynn, I Larsson, G Rathbun, L Davidson, S Muller, AB Kantor, LA Herzenberg. Defective B cell development and function in Btk-deficient mice. *Immunity* 3: 283–299, 1995.

131. JD Kerner, MW Appleby, RN Mohr, S Chien, DJ Rawlings, CR Maliszewski, ON Witte, RM Perlmutter. Impaired expansion of mouse B cell progenitors lacking Btk. *Immunity* 3: 301–312, 1995.

132. D Wang, J Feng, R Wen, JC Marine, MY Sangster, E Parganas, A Hoffmeyer, CW Jackson, JL Cleveland, PJ Murray, JN Ihle. Phospholipase Cgamma2 is essential in the functions of B cell and several Fc receptors. *Immunity* 13: 25–35, 2000.

133. J Ruland, GS Duncan, A Elia, BB del, I, L Nguyen, S Plyte, DG Millar, D Bouchard, A Wakeham, PS Ohashi, TW Mak. Bcl10 is a positive regulator of antigen receptor-induced activation of NF-kappaB and neural tube closure. *Cell* 104: 33–42, 2001.

134. PC Lucas, M Yonezumi, N Inohara, LM McAllister-Lucas, ME Abazeed, FF Chen, S Yamaoka, M Seto, G Nunez. Bcl10 and MALT1, independent targets of chromosomal translocation in malt lymphoma, cooperate in a novel NF-kappa B signaling pathway. *J Biol Chem* 276: 19012–19019, 2001.

135. T Yoneda, K Imaizumi, M Maeda, D Yui, T Manabe, T Katayama, N Sato, F Gomi, T Morihara, Y Mori, K Miyoshi, J Hitomi, S Ugawa, S Yamada, M Okabe, M Tohyama. Regulatory mechanisms of TRAF2-mediated signal transduction by Bcl10, a MALT lymphoma-associated protein. *J Biol Chem* 275: 11114–11120, 2000.

136. I Aifantis, F Gounari, L Scorrano, C Borowski, H von Boehmer. Constitutive pre-TCR signaling promotes differentiation through Ca^{2+} mobilization and activation of NF-kappaB and NFAT. *Nat Immunol* 2: 403–409, 2001.

137. RE Voll, E Jimi, RJ Phillips, DF Barber, M Rincon, AC Hayday, RA Flavell, S Ghosh. NF-kappa B activation by the pre-T cell receptor serves as a selective survival signal in T lymphocyte development. *Immunity* 13: 677–689, 2000.

138. TD Gilmore. Multiple mutations contribute to the oncogenicity of the retroviral oncoprotein v-Rel. *Oncogene* 18: 6925–6937, 1999.

139. B Rayet, C Gelinas. Aberrant rel/nfkb genes and activity in human cancer. *Oncogene* 18: 6938–6947, 1999.

140. HL Pahl. Activators and target genes of Rel/NF-kappaB transcription factors. *Oncogene* 18: 6853–6866, 1999.

141. TR Newton, NM Patel, P Bhat-Nakshatri, CR Stauss, RJ Goulet, Jr., H Nakshatri. Negative regulation of transactivation function but not DNA binding of NF-kappaB and AP-1 by IkappaBbeta1 in breast cancer cells. *J Biol Chem* 274: 18827–18835, 1999.

2

Nuclear Receptors and Dietary Ligands

MIRIAM N. JACOBS

INTRODUCTION

Steroid, heme, and xenobiotic metabolism are controlled by a network of interrelated but unique nuclear receptors (NRs). At the molecular level, the activation of nuclear receptors by metabolites underlies most of the genetic responses controlling fat, glucose, cholesterol, bile acid, and xenobiotic/endobiotic metabolism.

This chapter examines the molecular mechanisms of interrelated steroid hormone and nuclear receptors, giving selected examples with particular reference to the endocrine system. The estrogen receptor (ER) is one of the earliest receptors observed to be affected by dietary intakes and non-nutrients including synthetic chemicals and pharmaceuticals, and was one of the first steroid hormone receptors (SHRs) to be crystallized. Thus, the starting point of this chapter is sex hormones, and especially the estrogen-mediated effects on the steroid hormone and nuclear receptors in humans where the ligands are delivered through the diet.

The Endocrine System and Endocrine Disruption

The endocrine system is complex with many organs producing different hormones, contributing to a multifaceted feedback regulatory system, which is deficient in the developing fetus and infant. Endocrine disrupting compounds (EDCs) are substances that can cause adverse effects by interfering in some way with the body's hormones or chemical messengers that constitute the endocrine system. Under the control of the central nervous system (CNS), hormones are secreted by the endocrine glands and exert control on other cells of the body. Endocrine-disrupting chemicals are differentiated from classical toxicants such as carcinogens, neurotoxicants, and heavy metals because they can interfere with normal blood hormone levels or the subsequent action of those hormones but do not have a classical toxic effect. The effects can influence and disrupt the hormonal regulation and hormonal imprinting (in the fetus) of normal cell differentiation, growth, development, metabolism, and reproduction throughout life. Endocrine disruption can occur at levels far lower than those of traditional concern to toxicologists. Sometimes high doses shut off effects that occur at low levels, and sometimes low and intermediate doses produce greater effects than those observed at high doses. Human populations are exposed to complex mixtures of dietary, environmental, and endogenous agents, which may act together or modulate one another to produce biological effects. Exogenous endocrine modulators can involve any hormonal system and be affected by normal physiological states, diet, stress, and other lifestyle factors.

Other members of the SHR family and other NRs of unknown functions (not discussed here) are also key to maintaining endocrine equilibrium.[1] For example, EDCs are known to affect the thyroid, adrenal glands, and pancreas; a clear link between DDT and pancreatic cancer was recently reported;[2] and the persistent DDT metabolite *p, p*-DDE has been shown to have antiandrogenic potential.[3] Food contaminant EDCs such as phthalic acid and nonylphenol are now known to activate the pregnane X receptor (PXR), both *in vitro* and *in vivo*, as observed in the ERs, whereas bisphenol A has been both reported to have no effect on PXR-mediated

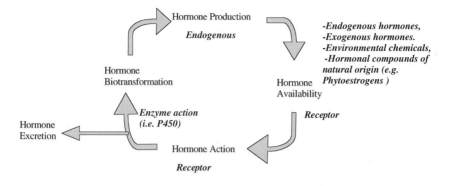

Figure 2.1 A simplified model of hormone dynamics. (Adapted from D Crain, A Rooney, E Orlando, LJ Guillette. Endocrine disrupting contaminants and hormone dynamics: lessons from wildlife. In L Jr Guillette, DA Crain, Eds., *Environmental Endocrine Disruptors, An Evolutionary Perspective.* Taylor and Francis, New York. 2000. pp 1–22.)

transcription,[4] and, later, to be a ligand for PXR[5] while also significantly enhancing ER-mediated transcription.[4]

Hormone Dynamics

Understanding a dietary chemical's potential to interact with the receptor-mediated aspects of endocrine system cell signaling depends upon adequate consideration of the entire hormone dynamic pathway, together with changes in hormone activity. Figure 2.1 presents a simplistic model of hormone dynamics, following hormones from production to excretion.

After a hormone is produced, circulating and intracellular-binding proteins regulate the hormone's bioactivity. Then, the hormone triggers action by binding to a specific cellular receptor. The hormone is next either excreted in the urine after conjugation reactions in the liver or biotransformed into another hormone, which will begin the cycle of bioavailability, action, and excretion and biotransformation.

Hormones and hormone-mimicking chemicals can potentially affect each point in the hormone dynamic cycle, each point in the steroid hormone pathway, and the enzyme families associated with the pathway. Assessing the mode of action of a nutrient or chemical is further complicated by many feedback mechanisms within and between the different hormone systems, as well as interconnections with the nervous and immune systems. Current scientific knowledge of these systems is fragmentary, but the emerging picture suggests an interlinked fabric of hormone dynamics in which the delicately balanced compensatory systems are easily perturbed. Hormone dynamics have evolved over a long period of time to deal with hormone and dietary phytochemical exposure. At the molecular level, receptors mediate alterations of hormone availability, action, excretion, and biotransformation, in concert with the cytochrome P450 enzyme system. Many receptors have been identified, and the recognition of their specific ligands and functions is awaited. It is likely that more receptors will be discovered in the future. Each type of receptor has the potential to regulate a distinct endocrine signaling pathway, as part of an evolved flexible system of great intricacy and complexity. Thornton suggests that the terminal ligand is the first for which a receptor evolves; selection for this hormone also selects for the synthesis of intermediates, despite the absence of receptors, and duplicated receptors then evolve affinity for these substances.[6] This might even be a preemptive strategy for vertebrates in combating plant defenses.[7] Although originally developing to facilitate an increasingly complex endocrine system in vertebrates, NRs have also been required to respond to a wide variety of xenobiotic challenges. They have a well-developed role as the sensory component of the defense system that includes regulation of Phase I and Phase II enzymes and transporter proteins in response to xenobiotics.

MOLECULAR MECHANICS OF ENDOCRINE ACTION: RECEPTOR-BASED MECHANISMS

There are many different steroid hormones and so many different SHRs and NRs. All cells have the ability to synthesize all receptors, but variations in the extent of gene expression means that an individual cell or tissue probably does not synthesize receptors for more than three or four hormones.[9]

This is considered to be part of the selectivity of the hormone mechanism to produce effects on certain target tissues.

The NR family has structural features in common. These include a central highly conserved DNA-binding domain (DBD) that targets the receptor to specific DNA sequences, termed hormone response elements (HREs). A C-terminal portion of this receptor includes the ligand-binding domain (LBD) that interacts directly with the hormone. Embedded within the LBD is a hormone-dependent transcriptional activation domain. The LBD acts as a molecular switch that recruits coactivator proteins and activates the transcription of target genes when flipped into the active conformation by hormone binding. The currently accepted theory of steroid hormone binding suggests that in the absence of the hormone, each receptor is associated with certain "chaperone" proteins, such as heat-shock complexes (these are other proteins that protect and aid the receptor).[10] Binding of the steroid hormone with the receptor protein causes a conformational change. This molecular switch results in the removal of the heat shock complex and allows the receptors to dimerize. Then DNA binding to a HRE occurs to produce a complex that can trigger or suppress the transcription of a selected set of genes.[10,11]

The Estrogen Receptors (ERs)

The ERs, most studied of all the NRs, may also be the ancestral origin for SHRs.[6] Based on phylogenetic analyses, and sequence reconstruction and knowledge of functional attributes of ancestral proteins, Thornton has shown that the first steroid receptor was an ER followed by a progesterone receptor. He suggests that NRs have developed through ligand exploitation and serial genome expansions. The full complement of present-day mammalian steroid receptors evolved from these ancient receptors by two large-scale genomic expansions, before and after the advent of jawed vertebrates about 450 million years ago.[6] Specific regulation of physiological processes by androgens and corticoids emerged after these events, and the classical ERα subtype[12] (see Figure 2.2) and ERβ receptors and isoforms[13–15] diverged at this

Figure 2.2   17β estradiol docked in the LBD of the ERα crystal structure. The ribbon indicates the protein backbone of the receptor. Key amino acids for binding (Arg 394, Glu 353, and His 524) are indicated in the LBD.

time,[16] suggesting that although they have evolved in parallel, this ancient duplication was to facilitate unique roles in vertebrate physiology and reproduction. The ERs are known to differ in tissue distribution and relative ligand-binding affinities,[13,15] which may help explain the selective action of estrogens in different tissues.[17] Indeed, there is evidence that ERβ plays an important role in the organization and adhesion of epithelial cells in breast tissue and is therefore instrumental in the differentiation of tissue morphology, particularly if stimulated at an early age.[18] This has important disease implications because different spliced variants are observed in malignant as opposed to normal tissues that lead to poor patient prognosis related to estrogen refractoriness.[18–21]

Tissue Differences in ER Distribution in Adults

Whereas women have ERα and ERβ in breast, uterine, and ovarian tissue, men have ERβ in the prostate and ERα in the

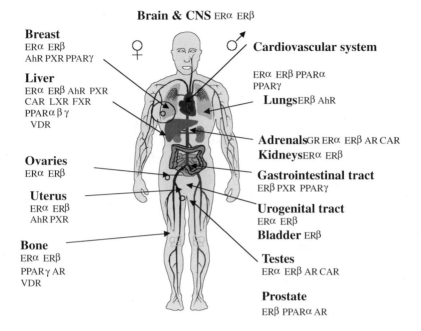

Brain & CNS ERα ERβ

Breast
ERα ERβ
AhR PXR PPARγ

Cardiovascular system
ERα ERβ PPARα
PPARγ

Liver
ERα ERβ AhR PXR
CAR LXR FXR
PPARα β γ
VDR

Lungs ERβ AhR

Adrenals GR ERα ERβ AR CAR
Kidneys ERα ERβ

Ovaries
ERα ERβ

Gastrointestinal tract
ERβ PXR PPARγ

Uterus
ERα ERβ
AhR PXR

Urogenital tract
ERα ERβ
Bladder ERβ

Bone
ERα ERβ
PPARγ AR
VDR

Testes
ERα ERβ AR CAR

Prostate
ERβ PPARα AR

Figure 2.3 Schematic diagram showing gender differences in tissue distribution of nuclear receptors (collated from the literature).

testes. Both women and men have ERα in the adrenals and kidneys but ERβ in the brain, thymus, lung, vascular system, bladder and bone, and skin.[13,23–25] Figure 2.2 shows the endogenous ligand 17β oestradiol docked into the ligand-binding site of ERα.[22] ERα dominates specifically in the reproductive tissues, whereas ERβ plays an important role in the physiology of several tissues as indicated in Figure 2.3. Female and male sex hormones can be understood to act as functional antagonists, such that an excess of estrogenic hormones may depress male development or male functions. There are also situations in which male and female hormones may act synergistically and show effects upon bone density or the promotion of liver tumors, for instance. This complexity is the consequence of the multiple targets of these hormones within mammals, including target tissue other than the sex organs, as the tissue distribution of the ERs in adults indicates.

ER and the Brain: Relationships with Neurodegenerative Conditions

The brain, and the role of ERβ in the brain, is attracting increasing interest. The expression of ERs in the adult cortex and hippocampus, areas associated with learning and memory, have been observed to be responsive to neuronal injury, suggesting a link may exist between early onset of senile dementia or Alzheimer's disease in postmenopausal women with significantly reduced estrogenic activity, compared with normal postmenopausal levels.[26,27] During fetal and postnatal development, peak neuronal cell proliferation occurs within specific brain regions, including the hippocampus and cortex. Unlike other organ systems, the brain and CNS development have a limited capacity to compensate for cell loss, and environmentally induced cell death can lead to a permanent reduction in cell number. Combined with higher relative cerebral blood flow and the immaturity of the blood–brain barrier, this can give rise to increased exposure of the brain to potential neurotoxins and EDCs.[28,29] The involvement of ERα and ERβ in the brain learning and memory centers suggests a logical mechanism of action of EDCs in the brain that could trigger learning difficulties, as reported in male ERβ-deficient mice. Although this is speculation, *in vivo* work on ER-deficient and aromatase (CYP 17A1/19A1)-deficient mice[30] may shed more light on the mechanism of the disruption of ERs and the effect of aromatase on brain function in mammals.

Receptors and "Ligand Cross Talk"

Receptors often have ligands in common (albeit with different binding affinities), and there is also a great deal of "cross talk" and "ligand promiscuity." That is, the same chemical may be able to bind with different receptors, and if they do, the binding strength may differ greatly between different receptors. This can occur for endogenous hormones, exogenous hormones, and environmental chemicals. For example, forms of the estrogenic (estradiol) and androgenic (androstanes) hormones are both ligands for ERβ. Estradiol is less potent in ERβ than with ERα, whereas the natural ligands for ERβ may actually be androgens, i.e., 5α androstane diol and 3β androstane diol.[31] The organochlorine pesticides transnonachlor and chlordane are known to activate both known ERs and PXR, but with different affinities.[13,32–34] Because these receptors are present in different ratios in different cell types and tissues, the response on a cellular, tissue, and systemic level may be quantitatively very different and may vary over time. Receptor modulation has been seen with lactation, when a form of ERβ has been observed to increase in the rat mammary glands,[31,35] and in breast tissue hyperplasia, where a frequent mutation in the ERα gene shows increased sensitivity to estrogen, compared with wild-type ERα, by affecting the border of the hinge and hormone-binding domains (HBD).[36]

Phytochemicals and EDCs may act on some but not all of the receptors and their isoforms in the tissues of these organs or act with different affinities, as methoxychlor and its analogue DDT do in ERα and ERβ. More specifically, ERβ is dependent on pure agonists for the activation of transcription from its target promoter, whereas ERα can be activated by agonists, partial agonists (such as tamoxifen, which is used in the treatment of ER-positive breast cancer), and ligand-independent mechanisms. There is also evidence that isoforms of different receptors modulate each other at a functional level, attempting to retain a balance. Different ERβ isoforms have been reported to modulate each other at low[37] and high hormone levels;[15] this may enable a tissue to govern its own responsiveness to estradiol, estradiol metabolites, and related hormones such as progestins, and estrogen mimics or EDCs. This speculation is supported by what appears to be an emerging pattern in nuclear receptor signaling, as similar balancing acts have been observed also in the α and β forms of the human glucocorticoid receptor (GR), the progesterone receptor (PR), and more recently between the PXR and the constitutive androstane receptor (CAR).[38]

Disease Scenarios: A Question of Receptor Balance or Imbalance?

The proliferative role of ERα in breast cancer is well established, but there is increasing evidence that ERβ is part of the disease picture through antiproliferative activity, which has therapeutic implications. The differences in activational mechanisms between the ERs and their isoforms will have a direct effect on possible disease scenarios such as ER-positive and ER-negative breast cancers, testicular and prostate cancers, and perhaps also brain injury.

EDCs may act on some but not all of the receptors and receptor isoforms in the tissues of these organs or act with different affinities, as many phytoestrogens (e.g., coumestrol and genistein) and synthetic chemicals such as bisphenol A[39] and organochlorine pesticides (e.g., methoxychlor and its analogue DDT) do in ERα and ERβ.

These differences between the two isoforms are of great importance with respect to the endocrine-disrupting ability of a ligand. The ability of ERβ to function both as an inhibitor or activator, depending on the agonist concentration, suggests that totally different patterns of gene expression may be observed at different hormone levels, and may be a mechanism by which cellular sensitivity to hormones is controlled.

In tissues where ERα and ERβ colocalize, fluctuations in the bioavailability of receptor-activating ligands may have a greater impact on the interactions of EDCs with both ERα and ERβ, as well as other NRs (see Figure 2.3). There can also be additional nonhormonal pathways influenced by specific compounds. A substance may act in a synergistic way on one target and in an antagonistic way on another one.[23] Such opposing effects may also occur at different dose levels of the same substance or as combinations of phytoestrogens and polychlorinated biphenyls (PCBs),[40] and other EDCs,[41] but there have been difficulties in replicating these findings, and consequently one study has been retracted.[42–44] The *in vitro* evidence of additive potency of combined EDC pesticides in ERα is far stronger[41,45,46] and has been noted for other receptors also.[47]

Phytoestrogens appear to have a greater affinity for ERβ than ERα,[13] but they have an ERα-selective efficacy,[31,48] whereas EDCs such as the hydroxylated metabolites of methoxychlor appear to be an ERα-specific agonist and ERβ antagonist[35] but with about the same affinity for both isoforms,[13] as indicated in Table 2.1.

Table 2.1   Relative Binding Affinities (RBA)
of Suspected Endocrine Disrupters for ERα and ERβ,
Adapted from Solid Phase Competition Experiments

Compound	RBA ERα	RBA ERβ
17β Estradiol	100	100
Isoflavones		
Coumestrol	20	140
Genistein	4	87
Daidzein	0.1	0.5
Pesticides		
o,p-DDT	0.01	0.02
Chlordecone	0.06	0.1
Endosulphan	<0.01	<0.01
Methoxychlor[a]	<0.01	<0.01

Note: The RBA of each competitor was calculated as a ratio of concentrations of estradiol and competitor required to reduce the specific radioligand binding by 50% (= ratio of IC_{50} values); RBA value for estradiol was arbitrarily set at 100.

[a] The metabolite of methoxychlor, 2,2-bis-(*p*-hydroxyphenyl)-1,1,1-trichloroethane (HPTE) is approximately 100-fold more active at ERα than methoxychlor. (From K Gaido, L Leonard, S Maness, J Hall, D McDonnell, B Saville, S Safe. Differential interaction of the methoxychlor metabolite 2,2-bis-(*p*-hydroxyphenyl)-1,1,1-tricloroethane with estrogen receptors α and β. *Endocrinol* 140: 5746–5753, 1999.)

Source: Adapted from G Kuiper, B Carlsson, K Grandien, E Enmark, J Haggblad, S Nilsson, JÅ Gustafsson. Comparison of the ligand binding specificity and transcript tissue distribution of estrogen receptors alpha and beta. *Endocrinol* 138: 863–870, 1997.

ERβ and Men

Perinatal exposure of the male fetus to potent estrogens is known to increase the incidence of cryptochordism and hypospadias at birth, and small testes, reduced sperm counts, epididymal cysts, prostate abnormalities, and testicular cancer in adult animals. ERβ appears to be preferentially expressed at significant levels in spermatocytes at various developmental stages in the testes, specifically the gonocytes, spermatogonia, and spermatocytes. This suggests that EDCs with an affinity for ERβ may find their way into precursors of sperm and cause disturbances in their function, disrupting male reproductive functions.[26,49–54]

Timing is all-important. During fetal testicular development, if one phase of development is out of phase with the following step (Müllerian duct regression, for instance), developmental problems ensue in the adult that may not become apparent until later developmental stages, such as puberty, have been completed. This is reflected particularly in the reports of falling sperm counts,[55] incidence of testicular cancer,[56] and decreased *in vitro* fertilization rates in couples with paternal pesticide exposure.[57] Coupled with the likelihood that the endogenous ligand for ERβ is probably not estradiol but an androgenic metabolite, investigation of the androgenic interactions with ERβ is critical for understanding the increasing adverse reproductive effects seen in men in relation to dietary exposure to hormonally active compounds.

ERβ is reported to play a central role in estrogen signaling in normal and malignant prostate epithelial cells;[58] whereas the roles played by estrogens in the transformation from a healthy prostate cell to a cancerous one remain controversial, the role of androgens in this transformation is clear. However, ERβ, AR, androstane, and CYP7B1 are known to regulate an endocrine pathway in the prostate.[59] Peroxisome proliferator–activated receptor α (PPARα) is functionally present in human prostate but is downregulated by androgens and overexpressed in advanced prostate cancer.[60]

The Many Biological Roles of ERβ

Data addressing the various biological roles of ERβ are being generated from *in vivo* studies in developing mice with a deleted ERβ gene.[18,61] These ERβ knockout (BERKO) mice display significantly reduced fertility in the female and show the essentiality of ERβ for normal functioning ovaries. At the beginning of the estrus cycle, in particular, ERβ expression is high, but after the luteinizing hormone (LH) surge, ERβ is rapidly downregulated; the ratio between ERα and ERβ in the ovary is about 1:9. Another phenotypic characteristic of BERKO animals of both sexes is that the bladder epithelium, the epithelium of the dorsal prostate, the coagulation glands, and the urethra show signs of hyperproliferation, suggesting that the growth control of these tissues is impaired if ERβ levels are compromised and that ERβ may have a protective role against hyperproliferation and carcinogenesis. For both male and female BERKO mice, reproductive behavior appears normal, although the males appear to be more aggressive under certain conditions (compared with wild-type mice).

Kuiper, Gustafsson, and co-workers are continuing to investigate the phenotypic characteristics of BERKO animals with particular reference to the cardiovascular system, bone, the immune system, sexually differentiated liver metabolism, and reproductive and nonreproductive behavior. Research into defects in ERβ expression and activity may also yield useful data for disease syndromes involving excess androgens in women, particularly as seen in women with polycystic ovarian syndrome (PCN). PCN sufferers exhibit symptoms such as secondary male characteristics, menstrual disruption, and difficulty in conceiving.[62,63]

ERβ and Ovarian Cancer

SHR expression in ovarian surface epithelial cells (the tissue of origin for over 90% of ovarian cancers) has been observed to be disrupted in ovarian cancer cells taken from postmenopausal women.[58] In the cancerous cells, the normal healthy coexpression of ERα and ERβ mRNA (along with other receptors) were disrupted with an ensuing loss of ERα, PR, and AR mRNA, suggesting that these receptors may be responsible for the neoplastic transformation of this cell type but not ERβ, whose mRNA levels appear to be unaffected by this malignant state. The researchers also suggest that ERβ action may depend upon functional ERα levels. Taken together, these findings implicate the regulation of normal ovarian cells by estrogens,[64] androgens, and progestins. Estrogen has been shown to contribute greatly to growth and development in endometrial cancer. The emergence of sex hormone resistance via downregulation or mutational inactivation of receptors (including the receptors ER, AR, CAR, and PXR described below) may be a key feature of ovarian epithelial transformation, which may lead to diseases such as endometriosis, some cases of sterility, and uterine and ovarian cancers. Recently, various levels of PXR expression have been found in endometrial cancer tissues and endometrial tissues of patients taking oral contraceptives, but not in normal tissues.[65]

Dietary Exposure: Synthetic Estrogens vs. Phytoestrogens

Phytoestrogens and lignans are found in the diet and are widely available from plant foods, including nitrogen-fixing plants and legumes such as clover, lentils, soya, grains such as rye, and also linseeds, berries and hops.[66–69] They do not rapidly accumulate and are water soluble. They are readily excreted in urine (6 to 8 h), and are probably the source of greatest dietary exposure to environmentally derived estrogen mimics. Phytoestrogens can be regarded as a defense mechanism for the plant to reproductively impair herbivore and omnivore endocrine systems and thus enable an effective strategy to reduce local herbivore populations. This was first noted in the 1950s when Australian clover was found to be the feminizing agent responsible for impairing sexual performance in rams, with a dramatic effect in lambing.[70] However, on the other hand, scientific evidence indicates that counter-defenses have evolved so that adult diets rich in phytoestrogens are associated with a reduced incidence of cardiovascular disease, breast cancer, prostate cancer, and osteoporosis. Asian women and vegetarians have a lower than average breast cancer risk, together with a relatively higher excretion of urinary phytoestrogens.[66,67,69,71–73]

Some of the possible SHR mechanisms of action of EDCs and phytoestrogens such as genistein are structure specific (Figure 2.4a), but others combine with structurally diverse ligands — for example, hydroxylated PCB 153 (Figure 2.4b).

When the structures of many phytoestrogen compounds are overlaid with the estrogen structure, they are virtually superimposable; the distances between the hydroxyl groups at each end of the molecules are almost identical. These distances determine hydrogen-bond interaction with the receptor. The more structurally diverse synthetic compounds have far greater flexibility, so they can move to, bind with, and

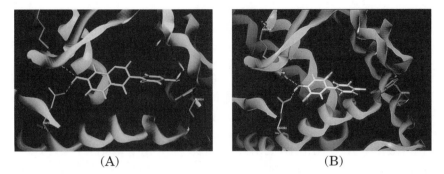

(A)   (B)

Figure 2.4 (A) Daidzein docked in the LBD of the hERα crystal structure (left) and (B) hydroxy PCB 153 docked in the LBD of the hERα crystal structure (right). The dashed lines indicate hydrogen bonding of the ligand with key amino acids (From MN Jacobs. Molecular modelling of oestrogenic compounds in the human oestrogen receptor and investigation of the activity of oestrogenic compounds in breast cancer cell lines. MSc thesis, University of Surrey, U.K., 1998.)

activate a wider range of receptors. The less flexible phytoestrogens such as genistein and daidzein, however, have suboptimal interactions in the LBD of the ERs (see Figure 2.4a in comparison with Figure 2.4b), especially in ERβ, giving a partial agonist activity compared with antagonists such as raloxifene.[48]

Synthetic hormone mimics, such as PCBs, dioxins, polybrominated flame retardants (PBDEs), and organochlorine pesticides, accumulate within the food chain and, due to their fat solubility and long half-lives, persist and bioaccumulate in adipose tissue and bone for many years, unlike phytoestrogens. DDT, for example, has a half-life of 10 years. Although dietary intakes are in far smaller doses compared with phytoestrogen intakes, they remain available as ligands to the SHR system for a far longer period of time, and the metabolized forms, particularly when hydroxylated, have greater affinity for the ERs.[22,74] The ubiquitous EDCs represent potential risk factors for a number of human cancers and other detrimental health effects. PCBs, DDT metabolites, dioxins, and PBDEs are the predominant persistent organic pollutant contaminants (POPs) of current concern in fatty animal-based foods such as oily fish, fish oils, meat, and dairy products,[75–81] as well as human breast milk.[82,83]

Phytoestrogens may influence carcinogenesis via their hormonal, antihormonal, and antioxidant actions in a beneficial way when administered throughout life. Substantial evidence indicates that diets high in plant-based foods may explain epidemiological variance of many hormone-dependent diseases that are major causes of mortality and morbidity in Western populations. Extracted phytoestrogens such as genistein (from soya) have been shown to stimulate cell proliferation and bind to ERs at relatively low levels (i.e., a low affinity), both *in vitro*[40,71–73] and *in vivo,*[67] and have been reported to have a protective effect if given before puberty[84] and an inhibitive effect on the growth of cancerous prostate cells,[85] but a detrimental effect with increased risk of uterine cancer if administered as a high-purity chemical in infancy.[86]

EDC pesticides have a more flexible molecular structure compared with phytoestrogens and thus greater affinity with the cellular receptors. It is frequently argued in the literature that the EDC pesticides bind very weakly to the estrogen receptor compared with natural estrogens and some phytoestrogens.[87–90] Quantitative structure activity relationship (QSAR) and molecular modelling studies have revealed more fully the importance of understanding differences in ligand binding within the ER,[22,91] and these differences need to be related to differences in serum levels of hormones at all stages of human development. EDC pesticides' binding affinity in the classical estrogen receptor (ERα) differs markedly from those of ERβ and other receptors (e.g., PXRs), and there are additional modes of action taking place. For example, organochlorine pesticides such as *o, p* DDT, and alachlor have been reported to partially mimic estradiol and function to suppress apoptosis (programmed cell death) in ER-responsive cells. Apoptosis appears to play a critical role in the generation and progression of cancer and is probably regulated by steroid hormones in hormone-responsive tissues.[92] In ER-negative breast cancer cells *p, p* DDT has been found to be capable of activating cellular signaling events,[93,94] so it is highly likely that some organochlorine pesticides may function through other signaling pathways.

Chronic exposure to large quantities of phytoestrogens in foods might have a direct binding effect upon the ER and other hormone receptors. Indeed, coumestrol, a very potent phytoestrogen, has uterotropic activity in the immature rat that is typical of the activity of the endogenous estradiol.[95] Newbold and coworkers[86] report that *in vivo* genistein exposure in newborn mice, at a time when the developing organism would normally be using natural estrogen signals to guide development, increases the risks of uterine cancer in adult life. The amount of genistein used by Newbold et al. was slightly higher than the amount consumed by infants but was within one order of magnitude of the level of human exposure (approximately 27 mg of genistein per day for infants feeding on formula vs. 50 mg/d in the experiment). Outbred female CD-1 mice were treated on days 1 to 5 with equivalent estrogenic doses of diethylstilbestrol (0.001 mg/kg/d) or genistein (50 mg/kg/d). At 18 months, the incidence of uterine adenocarcinoma was 35% for genistein and 31% for DES. These data suggest that genistein is carcinogenic if exposure occurs during critical periods of differentiation. Other impacts observed from genistein exposure included reductions in fertility during adulthood following exposure as a newborn.

Inappropriate estrogens can alter development by changing the intensity of the estrogen signal, whether they are from natural or synthetic sources. In the adult, such an intake may also have indirect modulating effects upon associated and related factors (e.g., heat shock proteins), such that even while acting as an estrogen at certain doses, phytoestrogens such as genistein may be also be acting upon multiple sites, having indirect (and nonreceptor-mediated) antiestrogenic and anticancer effects.

A significant but neglected dietary intake avenue of estrogens occurs in hormone residues from meat treated with sex steroids for growth promotion.[96] Although the Joint Food and Agriculture Organisation/World Health Organisation (FAO/WHO) expert committee on food additives (JECFA) and the U.S. Food and Drug Administration (FDA) consider that the residues found in meat from treated animals are safe for consumers, they have not considered the sensitivity of healthy prepubertal children to low levels of estradiol[96] or estrogen-mimicking compounds,[97] and pre- and postnatal infants to gene imprinting,[98] from this dietary source of estrogens. Possible adverse effects on human health by consumption of meat from hormone-treated animals cannot be excluded.

RECEPTORS INTERACTING WITH CYTOCHROME P450 ENZYMES: STEROIDOGENIC PATHWAY AND XENOBIOTIC METABOLISM

The NR and chaperone cofactors also mediate hormonal homeostasis by the coordinated release and degradation of bioactive hormones. Steroid hormones, their metabolites, ingested plant and animal steroids, and bioactive xenobiotic compounds are primarily metabolized by reduction and oxidation through cytochrome P450 (P450s) enzymes in the liver. Many P450s have broad substrate specificity and appear to be integrated into a coordinated metabolic pathway, such that while some receptors are ligand specific, others may have a broader specificity and low ligand affinity. In this way they can monitor aggregate levels of inducers to trigger production of metabolizing enzymes and thereby mount a defense against toxic compounds in the diet. This hypothesis is strongly supported by the expression of a receptor-sensing system in the digestive tissue.[34] P450 induction by nutrients and xenobiotics may therefore lead to alterations of endogenous regulatory pathways, with associated physiological consequences beneficial, or harmful, to health.

Aromatase (CYP17/19A1) is a key P450 in the production of estrogen, catalyzing the conversion of androgens, androstenedione, and testosterone via three hydroxylation steps to estrone and estradiol as part of the steroidogenic P450-signaling pathway[99,100] (Figure 2.5). Aromatase is expressed in many tissues, including ovaries, testes, placenta, brain, and adipose tissue of the breasts, abdomen, thighs, buttocks, and bone osteoblasts. Adipose tissue is the major site of estrogen biosynthesis in postmenopausal women, with the local production of estrogen in breast adipose tissue implicated in the development of ER-positive breast cancer tumor sites.[100,101]

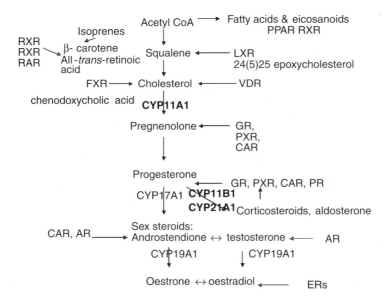

Figure 2.5 The key known receptors and P450s involved in the steroidogenic pathway. Reprinted from *Journal of Toxicology,* 205, Jacobs, M.N., *In silico* tools to aid risk assessment of endocrine disrupting chemicals, 43–53, 2004. With permission from Elsevier.

Adipose tissue is also the most significant storage site in the human body for environmental pollutants such as PCBs and DDT, compared with blood and the liver. IIDA and coworkers report measured values for PCB 118, as follows: blood, 59.2 pg/g tissue; liver, 1692 pg/g tissue; and fat, 13,896 pg/g tissue.[102]

Aromatase is also stimulated through other biological pathways. For instance, elevated levels of the cyclooxygenases COX-1 and COX-2 appear to increase aromatase activity. This can increase oestradiol biosynthesis,[103] as well as increasing prostaglandin secretion. Both provide an appropriate supply of ligands, may increase activation of peroxisome proliferators-activated receptors (PPARs) in local tissues. Inhibition of this pathway is one method that is exploited pharmacologically in ER-positive breast cancer treatments to inhibit estrogen production. Another approach is to inhibit estrogen action by antiestrogens, which interact with the estrogen receptor in the

ER positive tumors. Phytoestrogens are potent inhibitors of aromatase.[104]

Recently, rosiglitazone and troglitazone, compounds known to activate another receptor, PPARγ (a key factor in adipocyte differentiation), used in the control of insulin-resistant diabetes, have been found to inhibit aromatase expression in human breast adipose stromal cells.[30,105] They have also been observed to stimulate proliferation in breast cancer cell lines.[106] Developmental inhibition of the aromatase pathway can give rise to PCN.[62] EDCs that may affect this pathway include those that have been observed to bind with the ER. Figure 2.3 indicates current knowledge of the gender differences in the distribution of steroid hormone and nuclear receptors. There are also polymorphisms to consider, as interindividual variations in gene sequences[107] suggest that individuals may vary in terms of amount and function of all the receptors discussed herein. The hierarchy of ligand activation differs between the receptors, as well as for receptors isolated from different species. In many instances, molecules that were previously regarded as metabolic intermediates are in fact "intracrine signaling" molecules within tightly coupled metabolic pathways for altering gene expression.

The ER, the aryl hydrocarbon receptor (AhR), PXR, and CAR receptors mediate the steroidogenic pathway. PPARα regulates the catabolism of fatty acids and impacts upon several steps of the reverse cholesterol pathways. Other receptors not discussed here, such as the liver X receptor (LXR), farnesoid X receptor (FXR), and hepatic nuclear factor-4 (HNF-4), are part of the endobiotic bile regulatory pathways. The combined effects of the ligands for the nuclear receptors are vast, influencing virtually every fundamental biological process from development and homeostasis to proliferation and differentiation. The supply of ligands is a further regulatory route for the receptors. The cognate ligand for each receptor is usually dependent on the supply of a specific dietary component for its synthesis, e.g., iodine and tyrosine for thyroid hormone; cholesterol for the vitamin D receptor (VDR), GR and receptor-mediated steroidogenic pathway; fatty acids for the PPARs; and vitamin A as the ligand for

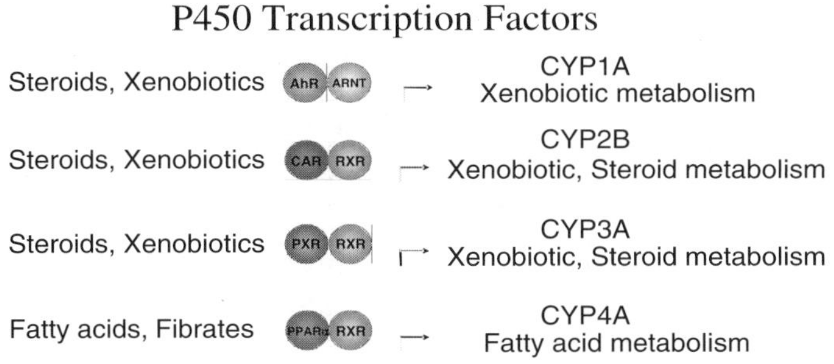

P450 Transcription Factors

Steroids, Xenobiotics (AhR ARNT) → CYP1A
Xenobiotic metabolism

Steroids, Xenobiotics (CAR RXR) → CYP2B
Xenobiotic, Steroid metabolism

Steroids, Xenobiotics (PXR RXR) → CYP3A
Xenobiotic, Steroid metabolism

Fatty acids, Fibrates (PPARα RXR) → CYP4A
Fatty acid metabolism

Figure 2.6 NRs and P450 transcription factors. Reprinted from *Journal of Toxicology,* 205, Jacobs, M.N., *In silico* tools to aid risk assessment of endocrine disrupting chemicals, 43–53, 2004. With permission from Elsevier.

all the *trans*-retinoic acid receptors, including the multiple-receptor heterodimerization partner retinoid-X-receptor (RXRα) (Figure 2.6). Table 2.2 provides a summary of the selected receptors and their known ligands.

Key Nuclear Receptors: Multifaceted Communication Systems

Members of the same nuclear receptor family share a common heterodimerization partner, RXR, and there is cross talk with other nuclear receptors and with a broad range of intracellular signaling pathways. There may be competition for RXR for the dimerization stage of receptor activation of DNA. There may even be a cascade effect, in which metabolites produced through the activities of one receptor are specific signaling molecules (and ligands) to modulate the next receptor, a link in the nuclear receptor intercommunication web of the body (Figure 2.3).

Activation of an NR may either increase or decrease the synthesis of that or another receptor. A given receptor population is not constant and may be influenced by the concentrations of circulating ligands and also the state of the receptor population.

The state of the receptor population can affect the receptors that require a heterodimerization partner. Whereas SHRs

Table 2.2   Selected Receptors and Their Ligands

Receptor	Activation Compounds	References
ERβ	5α androstane diol and 3β androstane diol, 17 β estradiol, coumestrol, genistein, daidzein, *trans*-nonachlor, endosulphan, *o,p'* DDT	13,14,31,35
PXR	Rifampicin, RU 486, SR 12813, androstanol, coumestrol, PB, TCPOBOP, pregnenolone 16α-carbonitrile (PCN), hyperforin (active constituent of St. Johns Wort), 17 β estradiol, β pregnane-3,20,dione, vitamin E, kava kava, bisphenol A, paclitaxel, guggulsterone, *trans*-nonachlor, triclosan, DDE (in rPXR), enterolactone	5,91,156–163
CAR	Androstenol, androstanol, clotrimazole, TCPOBOP (in mCAR), DDE (in rCAR) PB, 5β pregnane-3,20,dione, PCN, bilirubin	38,158, 163–165
AhR	Halogenated aromatic hydrocarbons (HAHs), e.g., polychlorinated dibenzo-*p*-dioxins (dioxins), dibenzofurans, and biphenyls (PCBs), polycyclic aromatic hydrocarbons (PAHs), e.g., benzo(*a*)pyrene, 3-methylcholanthrene, β-naphthaflavone, benzoflavones, carbaryl diaminotoluene, omeprazole, brevetoxin, indole carbinols (found in cruciferous vegetables), endogenous ligands, e.g., bilirubin, biliverdin, water-soluble metabolites of tryptophan, tryptamine, indole acetic acid, 2-(1H-indole-3carbonyl)-thiazole-4-carboxylic acid methyl ester, retinoids, antioxidants, Lipoxin A4, arachidonic acid, prostaglandins, indirubin, indigo	112,118,119, 121–123, 125, 166–169
PPARγ	Fatty acids, eicosanoids, prostaglandins, metabolites of arachidonic acid and linoleic acid, antidiabetic thiazolidinedione (TZD) drugs including NSAIDs	146,149,150, 152,170, 171
PPARα	Long-chain polyunsaturated fatty acids (LCPUFA), eicosanoids, fibrates, phthalate ester plasticizers, pristinic acid, phytanic acid	149,150,152, 154, 170–173

such as the ER, the GR, progesterone (PR), androgen (AR), and mineralocorticoid receptors (MR) bind to DNA as homodimers and recognize a palindromic element, the xenobiotic receptors PXR and CAR, the PPARs, thyroid (TR), retinoid (RAR), VDR, LXR, and HNF-4 bind to DNA as a heterodimer with RXR. The cytosolic Ah receptor differs by requiring the aryl hydrocarbon receptor nuclear transporter protein (ARNT) as the heterodimer to enable DNA binding (Figure 2.6).

To demonstrate how important it is to consider the spectrum of receptors when assessing dietary exposure to nutrients, phytochemicals, xenobiotics, and EDCs, a brief description of selected receptors and their known interactions with ERs follows. The interactions are far more complex than the brief descriptions may imply.

Aryl Hydrocarbon Receptor (AhR)

The AhR, a member of the PAS (Per-Arnt-Sim) family of nuclear regulatory basic helix-loop-helix proteins, has been detected in nearly all vertebrate groups examined[108,109] and appears to have a fundamental role in cellular physiology, neurodevelopment, and circadian rhythmicity.[110] AhR is predominantly found in hepatocytes, but also in breast cancer cells,[111] and the lung.[111,112] AhR regulates the expression of a number of genes, including cytochrome P450 1A1, 1A2, 1B1, and glutathione S transferase M (GSTM) in a ligand-dependent manner. AhR is also upregulated during cell division and in patients with cardiomyopathy.[113] Exposure to 2,3,7,8-tetrachlorodibenzo-*p*-dioxin (TCDD), the most potent AhR ligand known, results in a wide variety of species- and tissue-specific toxic and biological responses. Animals treated with 2,3,7,8-TCDD have developed abnormalities in several organs including the thyroid, thymus, lung and liver, and in immune and endocrine function. Wasting and lethality and induction of gene expression have also been shown to be AhR dependent.[114,115]

Within the cytosol of the cell, the AhR is associated with a heterodimeric transporter protein partner, ARNT. The unliganded AhR may also act through other mechanisms by being phosphorylated to key regulatory proteins such as hsp90, which appears to be required for maintaining the receptor in a nonactivated ligand-binding conformation,[110] as with the ERs and other proteins such as p37, AIP, XAP2, *src, rel*, and *Rb*.[116,117] AhR knockout mice display reduced fertility, reduced viability, and liver and immune deficits[115] in some independently generated lines but not others, as discussed in previous reviews.[110]

The best-characterized high affinity ligands for the AhR include a variety of ubiquitous manmade toxic and hydrophobic chemicals, including halogenated aromatic hydrocarbons (HAHs) such as the polychlorinated dibenzo-*p*-dioxins, dibenzofurans and biphenyls, and polycyclic aromatic hydrocarbons (PAHs) such as benzo(*a*)pyrene, 3-methylcholanthrene, benzoflavones, and other chemicals. Certain dietary indole derivatives such as indolo(3,2-*b*)carbazole appear to bind with the same affinity as that of 2,3,7,8-TCDD,[118] but a physiological receptor ligand has not been identified yet. Weaker ligands include diaminotoluene, omeprazole, brevetoxin, indole carbinols (found in cruciferous vegetables); and endogenous weak ligands such as bilirubin, biliverdin;[119,120] metabolites of tryptophan such as tryptamine and indole acetic acid;[119] equol, a daidzein metabolite, and compounds produced endogenously in the prostaglandin pathways, including arachidonic acid, prostaglandins, and lipoxin A4.[122,123] More recently, another endogenous ligand, with an indole thiazole structure, ITE (2-(1H- indole-3carbonyl)-thizole-4-carboxylic acid methyl ester), has been proposed after isolation from porcine lung. ITE has been shown to have a relatively high affinity with the AhR, with similar activity to 3βNF.[112] However, it is possible that this compound is not endogenous but an artifact formed during extraction and purification. Clearly, there is a wide gamut of structural diversity in AhR ligands, many of which are ligands for other NRs.

QSAR studies of AhR dioxin-like ligands indicate the HOMO energy to be an important descriptor with molecular planarity, together with the overall rectangularity as measured by the length to width ratio, planarity, and length and energy of the highest-occupied MO.[124] These descriptors have also been described for synthetic retinoids, with structural similarities to

2,3,7,8 TCDD and the coplanar PCBs, for the retinoic acid receptor (RAR), and the AhR–ARNT pathway.[125] *In silico* models indicate good steric complimentarity for TCDD and closely related compounds such as PCB 126.[126] The LBD of the AhR appears also to be able to accommodate many other recently identified dietary ligands, suggesting a dynamic plasticity in the LBD that may be similar to PXR, *in vitro* and *in vivo*.

Many of these ligands activate other receptors. For example, TCDD has antiestrogenic activity in the ER, whereas equol is strongly estrogenic. The endogenous ligands hint further at the developmental role of AhR, as persistent CYP1A1 and CYP1A2 gene expression has been observed by Lorenzen and Kennedy (1993) in congenitally jaundiced Gunn rats.[120] Both enzymes play a role in the oxidative metabolism of bilirubin, decreasing its toxicity and enhancing its elimination. Increased AhR induction is observed with increased oxidative stress, and it is probable that a route for antioxidant activity is through the AhR to downregulate oxidative signaling.

Pregnane X Receptor (PXR)

An important requirement for homeostasis is the detoxication and removal of endogenous hormones and xenobiotic compounds with biological activity. PXR[127–129] is involved in activating the expression of several P450 detoxifying enzymes, including CYP3A4 in the adult and CYP3A7 in the fetus in response to xenobiotics and steroids.[130] CYP3A4 is the major human hepatic P450, and has been suggested as being involved in the metabolism of over 60% of drugs in clinical use.[131] Wide species differences exist in PXR activation due to differences in the LBD.[34]

The major site of PXR expression is in liver cells and gastrointestinal tissues. They are also present in both normal and neoplastic breast tissue and endometrial cancer tissue, but not in normal endometrial tissue. In breast tissue a statistically inverse relationship has been observed between the level of PXR mRNA expression and ER status ($p = .04$).[132] With endometrial cancer tissue, those with high PXR expression showed significantly high expression of CYP3A4/7 and low expression of ER ($p < .01$), compared with levels in normal endometrial tissues showing low PXR expression (CYP3A4/7/GAPDH mRNA = 0.02 ± 0.01).[65] PXR can be activated by a variety of chemically distinct ligands, in a species-dependent manner, including endogenous hormones such as estrogen, pregnenolone, and progesterone; their synthetic derivatives such as pregnenolone 16α carbonitrile (PCN); bile acids; organochlorine pesticides; phthalic acid; nonylphenol; rifampicin; dexamethasone; corticosterone; spironolactone; phenobarbital; vitamin D_3 (1,25-(OH)2-D_3); and plant-sourced ligands such as hyperforin, coumestrol and genistein.[91,133,134] Many of these compounds activate other receptors, including liver receptors such as LXR and FXR, the ERs, AhR, and the vitamin D receptor. The percentage of inhibition in competition-binding assays for the pesticide *trans*-nonachlor is close to 100%, with 7 times greater activation than by PCN in human PXR transfection assays.[34] PXR is also essential in mediating transcriptional activation of *CYP3A* by environmental contaminants.[135]

It appears that there is a specific regulatory pathway where the accumulation of steroidal PXR ligands, including xenobiotics, results in increased CYP3A transcription and steroid catabolism, possibly providing the route for excess steroids to be eliminated from the body. So not only is PXR a xenobiotic sensor, it is also a key player in the regulation of steroid homeostasis by involvement in the expression of steroid hydroxylases and detoxication.[136] By implication, EDCs affecting PXR may also have indirect effects upon the regulation of steroid homeostasis.

Studies using the recently published PXR crystal structure, both alone and complexed with the high-affinity ligand SR 12813, indicate the five key polar residues critical for establishing the activation of PXR in the LBD. These are Ser 208, Ser 247, His 407, Arg 410, and Gln 285. The ligand-binding pocket is highly flexible and may be able to expand the number of possible hydrophobic contacts by enlarging via a pore.[137] Two of these amino acids, Gln 285 and His 407, have been shown by Östberg et al.[138] to be essential for the species-specific activation of PXR in mouse/human PXR receptor studies. Mutating the glutamine residue altered the activation of the receptor from rifampicin, the positive *in vitro* control for human PXR to PCN, the positive *in vitro* control used for rodent PXR species. The histidine residue, on the other hand, appears to be important for the basal level of activation of PXR.[138]

Some steroid hormone receptors have quite different, unanticipated mechanisms of action.

The Constitutive Androstane Receptor (CAR)

In a pattern similar to that of the ERs, based upon phylogenetic analyses it has been suggested that PXR and CAR are closely related to each other.[139] CAR is also present largely in the liver (it has also been detected by one group in the intestine, kidneys, lungs, heart, and muscle),[140] but not by Negishi's group[141] (Figure 2.3). It interacts with and is inhibited by two endogenous testosterone metabolites, androstanol and androstenol, via a mechanism that involves a widely expressed nuclear receptor coactivator, steroid hormone receptor coactivator-1 (SRC-1).[142]

Unlike most nuclear receptors, including PXR and ER, the steroidal ligand for CAR inhibits receptor-dependent gene transcription by way of a ligand-independent recruitment of transcriptional coactivators.[142] CAR functions in a manner opposite to that of the conventional nuclear-receptor pathways and can be considered a "repressed" nuclear receptor in the presence of androstane metabolites.[143] In cell-based reporter-gene assays, exogenously expressed CAR enters the nucleus and regulates the expression of target genes;[140–142] it is not present in the nucleus but is sequestered in the cytoplasm, unlike most of the other receptors discussed here. There are significant sex differences in plasma androstane levels, and it has been recently implicated as a transcriptional regulator of the gene governing the steroid hydroxylase CYP2B after binding with its cognate DNA response elements as a heterodimer with RXR.[142] There appear to be additional mechanisms for the regulation of CAR activity, including phosphorylation by phenobarbital (PB). The effects of PB on

CYP2B expression are blocked by the phosphatase inhibitor okadaic acid,[144] suggesting that dephosphorylation of CAR, rather than direct ligand binding, is involved in its translocation into the nucleus. CAR is highly responsive to PB-type compounds, including certain PCBs, pesticides, drugs, solvents, and other xenobiotics. Endogenous *CYP2B* is responsive to PB and other inducers only in the presence of suppressed CAR. At the systemic level, hormonal imbalances that affect androstanol and androstenol levels may impart a perturbation at the molecular level, preventing or inducing the inhibition of CAR and consequently the steroid hydroxylase CYP2B. Bilirubin also activates CAR but does so via binding to the CAR LBD.[38]

This receptor is a coordinate regulator of hepatic gene expression in defense against chemical toxicity. It also suggests a new area of androgen physiology whose significance is unknown as yet.[129] Negishi has presented a model of CAR where a cytoplasmic CAR retention protein (CCRP), bound with hsp90 (a cofactor held in common with the ERs and AhR), folds over to enclose the CAR. It appears that potent drugs can trigger translocation of CAR from this protein bundle, but they are not necessarily direct activators.[141] CAR also has PB-like PCB ligands in common with AhR. Both CAR and PXR m RNAs are also markedly reduced by the cytokine interleukin-6 (IL-6), but this has not been observed for the AhR or GR.[145]

Human Peroxisome Proliferator-Activated Receptor

The peroxisome proliferator-activated receptors (hPPARs) are a family of orphan receptors with fundamental roles in regulating energy balance[146–150] and gene expression by dietary fatty acids.[151] A number of prevalent metabolic disorders such as obesity, atherosclerosis, and type 2 diabetes are associated with a shift in this balance. The PPARs are activated by chemicals that elicit increases in the number and size of peroxisomes when administered to rodents.[136]

Three closely related PPARs, α, β/δ, and γ, are found in the liver, kidney, heart, and hematopoietic and adipose tissue but with different expression patterns. PPAR α is found in liver, kidney, heart, and muscle; PPAR β/δ is expressed in nearly all tissues, and PPAR γ is expressed in fat cells, the large intestine, and monocyte lineage cells.[152] They each play key roles in lipid metabolism and homeostasis; they are responsible for CYP4A induction, peroxisomal enzyme induction, and hepatic peroxisome proliferation. PPARα has a central role in hepatogenesis, and PPARγ, a central regulatory role in adipogenesis.[149,153] PPARα regulates key steps in lipid and fibrate metabolism, and this receptor is the molecular target for naturally occurring plant fatty acids (pristinic acid and phytanic acid);[154] long-chain polyunsaturated fatty acids (LCPUFA), eicosanoids;[148] peroxisome proliferators, which include drugs such as the fibrates (used widely to lower high triglyceride levels); synthetic chemicals, such as the phthalate ester plasticizers, and pesticides.[136] PPARγ ligands include fatty acids, prostaglandins, and the antidiabetic thiazolidinedione (TZD) drugs.[149,152] Pristinic acid and phytanic acid are branch-chained fatty acids obtained through the diet from the chlorophyll in plants. Present at micromolar concentrations in healthy individuals, they can accumulate in a variety of inherited disorders. Potent binding of pristinic acid and phytanic acid in PPARα[154] indicates a primary mechanism for metabolizing these dietary fatty acids. The PPARs have a far larger ligand-binding pocket than the receptors so far discussed, and there are differences in the shape of each PPAR ligand-binding pocket,[126] giving broad ligand specificity on a structural basis. For instance, docking studies of rosiglitazone occupied a fraction of the available LBD space in PPARα and less than that in PPARγ, particularly the rosiglitazone TZD head group, and thus comparatively reduced selectivity was observed. This has been observed for different ligands in the PPAR family, and is a clear descriptor for PPAR selectivity.

Another factor to be considered is modulation through cross talk between PPAR and other nuclear receptors and signaling molecules. For example, thyroid hormone suppresses hepatic PP responses and exhibits inhibitory cross talk with PPARα, due in part to competition between the thyroid receptor and PPAR for their common heterodimerization partner RXR.[155]

COMODULATORS THAT ENHANCE OR
SUPPRESS TRANSCRIPTIONAL ACTIVITY

SHR and NR interact with a group of novel nuclear proteins, including SRC-1, receptor-interacting protein 140 (RIP140), and steroidogenic factor 1 (SF-1).[174] In a yeast two-hybrid protein-interaction assay, PXR has been observed to interact with SRC1 and RIP140.[134] SRC-1, when bound to the surface of PXR, limits the flexibility of PXR, thus aiding a single active conformation of the high-affinity ligand SR12813.[175] RIP140 has been shown to interact with the ER LBD in the presence of estrogen-amplifying and potentiating ER-dependent transcriptional activity.[176] SF-1 is a key regulator of the tissue-specific expression of the cytochrome P450 steroid hydroxylases, and is essential for the embryonic survival of the primary steroidogenic organs and the regulation of reproductive function at all three levels of the hypothalamic–pituitary–gonadal axis from the earliest stages of gonadogenesis.[174] *In vitro* studies suggest that exposure to *o,p*-DDT (and other estrogenic compounds) at sufficient concentrations or in the presence of an ER coactivator could have a deleterious effect on normal cell function due to the untimely activation of estrogen-regulated genes.[176] Similarly, cellular efflux transporter proteins present in the intestine and liver, such as organic anion transporting polypeptides (e.g., OATP-C, OATP-1), P-glycoprotein (P-gp), and multidrug-resistant proteins (e.g., MDR1, MDR2), can have major effects on PXR ligand availability by interacting with PXR pathways. It has also been suggested that the differences in potency of phytoestrogen ligands in binding studies for ERs, especially ERβ, compared with the lower potency detected in whole-cell assays, may be a consequence of interactions with binding proteins.

Perturbation of the endocrine system hormones can also affect certain transport proteins, many of which are gender specific. It is relevant to note that sex steroids are known to regulate certain excretory transport proteins, such as the mRNA levels of rat organic anion transporters (OAT) in *in vivo* rat models, in a gender-specific manner. Androgens have been reported to increase rat organic anion transporter 1 (OAT1) mRNA in kidney, whereas the female GH (growth hormone) secretion pattern increases a different OAT, OAT2, with both androgens, and GH-regulating OAT3 mRNA levels in the liver.[177]

The implications of the interactions with transport proteins, such as OATP, P-gp, MRD1, and MRD2 can be significant with respect to ligand availability.

CONCLUSION

Receptor–Ligand Interactions

The NRs discussed here act as steroid, metabolic, and toxicological sensors that have evolved to provide a fast, highly adaptive response system to environmental changes by inducing the appropriate metabolic pathways that will provide the optimum protection from ligand overexposure. For the xenobiotic receptor sensors, these compounds can be present at high concentrations (μM range), unlike the classical SHRs (e.g., ERs), where the endocrine signaling molecules are generated in minute quantities (nM range). Xenobiotic NRs have generally evolved in response to intermediates from metabolic pathways, phytochemicals, and metabolites of dietary origin. Now, however, humans are chronically exposed to a wider range of ligands for these receptors, due to a Westernized lifestyle characterized by high fat and processed food intakes, and exposure to drugs and xenobiotics.

Hydrogen-bonding properties of the ligands have been a key part of the evolutionary development of NR. Hydrogen bonding has long been recognized as an important type of molecular interaction that is ubiquitous in nature, being found in systems ranging from water to cellular proteins and DNA.[178,179] The variability in forms of hydrogen bonding[180,181] give rise to an even more elaborate communication system, where it may only be a difference between a few key residues in the LBD that produces species variability and within-species receptor differences in response.

Ligands bind with, activate, or suppress receptors in a wide variety of ways, but conformational change that leads to DNA interaction is not always induced. For instance, recent work by Henry and Gasiewicz[182] shows that agonists, but not antagonists (such as dietary and synthetic flavones), cause a conformational change to the AhR DRE-binding conformation *in vitro*. The mutation of one key residue in the androgen receptor LBD ablates the ligand binding of testosterone,[183] whereas the mutation of four specific LBD residues (or two different residues in mouse and human each) causes a species specificity change in PXR.[137,138]

Role of Protein and Enzyme Coregulators

Molecular mechanisms of ligand–receptor binding and gene activation include a large number of modulatory protein transcription factors and transporters as part of the receptor activation process. There is an intricate network of these cofactors embedded into the NR communication web, each one having functionally significant interrelated roles with specific receptors, sometimes several receptors, and other proteins. Many cofactors act in concert with each receptor to facilitate transport, activation, or repression. Common intermediaries are shared in receptor and other signaling pathways. Some of the cofactors are known to have functionally significant roles for the receptors, particularly for PXR, in which SRC-1 promotes the specific interaction between ligand and receptor.[175] Receptors are also known to regulate cofactors; thus, PXR and CAR not only coregulate xenobiotic metabolism but also drug transport.[184] As previously discussed, the key cofactors that are part of the cross-talk pathways between the AhR, ERs, PXR, CAR, and PPARs, shown in Figure 2.7, emphasize the influence that cofactors and receptors exert upon each other to maintain equilibrium.

Receptor–Receptor and Receptor–Enzyme Interactions

Receptor–receptor and receptor–enzyme interactions constitute a very large field of study, and this chapter has attempted to describe some of the interactions between the ERs, AhR, PXR, CAR, the PPARs (to a lesser degree), and the P450s these receptors induce. Here, three further examples are given of the many varied signaling pathways between related NRs.

The AhR and ER have a well-established intricate relationship which may be further modulated or regulated by

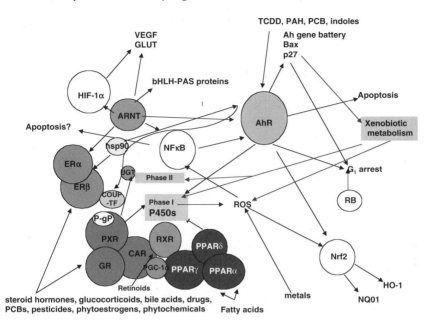

Figure 2.7 Functionally significant cross-talk pathways between nuclear receptors and selected cofactors. (Adapted from D Carpenter, K Arcaro, D Spink. Understanding the human health effects of chemical mixtures. *Environ Health Perspect* 110: 25–42, 2002.)

Abbreviations not included in the text: Glut1, glucose transporter 1; HIFα, Hypoxia Inducible Factor α; Nrf2, nuclear factor drythrod 2-related factor; RB, retinoblastoma protein; VEGF, vascular endothelial growth factor.

COUP-TF1. COUP-TF, known to be part of the CYP3A signaling pathway,[185] has been shown by Klinge and colleagues to regulate AhR action *in vitro* by direct competitive DNA binding and protein–protein interactions.[186]

As part of the CYP3A signaling pathway, Drocourt et al.[187] have recently shown in primary human hepatocytes that, together with PXR and CAR, the VDR has a regulatory (but not primary) role, controlling the basal and inducible expression of *CYP3A4, CYP2B6,* and *CYP2C9* (which is primarily a GR-responsive gene). They suggest that the expression of VDR-controlled genes might be affected by xenobiotics, such as rifampicin, through the PXR and CAR pathway.

In vivo studies in an immature gonadotrophin-primed female rat model have shown that in ovaries (an estrogen-responsive tissue), COX-2 is induced the morning after ovulation, but greatly inhibited by TCDD via AhR signaling.[188] GR was also observed to be downregulated. Blockage or reduction of COX-2 expression is known to impair ovulation. There appears to be a feedback loop in the ovary between AhR signaling and reduced COX-2 expression that may depend upon a reduced leutinising hormone (LH) surge due to the stimulation of corticoids, via GR autoregulation, and the inhibition of gonadotrophin surges.[188]

Heterodimerization Partners

Belonging to the steroid hormone nuclear-receptor superfamily, ubiquitous RXRα is the necessary heterodimerization partner for many receptors, including the thyroid hormone receptor, ERs, PXR, CAR,[189] LXR, FXR, PPARα,[190] and PPARγ,[191] to bind to DNA. The crystal structure data of the PPARγ-RXRα heterodimer shows the asymmetric conserved heterodimerization interfaces of both receptors.[192] RXRα dimerizes through a 40-amino-acid subregion within the LBD, known as the "identity box." Mutation of two important determinants (alanine 416 and arginine 421) within this box has been shown to impair the actions of receptor dimerization partners.[193] RXRα is established as a heterodimeric integrator of multiple physiological processes in the liver and is a regulatory component of cholesterol,[194–196] fatty acid,[197] bile acid,[198,199] steroid and xenobiotic metabolism, and homeostasis.[200] The retinoid ligands of RXR have distinct effects in different contexts and have been reported to significantly alter the response of the CAR-RXR heterodimer to CAR ligands.[201]

Suppression of RXRα has a concomitant effect upon the heterodimerization partner. For example, LXR is reported in this way to inhibit PPARα signaling in the nutritional regulation of fatty acid metabolism. PPARα has a counterinhibitory action repressing LXR/RXR binding through the sterol regulatory-element binding protein −1c (SREBP-1c).[202,203]

RXRα also has cofactors in common with the NR; for instance, Weibel et al. (1999) have shown a competitive element between RIP140 and SRC-1 in binding OR-1 with RXR to heterodimers of a novel orphan receptor. RIP140 is also implicated in the potentiation of EDCs in yeast ER *in vitro* studies.[176] SRC-1 RXR phosphorylation can also be induced through stress pathway activation,[193] which would reduce RXR heterodimerization availability for other receptor partners. ARNT, the heterodimerization partner of AhR,[204] is also a potent coactivator of ER-dependent transcription that requires the estradiol-activated LBD of ERα or ERβ,[205] but, unlike AhR, does not also interact directly with orphan receptor COUP-TF.[186]

Final Remarks

NRs are dynamic protein structures and will adapt in different ways to given situations. Thus, the *in silico* models, the *in vitro* host cell systems, and even the receptor null *in vivo* studies cannot give precise information on the real-life situation. At best, they can only suggest possible molecular mechanisms of activation, predict cellular mechanisms *in vitro,* and indicate the possible physiological pathways *in vivo.* Other receptors, signaling proteins and transporters will expand their metabolic activities to fulfill the role required of the missing receptor to adapt and optimize the receptor signaling pathways in living systems. However, the utilization of *in silico* and *in vitro* tools can help expand our understanding of how the receptors function as part of hormone dynamics and xenobiotic metabolism. Applying these techniques to an extended range of receptors, and new crystal structures as they become available, is a primary avenue of future research. NRs are important dietary targets for preventative and restorative health care, as well as drug targets for intervention in disease processes.

Exogenous compounds that target these receptors can disrupt both normal and abnormal functioning of these key metabolic pathways. While environmental hormone mimics contribute to detrimental health effects by activating certain receptors and disturbing normal function, there are therapeutic uses from both dietary and pharmacolocial treatment for abnormal functioning of the hormone pathways and hormone dependent diseases.

REFERENCES

1. M Parker. Steroid and related receptors. *Current Opinion in Cell Biology* 5: 499–504, 1993.

2. M Porta, N Malats, M Jariod, J Grimault, J Rifa, A Carrato, L Guarner, A Salas, M Santiago-Silva, J Vorominas, M Andreu, F Real. Serum concentrations of organochlorine compounds and K-*ras* mutations in exocrine pancreatic cancer. *Lancet* 354: 2125–2129, 1999.

3. R Kelce, R Stone, C Laws, E Gray, A Kemppainen, M Wilson. Persistent DDT metabolite *p,p'*-DDE is a potent androgen receptor antagonist. *Nature* 373: 581–585, 1995.

4. H Masuyama, Y Hiramatsu, M Kunitomi, T Kudo, P MacDonald. Endocrine disrupting chemicals, phthalic acid and nonylphenol, activate pregnane X receptor-mediated transcription. *Mol Endocrinol* 14: 421–428, 2000.

5. A Takeshita, N Koibuchi, J Oka, M Taguchi, Y Shishiba, Y Ozawa. Bisphenol-A, an environmental estrogen, activates the human orphan nuclear receptor, steroid and xenobiotic receptor-mediated transcription. *Eur J Endocrinol* 145: 513–517, 2001.

6. J Thornton. Evolution of vertebrate steroid receptors from an ancestral estrogen receptor by ligand exploitation and serial genome expansions. *Proc Nat Acad Sci USA* 98: 5671–5676, 2001.

7. K Wynne-Edwards. Evolutionary biology of plant defenses against herbivory and their predictive implications for endocrine disruptor susceptibility in vertebrates. *Environ Health Perspect* 109: 443–448, 2001.

8. D Crain, A Rooney, E Orlando, LJ Guillette. Endocrine disrupting contaminants and hormone dynamics: lessons from wildlife. In LJ Guillette, DA Crain, Eds., *Environmental Endocrine Disruptors, An Evolutionary Perspective.* Taylor and Francis, New York. 2000. pp 1–22.

9. P Gard. Human Endocrinology. Taylor and Francis, London, 2001.

10. N Weigel. Steroid hormone receptors and their regulation by phosphorylation. *Biochem J* 319: 657–667, 1996.

11. B Alberts, D Bray, A Johnson, J Lewis, M Raff, K Roberts, P Walter. *Essential Cell Biology*, Garland Publishing, New York, 1998.

12. A Brzozowski, A Pike, Z Dauter, R Hubbard, T Bonn, O Engstrom, L Ohman, G Greene, JÅ Gustafsson, M Carlqvist. Molecular basis of agonism and antagonism in the oestrogen receptor. *Nature* 389: 753–758, 1997.

13. G Kuiper, B Carlsson, K Grandien, E Enmark, J Haggblad, S Nilsson, JÅ Gustafsson. Comparison of the ligand binding specificity and transcript tissue distribution of estrogen receptors alpha and beta. *Endocrinol* 138: 863–870, 1997.

14. G Kuiper, J Lemmen, B Carlsson, J Corton, S Safe, P van der Saag, B van der Burg, JÅ Gustaffson. Interaction of estrogenic chemicals and phytoestrogens with estrogen receptor β. *Endocrinol* 139: 4252–4263, 1998.

15. D Petersen, G Tkalcevic, P Koza-Taylor, T Turi, T Brown. Identification of estrogen receptor beta (2), a functional variant of estrogen receptor beta expressed in normal rat tissues. *Endocrinol* 139: 1082–1092, 1998.

16. S Kelley and V Thackray. Phylogenetic analyses reveal ancient duplication of estrogen receptor isoforms. *J Mol Evol* 49: 609–614, 1999.

17. JÅ Gustafsson. Estrogen receptor β — a new dimension in estrogen mechanism of action. *J Endocrinol* 163: 379–383, 1999.

18. C Forster, S Makela, A Warri, S Kietz, D Becker, K Hultenby, M Warner, JÅ Gustafsson. Involvement of estrogen receptor β in terminal differentiation of mammary gland epithelium. *Proc Nat Acad Sci USA* 99: 15578–15583, 2002.

19. J Fujimoto, H Sakaguchi, I Aoki, T Tamaya. Steroid receptors and metastatic potential in endometrial cancers. *Eur J Cancer* 36: S33, 2000.

20. J Fujimoto, R Hirose, H Sakaguchi, T Tamya. Clinical significance of expression of estrogen receptor α and β mRNAs in ovarian cancers. *Oncology* 58: 334–341, 2000.

21. Y Omoto, S Kobayashi, S Inoue, S Ogawa, T Toyama, H Yamashita, M Muramatsu, JÅ Gustafsson, H Iwase. Evaluation of oestrogen receptor β wild-type and variant protein expression, and relationship with clinicopathological factors in breast cancers. *Eur J Cancer* 38: 380–386, 2002.

22. MN Jacobs. Molecular modelling of oestrogenic compounds in the human oestrogen receptor and investigation of the activity of oestrogenic compounds in breast cancer cell lines. MSc thesis, University of Surrey, U.K., 1998.

23. K Paech, P Webb, G Kuiper, S Nilsson, JÅ Gustafsson, P Kushner, T Scanlan. Differential ligand activation of estrogen receptors ER and ER at AP1 sites. *Science* 277: 1508–1510, 1997.

24. M Tetsuka, R Anderson, S Hillier. Distribution of messenger RNA's encoding oestrogen receptor isoforms in female reproductive tissues. *J Endocrinol* 152: OC20, 1997.

25. M Thornton, A Taylor, K Mulligan, F Al-Azzawi, C Lyon, J O'Driscoll, A Messenger. Oestrogen receptor beta is the predominant oestrogen receptor in human scalp skin. *Exp Dermatology* 12 (2): 181–190, 2003.

26. P Shughrue, I Merchenthaler. Estrogen is more than just a "sex hormone": Novel sites for estrogen action in the hippocampus and cerebral cortex. *Frontiers Neuroendocrinol* 21: 95–101, 2000.

27. P Wise, D Dubal, M Wilson, S Rau, M Bottner. Minireview: Neuroprotective effects of estrogen — new insights into mechanisms of action. *Endocrinol* 142: 969–973, 2001.

28. E Faustman, S Silbernagel, R Fenske, T Burbacher, R Ponce. Mechanisms underlying children's susceptibility to environmental toxicants. *Environ Health Perspect* 108: 13–21, 2000.

29. GW ten Tusscher, JG Koppe, Perinatal dioxin exposure and later effects — a review. *Chemosphere* 54: 1329–1336, 2004.

30. E Simpson, G Rubin, C Clyne, K Robertson, L O'Donnell, M Jones, S Davis. The role of local estrogen biosynthesis in males and females. *Trends Environ Med* 11: 184–188, 2000.

31. JÅ Gustafsson. The Role of the ERβ in oestrogen mediated carcinogenesis. *Toxicol Lett* 116: 1, 2000.

32. T Barkhem, B Carlsson, Y Nilsson, E Enmark, JÅ Gustafsson, S Nilsson. Differential response of estrogen receptor α and estrogen receptor to partial estrogen agonists/antagonists. *Mol Pharmacol* 54: 105–112, 1998.

33. D Waxman. P450 gene induction by structurally diverse xenochemicals: central role of nuclear receptors CAR, PXR and PPAR. *Arch Biochem Biophys* 369: 11–23, 1999.

34. S Jones, L Moore, J Shenk, G Wisely, G Hamilton, D McKee, N Tomkinson, E LeCluyse, M Lambert, T Wilson, S Kliewer, J Moore. The pregnane X receptor: a promiscuous xenobiotic receptor that has diverged during evolution. *Mol Endocrinol* 14: 27–39, 2000.

35. K Gaido, L Leonard, S Maness, J Hall, D McDonnell, B Saville, S Safe. Differential interaction of the methoxychlor metabolite 2,2-bis-(p-hydroxyphenyl)-1,1,1-tricloroethane with estrogen receptors α and β. *Endocrinol* 140: 5746–5753, 1999.

36. S Fuqua, C Wiltschke, Q Zhang, A Borg, C Castles, W Friedrichs, T Hopp, S Hilsenbeck, S Mohsin, P O'Connell, D Allred. A hypersensitive estrogen receptor alpha mutation in premalignant breast lesions. *Cancer Res* 60: 4026–4029, 2000.

37. J Hall, D McDonnell. The estrogen receptor β-isoform (ERβ) of the human estrogen receptor modulates ER α transcriptional activity and is a key regulator of the cellular response to estrogens and antiestrogens. *Endocrinol* 140: 5566–5578, 1999.

38. P Honkakoshi, T Sueyoshi, M Negishi. Drug-activated nuclear receptors CAR and PXR. *Annals of Medicine* 35: 172–182, 2003.

39. J Matthews, K Twomey, T Zacharewski. *In vitro* and *in vivo* interactions of bisphenol A and its metabolite, bisphenol a glucuronide, with estrogen receptors α and β. *Chem Res Toxicol* 14: 149–157, 2001.

40. MN Jacobs, S Starkey, R Hoffman. Combination effects of different ratios of PCB 153 and the phytoestrogen genistein upon DNA synthesis in an MCF-7 breast cancer cell line. *Organohalogen Comp* 42: 73–77, 1999.

41. J Payne, M Scholze, A Kortenkamp. Mixtures of four organochlorines enhance human breast cancer cell proliferation. *Environ Health Perspect* 109: 391–397, 2001.

42. S Arnold, D Klotz, B Collins, P Vonier, L Guillette Jr, J McLachlan. Synergistic activation of estrogen receptor with combinations of environmental chemicals. *Science* 272: 1489–1492, 1996.

43. S Arnold, P Vonier, B Collins, D Klotz, L Guillette Jr, J McLachlan. *In vitro* synergistic interaction of alligator and human estrogen receptors with combinations of environmental chemicals. *Environ Health Perspect* 105: 615–632, 1997.

44. J McLachlan. Synergistic effect of environmental oestrogens: report withdrawn. *Science* 277: 463–464, 1997.

45. K Ramamoorthy, K Wang, IC Chen, S Safe, J Norris, D McDonnell, K Gaido, W Bocchinfuso, K Korach. Potency of combined estrogenic pesticides. *Science* 275: 405–406, 1997.

46. E Silva, N Rajapanske, A Kortenkamp. Something from "nothing" — eight weak estrogenic chemicals combined at concentrations below NOECs produce significant mixture effects. *Environ Sci Technol* 36: 1751–1756, 2002.

47. D Carpenter, K Arcaro, D Spink. Understanding the human health effects of chemical mixtures. *Environ Health Perspect* 110: 25–42, 2002.

48. A Pike, A Brzozowski, R Hubbard, T Bonn, AG Thorsell, O Engstrom, J Ljunggren, JÅ Gustafsson, M Carlquist. Structure of the ligand-binding domain of oestrogen receptor beta in the presence of a partial agonist and a full agonist. *EMBO J* 18: 4608–4618, 1999.

49. A Giwercman, E Carlsen, N Keiding, N Skakkebaek. Evidence for the increasing incidence of abnormalities of the human testis: A review. *Environ Health Perspect* 102: 56–71, 1993.

50. R Hess, D Bunick, KH Lee, J Bahr, J Taylor, K Korach, D Lubahn. A role for oestrogens in the male reproductive system. *Nature* 390: 509–512, 1997.

51. R Sharpe. Do males rely on female hormones? *Nature* 390: 447–448, 1997.

52. R Sharpe, N Atanaessova, C McKinnell, P Parte, K Turner, J Fisher, J Kerr, N Groome, S Macpherson, M Millar, P Saunders. Abnormalities in functional development of the Sertoli cells in rats treated neonatally with diethylstilbestrol: A possible role for estrogens in Sertoli cell development. *Biology of Reproduction* 59: 1084–1094, 1998.

53. R Sharpe. Hormones and testis development and the possible adverse effects of environmental chemicals. *Toxicol Lett* 120: 221–232, 2001.

54. J Richtoff, L Rylander, B Jönsson, H Åkesson, L Hagmar, P Nilsson-Ehle, M Stridsberg, A Giwercman. Serum levels of 2,2′,4,4′,5,5′-hexachlorobiphenyl (CB-153) in relation to markers of reproductive function in young males from the general Swedish population. *Environ Health Perspect* 111(4): 409–413, 2003.

55. J Auger, J Kunstmann, F Czyglik, P Jouannet. Decline in semen quality among fertile men in Paris during the past 20 years. *New Eng J Med* 332: 281–285, 1995.

56. HO Adami, R Bergstrom, M Mohner, W Zatonski, H Storm, A Ekbom, S Tretli, L Teppo, H Zeigler, M Rahu, R Gurevicius, A Stengrevics. Testicular cancer in nine northern European countries. *Int J Cancer* 59: 33–38, 1994.

57. E Tielemans, R van Kooij, E Te Velde, A Burdoff, D Heederik. Pesticide exposure and decreased fertilisation rates *in vitro*. *Lancet* 354: 484–485, 1999.

58. KM Lau, S Mok, SM Ho. Expression of human estrogen receptor α and, progesterone receptor and androgen receptor mRNA in normal and malignant ovarian epithelial cells. *Proc Nat Acad Sci USA* 96: 5722–5727, 1999.

59. Z Weihua, R Lathe, M Warner, JÅ Gustafsson. An endocrine pathway in the prostate, ERβ, AR, 5α-androstane-3β, 17β-diol, and CYPB1, regulates prostate growth. *Proc Nat Acad Sci USA* 99: 13589–13594, 2002.

60. G Collett, A Betts, M Johnson, A Pulimood, S Cook, D Neal, C Robson. Peroxisome proliferator-activated receptor a is an androgen-responsive gene in human prostate and is highly expressed in prostatic adenocarcinoma. *Clin Cancer Res* 6: 3241–3248, 2000.

61. JÅ Gustafsson. Estrogen receptor — a new dimension in estrogen mechanism of action. *J Endocrinol* 163: 379–383, 1999.

62. S Franks, N Gharani, C Gilling-Smith. Polycystic ovary syndrome: evidence for a primary disorder of ovarian steroidogenesis. *J Steroid Biochem Mol Biol* 69: 269–272, 1999.

63. D Abbott, D Dumesic, S Franks. Developmental origin of polycystic ovarian syndrome — a hypothesis. *J Endocrinol* 174: 1–5, 2002.

64. K Britt, J Finday. Estrogen activity in the ovary revisited. *J Endocrinol* 175: 269–276, 2002.

65. H Masuyama, Y Hiramatsu, Y Kodama, T Kudo. Expression and potential roles of pregnane X receptor in endometrial cancer. *J Clin Endocrinol Metab* 88(9): 4446–4454, 2003.

66. H Aldercreutz. Phytoestrogens and human health. British Toxicological Society Annual Congress, University of Surrey, U.K., 19–22 April 1998, pp 45.

67. S Milligan, A Balasubramanian, J Kalita. Relative potency of xenobiotic estrogens in an acute *in vivo* mammalian assay. *Environ Health Perspect* 106: 23–26, 1998.

68 S Milligan, J Kalita, A Heyerick, H Rong, L De Coomsan, D De Keukeleire. Identification of a potent phytoestrogen in hops (*Humulus lupulus* L.) and beer. *J Clin Endocrinol Metab* 84: 2249–2252, 1999.

69 A Cassidy, M Faughnan. Phyto-oestrogens through the life cycle. *Proc Nutr Soc* 59: 489–496, 2000.

70. R Bradbury, D White. Estrogens and related substances in plants. *Vitamins and Hormones* 12: 207–233, 1954.

71. A Cassidy. Physiological effects of phyto-oestrogens in relation to cancer and other human health risks. *Proc Nutr Soc* 55: 399–417, 1996.

72. S Bingham, C Atkinson, J Liggins, L Bluck, A Coward. Phyto-oestrogens: Where are we now? *Br J Nutr* 79: 393–406, 1998.

73. A Cassidy, S Milligan. How significant are environmental oestrogens to women? *Climeractic* 1: 229–242, 1998.

74. I Meerts, R Letcher, S Hoving, G Marsh, A Bergman, J Lemmen, B van der Berg, A Brower. *In vitro* estrogenicity of polybrominated diphenyl ethers, hydroxylated PBDEs and polybrominated bisphenol A compounds. *Environ Health Perspect* 109: 399–407, 2001.

75. A Liem, R Theelen. Dioxins: Chemical analysis, exposure and risk assessment. PhD thesis. National Institute of Public Health and the Environment, Bilthoven, The Netherlands, 1997.

76. MN Jacobs, D Santillo, P Johnston, C Wyatt. Organo-chlorine residues in fish oil dietary supplements: comparison with industrial grade oils. *Chemosphere* 37: 1709–1731, 1998.

77. A Liem, P Furst, C Rappe. Exposure of populations to dioxins and related compounds. *Food Addit Contam* 17: 241–259, 2000.

78. MN Jacobs, A Covaci, P Schepens. Investigation of selected persistent organic pollutants in farmed atlantic salmon (*Salmo salar*), salmon aquaculture feed, and fish oil components of the feed. *Environ Sci Technol* 36: 2797–2805, 2002.

79. MN Jacobs, J Ferrario, C Byrne. Investigation of polychlorinated dibenzo-*p*-dioxins, dibenzo-*p*-furans and selected coplanar biphenyls in Scottish farmed Atlantic salmon (*Salmo salar*). *Chemosphere* 46: 183–191, 2002.

80. MN Jacobs, A Covaci, A Ghoerghe, P Schepens. A time trend investigation of PCBs, PBDEs and organochlorine pesticides in selected *n*-3 polyunsaturated fatty acid rich dietary fish oil and vegetable oil supplements; nutritional relevance for human essential *n*-3 fatty acid requirements. *J Agric Food Chem* 52: 1780–1788, 2004.

81. R Hites, JA Foran, DO Carpenter, MC Hamilton, BA Knuth, SJ Schwager. Global assessment of organic contaminants in farmed salmon. *Science* 303: 226–229, 2004.

82. G Solomon, P Weiss. Chemical contaminants in breast milk: time trends and regional variability. *Environ Health Perspect* 110: A339–A347, 2002.

83. G Solomon, A Huddle. Low levels of persistent organic pollutants raise concerns for future generations. *J. Epidemiol Community Health* 56: 826–827, 2002.

84. C Lamartinière, JHM Moore, S Barnes. Neonatal genistein chemoprevents mammary carcinogenesis. *Proc Soc Exp Biol Med* 208: 120–123, 1995.

85. G Hillman, J Forman, O Kucuk, M Yudelev, R Maughan, J Rubio, A Layer, S Tekyi-Mensah, J Abrams, F Sarker. Genistein potentiates the radiation effect on prostate carcinoma cells. *Clin Cancer Res* 7: 382–390, 2001.

86. R Newbold, E Padilla Banks, B Bullock, W Jefferson. Uterine adenocarcinoma in mice treated neonatally with genistein. *Cancer Res* 61: 4325–4328, 2001.

87. S Safe. Environmental and dietary estrogens and human health: Is there a problem? *Environ Health Perspect* 103: 346–351, 1995.

88. S Safe, K Connor, K Ramamoorthy, K Gaido, S Maness. Human exposure to endocrine-active chemicals: hazard assessment problems. *Reg Toxicol Pharmacol* 26: 52–58, 1997.

89. S Safe. Xenoestrogens and breast cancer. *New Eng J Med* 337: 1303–1304, 1997.

90. S Safe. Is there an association between exposure to environmental estrogens and breast cancer? *Environ Health Perspect* 105: 675–678, 1997.

91. MN Jacobs. Investigations of ligand interactions with nuclear receptors and implications for human health. PhD thesis. University of Surrey, U.K., 2003.

92. ME Burow, Y Tang, BM Collins-Burow, S Krajewski, JC Reed, JA McLachlan, BS Beckman. Effects of environmental estrogens on tumor necrosis factor-mediated apoptosis in MCF-7 cells. *Carcinogenesis* 20: 2057–2061, 1999.

93. K Shen, R Novak. Differential effects of aroclors and DDT on growth factor gene expression and receptor tyrosine kinase activity in human breast epithelial cells. *Adv Exp Med Biol* 407: 295–302, 1997.

94. K Shen, R Novak. DDT stimulates c-erbB2, c-met, and STATS Tyrosine phosphorylation, Grb2-sos association, MAPK phosphorylation, and proliferation of human breast epithelial cells. *Biochem Biophys Res Comm* 231: 17–21, 1997.

95. J Ashby, H Tinwell, A Soames, J Foster. Induction of hyperplasia and increased DNA content in the uterus of immature rats exposed to coumestrol. *Environ Health Perspect* 107: 819–822, 1999.

96. AM Andersson, N Skakkebaek. Exposure to exogenous estrogens in food: possible impact on human development and health. *Eur J Endocrinol* 140: 477–485, 1999.

97. K Howdeshell, A Hotchkiss, K Thayer, J Vandenbergh, F von Saal. Exposure to bisphenol A advances puberty. *Nature* 401: 763–765, 1999.

98. J McLachlan, M Burow, TC Chiang, LS Fang. Gene imprinting in developmental toxicology: a possible interface between physiology and pathology. *Toxicol Lett* 120: 161–164, 2001.

99. C Martucci, J Fishman. P450 enzymes of estrogen metabolism. *Pharmacol Ther* 57: 237–257, 1993.

100. A Brodie, Q Lu, B Long. Aromatase and its inhibitors. *J Steroid Biochem Molec Biol* 69: 205–210, 1999.

101. D Liao, R Dickson. Roles of androgens in development, growth and carcinogenesis of the mammary gland. *J Steroid Biochem Molec Biol* 80: 175–189, 2002.

102. T Iida, H Hirakawa, T Matsueda, J Nagayama, T Nagata. Polychlorinated dibenzo-*p*-dioxins and related compounds: correlations of levels in human tissues and in blood. *Chemosphere* 38: 2767–2774, 1999.

103. J Richards, T Petrel, R Brueggemeier. Signaling pathways regulating aromatase and cycloxygenases in normal and malignant breast cells. *J Steroid Biochem Molec Biol* 80: 203–212, 2002.

104. P Whitten, H Patisaul. Cross-species and interassay comparisons of phytoestrogen action. *Environ Health Perspect* 109: 5–20, 2001.

105. G Rubin, Y Zhao, A Kalus, E Simpson. Peroxisome proliferator-activated receptor g ligands inhibit estrogen biosynthesis in human breast adipose tissue: possible implications for breast cancer therapy. *Cancer Res* 60: 1604–1608, 2000.

106. E Mueller, P Sarraf, P Tontonoz, R Evans, K Martin, M Zhang, C Fletcher, S Singer, B Spiegelman. Terminal differentiation of human breast cancer through PPAR gamma. *Mol Cell* 1: 465–470, 1998.

107. N Masahiko, P Honkakoski. Induction of drug metabolism by nuclear receptor CAR: molecular mechanisms and implications for drug research. *Eur J Pharm Sci* 11: 259–264, 2000.

108. M Hahn, B Woodin, J Stegeman, D Tillitt. Aryl hydrocarbon receptor function in early vertebrates: inducibility of cytochrome P450 1A in agnathan and elasmobranch fish. *Comp Biochem Physiol* Part C 120: 67–75, 1998.

109. M Hahn. Aryl hydrocarbon receptors: diversity and evolution. *Chemico-Biol Interact* 141: 131–160, 2002.

110. L Poellinger. Mechanistic aspects-the dioxin (aryl hydrocarbon) receptor. *Food Addit Contam* 17: 261–266, 2000.

111. T Nguyen, D Hoivik, JE Lee, S Safe. Interactions of nuclear receptor coactivator/corepressor proteins with the aryl hydrocarbon receptor complex. *Arch Biochem Biophys* 367: 250–257, 1999.

112. J Song, M Clagett-Dame, R Peterson, M Hahn, W Westler, R Sicinski, H DeLuca. A ligand for the aryl hydrocarbon receptor isolated from lung. *Proc Nat Acad Sci USA* 99: 14694–14699, 2002.

113. M Mehrabi, G Steiner, C Dellinger, A Kofler, K Schaufler, F Tamaddon, K Plesch, CMG Ekmekcioglu, H Glogar, T Thalhammer. The arylhydrocarbon receptor (AhR), but not the AhR-nuclear translocator (ARNT), is increased in hearts of patients with cardiomyopathy. *Virchows Arch* 441: 481–489, 2002.

114. B Abbott, J Schmid, J Pitt, A Buckalew, C Wood, G Held, J Diliberto. Adverse reproductive outcomes in the transgenic Ah receptor–deficient mouse. *Toxicol Appl Pharmacol* 155: 62–70, 1999.

115. J Diliberto, B Abbot, L Birnbaum. Use of AhR knockout (AhR–/–) mice to investigate the role of the Ah receptor on the disposition of TCDD. *Organohalogen Comp* 49: 121–123, 2000.

116. L Birnbaum, J Tuomisto. Non-carcinogenic effects of TCDD in animals. *Food Addit Contam* 17: 275–288, 2000.

117. L Birnbaum. Health effects of dioxins: People are animals and vice versa! *Organohalogen Comp* 49: 101–103, 2000.

118. M Gillner, J Bergman, C Ambilleau, B Ferustrom, JÅ Gustafsson. Interactions of indolo(3-2-6) carbazoles and related polycyclic aroma hydrocarbons with specific binding sites for 2,3,7,8-tetrachlorodobenzo-p-dioxin in rat liver. *Mol Pharmacol* 28: 336–345, 1993.

119. C Sinal, R Bend. Aryl hydrocarbon receptor-dependent induction of Cyp1A1 by bilirubin in mouse hepatatoma hepa 1c1c7 cells. *Mol Pharmacol* 52: 590–599, 1997.

120. D Phelan, G Winter, J Lam, M Denison. Activation of the Ah receptor signal transduction pathway by bilirubin and biliverdin. *Arch Biochem Biophys* 357: 155–163, 1998.

121. S Heath-Pagliuso, W Rogers, K Tullis, S Seidel, P Cenijn, A Brouwer, M Denison. Activation of the Ah receptor by tryptophan and tryptophan metabolites. *Biochem* 37: 11508–11515, 1998.

122. C Schaldach, J Riby, L Bjeldanes. Lipoxin A$_4$: A new class of ligand for the Ah receptor. *Biochem* 38: 7594–7600, 1999.

123. M Denison, A Pandini, S Nagy, E Baldwin, L Bonati. Ligand binding and activation of the Ah receptor. *Chemico-Biologic Interact* 141: 3–24, 2002.

124. DFV Lewis, MN Jacobs. A QSAR study of some PCBs ligand-binding affinity to the cytosolic Ah Receptor. *Organohalogen Comp* 41: 537–540, 1999.

125. C Gambone, J Hutcheson, J Gabriel, R Beard, RAS Chandraaratna, R Soprano. Unique property of some synthetic retinoids: Activation of the Aryl hydrocarbon receptor pathway. *Mol Pharmacol* 61: 334–342, 2002.

126. MN Jacobs, M Dickins, DFV Lewis. Homology modelling of the nuclear receptors: human oestrogen receptor β (hERβ), the human pregnane-X-receptor (PXR), the Ah receptor (AhR) and the constitutive androstane receptor (CAR) ligand binding domains from the human oestrogen receptor α(hERα)crystal structure, and the human peroxisome proliferator-activated receptor α (PPARα) ligand binding domain from the human PPARγ crystal structure. *J Steroid Biochem Mol Biol* 84: 117–132, 2003.

127. J Lehmann, D McKee, M Watson, T Willson, J Moore, S Kliewer. The human orphan nuclear receptor PXR is activated by compounds that regulate CYP3A4 gene expression and cause drug interactions. *J Clin Invest* 102: 1016–1023, 1998.

128. S Kliewer, J Moore, L Wade, J Staudinger, M Watson, S Jones, D McKee, B Oliver, T Willson, R Zetterstrom, T Perlmann, J Lehmann. An orphan nuclear receptor activated by pregnanes defines a novel steroid signalling pathway. *Cell* 92: 73–82, 1998.

129. B Blumberg, W Sabbagh, H Juguilon, J Bolado Jr, C van Meter, E Ong, R Evans. SXR, a novel steroid and xenobiotic sensing nuclear receptor. *Genes and Development* 12: 3195–3205, 1998.

130. JM Pascussi, Y Jounaidi, L Drocourt, J Domergue, C Balabaud, P Maurel, MJ Vilarem. Evidence for the presence of a functional pregnane X receptor response element in the CYP3A7 promoter gene. *Biochem Biophys Res Comm* 260: 377–381, 1999.

131. P Maurel. The CYP 3 family, in C Ionnides, (Ed.), Cytochromes P450: metabolic and toxicological aspects. CRC Press, Boca Raton, FL. 1996. pp 241–270.

132. H Dotzlaw, E Leygue, P Watson, L Murphy. The human orphan receptor PXR messenger RNA is expressed in both normal and neoplastic breast tissue. *Clin Cancer Res* 5: 2103–2107, 1999.

133. L Moore, B Goodwin, S Jones, G Wisely, T Willson, J Collins, S Kliewer. St. John's wort induces hepatic drug metabolism through activation of the pregnane X receptor. *Proc Natl Acad Sci USA* 97: 7500–7502, 2000.

134. H Masuyama, Y Hiramatus, M Kunitomi, T Kudo, P MacDonald. Endocrine disrupting chemicals, phthalic acid and nonylphenol activate pregnane X receptor-mediated transcription. *Mol Pharmacol* 14: 421–428, 2000.

135. E Schuetz, C Brimer, J Schuetz. Environmental xenobiotics and the antihormones cyproterone acetate and spironolactone use the nuclear hormone pregnenolone X receptor to activate the CYP3A23 hormone response element. *Mol Pharmacol* 54: 1113–1117, 1998.

136. S Kliewer, J Lehmann, T Willson. Orphan nuclear receptors: shifting endocrinology into reverse. *Science* 284: 757–760, 1999.

137. R Watkins, G Wisely, L Moore, J Collins, M Lambert, S Williams, T Willson, S Kliewer, M Redinbo. The human nuclear xenobiotic receptor PXR: structural determinants of directed promiscuity. *Science* 292: 2329–2333, 2001.

138. T Ostberg, G Bertilsson, L Jenderberg, A Berkenstam, J Uppenberg. Identification of residues in the PXR ligand binding domain critical for specifes specific and constitutive activation. *Eur J Biochem* 269: 4896–4904, 2002.

139. L Moore, D Parks, S Jones, R Bledsoe, T Consler, J Stimmel, B Goodwin, C Liddle, S Blanchard, T Willson, J Collins, S Kliewer. Orphan nuclear receptors, constitutive androstane receptor and pregnane X receptor share xenobiotic and steroid ligands. *J Biol Chem* 275: 15122–15127, 2000.

140. M Baes, T Gulick, HS Choi, M Martinoli, D Simha, D Moore. A new orphan member of the nuclear hormone receptor superfamily that interacts with a subset of retinoic acid response elements. *Mol Cell Biol* 14: 1544–1552, 1994.

141. M Negishi, R Moore, T Seuyoshi. Transient humanization of nuclear receptor CAR in mouse liver. British Toxicology Society Annual Congress, abstracts, 51. 2003.

142. B Forman, I Tzameli, HS Choi, J Chen, D Simha, W Seol, R Evans, D Moore. Androstane metabolites bind to and deactivate the nuclear receptor CAR-β. *Nature* 395: 612–615, 1998.

143. I Tzameli, D Moore. Role reversal: new insights from new ligands for the xenobiotic receptor CAR. *Trends Endocrinol Metabol* 12: 7–10, 2001.

144. T Kawamoto, T Seuyoshi, I Zelko, R Moore, K Washburn, M Negishi. Phenobarbital-responsive nuclear translocation of the receptor CAR in induction of the CYP2B gene. *Mol Cell Biol* 19: 6318–6322, 1999.

145. JM Pascussi, S Gerbal-Chaloin, L Pichard-Garcia, M Daujat, JM Fabre, P Maurel, MJ Vilarem. Interleukin-6 negatively regulates the expression of pregnane X receptor and constitutively activated receptor in primary human hepatocytes. *Biochem Biophys Res Comm* 274: 707–713, 2000.

146. T Johnson, M Holloway, R Vogel, S Rutledge, J Perkins, G Rodan, A Scmidt. Structural requirements and cell type specifity for ligand activation of peroxisome proliferator activated receptors. *J Steroid Biochem Mol Biol* 63: 1–8, 1997.

147. B Blumberg, R Evans. Orphan nuclear receptors — new ligands and new possibilities. *Genes and Development* 12: 3149–3155, 1998.

148. R Uauy, P Mena, C Rojas. Essential fatty acids in early life: structural and functional role. *Proc Nutr Soc* 59: 3–15, 2000.

149. T Willson, P Brown, D Sternbach, B Henke. The PPARs: from orphan receptors to drug discovery. *J Med Chem* 43: 527–550, 2000.

150. J Bar-Tana. Peroxisome proliferator-activated receptor gamma (PPARγ) activation and its consequences in humans. *Toxicol Lett* 120: 9–19, 2001.

151. S Khan, J Vanden Heuvel. Role of nuclear receptors in the regulation of gene expression by dietary fatty acids (review). *J Nutr Biochem* 14: 554–567, 2003.

152. R Memon, L Tecott, K Nonogaki, A Beigneux, A Moser, C Grunfeld, K Feingold. Up-regulation of peroxisome proliferator-activated receptor α (PPARα) and PPAR messenger ribonucleic acid expression in the liver in murine obesity: Troglitazone induces expression of PPAR responsive adipose tissue-specific genes in the liver of obese diabetic mice. *Endocrinol* 141: 4021–4031, 2000.

153. S Kato, M Suzawa, I Takada, K Takeyama, J Yanagizawa, R Fujiki, H Kitagawa. The function of nuclear receptors in bone tissues. *J Bone Miner Metab* 21: 323–336, 2003.

154. A Zomer, B van der Burg, G Jansen, J Wanders, B Poll-The, P van der Saag. Pristanic acid and phytanic acid: naturally occurring ligands for the nuclear receptor peroxisome proliferator-activated receptor α. *J Lipid Res* 41: 1801–1807, 2000.

155. T Miyamoto, A Keneko, T Kakizawa, H Yajima, K Kamijo, R Sekine, K Hiramatsu, Y Nishii, T Hashimoto, K Hashizume. Inhibition of peroxisome proliferator signaling pathways by thyroid hormone-receptor-competitive binding to the response element. *J Biol Chem* 272: 7752–7758, 1997.

156. J Lehmann, D McKee, M Watson, T Willson, J Moore, SA Kliewer. The human orphan nuclear receptor PXR is activated by compounds that regulate CYP3A4 gene expression and cause drug interactions. *J Clin Invest* 102: 1016–1023, 1998.

157. L Moore, B Goodwin, S Jones, G Wisely, T Willson, J Collins, S Kliewer. St. John's Wort induces hepatic drug metabolism through activation of the pregnane X receptor. *Proc New York Acad Sci USA* 97: 7500–7502, 2000.

158. L Moore, D Parks, S Jones, R Bledsoe, T Cousler, J Stimmel, B Goodwin, C Liddle, S Blanchard, T Willson, J Collins, S Kliewer. Orphan nuclear receptors constitutive antrostane receptor and pregnane X receptor share xenobiotic and steroid ligands. *J Biol Chem* 275: 15122–15127, 2000.

159. N Landes, P Pfluger, D Kluth, M Birringer, R Ruhl, GF Böl, H Glatt, R Brigelius-Flohé. Vitamin E activates gene expression via the pregnane X receptor. *Biochem Pharmacol* 65: 269–273, 2003.

160. SC Nallani, B Goodwin, JM Maglich, DJ Buckley, AR Buckley, PB Desai. Induction of cytochrome P450 3A by paclitaxel in mice: pivotal role of the nuclear xenobiotic receptor, pregnane X receptor, *Drug Metab Dis* 31: 681–684, 2003.

161. E Owsley, J Chiang. Gugglesterone antagonizes farnesoid X receptor induction of bile salt pump but activates pregnane X receptor to inhibit cholesterol 7α–hydroxylase gene. *Biochem Biophys Res Comm* 304: 191–195, 2003.

162. JL Raucy. Regulation of CYP3A4 Expression in human hepatocytes by pharmaceuticals and natural products. *Drug Metab Dis* 31: 533–539, 2003.

163. M Wyde, E Bartolucci, A Ueda, H Zhang, B Yan, M Negishi, L You. The environmental pollutant 1,1-dichloro-2,2-bis(p-chlorophenyl)ethylene induces rat hepatic cytochromeP4502B and 3A expression through the constitutive androstane receptor and pregnane X receptor. *Mol Pharmacol* 64 (2): 474–481, 2003

164. JM Maglich, DJ Parks, LB Moore, JL Collins, B Goodwin, AN Billin, CA Stoltz, SA Kliewer, MH Lambert, TM Willson, JT Moore. Identification of a novel human constitutive androstane receptor (CAR) agonist and its use in the identification of CAR target genes. *J Biol Chem* 278: 17277–17283, 2003.

165. W Huang, J Zhang, SS Chua, M Qatanani, Y Han, R Granata, DD Moore. Induction of bilirubin clearance by the constitutive androstane receptor. *Proceedings of the National Academy of Science USA* 100 (7): 4156–4161, 2003.

166. M Denison, J Whitlock. Xenobiotic-inducible transcription of cytochrome P450 genes. *J Biol Chem* 270: 18175–18178, 1995.

167. M Denison, D Phelan, G Winter, M Ziccardi. Carbaryl, a carbamate insecticide is a ligand for the hepatic Ah (dioxin) Receptor. *Toxicol Appl Pharmacol* 152: 406–414, 1998.

168. D Phelan, G Winter, W Rogers, J Lam, M Denison. Activation of the Ah receptor signal transduction pathway by bilirubin and biliverdin. *Arch Biochem Biophys* 357: 155–163, 1998.

169. J Adachi, Y Mori, S Matsui, H Takigami, J Fujino, H Kitagawa, C Miller, T Kato, K Sacki, T Matsuda. Indirubin and indigo are potent aryl hydrocarbon receptor ligands present in human urine. *J Biol Chem* 276: 31475–31478, 2001.

170. A Chawla, J Repa, R Evans, D Mangelsdorf. Nuclear receptors and lipid physiology: Opening the X files. *Science* 294: 1866–1870, 2001.

171. JB Nixon, H Kamitani, SJ Baek, TE Eling. Evaluation of eicosanoids and NSAIDs as PPAR(gamma) ligands in colorectal carcinoma cells. *Prostaglandins, Leukotrienes and Essential Fatty Acids* 68: 323–330, 2003.

172. P Cronet, J Petersen, R Folmer, N Blomberg, K Sjoblom, U Karlsson, EL Lindstedt, K Bamberg. Structure of the PPAR α and β-ligand biding domain in complex with AZ 242; ligand selectivity and agonist activation in the PPAR family. *Structure* 9: 699–706, 2001.

173. HE Xu, MH Lambert, VG Montana, DJ Parks, SG Blanchard, PJ Brown, DD Sternbach, JM Lehmann, GB Wisely, TM Willson, SA Kliewer, MV Miburn. Molecular recognition of fatty acids by peroxisome proliferator activated receptors. *Mol Cell* 3: 397–403, 2003.

174. X Luo, Y Ikeda, D Lala, D Rice, M Wong, K Parker. Steroidogenic factor-1 (SF-1) is essential for endocrine development and function. *J Steroid Biochem Mol Biol* 69: 13–18, 1999.

175. R Watkins, P Davis-Searles, M Lambert, M Redinbo. Coactivator binding promotes the specific interaction between ligand and the pregnane X receptor. *J Mol Biol* 331: 815–828, 2003.

176. C Sheeler, M Dudley, S Khan. Environmental estrogens induce transcriptionally active estrogen receptor dimers in yeast: Activity potentiated by the coactivator RIP140. *Environ Health Perspect* 108: 97–103, 2000.

177. SCN Buist, NJ Cherrington, CD Klaassen. Endocrine regulation of rat organic anion transporters. *Drug Metab Dis* 31: 559–564, 2003.

178. G Pimentel, A McClellan. *The Hydrogen Bond*. L Pauling, WH Freeman, New York, 1960.

179. J Del Bene. Hydrogen Bonding: 1. P von Ragué Schleyer, NL Allinger, T Clark, J Gasteiger, PA Kollman, HF Schaefer, PR Schreiner. *Encyclopedia of Computational Chemistry* E-L. 2, 1263–1271. Wiley, Chichester, U.K., 1998.

180. MH Abraham, A Ibrahim, AM Zissimos, YH Zhao, J Comer, DP Reynolds. Application of hydrogen bonding calculations in property based drug design, *Drug Discovery Today* 7: 1056–1063, 2002.

181. MH Abraham, K Enomoto, ED Clarke, G Sexton. Hydrogen bond basicity of the chlorogroup; hexachlorohexanes as strong hydrogen bond bases. *J Organic Chem* 67: 4782–4786, 2002.

182. EC Henry, TA Gasiewicz. Agonist but not antagonist ligands induce conformational change in the mouse aryl hydrocarbon receptor as detected by partial proteolysis. *Mol Pharmacol* 63: 392–400, 2003.

183. F Rabenoelina, P Nirdé, N Servant, C Sultan, B Terouanne, G Auzou. A single amino acid substitition in the ligand binding domain of the human androgen receptor modulates the receptor transactivation capacity. P128. 15th International Symposium of the Journal of Steroid Biochemistry and Molecular Biology, Munich, Germany, May 17–20, 2002.

184. JL Staudinger, A Madan, KM Carol, A Parkinson, Regulation of Drug Transporter Gene Expression by Nuclear Receptors. *Drug Metab Dis* 31: 523–527, 2003.

185. K Swales. An investigation of host cell effects on the xenobiotic induction of cytochrome P450 3A. PhD thesis, University of Surrey, U.K., 2002.

186. C Klinge, K Kaur, H Swanson. The aryl hydrocarbon receptor interacts with estrogen receptor alpha and orphan receptors COUP-TF1 and ERRα 1. *Arch Biochem Biophys* 373 (1): 163–174, 2000.

187. L Drocourt, JC Ourlin, JM Pascussi, P Maurel, MJ Vilarem. Expression of CYP3A4, CYP2B6, and CYP2C9 is regulated by the vitamin D receptor pathway in primary human hepatocytes. *J Biol Chem* 277: 25125–25132, 2002.

188. K Mizuyachi, DS Son, K Rozman, P Terranova. Alteration in ovarian gene expression in response to 2,3,7,8-tetrachlorodibenzo-*p*-dioxin: reduction of cyclooxygenase-2 in the blockage of ovulation. *Repro Toxicol* 16: 299–307, 2002.

189. S Kakizaki, S Karami, M Negishi. Retinoic acids repress constitutive active receptor-mediated induction by 1,4-bis[2-(3,5-dichlorpyridyloxy)] benzene of the CYP2B10 gene in mouse primary hepatocytes. *Drug Metab Dis* 30: 208–211, 2002.

190. Y Cai, T Konishi, G Han, K Campwala, S French, YJY Wan. The role of hepatocyte RXRa in xenobiotic-sensing nuclear receptor-mediated pathways. *Eur J Pharm Sci* 15: 89–96, 2002.

191. L Dubuquoy, S Dharancy, S Nutten, S Pettersson, J Auwerx, P Desreumaux. Role of peroxisome proliferator-activated receptor γ and retinoid X receptor heterodimer in hepatogastroenterological diseases. *Lancet* 360: 1410–1418, 2002.

192. R Gampe Jr, V Montana, M Lambert, A Miller, R Bledsoe, M Milburn, S Kliewer, T Willson, H Xu. Asymmetry in the PPARγ/RXRα crystal structure reveals the molecular basis of heterodimerization among nuclear receptors. *Mol Cell* 5: 545–555, 2000.

193. HY Lee, YA Suh, M Robinson, J Clifford, W Hong, J Woodget, M Cobb, D Mangelsdorf, J Kurie. Stress pathway activation induces phosphorylation of retinoid X receptor. *J Bio Chem* 275: 32193–32199, 2000.

194. P Edwards, J Ericsson. Sterols and isoprenoids: signalling molecules derived from the cholesterol biosynthetic pathway. *Ann Rev Biochem* 68: 157–185, 1999.

195. D Russell. Nuclear orphan receptors control cholesterol catabolism. *Cell* 97: 539–542, 1999.

196. C Song, S Liao. Cholestenoic acid is a naturally occurring ligand for liver X receptor α. *Endocrinol* 141: 4180–4184, 2000.

197. P Edwards, H Kast, A Anisfeld. BAREing it all: the adoption of LXR and FXr and their roles in lipid homeostasis. *J Lipid Res* 43: 2–12, 2002.

198. H Wang, J Chen, K Hollister, L Sowers, B Forman. Endogenous bile acids are ligands for the nuclear receptor FXR/BAR. *Mol Cell* 3: 543–553, 1999.

199. M Makishima, A Okamoto, J Repa, H Tu, M Learned, A Luk, M Hull, K Lustig, D Mangelsdorf. Identification of a nuclear receptor for bile acids. *Science* 284: 1362–1368, 1999.

200. YJY Wan, D An, Y Cal, J Repa, THP Chen, M Flores, C Postic, M Magnuson, J Chen, K Chien, S French, D Mangelsdorf, H Sucov. Hepatocyte-specific mutation establishes retinoid X receptor a as a heterodimeric integrator of multiple physiological processes in the liver. *Mol Cell Biol* 20: 4436–4444, 2000.

201. I Tzameli, S Chua, B Cheskis, D Moore. Complex effects of rexinoids on ligand-dependent activation or inhibition of the xenobiotic receptor CAR. *Nucl Recept* 1:2, 2003.

202. T Ide, H Shimano, T Yoshikawa, N Yahagi, M Amemiya-Kudo, T Matsuzaka, M Nakakuki, S Yatoh, Y Iizuka, S Tomita, K Ohashi, A Takahashi, H Sone, T Gotoda, J Osuga, S Ishibashi, N Yamada. Cross talk between peroxisome proliferator-activated receptor (PPAR)α and liver X receptor (LXR) in nutritional regulation of fatty acid metabolism. II. LXRs suppress lipid degradation gene promoters through inhibition of PPAR signaling. *Mol Endocrinol* 17: 1255–1267, 2003.

203. T Yoshikawa, T Ide, H Shimano, N Yahagi, M Amemiya-Kudo, T Matsuzaka, S Yatoh, T Kitamine, H Okazaki, Y Tamura, M Sekiya, A Takahashi, A Hasty, R Sato, H Sone, J Osuga, S Ishibashi, N Yamada. Cross talk between peroxisome proliferator-activated receptor (PPAR)α and liver X receptor (LXR) in nutritional regulation of fatty acid metabolism. I.PPARs suppress sterol regulatory element binding protein-1c promoter through inhibition of LXR signaling. *Mol Endocrinol* 17: 1240–1254, 2003.

204. C Elbi, T Misteli, GL Hager. Recruitment of dioxin receptor to active transcription sites. *Mol Biol Cell* 13: 2001–2015, 2002.

205. S Brunnberg, K Pettersson, E Rydin, J Matthews, A Hanberg, I Pongratz. The basic helix-loop-helix-PAS protein ARNT functions as a potent coactivator of estrogen receptor-dependent transcription, *Proc Nat Acad Sci USA* 100: 6517–6522, 2003.

3

Retinoid Signaling

DIANNE ROBERT SOPRANO, PU QIN, AND
KENNETH J. SOPRANO

INTRODUCTION

The term *retinoid* refers to a class of compounds including vitamin A, its natural derivatives, and more than 10,000 synthetic analogs. The structure of retinoids can be divided into three moieties: a β-ionone ring (cyclohexenyl), a conjugated side chain, and a polar terminal group (Figure 3.1). Natural retinoids include all-*trans*-retinol, all-*trans*-retinal, all-*trans*-retinoic acid (RA), 9-*cis*-RA, and 13-*cis*-RA. Furthermore, organic chemists have synthesized a large number of synthetic retinoids as potential pharmacological agents.

The Recommended Daily Allowance for vitamin A ranges from 375 μg retinol equivalents/d for infants to 1000 μg retinol equivalents/d for male adults. Vitamin A can be obtained from green leafy vegetables, carrots, and sweet potatoes as carotenoids (including β-carotene) and milk, meat, and liver as retinyl esters. It is an important nutrient for a diverse group of biological functions including growth, vision, epithelial differentiation, immune function, embryonic development, and reproduction. With the exception of night vision (mediated by 11-*cis*-retinal), each of these functions are mediated by RA.

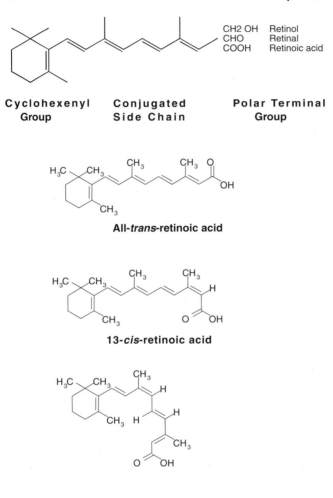

Figure 3.1   Structural features of retinoids.

In developing countries, deficiency among children remains a leading cause of severe visual impairment and night blindness. Vitamin A deficiency is also likely to cause vulnerability to other illnesses in both adults and children, such as iron-deficiency anemia and measles. Deficiency during pregnancy leads to placental dysfunction, fetal loss, and congenital malformations. On the other hand, excessive intake of vitamin A can lead to reduced bone mineral density, liver abnormalities, nausea, vomiting, and even death, whereas exposure of developing embryos to excessive amounts of vitamin A (particularly, RA) can cause birth defects.

TRANSPORT OF VITAMIN A

In the small intestine, dietary retinyl esters are hydrolyzed to retinol and β-carotene is cleaved to all-*trans*-retinal by carotene 15-15 dioxygenase (also referred to as carotene cleavage enzyme). Retinal is further reduced to retinol by a microsomal retinal reductase in the intestinal mucosa cells. Retinol is then complexed with cellular retinol binding protein II (CRBPII). The CRBPII bound retinol is subsequently converted to retinyl esters by lecithin:retinol acyltransferase (LRAT). Retinyl esters are incorporated into chylomicrons along with triacylglycerol, cholesteryl esters, and other fat-soluble vitamins and delivered to various tissues, principally to the liver in addition to other tissues such as bone marrow, peripheral blood cells, spleen, adipose tissue, and kidney.

Vitamin A is mainly stored in the stellate cells of the liver in the form of retinyl esters. When needed, retinyl esters are hydrolyzed to retinol, complexed with retinol-binding protein (RBP), and transported to different tissues (for review, see Reference 1).[1] In plasma, the 21-kDa RBP circulates as a complex with transthyretin (TTR, 80 kDa) to avoid glomerular filtration in the kidney. The RBP-bound retinol in plasma is easily taken up by target cells, reversibly oxidized to retinal (by retinol dehydrogenases) and then irreversibly oxidized to all-*trans*-RA (by retinal dehydrogenases) (Figure 3.2).

In the cytoplasm of target cells, retinol and RA bind to their binding proteins, termed *cellular retinol-binding protein* (CRBP-I and CRBP-II) and *cellular retinoic acid-binding protein* (CRABP-I and CRABP-II), respectively (for a review, see Reference 2).[2] CRBP and CRABP are able to facilitate the intracellular transport and metabolism of retinoids. Moreover, CRABP-II can deliver RA into the nuclei of cells, where it can bind to the nuclear receptors, retinoic acid receptors (RARs),

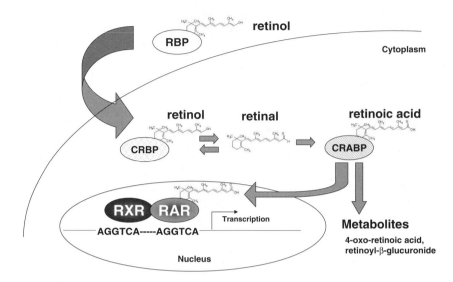

Figure 3.2 Vitamin A signal transduction pathway. Retinol is delivered to target tissues bound to RBP. Retinol taken up by target cells binds to CRBP. Retinol is oxidized to retinoic acid, which binds to CRABP. Retinoic acid modulates transcription of target genes via RAR/RXR. Retinoic acid can also be further metabolized to inactivate products. Abbreviations: CRABP, cellular retinoic acid-binding protein; CRBP, cellular retinol-binding protein; RAR, retinoic acid receptor; RBP, retinol-binding protein; RXR, retinoid X receptor.

and retinoid X receptors (RXRs) and regulate the transcription of target genes (Figure 3.2).

METABOLISM OF VITAMIN A

Regulation of retinoid synthesis and catabolism are both important ways in which distinct spatiotemporal patterns of active retinoids can be maintained in the developing embryo and the adult (for review, see References 3–5).[3–5] A particular cell will respond to RA only if (1) it expresses RARs and RXRs, and (2) the RA concentration lies within a range that is appropriate for its response. Thus, the precise control of RA distribution and concentration is critical.

Studies utilizing transgenic mice carrying an RA-activated reporter gene have showed that RA is not homogeneously distributed throughout the whole embryo but is localized to specific territories within a restricted number of developing organs in the embryo.[6,7] RA synthesizing and catabolizing enzymes are localized in subregions of developing tissues in a pattern that resembles that of RA reporter gene expression.[8–10] The most detailed description of this localization is available for the embryonic retina[11] and spinal cord.[11]

As described earlier, retinol is reversibly oxidized to retinal and further irreversibly oxidized to RA (Figure 3.2). Members of the medium-chain alcohol dehydrogenase family (ADH1 and ADH4) and short-chain dehydrogenase family are proposed to play a critical role in the oxidation of retinol to retinal. Ethanol can inhibit the enzymatic activity of ADH1 but not ADH4. In adults, ethanol inhibits RA synthesis in several tissues, including testes,[12] liver,[13] and cornea.[14] Moreover, the alcohol dehydrogenase specific inhibitor 4-methylpyrazole has been shown to inhibit retinal synthesis.[12] These data as well as the similarities between the embryonic abnormalities in RA deficiency and in fetal alcohol syndrome suggest that fetal alcohol syndrome could result from the inhibition of RA synthesis by ethanol. RA synthesis in both the ADH1- and ADH4-null mutant mice is reduced compared with that in wild-type animals, which suggests a contributory role of both enzymes to RA synthesis.[15] On the other hand, the fact that both ADH1- and ADH4-null mutant mice are normal and fertile argues for a functional redundancy between at least these two enzymes. At least five short-chain dehydrogenases are known to be expressed in the adult although their embryonic distribution is largely unknown.[16]

The second step in RA synthesis is the conversion of retinal to RA. The synthesis of RA from retinal is catalyzed by retinal oxidase and a group of enzymes that belong to the retinal dehydrogenase (RALDH) family. The work of Drager and her coworkers supports the role of RALDHs in the synthesis of RA in the mouse embryo.[8,9,11] Four different RALDHs have been identified which can convert all-*trans*-retinal and 9-*cis*-retinal to RA. These enzymes all belong to the aldehyde

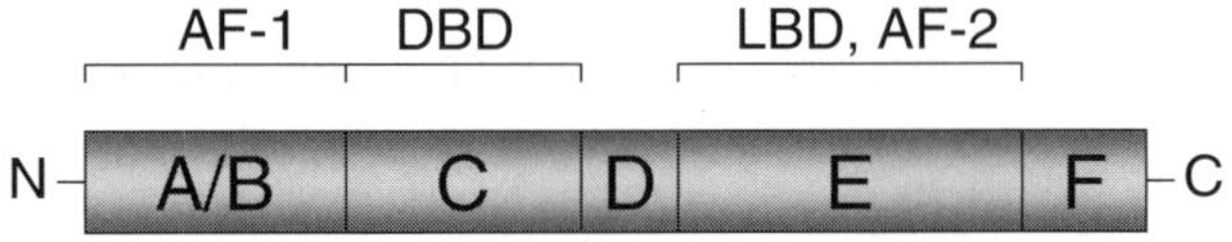

	DBD	LBD
RARα	100%	100%
RARβ	95%	82%
RARγ	95%	85%
RXRα	61%	27%
RXRβ	56%	25%
RXRγ	58%	24%

Figure 3.3 Molecular structure of retinoid receptors. Six domains termed A-F are shown. The percentage amino acid identity for the DNA-binding and ligand-binding domains is indicated in relation to RARα. Abbreviations: AF-1, activation function-1; AF-2, activation function-2; DBD, DNA-binding domain; LDB, ligand-binding domain.

dehydrogenase (ALDH) superfamily, which consists of at least 86 members.[17] RALDH1 catalyzes the conversion of both all-*trans*-retinal and 9-*cis*-retinal to RA, with similar efficiencies.[18] RALDH2 also catalyzes both reactions but is fourfold less efficient with 9-*cis*-retinal than with all-*trans*-retinal.[3] RALDH3 only converts all-*trans*-retinal to all-*trans*-RA and does not work on 9-*cis*-retinal.[19] RALDH4, the newest member of this family, only catalyzes the dehydrogenation of 9-*cis*-retinal to 9-*cis*-RA but not that of all-*trans*-retinal.[20] RALDH2 expression appears early during mouse embryogenesis and is essential for normal development. RALDH2 knockout mice die on gestation day 10.5 with multiple malformations such as shortened trunk and opened neural tube.[21]

Catabolism of RA involves oxidation, isomerization, and formation of glucuronide and taurine conjugates[22–25] (Figure 3.2). The predominant metabolites include 4-hydroxy-RA, 18-oxo-RA, 4-oxo-RA, 5,8-epoxy-RA, and retinoyl-β-glucuronide. Recently, a new member of the cytochrome P-450 family,

CYP26,[26] has been identified as an enzyme that specifically mediates RA oxidation. It was first cloned in zebra fish and later in humans, mice, and rats.[27–30] The expression of CYP26 in embryos suggests that it also regulates RA distribution.[26] The promoter of the CYP26 gene contains a retinoic acid response element (RARE), which has been shown to regulate CYP26 expression by RA, probably by a feedback loop regulation responding to high concentration of RA.[31,32] Exposure of the mouse embryos to RA induces CYP26 expression in a number of regions.[26] Additionally, CYP26 is regulated by RA in a number of cell lines including embryonal carcinoma cells, breast cancer cells, human tracheobronchial epithelial cells, human colon carcinoma cells, and head and neck squamous carcinoma cells.[33–37] As a matter of fact, CYP26 has been shown to be upregulated in acute promyelocytic leukemia patients after all-*trans*-RA treatment and has been suggested to contribute to resistance to all-*trans*-RA therapy.[38,39]

In conclusion, both RA synthesis and catabolism enzymes play important roles in regulating the distribution of RA during embryonic development and in adults. Disruption of the expression or functional activity of any of these enzymes can result in abnormal RA signaling.

General Features of RARs and RXRs

Retinoic acid receptors (RARs) were initially discovered by Pierre Chambon's group and Ronald Evans' group simultaneously in 1987.[40,41] Retinoid X receptors (RXRs) were discovered a few years later.[42] RARs and RXRs belong to the superfamily of steroid and thyroid hormone receptors, which includes estrogen receptors, androgen receptors, thyroid hormone receptors, glucocorticoid receptors and vitamin D receptors. These receptors are ligand-dependent nuclear receptors that form heterodimers or homodimers (RAR-RXR or RXR-RXR, respectively) to activate transcription of their target genes upon binding of ligand (for review, see Reference 43).[43] RAR and RXR each have 3 subtypes (α, β, γ), which are encoded by six different genes located at distinct chromosomal loci (Table 3.1). In addition, multiple isoforms of each subtype

Table 3.1 RARs and RXRs

Gene	Isoforms	Chromosomal Location		Ligand
		Human	Mouse	
RARα	α1 - α7	17q21.1	11	ATRA, 9CRA
RARβ	β1 - β4	3p24	14	ATRA, 9CRA
RARγ	γ1 - γ7	12q13	15	ATRA, 9CRA
RXRα	α1, α2	9q34.3	2	9CRA
RXRβ	β1, β2	6q21.3	17	9CRA
RXRγ	γ1, γ2	1q22-q23	1	9CRA

Note: ATRA = all-*trans*-RA; 9CRA = 9-*cis*-RA.

are generated as a result of different promoter usage, alternative splicing, and the use of an internal CUG codon for the initiation of translation (Table 3.1). Four major isoforms of RARβ (β1, β2, β3, and β4) and two major isoforms of both RARα (α1 and α2) and RARγ(γ1 and γ2) have been extensively studied (44-49). In addition, five additional minor isoforms for both RARα (α3 to α7) and RARγ (γ3 to γ7) have been reported. Each of the three RXR subtypes has two isoforms (α1, α2, β1, β2, γ1, and γ2).[50–52]

Comparison of the amino acid sequences of the RAR family and RXR family of nuclear receptors demonstrates only weak homology over the entire length of the proteins, with the highest degree of similarity found in the DNA-binding domain (approximately 60%), whereas the ligand-binding domain displays only approximately 25% amino acid identity (Figure 3.3). Interestingly, there is an RXR homolog in drosophila called *ultraspiracle* that dimerizes with the ecdysone receptor, but there are no RAR homologs identified in drosophila.[53] From these findings it has been suggested that RARs and RXRs arose separately during evolution and that receptor dimerization (see following discussion) has been well conserved evolutionarily.

RARs bind both all-*trans*-RA and 9-*cis*-RA with equal affinity (K_d = 1 to 5 nM), whereas RXRs have been shown to bind *in vitro* only 9-*cis*-RA with high affinity (K_d = 1 to 10 nM).[54,55] Although both RARs and RXRs display high affinity for 9-*cis*-RA using *in vitro* binding assays, it has been difficult to detect similar quantities *in vivo*. This has raised the possibility that biologically important RXR-specific ligands that are yet unknown may exist.[56,57] Furthermore, much effort has been directed toward the design and synthesis of selective retinoids that can target an individual RAR or RXR subtype or isoform. To date, a number of RAR isoform-selective retinoids along with several RXR subtype-selective retinoids have been developed. In addition to the development of these selective agonists, other retinoids that display properties of antagonists, inverse agonists, and inverse antagonists have also been described.[58–60] Retinoids such as these have the potential of being of great clinical benefit for the treatment of a variety of disorders in the areas of dermatology, oncology, type 2 diabetes, atherosclerosis, and obesity.

FUNCTIONAL DOMAINS OF RARS AND RXRS

The RARs and RXRs exhibit the conserved modular structure shared by all members of the steroid-thyroid hormone superfamily (for review, see Reference 43).[43] Their amino acid sequences can be divided into five or six functionally distinct regions (termed domains A to F), based on homology among themselves and with other members of the steroid and thyroid hormone superfamily members (Figure 3.3).

DNA-Binding Domain

Domain C contains the DNA-binding domain and is the most highly conserved region of the nuclear receptors (95% homology between three RAR subtypes and 92 to 95% homology between three RXR subtypes) (Figure 3.3). The DNA binding domain contains a 66-amino-acid-long central region, which is responsible for the interaction with the DNA responsive elements in the promoter region of target genes. This region of the DNA binding domain contains a pair of zinc-binding motifs, termed *zinc fingers*, which mediate the interaction between the nuclear receptor and the major groove of DNA. In addition, these zinc fingers are responsible for the discrimination of the DNA response element half-site sequence and spacing. The

DNA-binding domain allows the RAR-RXR heterodimer or the RXR homodimer to come in close proximity to the transcription machinery to regulate mRNA transcription. The RAR-RXR heterodimer binds to the retinoic acid responsive element (RARE), whereas the RXR homodimer binds to the retinoid X responsive element (RXRE).

RAREs and RXREs consist of a half-site with the consensus sequence AGGTCA.[61–64] Comparison of a number of RAREs and RXREs in RA target genes (Table 3.2) demonstrates that the half-site sequence displays considerable pliancy for nucleotide substitution. Furthermore, the half-site sequence can form a direct repeat, a palindrome, or can be found without any consensus structure. However, a direct repeat is seen most commonly. The RAR-RXR heterodimer binds tightly to the half-site sequence separated by two or five nucleotides (RARE-DR2 and RARE-DR5, respectively), with the RARE-DR5 seen most commonly. On the other hand, the RXR-RXR homodimer typically binds to direct repeats of the half-site sequence containing a one-nucleotide spacer between the repeats (RXRE-DR1). In most RA responsive genes, the RARE or RXRE is located in the promoter region 5′ to the start site of transcription. However, both HOXB1 and HOXA1 have RARE enhancers 3′ to the gene. RAR-RXR has been demonstrated to bind to those regions and regulate the expression of HOXB1 and HOXA1.[65–67]

Ligand-Binding Domain

Domain E is a functionally complex region composed of approximately 220 amino acid residues. It contains the ligand-binding domain (LBD), the ligand-dependent activation function 2 (AF-2), and a dimerization surface for RAR-RXR or RXR-RXR interaction.[68] The amino acid sequence of domain E is also highly conserved between the three RAR subtypes (82 to 85%) and the three RXR subtypes (86 to 90%) (Figure 3.3).

Over the last several years, the x-ray crystal structure of the LBD of *apo*-RXRα, *holo*-RARγ, *holo*-RXRα, and a heterodimer between antagonist-bound RARα and the constitutively active RXRαF318A have been reported.[69–72] These crystal structures have demonstrated that each of these LBDs share a novel protein fold, termed an *antiparallel α-helical sandwich*, composed of 12 α-helices (H1 to H12) and a β-turn residing between helices H5 and H6.[69–73] RA within the ligand-binding pocket of the LBD comes in close contact with a large number of amino acid residues located on α-helices H1, H3, H5, H11, and H12, loops 6 to 7 and 11 to 12, and β-sheet s1.

The direct comparison of the x-ray structures of the LBD of RARs and RXR with and without bound agonists or antagonists has clearly demonstrated that there are major conformational changes that occur upon binding of ligand. Binding of 9-*cis*-RA in the ligand-binding pocket of RXRα causes a major conformational change in helix H12, along with more minor changes in other helices, including H3, H6, and H11. In the *apo*-form, helix H12 protrudes from the protein core and is exposed to solvent, whereas in the *holo*-form, helix H12 rotates and folds back toward the ligand-binding pocket. In the *holo*-form, the LBD is more compact, and there is a formation of a new surface that binds coactivators involved in mediating transcriptional activity (see following discussion). Interestingly, helix H12 has been shown to exist in at least two distinct conformational states, depending upon whether the bound ligand is an agonist or an antagonist. It has been suggested that this conformational difference in helix H12 is at least partially responsible for mediating the different transcriptional activities of agonist-bound and antagonist-bound receptors.

In addition to the crystal structures, site-directed mutagenesis studies have demonstrated that multiple amino acid residues in the ligand-binding pocket of RARs associate with RA 9.[74–79] The homologous Arg residue (αArg 276, βArg 269 and γArg 278) in each of the RARs has been demonstrated to be critical for the binding of RA, by interaction with the carboxylate group of RA. This is consistent with the crystal structure of *holo*-RARγ, in which γArg 278 appears to form a salt bridge with the carboxylate O-22 of RA. In both RARα and RARγ, this conserved Arg residue appears to function relatively independently in coordinating with the carboxyl

Table 3.2 Typical RAREs and RXREs

Gene[a]	Location[b]	Sequence[c]	Category
DR5	Synthetic	TCAGGTCACCAGGAGGTCAGA	RARE-DR5
mRARβ2	−55 ~ −35	AGGGTTCACCGAAAGTTCACT	RARE-DR5
mRARα2	−57 ~ −35	CGAGTTCAGCGAGAGTTCAGC	RARE-DR5
hRARα2	−56 ~ −36	CGAGTTCAGCGAGAGTTCAGC	RARE-DR5
hADH3	−282 ~ −302	AGGGGTCATTCAGAGTTCAGT	RARE-DR5
DR2	Synthetic	TCAGGTCATCAGGTCACG	RARE-DR2
mHOXB1	3′ end	AGAGGTAAAAAGGTCAGC	RARE-DR2
mCRBPI	−1013 ~ −996	GTAGGTCAAAAGGTCAGA	RARE-DR2
hOST	−512 ~ −497	CTAGGTGACTCACCGGG	RARE-palindrome

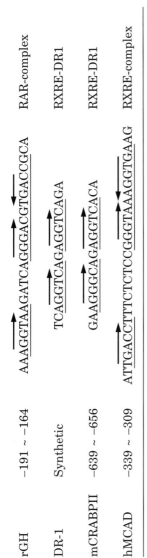

rGH	-191 ~ -164	AAAGGTAAGATCAGGGACGTGACCGCA	RAR-complex
DR-1	Synthetic	TCAGGTCAGAGGTCAGA	RXRE-DR1
mCRABPII	-639 ~ -656	GAAGGGCAGAGGTCACA	RXRE-DR1
hMCAD	-339 ~ -309	ATTGACCTTTCTCTCCGGGTAAAGGTGAAG	RXRE-complex

Note: [a]Abbreviations used for gene names: m(h)RAR(α, β)2, mouse (human)RAR(α, β)2; hADH, human alcohol dehydrogenase 3; mCRBPI, mouse cellular retinol binding protein typeI; hOST, human osteocalcin; rGH, rat growth hormone; mCRABPII, mouse cellular retinoic acid binding protein type II; hMCAD, human medium chain acyl-coenzyme A dehydrogenase; mHOXB1, mouse homeobox protein B1. [b]Except synthetic response elements, numbers indicate relative location of the response element from transcription start site. In the case of mHOXB1, the RARE is located at the 3 end of the gene. [c]Arrows indicate orientation of a common half-site hexamer 5-AGGTCA-3.

Source: Adapted from DJ Mangelsdorf, K Umesono, RM Evans. The retinoid receptors. In: MB Sporn, AB Roberts, DS Goodman, Eds. *The Retinoids: Biology, Chemistry and Medicine.* 2nd ed. Raven Press, NY. 1994, pp 319–349.

group of the bound RA molecule. On the other hand, Arg 269 in RARβ appears to function in conjunction with two additional amino acid residues (Lys 220 and Ser 278) in the binding of RA. This suggests that the orientation of this amino acid residue in the ligand-binding site or the electronic environment associated with this Arg in each RAR subtype is different.

In addition to the conserved Arg residue, three nonconserved amino acid residues located with the ligand-binding pocket of RARs have been demonstrated to be important for RAR-subtype ligand specificity.[80–82] The mutation of Ile 263 in RARβ to a Met or the mutation of both Ile 263 and Val 388 in RARβ to a Met and an Ala, respectively, creates a mutant RARβ that binds the RARγ-selective retinoid CD437 with high affinity. Similarly, the mutation of Ala 225 in RARβ to a Ser results in a mutant RARβ that binds the RARα-selective retinoid AM580 with high affinity. These data clearly demonstrate that these three amino acid residues play a critical role in distinguishing these RAR subtype selective ligands. Further studies involving the examination of the interrelationship between the Arg residue responsible for interaction with the carboxyl group of retinoids and these three amino acid residues involved in subtype ligand specificity suggest that these two regions of the ligand-binding pocket function independently.[83]

A and B Domains

The amino-terminal A/B domain is of variable length and is the least homologous between the receptors. The different isoforms of each of the RAR and RXR subtypes vary in the amino acid sequence of the amino-terminal A domain, whereas the amino acid sequence of the B to F domains is identical for each of the isoforms within each subtype. This region has been shown to contain a second transcription activation function, termed AF-1, which is ligand independent and has been demonstrated to synergize with the AF-2 region in the LBD.[68]

In the case of RARα1, the synergy between AF-1 and AF-2 is likely to be mediated by TIF2 which forms a bridge between these two domains.[84] In addition, RARα is phosphorylated at a specific serine residue (Ser 77 of the A/B domain) by cdk7 of the general transcription factor TFIIH, which is important for the transactivation function of RARα.[85] Recently, it was found that the A/B domain of RARγ2 is phosphorylated by p38MAPK. This leads to the proteasome-dependent degradation of this receptor. The degradation process is also associated with the transactivation function of RARγ.[86] The phosphorylation of RARα and RARγ is closely related to the RA-dependent and cAMP-dependent induction of differentiation of mouse teratocarcinoma F9 cells.

D and F Domains

Region D is the hinge region that connects domains C and E. The high conservation in the amino acid sequence of the D domain among RAR subtypes and different animal species strongly suggests that this domain may have some yet unknown function. The function of region F in RARs is unknown, whereas RXRs does not have this region.

MECHANISM OF TRANSCRIPTIONAL REGULATION BY RARS AND RXRS

Dimerization

In addition to ligand-binding, heterodimerization of RXR and RAR is required for tight binding to DNA and transactivation of gene expression through a RARE. RAR and RXR dimerize weakly in solution but this interaction is strengthened once they bind DNA. The heptad repeat motif located in the E domain and the zinc finger D box located in the C domain are both required for dimerization and recognition of the RARE by the RAR-RXR dimer. The RXR in the RAR-RXR heterodimer binds to the 5 upstream half-site of the RARE, whereas the RAR binds to the 3 downstream half-site (Figure 3.2 and Figure 3.4).

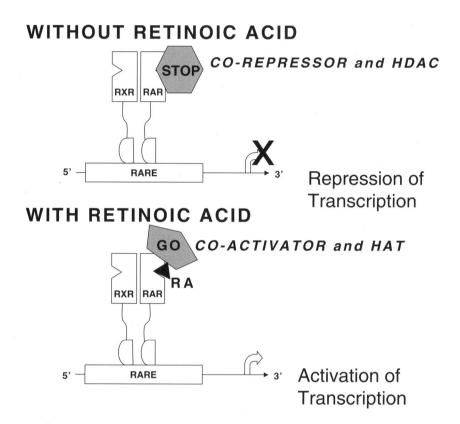

Figure 3.4 Model of the Ligand-Dependent Transactivation by RAR-RXR Heterodimer. RAR and RXR heterodimerize and bind to RARE. In the absence of ligand, they recruit transcription corepressors such as N-CoR or SMRT and repress the transcription of their target genes. When RA is present, it binds to the LBD of RAR and induces a conformational change in the RAR. Subsequently, corepressors are released and coactivators are recruited causing activation of transcription. Abbreviations: HAT, histone acetyltransferase; HDAC, histone deacetylase complex; RAR, retinoic acid receptor; RARE, retinoic acid receptor response element; RXR, retinoid X receptor.

Besides interacting with RARs, RXRs can homodimerize and bind to the DR-1 RARE. In addition, RXR can form a heterodimer with a variety of other nuclear receptors such as vitamin D receptor, thyroid hormone receptor, chicken ovalbumin upstream promoter transcription factor (COUP-TF), and peroxisome proliferator activated receptor (PPAR) to regulate transcription of these receptors' target genes.[87-91]

Because RXR does not have to be liganded with 9-*cis*-RA during the transactivation process mediated by RAR-RXR heterodimer, it was originally thought to be a silent partner. However, concomitant addition of suboptimal concentrations of RAR and RXR selective ligands results in synergistic activation of RA responsive genes.[92,93] Thus, RXRs can synergize with RARs in activating transcription when it is also liganded.

Coactivactor and Corepressors

In unliganded RARs, transcriptional corepressors such as N-CoR (nuclear receptor corepressors) and SMRT (silencing mediator for RAR and TR) are associated with region E of RAR, resulting in the silencing of the associated promoter[94-96] (Figure 3.4). Both of these corepressors mediate their negative transcriptional effects by the recruitment of histone deacetylase complexes (HDACs).[97,98] HDACs catalyze the removal of acetyl groups from histone proteins, causing a compaction of chromatin and making the DNA inaccessible to the transcription machinery.

Upon RA binding, the LBD undergoes a major conformational change that involves the folding back of helix H12. The corepressors are released, and the AF-2 activation domain recruits transcriptional coactivators including TIF1 (transcriptional mediators/intermediary factor 1),[99,100] TIF2,[101] SRC-1 (steroid receptor coactivator-1),[102] and CBP/p300 (CREB-binding protein),[103] causing transcriptional activation (Figure 3.4). Some coactivators interact directly with the basal transcriptional machinery to enhance transcriptional activation. Others possess an intrinsic histone acetyltransferase (HAT) activity. HAT acetylates histone proteins, causing the opening up of chromatin and greater accessibility of the DNA to the basal transcription machinery.

FUNCTIONAL ROLE OF RETINOIC ACID RECEPTORS

One of the major unanswered questions is why six distinct RAR and RXR subtypes, along with many isoforms of each subtype, exist. Studies examining both the expression pattern of the various RAR or RXR subtypes and the effect of inactivation of specific RAR or RXR subtypes suggest that individual RAR or RXR subtypes mediate at least some specific, nonoverlapping functions.

Expression Pattern of RARs and RXRs

In situ hybridization studies have demonstrated that different RAR and RXR subtypes are widely expressed during embryogenesis, and each subtype (even different isoforms of a subtype) has its own expression pattern which may or may not overlap with that of other subtypes (for review, see References 104 and 105).[104,105] RARα is expressed in a general fashion during mouse embryogenesis, whereas the expression of RARβ and RARγ is more restricted.[106–108] Among the RXR subtypes, RXRβ is expressed in a general fashion, whereas RXRα and RXRγ have more restricted expression patterns.[53,109]

In adult tissues, RARα is expressed ubiquitiously, with the highest levels of mRNA observed in specific regions of the brain including the hippocampus and the cerebellum.[40,110,111] RARα1 mRNA is more abundant in the brain, skin, muscle, heart, and kidney. On the other hand, RARα2 mRNA is present in slightly higher levels in the lung, liver, and intestines.[46] RARβ expression is variable among adult tissues. Low levels of mRNA are observed in the breast and eye, average levels in the liver, spleen, uterus, ovary, and testes, and high levels in the kidney, prostrate, spinal cord, cerebral cortex, pituitary gland, and adrenal gland.[110,112] RARβ2 and RARβ4 mRNA levels are observed in the kidney, liver, lung, skin, heart, muscle, intestine, and brain, whereas RARβ1 and RARβ3 mRNA levels are expressed abundantly in the brain, skin, and lung.[47,49] Finally, RARγ transcripts are the most restricted in their expression, with both RARγ1 and RARγ2 found predominately in the skin and lung.[45,113,114] The three RXR mRNAs are widely expressed in adult tissues, with at least one RXR subtype in every tissue examined.[42,53]

RAR and RXR Knockout Studies

To identify the specific functions of these receptors during embryonic development, major efforts have been made to study the phenotype of mice lacking the expression of a specific RAR or RXR subtype or isoform. In addition, the study of the effect of functional inactivation of specific RAR and RXR subtypes on RA-dependent gene expression in F9 teratocarcinoma cells has also provided important information.

Generally, although different isoforms of each receptor subtype have unique distributions during development, inactivation of one specific isoform does not appear to affect normal embryogenesis. For example, mice homozygous for either RARα1, RARβ2, or RARγ2 inactivation were viable and did not display any phenotypic abnormalities.[115–118] Interestingly RARβ2-RARγ2 compound mutant mice did display retinal dysplasia.[119]

Inactivation of a single RAR or RXR subtype results in a number of specific developmental defects in mice. RARα−/− mice exhibit early postnatal lethality and testis degeneration, whereas RARγ−/− mice display homeotic transformations of the cervical vertebra and occipital region of the skull, fusion of the first and second ribs, and abnormalities of the tracheal rings.[117,120] RXRβ−/− mice are morphologically normal except male mice are sterile, RXRα−/− mice die between GD 13.5 and 16.5 due to hypoplastic development of the ventricular chambers of the heart, and RXRγ−/− knockout mice display abnormalities in the eyes.[121–123] These studies indicate the association between certain aberrations and the lack of specific retinoid receptor subtype. However, none of the phenotypes observed in the single retinoid receptor knockout mice could be predicted on the basis of the expression patterns of the specific retinoid receptors during embryogenesis.

On the other hand, RAR double subtype mutants die either *in utero* or shortly after birth, with many defects that are characteristic of fetal vitamin A deficiency.[124–129] Analysis of RARα-RARγ and RARα-RARβ mutant mice suggests that RARα or RARγ mediate events involved in specifying the prospective rhombomere 5/rhombomere 6 territory in the developing brain, whereas RARβ is important in setting the caudal boundary of this region later in embryonic development.[130] RXRβ-RXRγ double knockout mice and even RXRα[+/–]-RXRβ[–/–]-RXRγ[–/–] are viable without obvious congenital or even postnatal defects except a marked growth deficiency and male sterility, which are due to loss of RXRβ.[121] Therefore, it appears that one copy of RXR is sufficient to perform most of the functions of RXR. Finally, the study of various RAR subtype-RXRα compound mutant mice demonstrates that the RXRα-RAR heterodimer mediates retinoid signaling *in vivo*.[131]

More recently, investigators have focused on the functional inactivation of RXRα within a specific cell type (adipocytes, skin, and liver) at various times after birth, utilizing the Cre/lox system.[132–138] RXRα has been demonstrated to be important in the mechanisms associated with the life span of hepatocytes, in hair follicle growth in skin, and adipogenesis in adipocytes. This approach shows great promise in elucidating specific roles of retinoid receptors in individual tissues during both embrogenesis and in adult animals.

In conclusion, studies utilizing mutant mice in which a specific RAR or RXR subtype is functionally inactivated during the entire life span of a mouse demonstrates that there is a degree of forced redundancy between the individual RARs and RXRs. Similarly, functional redundancies have also been observed in a number of studies of retinoid receptor knockout F9 cells. However, several reports have provided evidence that gene knockout in F9 cells generates artefactual conditions within cells that unmask functional redundancies that actually do not normally occur in a promoter-dependent context.[139–141] Therefore, at this time, the exact role of each individual retinoid receptor subtype and isoform under wild-type physiological conditions remains unclear. Additional studies are necessary to resolve this important question.

ACKNOWLEDGMENTS

The National Institutes of Health (Grants DK44517, CA82770, CA64945, and DE13139) supported the work performed in the authors' laboratories.

References

1. ME Gottesman, L Quadro, WS Blaner. Studies of vitamin A metabolism in mouse model systems. *Bioessays* 23: 409–419, 2001.

2. N Noy. Retinoid-binding proteins: mediators of retinoid action. *Biochem J* 348: 481–495, 2000.

3. JL Napoli. Interactions of retinoid binding proteins and enzymes in retinoid metabolism. *Biochim Biophys Acta* 1440: 139–162, 1999.

4. JL Napoli. Retinoic acid: its biosynthesis and metabolism. *Prog Nucleic Acid Res Mol Biol* 63: 139–188, 1999.

5. G Duester. Families of retinoid dehydrogenases regulating vitamin A function: production of visual pigment and retinoic acid. *Eur J Biochem* 267: 4315–4324, 2000.

6. W Balkan, M Colbert, C Bock, E Linney. Transgenic indicator mice for studying activated retinoic acid receptors during development. *Proc Natl Acad Sci USA* 89: 3347–3351, 1992.

7. J Rossant, R Zirngibl, D Cado, M Shago, V Giguere. Expression of a retinoic acid response element-hsplacZ transgene defines specific domains of transcriptional activity during mouse embryogenesis. *Genes Dev* 5: 1333–1344, 1991.

8. JB Moss, J Xavier-Neto, MD Shapiro, SM Nayeem, P McCaffery, UC Drager, N Rosenthal. Dynamic patterns of retinoic acid synthesis and response in the developing mammalian heart. *Dev Biol* 199: 55–71, 1998.

9. K Niederreither, P McCaffery, UC Drager, P Chambon, P Dolle. Restricted expression and retinoic acid-induced down-regulation of the retinaldehyde dehydrogenase type 2 (RALDH-2) gene during mouse development. *Mech Dev* 62: 67–78, 1997.

10. D Zhao, P McCaffery, KJ Ivins, RL Neve, P Hogan, WW Chin, UC Drager. Molecular identification of a major retinoic-acid-synthesizing enzyme, a retinaldehyde-specific dehydrogenase. *Eur J Biochem* 240: 15–22, 1996.

11. P McCaffery UC Drager. Hot spots of retinoic acid synthesis in the developing spinal cord. *Proc Natl Acad Sci USA* 91: 7194–7197, 1994.

12. DH Van Thiel, J Gavaler, R Lester. Ethanol inhibition of vitamin A metabolism in the testes: possible mechanism for sterility in alcoholics. *Science* 186: 941–942, 1974.

13. E Mezey, PR Holt. The inhibitory effect of ethanol on retinol oxidation by human liver and cattle retina. *Exp Mol Pathol* 15: 148–156, 1971.

14. P Julia, J Farres, X Pares. Ocular alcohol dehydrogenase in the rat: regional distribution and kinetics of the ADH-1 isoenzyme with retinol and retinal. *Exp Eye Res* 42: 305–314, 1986.

15. L Deltour, MH Foglio, G Duester. Impaired retinol utilization in Adh4 alcohol dehydrogenase mutant mice. *Dev Genet* 25: 1–10, 1999.

16. J Su, X Chai, B Kahn, JL Napoli. cDNA cloning, tissue distribution, and substrate characteristics of a cis-Retinol/3alpha-hydroxysterol short-chain dehydrogenase isozyme. *J Biol Chem* 273: 17910–17916, 1998.

17. NA Sophos, A Pappa, TL Ziegler, V Vasiliou. Aldehyde dehydrogenase gene superfamily: the 2000 update. *Chem Biol Interact* 130–132: 323–337, 2001.

18. Z el Akawi, JL Napoli. Rat liver cytosolic retinal dehydrogenase: comparison of 13-cis-, 9-cis-, and all-trans-retinal as substrates and effects of cellular retinoid-binding proteins and retinoic acid on activity. *Biochemistry* 33: 1938–1943, 1994.

19. F Grun, Y Hirose, S Kawauchi, T Ogura, K Umesono. Aldehyde dehydrogenase 6, a cytosolic retinaldehyde dehydrogenase prominently expressed in sensory neuroepithelia during development. *J Biol Chem* 275: 41210–41218, 2000.

20. M Lin, M Zhang, M Abraham, SM Smith, JL Napoli. Mouse retinal dehydrogenase 4 (RALDH4), molecular cloning, cellular expression, and activity in 9-cis-retinoic acid biosynthesis in intact cells. *J Biol Chem* 278: 9856–9861, 2003.

21. K Niederreither, V Subbarayan, P Dolle, P Chambon. Embryonic retinoic acid synthesis is essential for early mouse post-implantation development. *Nat Genet* 21: 444–448, 1999.

22. T Arnhold, G Tzimas, W Wittfoht, S Plonait, H Nau. Identification of 9-cis-retinoic acid, 9,13-di-cis-retinoic acid, and 14-hydroxy-4,14-retro-retinol in human plasma after liver consumption. *Life Sci* 59: PL169–177, 1996.

23. G Genchi, W Wang, A Barua, WR Bidlack, JA Olson. Formation of beta-glucuronides and of beta-galacturonides of various retinoids catalyzed by induced and noninduced microsomal UDP-glucuronosyltransferases of rat liver. *Biochim Biophys Acta* 1289: 284–290, 1996.

24. R Kojima, T Fujimori, N Kiyota, Y Toriya, T Fukuda, T Ohashi, T Sato, Y Yoshizawa, K Takeyama, H Mano, S Masushige, S Kato. In vivo isomerization of retinoic acids. Rapid isomer exchange and gene expression. *J Biol Chem* 269: 32700–32707, 1994.

25. MA Shirley, YL Bennani, MF Boehm, AP Breau, C Pathirana, EH Ulm. Oxidative and reductive metabolism of 9-cis-retinoic acid in the rat. Identification of 13,14-dihydro-9-cis-retinoic acid and its taurine conjugate. *Drug Metab Dispos* 24: 293–302, 1996.

26. H Fujii, T Sato, S Kaneko, O Gotoh, Y Fujii-Kuriyama, K Osawa, S Kato, H Hamada. Metabolic inactivation of retinoic acid by a novel P450 differentially expressed in developing mouse embryos. *EMBO J* 16: 4163–4173, 1997.

27. JA White, B Beckett-Jones, YD Guo, FJ Dilworth, J Bonasoro, G Jones, M Petkovich. cDNA cloning of human retinoic acid-metabolizing enzyme (hP450RAI) identifies a novel family of cytochromes P450. *J Biol Chem* 272: 18538–18541, 1997.

28. JA White, YD Guo, K Baetz, B Beckett-Jones, J Bonasoro, KE Hsu, FJ Dilworth, G Jones, M Petkovich. Identification of the retinoic acid-inducible all-trans-retinoic acid 4-hydroxylase. *J Biol Chem* 271: 29922–29927, 1996.

29. WJ Ray, G Bain, M Yao, DI Gottlieb. CYP26, a novel mammalian cytochrome P450, is induced by retinoic acid and defines a new family. *J Biol Chem* 272: 18702–18708, 1997.

30. Y Yamamoto, R Zolfaghari, AC Ross. Regulation of CYP26 (cytochrome P450RAI) mRNA expression and retinoic acid metabolism by retinoids and dietary vitamin A in liver of mice and rats. *FASEB J* 14: 2119–2127, 2000.

31. O Loudig, C Babichuk, J White, S Abu-Abed, C Mueller, M Petkovich. Cytochrome P450RAI(CYP26) promoter: a distinct composite retinoic acid response element underlies the complex regulation of retinoic acid metabolism. *Mol Endocrinol* 14: 1483–1497, 2000.

32. SS Abu-Abed, BR Beckett, H Chiba, JV Chithalen, G Jones, D Metzger, P Chambon, M Petkovich. Mouse P450RAI (CYP26) expression and retinoic acid-inducible retinoic acid metabolism in F9 cells are regulated by retinoic acid receptor gamma and retinoid X receptor alpha. *J Biol Chem* 273: 2409–2415, 1998.

33. E Sonneveld, CE van den Brink, BM van der Leede, RK Schulkes, M Petkovich, B van der Burg, PT van der Saag. Human retinoic acid (RA) 4-hydroxylase (CYP26) is highly specific for all-trans-RA and can be induced through RA receptors in human breast and colon carcinoma cells. *Cell Growth Differ* 9: 629–637, 1998.

34. SY Kim, SJ Yoo, HJ Kwon, SH Kim, Y Byun, KS Lee. Retinoic acid 4-hydroxylase-mediated catabolism of all-trans retinoic acid and the cell proliferation in head and neck squamous cell carcinoma. *Metabolism* 51: 477–481, 2002.

35. YLR Mira, WL Zheng, YS Kuppumbatti, B Rexer, Y Jing, DE Ong. Retinol conversion to retinoic acid is impaired in breast cancer cell lines relative to normal cells. *J Cell Physiol* 185: 302–309, 2000.

36. SY Kim, H Adachi, JS Koo, AM Jetten. Induction of the cytochrome P450 gene CYP26 during mucous cell differentiation of normal human tracheobronchial epithelial cells. *Mol Pharmacol* 58: 483–490, 2000.

37. E Sonneveld, CE van den Brink, LG Tertoolen, B van der Burg, PT van der Saag. Retinoic acid hydroxylase (CYP26) is a key enzyme in neuronal differentiation of embryonal carcinoma cells. *Dev Biol* 213: 390–404, 1999.

38. B Ozpolat, K Mehta, AM Tari, G Lopez-Berestein. All-trans-Retinoic acid-induced expression and regulation of retinoic acid 4-hydroxylase (CYP26) in human promyelocytic leukemia. *Am J Hematol* 70: 39–47, 2002.

39. E Sonneveld, PT van der Saag. Metabolism of retinoic acid: implications for development and cancer. *Int J Vitam Nutr Res* 68: 404–410, 1998.

40. V Giguere, ES Ong, P Segui, RM Evans. Identification of a receptor for the morphogen retinoic acid. *Nature* 330: 624–629, 1987.

41. M Petkovich, NJ Brand, A Krust, P Chambon. A human retinoic acid receptor which belongs to the family of nuclear receptors. *Nature* 330: 444–450, 1987.

42. DJ Mangelsdorf, ES Ong, JA Dyck, RM Evans. Nuclear receptor that identifies a novel retinoic acid response pathway. *Nature* 345: 224–229, 1990.

43. P Chambon. A decade of molecular biology of retinoic acid receptors. *FASEB J* 10: 940–954, 1996.

44. V Giguere, M Shago, R Zirngibl, P Tate, J Rossant, S Varmuza. Identification of a novel isoform of the retinoic acid receptor gamma expressed in the mouse embryo. *Mol Cell Biol* 10: 2335–2340, 1990.

45. P Kastner, A Krust, C Mendelsohn, JM Garnier, A Zelent, P Leroy, A Staub, P Chambon. Murine isoforms of retinoic acid receptor gamma with specific patterns of expression. *Proc Natl Acad Sci USA* 87: 2700–2704, 1990.

46. P Leroy, A Krust, A Zelent, C Mendelsohn, JM Garnier, P Kastner, A Dierich, P Chambon. Multiple isoforms of the mouse retinoic acid receptor alpha are generated by alternative splicing and differential induction by retinoic acid. *EMBO J* 10: 59–69, 1991.

47. A Zelent, C Mendelsohn, P Kastner, A Krust, JM Garnier, F Ruffenach, P Leroy, P Chambon. Differentially expressed isoforms of the mouse retinoic acid receptor beta generated by usage of two promoters and alternative splicing. *EMBO J* 10: 71–81, 1991.

48. M Leid, P Kastner, P Chambon. Multiplicity generates diversity in the retinoic acid signalling pathways. *Trends Biochem Sci* 17: 427–433, 1992.

49. S Nagpal, A Zelent, P Chambon. RAR-beta 4, a retinoic acid receptor isoform is generated from RAR-beta 2 by alternative splicing and usage of a CUG initiator codon. *Proc Natl Acad Sci USA* 89: 2718–2722, 1992.

50. Q Liu, E Linney. The mouse retinoid-X receptor-gamma gene: genomic organization and evidence for functional isoforms. *Mol Endocrinol* 7: 651–658, 1993.

51. J Brocard, P Kastner, P Chambon. Two novel RXR alpha isoforms from mouse testis. *Biochem Biophys Res Commun* 229: 211–218, 1996.

52. T Nagata, Y Kanno, K Ozato, M Taketo. The mouse RXRb gene encoding RXR beta: genomic organization and two mRNA isoforms generated by alternative splicing of transcripts initiated from CpG island promoters. *Gene* 142: 183–189, 1994.

53. DJ Mangelsdorf, U Borgmeyer, RA Heyman, JY Zhou, ES Ong, AE Oro, A Kakizuka, RM Evans. Characterization of three RXR genes that mediate the action of 9-cis retinoic acid. *Genes Dev* 6: 329–344, 1992.

54. AA Levin, LJ Sturzenbecker, S Kazmer, T Bosakowski, C Huselton, G Allenby, J Speck, C Kratzeisen, M Rosenberger, A Lovey, J Grippo. 9-cis retinoic acid stereoisomer binds and activates the nuclear receptor RXR alpha. *Nature* 355: 359–361, 1992.

55. RA Heyman, DJ Mangelsdorf, JA Dyck, RB Stein, G Eichele, RM Evans, C Thaller. 9-cis retinoic acid is a high affinity ligand for the retinoid X receptor. *Cell* 68: 397–406, 1992.

56. AM de Urquiza, S Liu, M Sjoberg, RH Zetterstrom, W Griffiths, J Sjovall, T Perlmann. Docosahexaenoic acid, a ligand for the retinoid X receptor in mouse brain. *Science* 290: 2140–2144, 2000

57. JT Goldstein, A Dobrzyn, M Clagett-Dame, JW Pike, HF DeLuca. Isolation and characterization of unsaturated fatty acids as natural ligands for the retinoid-X receptor. *Arch Biochem Biophys* 420: 185–193, 2003.

58. RA Chandraratna. Future trends: a new generation of retinoids. *J Am Acad Dermatol* 39: S149–152, 1998.

59. A Johnson, RA Chandraratna. Novel retinoids with receptor selectivity and functional selectivity. *Br J Dermatol* 140 Suppl 54: 12–17, 1999

60. S Nagpal, RA Chandraratna. Recent developments in receptor-selective retinoids. *Curr Pharm Des* 6: 919–931, 2000.

61. K Umesono, V Giguere, CK Glass, MG Rosenfeld, RM Evans. Retinoic acid and thyroid hormone induce gene expression through a common responsive element. *Nature* 336: 262–265, 1988.

62. K Umesono, RM Evans. Determinants of target gene specificity for steroid/thyroid hormone receptors. *Cell* 57: 1139–1146, 1989.

63. K Umesono, KK Murakami, CC Thompson, RM Evans. Direct repeats as selective response elements for the thyroid hormone, retinoic acid, and vitamin D3 receptors. *Cell* 65: 1255–1266, 1991.

64. AM Naar, JM Boutin, SM Lipkin, VC Yu, JM Holloway, CK Glass, MG Rosenfeld. The orientation and spacing of core DNA-binding motifs dictate selective transcriptional responses to three nuclear receptors. *Cell* 65: 1267–1279, 1991.

65. JR Thompson, SW Chen, L Ho, AW Langston, LJ Gudas. An evolutionary conserved element is essential for somite and adjacent mesenchymal expression of the Hoxa1 gene. *Dev Dyn* 211: 97–108, 1998.

66. AW Langston, LJ Gudas. Identification of a retinoic acid responsive enhancer 3′ of the murine homeobox gene Hox-1.6. *Mech Dev* 38: 217–227, 1992.

67. D Huang, SW Chen, AW Langston, LJ Gudas. A conserved retinoic acid responsive element in the murine Hoxb-1 gene is required for expression in the developing gut. *Development* 125: 3235–3246, 1998.

68. S Nagpal, S Friant, H Nakshatri, P Chambon. RARs and RXRs: evidence for two autonomous transactivation functions (AF-1 and AF-2) and heterodimerization *in vivo. EMBO J* 12: 2349–2360, 1993.

69. W Bourguet, M Ruff, P Chambon, H Gronemeyer, D Moras. Crystal structure of the ligand-binding domain of the human nuclear receptor RXR-alpha. *Nature* 375: 377–382, 1995.

70. JP Renaud, N Rochel, M Ruff, V Vivat, P Chambon, H Gronemeyer, D Moras. Crystal structure of the RAR-gamma ligand-binding domain bound to all-trans retinoic acid. *Nature* 378: 681–689, 1995.

71. W Bourguet, V Vivat, JM Wurtz, P Chambon, H Gronemeyer, D Moras. Crystal structure of a heterodimeric complex of RAR and RXR ligand-binding domains. *Mol Cell* 5: 289–298, 2000.

72. PF Egea, A Mitschler, N Rochel, M Ruff, P Chambon, D Moras. Crystal structure of the human RXRalpha ligand-binding domain bound to its natural ligand: 9-cis retinoic acid. *EMBO J* 19: 2592–2601, 2000.

73. JM Wurtz, W Bourguet, JP Renaud, V Vivat, P Chambon, D Moras, H Gronemeyer. A canonical structure for the ligand-binding domain of nuclear receptors. *Nat Struct Biol* 3: 87–94, 1996.

74. N Tairis, JL Gabriel, KJ Soprano, DR Soprano. Alteration in the retinoid specificity of retinoic acid receptor-beta by site-directed mutagenesis of Arg269 and Lys220. *J Biol Chem* 270: 18380–18387, 1995.

75. N Tairis, JL Gabriel, M Gyda, KJ Soprano, DR Soprano. Arg269 and Lys220 of retinoic acid receptor-beta are important for the binding of retinoic acid. *J Biol Chem* 269: 19516–19522, 1994.

76. A Scafonas, CL Wolfgang, JL Gabriel, KJ Soprano, DR Soprano. Differential role of homologous positively charged amino acid residues for ligand binding in retinoic acid receptor alpha compared with retinoic acid receptor beta. *J Biol Chem* 272: 11244–11249, 1997.

77. ZP Zhang, CJ Gambone, JL Gabriel, CL Wolfgang, KJ Soprano, DR Soprano. Arg278, but not Lys229 or Lys236, plays an important role in the binding of retinoic acid by retinoic acid receptor gamma. *J Biol Chem* 273: 34016–34021, 1998.

78. CL Wolfgang, Z Zhang, JL Gabriel, RA Pieringer, KJ Soprano, DR Soprano. Identification of sulfhydryl-modified cysteine residues in the ligand binding pocket of retinoic acid receptor beta. *J Biol Chem* 272: 746–753, 1997.

79. ZP Zhang, M Shukri, CJ Gambone, JL Gabriel, KJ Soprano, DR Soprano. Role of Ser(289) in RARgamma and its homologous amino acid residue of RARalpha and RARbeta in the binding of retinoic acid. *Arch Biochem Biophys* 380: 339–346, 2000.

80. J Ostrowski, L Hammer, T Roalsvig, K Pokornowski, PR Reczek. The N-terminal portion of domain E of retinoic acid receptors alpha and beta is essential for the recognition of retinoic acid and various analogs. *Proc Natl Acad Sci USA* 92: 1812–1816, 1995.

81. J Ostrowski, T Roalsvig, L Hammer, A Marinier, JE Starrett, Jr., KL Yu, PR Reczek. Serine 232 and methionine 272 define the ligand binding pocket in retinoic acid receptor subtypes. *J Biol Chem* 273: 3490–3495, 1998.

82. M Gehin, V Vivat, JM Wurtz, R Losson, P Chambon, D Moras, H Gronemeyer. Structural basis for engineering of retinoic acid receptor isotype-selective agonists and antagonists. *Chem Biol* 6: 519–529, 1999.

83. ZP Zhang, JM Hutcheson, HC Poynton, JL Gabriel, KJ Soprano, DR Soprano. Arginine of retinoic acid receptor beta which coordinates with the carboxyl group of retinoic acid functions independent of the amino acid residues responsible for retinoic acid receptor subtype ligand specificity. *Arch Biochem Biophys* 409: 375–384, 2003.

84. M Bommer, A Benecke, H Gronemeyer, C Rochette-Egly. TIF2 Mediates the Synergy between RARalpha 1 Activation Functions AF-1 and AF-2. *J Biol Chem* 277: 37961–37966, 2002.

85. C Rochette-Egly, S Adam, M Rossignol, JM Egly, P Chambon. Stimulation of RAR alpha activation function AF-1 through binding to the general transcription factor TFIIH and phosphorylation by CDK7. *Cell* 90: 97–107, 1997.

86. M Gianni, A Bauer, E Garattini, P Chambon, C Rochette-Egly. Phosphorylation by p38MAPK and recruitment of SUG-1 are required for RA-induced RAR gamma degradation and transactivation. *EMBO J* 21: 3760–3769, 2002.

87. TH Bugge, J Pohl, O Lonnoy, HG Stunnenberg. RXR alpha, a promiscuous partner of retinoic acid and thyroid hormone receptors. *EMBO J* 11: 1409–1418, 1992.

88. XK Zhang, B Hoffmann, PB Tran, G Graupner, M Pfahl. Retinoid X receptor is an auxiliary protein for thyroid hormone and retinoic acid receptors. *Nature* 355: 441–446, 1992.

89. VC Yu, C Delsert, B Andersen, JM Holloway, OV Devary, AM Naar, SY Kim, JM Boutin, CK Glass, MG Rosenfeld. RXR beta: a coregulator that enhances binding of retinoic acid, thyroid hormone, and vitamin D receptors to their cognate response elements. *Cell* 67: 1251–1266, 1991.

90. SA Kliewer, K Umesono, RA Heyman, DJ Mangelsdorf, JA Dyck, RM Evans. Retinoid X receptor-COUP-TF interactions modulate retinoic acid signaling. *Proc Natl Acad Sci USA* 89: 1448–1452, 1992.

91. SA Kliewer, K Umesono, DJ Noonan, RA Heyman, RM Evans. Convergence of 9-cis retinoic acid and peroxisome proliferator signalling pathways through heterodimer formation of their receptors. *Nature* 358: 771–774, 1992.

92. B Roy, R Taneja, P Chambon. Synergistic activation of retinoic acid (RA)-responsive genes and induction of embryonal carcinoma cell differentiation by an RA receptor alpha (RAR alpha)-, RAR beta-, or RAR gamma-selective ligand in combination with a retinoid X receptor-specific ligand. *Mol Cell Biol* 15: 6481–6487, 1995.

93. R Taneja, B Roy, JL Plassat, cf. Zusi, J Ostrowski, PR Reczek, P Chambon. Cell-type and promoter-context dependent retinoic acid receptor (RAR) redundancies for RAR beta 2 and Hoxa-1 activation in F9 and P19 cells can be artefactually generated by gene knockouts. *Proc Natl Acad Sci USA* 93: 6197–6202, 1996.

94. R Kurokawa, M Soderstrom, A Horlein, S Halachmi, M Brown, MG Rosenfeld, CK Glass. Polarity-specific activities of retinoic acid receptors determined by a co-repressor. *Nature* 377: 451–454, 1995.

95. AJ Horlein, AM Naar, T Heinzel, J Torchia, B Gloss, R Kurokawa, A Ryan, Y Kamei, M Soderstrom, CK Glass. Ligand-independent repression by the thyroid hormone receptor mediated by a nuclear receptor co-repressor. *Nature* 377: 397–404, 1995.

96. JD Chen, RM Evans. A transcriptional co-repressor that interacts with nuclear hormone receptors. *Nature* 377: 454–457, 1995.

97. L Alland, R Muhle, H Hou, Jr., J Potes, L Chin, N Schreiber-Agu, RA DePinho. Role for N-CoR and histone deacetylase in Sin3-mediated transcriptional repression. *Nature* 387: 49–55, 1997.

98. T Heinzel, RM Lavinsky, TM Mullen, M Soderstrom, CD Laherty, J Torchia, WM Yang, G Brard, SD Ngo, JR Davie, E Seto, RN Eisenman, DW Rose, CK Glass, MG Rosenfeld. A complex containing N-CoR, mSin3 and histone deacetylase mediates transcriptional repression. *Nature* 387: 43–48, 1997.

99. B Le Douarin, C Zechel, JM Garnier, Y Lutz, L Tora, P Pierrat, D Heery, H Gronemeyer, P Chambon, R Losson. The N-terminal part of TIF1, a putative mediator of the ligand-dependent activation function (AF-2) of nuclear receptors, is fused to B-raf in the oncogenic protein T18. *EMBO J* 14: 2020–2033, 1995.

100. E vom Baur, C Zechel, D Heery, MJ Heine, JM Garnier, V Vivat, B Le Douarin, H Gronemeyer, P Chambon, R Losson. Differential ligand-dependent interactions between the AF-2 activating domain of nuclear receptors and the putative transcriptional intermediary factors mSUG1 and TIF1. *EMBO J* 15: 110–124, 1996.

101. JJ Voegel, MJ Heine, C Zechel, P Chambon, H Gronemeyer. TIF2, a 160 kDa transcriptional mediator for the ligand-dependent activation function AF-2 of nuclear receptors. *EMBO J* 15: 3667–3675, 1996.

102. SA Onate, SY Tsai, MJ Tsai, BW O'Malley. Sequence and characterization of a coactivator for the steroid hormone receptor superfamily. *Science* 270: 1354–1357, 1995.

103. TP Yao, G Ku, N Zhou, R Scully, DM Livingston. The nuclear hormone receptor coactivator SRC-1 is a specific target of p300. *Proc Natl Acad Sci USA* 93: 10626–10631, 1996.

104. P Chambon. The molecular and genetic dissection of the retinoid signaling pathway. *Recent Prog Horm Res* 50: 317–332, 1995.

105. DJ Mangelsdorf, C Thummel, M Beato, P Herrlich, G Schutz, K Umesono, B Blumberg, P Kastner, M Mark, P Chambon. The nuclear receptor superfamily: the second decade. *Cell* 83: 835–839, 1995.

106. P Dolle, E Ruberte, P Leroy, G Morriss-Kay, P Chambon. Retinoic acid receptors and cellular retinoid binding proteins. I. A systematic study of their differential pattern of transcription during mouse organogenesis. *Development* 110: 1133–1151, 1990.

107. E Ruberte, P Dolle, A Krust, A Zelent, G Morriss-Kay, P Chambon. Specific spatial and temporal distribution of retinoic acid receptor gamma transcripts during mouse embryogenesis. *Development* 108: 213–222, 1990.

108. C Mendelsohn, S Larkin, M Mark, M LeMeur, J Clifford, A Zelent, P Chambon. RAR beta isoforms: distinct transcriptional control by retinoic acid and specific spatial patterns of promoter activity during mouse embryonic development. *Mech Dev* 45: 227–241, 1994.

109. P Dolle, V Fraulob, P Kastner, P Chambon. Developmental expression of murine retinoid X receptor (RXR) genes. *Mech Dev* 45: 91–104, 1994.

110. H de The, A Marchio, P Tiollais, A Dejean. Differential expression and ligand regulation of the retinoic acid receptor alpha and beta genes. *EMBO J* 8: 429–433, 1989.

111. W Krezel, P Kastner, P Chambon. Differential expression of retinoid receptors in the adult mouse central nervous system. *Neuroscience* 89: 1291–1300, 1999.

112. D Benbrook, E Lernhardt, M Pfahl. A new retinoic acid receptor identified from a hepatocellular carcinoma. *Nature* 333: 669–672, 1988.

113. S Noji, T Yamaai, E Koyama, T Nohno, W Fujimoto, J Arata, S Taniguchi. Expression of retinoic acid receptor genes in keratinizing front of skin. *FEBS Lett* 259: 86–90, 1989.

114. A Zelent, A Krust, M Petkovich, P Kastner, P Chambon. Cloning of murine alpha and beta retinoic acid receptors and a novel receptor gamma predominantly expressed in skin. *Nature* 339: 714–717, 1989.

115. P Kastner, M Mark, P Chambon. Nonsteroid nuclear receptors: what are genetic studies telling us about their role in real life? *Cell* 83: 859–869, 1995.

116. E Li, HM Sucov, KF Lee, RM Evans, R Jaenisch. Normal development and growth of mice carrying a targeted disruption of the alpha 1 retinoic acid receptor gene. *Proc Natl Acad Sci USA* 90: 1590–1594, 1993.

117. D Lohnes, P Kastner, A Dierich, M Mark, M LeMeur, P Chambon. Function of retinoic acid receptor gamma in the mouse. *Cell* 73: 643–658, 1993.

118. C Mendelsohn, M Mark, P Dolle, A Dierich, MP Gaub, A Krust, C Lampron, P Chambon. Retinoic acid receptor beta 2 (RAR beta 2) null mutant mice appear normal. *Dev Biol* 166: 246–258, 1994.

119. JM Grondona, P Kastner, A Gansmuller, D Decimo, P Chambon, M Mark. Retinal dysplasia and degeneration in RARbeta2/RARgamma2 compound mutant mice. *Development* 122: 2173–2188, 1996.

120. T Lufkin, D Lohnes, M Mark, A Dierich, P Gorry, MP Gaub, M LeMeur, P Chambon. High postnatal lethality and testis degeneration in retinoic acid receptor alpha mutant mice. *Proc Natl Acad Sci USA* 90: 7225–7229, 1993.

121. W Krezel, V Dupe, M Mark, A Dierich, P Kastner, P Chambon. RXR gamma null mice are apparently normal and compound RXR alpha +/–/RXR beta –/–/RXR gamma –/– mutant mice are viable. *Proc Natl Acad Sci USA* 93: 9010–9014, 1996.

122. P Kastner, M Mark, M Leid, A Gansmuller, W Chin, JM Grondona, D Decimo, W Krezel, A Dierich, P Chambon. Abnormal spermatogenesis in RXR beta mutant mice. *Genes Dev* 10: 80–92, 1996.

123. HM Sucov, E Dyson, CL Gumeringer, J Price, KR Chien, RM Evans. RXR alpha mutant mice establish a genetic basis for vitamin A signaling in heart morphogenesis. *Genes Dev* 8: 1007–1018, 1994.

124. D Lohnes, M Mark, C Mendelsohn, P Dolle, D Decimo, M LeMeur, A Dierich, P Gorry, P Chambon. Developmental roles of the retinoic acid receptors. *J Steroid Biochem Mol Biol* 53: 475–486, 1995.

125. D Lohnes, M Mark, C Mendelsohn, P Dolle, A Dierich, P Gorry, A Gansmuller, P Chambon. Function of the retinoic acid receptors (RARs) during development (I). Craniofacial and skeletal abnormalities in RAR double mutants. *Development* 120: 2723–2748, 1994.

126. J Luo, HM Sucov, JA Bader, RM Evans, V Giguere. Compound mutants for retinoic acid receptor (RAR) beta and RAR alpha 1 reveal developmental functions for multiple RAR beta isoforms. *Mech Dev* 55: 33–44, 1996.

127. C Mendelsohn, D Lohnes, D Decimo, T Lufkin, M LeMeur, P Chambon, M Mark. Function of the retinoic acid receptors (RARs) during development (II). Multiple abnormalities at various stages of organogenesis in RAR double mutants. *Development* 120: 2749–2771, 1994.

128. MY Chiang, D Misner, G Kempermann, T Schikorski, V Giguere, HM Sucov, FH Gage, cf. Stevens, RM Evans. An essential role for retinoid receptors RARbeta and RXRgamma in long-term potentiation and depression. *Neuron* 21: 1353–1361, 1998.

129. V Dupe, NB Ghyselinck, O Wendling, P Chambon, M Mark. Key roles of retinoic acid receptors alpha and beta in the patterning of the caudal hindbrain, pharyngeal arches and otocyst in the mouse. *Development* 126: 5051–5059, 1999.

130. O Wendling, NB Ghyselinck, P Chambon, M Mark. Roles of retinoic acid receptors in early embryonic morphogenesis and hindbrain patterning. *Development* 128: 2031–2038, 2001.

131. P Kastner, M Mark, N Ghyselinck, W Krezel, V Dupe, JM Grondona, P Chambon. Genetic evidence that the retinoid signal is transduced by heterodimeric RXR/RAR functional units during mouse development. *Development* 124: 313–326, 1997.

132. D Metzger, P Chambon. Site- and time-specific gene targeting in the mouse. *Methods* 24: 71–80, 2001.

133. YJ Wan, Y Cai, W Lungo, P Fu, J Locker, S French, HM Sucov. Peroxisome proliferator-activated receptor alpha-mediated pathways are altered in hepatocyte-specific retinoid X receptor alpha-deficient mice. *J Biol Chem* 275: 28285–28290, 2000.

134. YJ Wan, D An, Y Cai, JJ Repa, T Hung-Po Chen, M Flores, C Postic, MA Magnuson, J Chen, KR Chien, S French, DJ Mangelsdorf, HM Sucov. Hepatocyte-specific mutation establishes retinoid X receptor alpha as a heterodimeric integrator of multiple physiological processes in the liver. *Mol Cell Biol* 20: 4436–4444, 2000.

135. T Imai, M Jiang, P Chambon, D Metzger. Impaired adipogenesis and lipolysis in the mouse upon selective ablation of the retinoid X receptor alpha mediated by a tamoxifen-inducible chimeric Cre recombinase (Cre-ERT2) in adipocytes. *Proc Natl Acad Sci USA* 98: 224–228, 2001.

136. T Imai, M Jiang, P Kastner, P Chambon, D Metzger. Selective ablation of retinoid X receptor alpha in hepatocytes impairs their lifespan and regenerative capacity. *Proc Natl Acad Sci USA* 98: 4581–4586, 2001.

137. M Li, H Chiba, X Warot, N Messaddeq, C Gerard, P Chambon, D Metzger. RXR-alpha ablation in skin keratinocytes results in alopecia and epidermal alterations. *Development* 128: 675–688, 2001.

138. M Li, AK Indra, X Warot, J Brocard, N Messaddeq, S Kato, D Metzger, P Chambon. Skin abnormalities generated by temporally controlled RXRalpha mutations in mouse epidermis. *Nature* 407: 633–636, 2000.

139. H Chiba, J Clifford, D Metzger, P Chambon. Distinct retinoid X receptor-retinoic acid receptor heterodimers are differentially involved in the control of expression of retinoid target genes in F9 embryonal carcinoma cells. *Mol Cell Biol* 17: 3013–3020, 1997.

140. J Plassat, L Penna, P Chambon, C Rochette-Egly. The conserved amphipatic alpha-helical core motif of RARgamma and RARalpha activating domains is indispensable for RA-induced differentiation of F9 cells. *J Cell Sci* 113: 2887–2895, 2000.

141. C Rochette-Egly, JL Plassat, R Taneja, P Chambon. The AF-1 and AF-2 activating domains of retinoic acid receptor-alpha (RARalpha) and their phosphorylation are differentially involved in parietal endodermal differentiation of F9 cells and retinoid-induced expression of target genes. *Mol Endocrinol* 14: 1398–1410, 2000.

142. DJ Mangelsdorf, K Umesono, RM Evans. The retinoid receptors. In: MB Sporn, AB Roberts, DS Goodman, Eds. *The Retinoids: Biology, Chemistry and Medicine*. 2nd ed.. Raven Press, NY. 1994, pp 319–349.

4

Regulation of Cytokines and Immune Function by 1,25-Dihydroxyvitamin D$_3$ and Its Analogs

EVELYNE VAN ETTEN, ANNAPAULA GIULIETTI,
CONNY GYSEMANS, ANNEMIEKE VERSTUYF,
LUT OVERBERGH, AND CHANTAL MATHIEU

ABBREVIATIONS

1,25(OH)$_2$D$_3$	1,25-dihydroxyvitamin D$_3$
APC	antigen-presenting cell
BB	Biobreeding
DC	dendritic cell
DRIP	vitamin D receptor interacting-protein
GM-CSF	granulocyte macrophage colony-stimulating factor
IFN	interferon
IL	interleukin
NOD	nonobese diabetic
RXR	retinoid X receptor
SRC	steroid receptor coactivator
Th	T helper
Treg	regulatory T lymphocyte
VDR	vitamin D receptor
VDRE	vitamin D responsive element

INTRODUCTION

1,25-Dihydroxyvitamin D_3 [1,25$(OH)_2D_3$], which was originally described as an essential player in bone and mineral
homeostasis, is the biologically active form of the secosteroid
vitamin D_3. Although vitamin D_3 can be derived from nutrition (from major dietary sources such as fortified dairy products, fatty fish, and fish liver oils), the main supply of vitamin
D_3 is through photosynthesis in the skin.[1] Exposure of skin
to UV light (270 to 300 nm) catalyzes the first step in the
vitamin D_3 biosynthesis, converting 7-dehydrocholesterol into
previtamin D_3, which is followed by a spontaneous and tem-
perature-dependent isomerization into vitamin D_3. To obtain
the biologically active metabolite 1,25$(OH)_2D_3$, vitamin D_3
must first be hydroxylated by D_3-25-hydroxylase (CYP2D25)
in the liver into 25-hydroxyvitamin D_3, the major circulating
form of the vitamin, which can then be further hydroxylated
by 25$(OH)D_3$-1-α-hydroxylase (CYP27B1) in the proximal con-
voluted tubule cells of the kidney into the active 1,25$(OH)_2D_3$
(reviewed in Reference 2).[2] In addition to the central function
of 1,25$(OH)_2D_3$ in mineral and bone homeostasis, important
effects on the differentiation, growth, and function of many
other target tissues, including normal (e.g., keratinocytes,
pancreatic islet cells) as well as malignant (e.g., breast, colon,
and prostate cancer cells) cell types, have been shown. These
biological effects of 1,25$(OH)_2D_3$ are mediated through the
vitamin D receptor (VDR), a member of the superfamily of
nuclear hormone receptors.[3] Ligand binding induces conformational changes in the VDR, promoting heterodimerization
with the retinoid X receptor (RXR) and recruitment of a number of nuclear receptor coactivator proteins, such as steroid
receptor coactivator family (SRC) members and the D-recep-
tor interacting protein (DRIP) coactivator complex. Thus, the
VDR functions as a ligand-activated transcription factor that
binds to vitamin D_3-response elements (VDREs) within the
promoter region of vitamin D_3-responsive genes, influencing
the rate of RNA polymerase II-mediated transcription.[4,5]
Besides this genomic mechanism of action, it has been suggested that 1,25$(OH)_2D_3$ is also able to exert its function on

target cells via rapid effects independent of *de novo* gene transcription (reviewed in Reference 6).[6] It is postulated that this nongenomic signaling pathway is initiated at the cell membrane, but whether the classical VDR or an alternative receptor is involved is still under investigation.

RECEPTORS AND METABOLISM OF 1,25(OH)$_2$D$_3$ IN THE IMMUNE SYSTEM

The discovery that the VDR is constitutively expressed by antigen-presenting cells (APCs) such as macrophages and dendritic cells (DCs) and inducibly expressed by lymphocytes following activation[7] has led to experiments using 1,25(OH)$_2$D$_3$ *in vitro* and *in vivo* in different settings, confirming a central role for 1,25(OH)$_2$D$_3$ in the immune system.[8] Moreover, the enzyme responsible for the final and rate-limiting hydroxylation step in the synthesis of 1,25(OH)$_2$D$_3$, 25(OH)D$_3$-1-α-hydroxylase, is expressed by activated macrophages,[9] making them able to synthesize and secrete 1,25(OH)$_2$D$_3$ in a regulated fashion. Although the enzyme found in macrophages is identical to the known renal form, its expression is regulated differently. Whereas renal 1-α-hydroxylase is mainly regulated by mediators of calcium and bone homeostasis (e.g., parathyroid hormone and 1,25(OH)$_2$D$_3$ itself), its macrophage version is predominantly under control of immune signals such as IFN-γ. Moreover, in macrophages no negative feedback by the end product, 1,25(OH)$_2$D$_3$ itself, could be observed, explaining the occasional hypercalcemia in situations of macrophage overactivation.[10]

Based on the widespread presence of VDR in the different cell types of the immune system and on the 25(OH)D$_3$-1-α-hydroxylase activity present in macrophages and regulated by immune signals, a paracrine role for 1,25(OH)$_2$D$_3$ in the immune system can be postulated.

SYNTHETIC ANALOGS OF 1,25(OH)$_2$D$_3$

The major obstacles in the widespread clinical application of 1,25(OH)$_2$D$_3$ for its immunomodulatory properties are the dose-limiting side effects including hypercalcemia, hypercalciuria, and increased bone resorption. Indeed, significant immunomodulatory effects appear only at high pharmacological doses of $1,25(OH)_2D_3$, at which the side effects on bone and calcium homeostasis are already detrimental. Worldwide efforts have led to the development of a whole array of synthetic structural analogs of $1,25(OH)_2D_3$, of which many maintain or amplify the nonclassical effects of $1,25(OH)_2D_3$ while displaying a decreased hypercalcemic potential.[11,12] The secosteroid $1,25(OH)_2D_3$, with its open B-ring and side chain, is indeed a very flexible molecule. Different parts of the molecule have been modified (A, seco B, C, and D rings and side chain) by addition or transposition of hydroxyl groups, introduction of unsaturated bonds and hetero atoms, inversion of the stereochemistry, and alterations in the length of the side chain. Moreover, nonsteroidal analogs have been created that lack the full CD-region of the parent molecule.[13,14] The recent elucidation of the crystal structure of the VDR bound to its natural ligand[15] will further facilitate the rational design of future $1,25(OH)_2D_3$ analogs with improved immunomodulatory potency, reduced calcemic activity, and increased tissue specificity.

Several mechanisms may contribute to the altered biological profile of these analogs when compared with $1,25(OH)_2D_3$. Differences in binding affinity to the vitamin D binding protein may result in altered availability and clearance rates in the various organs.[16] Metabolization of $1,25(OH)_2D_3$ analogs is probably different than for the parent compound. It is therefore conceivable that the metabolization rate and the occurrence of analog-specific metabolites contribute to the altered biological profile and tissue specificity.[17] Altered affinity to the VDR and modifications in the VDR conformation and stability due to interaction with an analog may affect the heterodimerization with RXR, the binding of the analog-VDR-RXR complex to a VDRE, and the interaction with coactivators and factors of the preinitiation complex, and thus lead to differences in gene transcription.[18,19] Moreover, the existence of tissue-specific forms of VDR has been postulated to explain the apparent tissue-specificity of certain analogs,[20] but this remains unconfirmed as yet.

EFFECTS OF 1,25(OH)$_2$D$_3$ AND ITS ANALOGS IN THE IMMUNE SYSTEM

T Lymphocytes

After the discovery of VDR expression in most cells of the immune system, a direct effect of 1,25(OH)$_2$D$_3$ on T lymphocytes was shown. Antigen- or lectin-stimulated human and murine T lymphocyte proliferation, cytokine secretion, and cell-cycle progression from G$_{1a}$ to G$_{1b}$ are inhibited by *in vitro* addition of 1,25(OH)$_2$D$_3$.[21,22] Indeed, the transcription of several key cytokines of T helper (Th) 1 lymphocytes, such as IFN-γ and IL-2, is a direct target of 1,25(OH)$_2$D$_3$. By inhibiting IFN-γ transcription, the major positive feedback signal for antigen-presenting cells, 1,25(OH)$_2$D$_3$ prevents further antigen presentation to and recruitment of T lymphocytes.[23] IL-2 is an autocrine growth factor for T lymphocytes, and inhibition of its expression by 1,25(OH)$_2$D$_3$ prevents their further activation and proliferation.[24,25] CD4$^+$ Th1 lymphocytes (IL-2 and IFN-γ secretion) are considered to be the key mediators in graft rejection and autoimmune diseases, whereas CD4$^+$ Th2 lymphocytes (IL-4, IL-5, IL-13 secretion) have a more regulatory function. While inhibiting the synthesis of the Th1 cytokines IL-2 and IFN-γ, and thus indirectly inducing CD4$^+$ T lymphocytes to polarize toward a Th2 phenotype and eventually attenuating autoimmunity and graft rejection, 1,25(OH)$_2$D$_3$ also affects the production of Th2 cytokines themselves. Boonstra et al. showed the development of a highly Th2-skewed lymphocyte population with increased IL-4, IL-5, and IL-10 production after *in vitro* 1,25(OH)$_2$D$_3$ treatment.[26] Furthermore, the 1,25(OH)$_2$D$_3$-induced effects on Th2 lymphocyte development were largely mediated via IL-4. This 1,25(OH)$_2$D$_3$-mediated upregulation of IL-4 and the subsequent protective effects in autoimmunity could be confirmed *in vivo*.[27,28] However, other reports suggest minimal or even inhibitory effects of 1,25(OH)$_2$D$_3$ treatment on Th2 development and IL-4 production.[29–31]

The synthesis of other T lymphocyte cytokines can also be influenced by 1,25(OH)$_2$D$_3$. Granulocyte-macrophage colony-stimulating factor (GM-CSF), which typically activates mature granulocytes and macrophages eliciting an inflammatory response to infection, is down-regulated after *in vitro* $1,25(OH)_2D_3$-treatment of normal mitogen-activated T lymphocytes or T lymphocytes from a cell line.[32] Besides influencing the proliferation and cytokine production of T lymphocytes, $1,25(OH)_2D_3$ is also able to affect their survival and death. The Fas/FasL system is the major pathway by which activation-induced cell death can be mediated in T lymphocytes. The expression of FasL by activated T lymphocytes can be inhibited by $1,25(OH)_2D_3$.[33]

Antigen-Presenting Cells: Monocytes, Macrophages, and Dendritic Cells

Most immunomodulatory agents that presently are used in the clinic focus their attention exclusively on T lymphocytes and, as a consequence, only a limited benefit can be obtained from a combination of different immunomodulators. A major asset of $1,25(OH)_2D_3$ as a regulator of immune functions is the fact that it not only interacts with T lymphocytes but primarily targets the pivotal cell in the immune cascade, the APC. Monocytes exposed to $1,25(OH)_2D_3$ have significantly reduced MHC class II expression and T-cell stimulatory capacity. In addition, the surface expression of costimulatory receptors (such as CD40, CD80, and CD86), determining the ability of APCs to provide secondary signals necessary for full T-cell activation, is also potently inhibited in $1,25(OH)_2D_3$-treated monocytes.[34–36] In contrast to the attenuation of the antigen-presenting function, the chemotactic and phagocytic capacity of monocytes and macrophages necessary for the tumor-cell cytotoxicity and microbacterial activity of these cells is enhanced by exposure to $1,25(OH)_2D_3$.[36,37]

More recently, $1,25(OH)_2D_3$ has been shown to potently inhibit DC maturation by a VDR-dependent mechanism.[38–42] DCs are by far the most potent among the APCs, and are the primary initiators of T-cell-mediated immune responses.[43,44] DCs are present within essentially all organs, mediating the uptake and transport of disease-associated antigens from peripheral tissues to lymph nodes in which cellular immune responses can be initiated. When fully mature, DCs express high levels of MHC class II, CD80, CD86, CD40, and an array of other immunostimulatory products. The process of migration from periphery to lymphoid tissue and the associated transition from low to high T-cell stimulatory capacity and from high to low antigen-uptake capacity is called maturation. Because inappropriate DC maturation may result in excessive or unnecessary T-cell-mediated immune responses and, eventually, autoimmunity, this maturation process is subject to multiple positive and negative regulatory mechanisms. In addition, in the process of tolerance induction to self- threatening or nonthreatening antigens, an active role has been attributed to immature DCs.[45] Due to this pivotal role that DCs play in controlling the immune responses, investigating the effects of 1,25(OH)$_2$D$_3$ on DCs is of major importance. The *in vitro* differentiation of DCs from its precursors (human peripheral-blood monocytes or murine bone marrow-derived precursors) is inhibited by 1,25(OH)$_2$D$_3$, and even deviated in some cases, giving rise to CD14-positive cells, with CD14 being a non-DC surface marker. Moreover, the antigen-presenting cell function of DCs is profoundly altered by 1,25(OH)$_2$D$_3$. Surface expressions of MHC class II, costimulatory molecules (CD86, CD80, and CD40), and other maturation-induced proteins (CD1a, CD83) are significantly reduced by treatment with 1,25(OH)$_2$D$_3$ during *in vitro* or *in vivo* maturation.

The secretion by APCs of cytokines, which are crucial for recruitment and activation of T lymphocytes, is also influenced by 1,25(OH)$_2$D$_3$. The immunostimulatory IL-12, the major cytokine determining the direction in which the immune system will be activated, is clearly inhibited by 1,25(OH)$_2$D$_3$ in DCs as well as in other APCs.[46,47] IL-12 stimulates the development of CD4$^+$ Th1 lymphocytes and inhibits the development of CD4$^+$ Th2 lymphocytes. Due to this inhibition of IL-12, observed *in vitro* as well as *in vivo*, 1,25(OH)$_2$D$_3$ interferes directly with a key event in the immune cascade, shifting the ongoing reaction toward a Th2 profile.[31] In addition, expression by DCs of the immunosuppressive cytokine IL-10, opposing the Th1-driving effects of IL-12, is increased by treatment with 1,25(OH)$_2$D$_3$.[38,39]

Other factors secreted by APCs are also under the influence of $1,25(OH)_2D_3$. Expression of the suppressive prostaglandin E_2 by monocytes is stimulated after *in vitro* $1,25(OH)_2D_3$ treatment.[48] Expression of the inflammatory cytokine TNF-α by monocytes or macrophages is also modulated by $1,25(OH)_2D_3$, although different studies yielded contradictory results. In general, the differentiation or maturation status of the cells determines their response to $1,25(OH)_2D_3$; whereas TNF-α production is increased in immature cells such as some tumor cell lines[49] or bone marrow cells,[50] a decreased TNF-α production can be observed after $1,25(OH)_2D_3$ treatment of more mature cells such as peripheral-blood mononuclear cells[51] or peritoneal macrophages.[52] Furthermore, expression of the inflammatory cytokines IL-1 and IL-6 can also be inhibited by $1,25(OH)_2D_3$ treatment.[34,51,53]

Functionally, the capacity of APCs to induce T lymphocyte activation, proliferation, and cytokine secretion is profoundly modified by $1,25(OH)_2D_3$ treatment. CD4+ T lymphocytes are, besides the direct effects of $1,25(OH)_2D_3$ on the T lymphocyte level, particularly in this indirect way influenced by $1,25(OH)_2D_3$. Besides Th1 and Th2 lymphocytes, the existence of a population of regulatory CD4+ T lymphocytes (Tregs), characterized by the combined expression of CD4 and CD25, the secretion of potentially inhibitory cytokines (IL-10 and TGF-β), and the ability to potently inhibit antigen-specific T-cell activation, has been reported.[54] A complete hyporesponsiveness (inhibition of proliferation as well as cytokine production) of naive T lymphocytes could be observed when cultured *in vitro* in the presence of $1,25(OH)_2D_3$-pretreated DCs with $1,25(OH)_2D_3$ absent.[39] Moreover, one study shows that a committed human autoreactive CD4+ T lymphocyte clone can be affected by $1,25(OH)_2D_3$-pretreated DCs.[38] After incubation with $1,25(OH)_2D_3$-conditioned DCs, again in the absence of $1,25(OH)_2D_3$ itself, IFN-γ production in these autoreactive T lymphocytes was decreased, suggesting a redirection toward a more protective Th2 profile. Also, $1,25(OH)_2D_3$ can induce DCs with tolerogenic properties *in vivo*, as demonstrated in models of allograft rejection and autoimmunity. After treatment with a combination of $1,25(OH)_2D_3$ and mycophenolate mofetil, *in vivo*-generated tolerogenic DCs are probably responsible for the increased proportion of $CD4^+CD25^+$-regulatory T lymphocytes found in regional lymph nodes. Those Tregs protect against insulin-producing islet allograft rejection, not only in treated animals but also after transfer of the Tregs to nontreated animals.[55] Moreover, in autoimmune nonobese diabetic (NOD) mice (having a defect in these regulatory T lymphocytes), a restoration of the regulatory T lymphocyte population could be observed after $1,25(OH)_2D_3$ treatment.[56,57] Again, transfer of the Treg population protected nontreated mice. However, when this regulatory T lymphocyte population was eliminated by cyclophosphamide, $1,25(OH)_2D_3$-treated mice were still protected from diabetes development.[58] These data confirm that the effects of $1,25(OH)_2D_3$ are not solely mediated through induction of regulatory T lymphocytes but also through elimination of effector T lymphocytes. Indeed, restoration of the apoptosis-sensitivity of T lymphocytes centrally in the thymus of NOD mice was achieved by treatment with $1,25(OH)_2D_3$, resulting in a better elimination of autoreactive diabetogenic T lymphocytes.[58,59] More recently, an arrest in the progression of autoimmune diabetes in NOD mice treated with an analog of $1,25(OH)_2D_3$ was shown to be associated with an enhanced frequency of $CD4^+CD25^+$-regulatory T lymphocytes in pancreatic lymph nodes.[60] In addition, a selective decrease of effector Th1 lymphocytes could be observed in pancreas and pancreatic lymph nodes without, however, a deviation toward the Th2 phenotype. These effects were postulated to be an indirect result of analog-mediated induction of tolerogenic DCs rather than a direct T lymphocyte effect.

Although it is commonly accepted that the $1,25(OH)_2D_3$-mediated induction of regulatory T lymphocytes occurs indirectly through the generation of tolerogenic DCs, one study reports that in an APC-free *in vitro* system a combination of $1,25(OH)_2D_3$ and dexamethasone could directly induce a Treg population producing IL-10 and no IL-4, IL-5, or IFN-γ, being able to prevent autoimmune demyelination in an antigen-specific manner.[61]

Effects of Vitamin D Deficiency on the Immune System

In addition to the pharmacological effects of $1,25(OH)_2D_3$ on the immune system, several data point toward a possible physiological immune function of $1,25(OH)_2D_3$, suggesting that vitamin D_3 deficiency could lead to immune malfunctioning. Subjects with severe vitamin D_3 deficiency are abnormally susceptible to infections[62] such as tuberculosis.[63] Different studies provide evidence for defects in macrophage functions, such as chemotaxis, phagocytosis, and the production of proinflammatory cytokines, in vitamin D_3-deficient conditions.[64,65] Also, localized antigen-specific cellular immune responses (delayed type hypersensitivity) are defective during vitamin D_3 deficiency.[66]

Also, the prevalence of autoimmune diseases has been correlated with vitamin D_3 status.[67] A positive correlation of diabetes incidence with latitude,[68,69] sunshine exposure, and season of birth[70] has been shown. In this context, we demonstrated that in NOD mice, vitamin D_3 deficiency in early life (*in utero* and during the first 100 d of life) leads to a more aggressive presentation of type 1 diabetes, with an earlier onset and a higher final disease incidence, both in females as well as in the more diabetes-resistant males.[70a] A limited vitamin D_3 deficiency was created in this mouse model, with only a measurable deficiency in blood but no impact on calcium levels in serum or bone calcium content, which was the equivalent of the vitamin D_3 status in many infants and small children. This vitamin D_3 deficiency, although mild, affected the immune system, causing an aberrant macrophage cytokine profile (low IL-1 and IL-6 and high IL-15) and a defect in $CD4^+CD62L^+$-regulatory T lymphocytes. Also in humans, increased diabetes incidence can be observed in a vitamin D_3-deficient setting. Children with vitamin D_3 deficiency or rickets in early life have a threefold increase in the prevalence of diabetes later in life.[71]

A negative correlation between supplements of vitamin D_3 in early life and the risk of type 1 diabetes later on has been demonstrated.[72–74] Children exposed to vitamin D_3 supplements in early infancy were protected from the development of type 1 diabetes. This effect of vitamin D_3 supplementations in early life is, however, much smaller than the effects of early-life vitamin D_3 deficiency, suggesting that mainly a vitamin D_3-deficient status in a genetically predisposed background might trigger autoimmunity, whereas for prevention of the disease in vitamin D_3-sufficient individuals, high doses of the active $1,25(OH)_2D_3$ or an analog are needed. A study in NOD mice showed that treatment of vitamin D_3-sufficient NOD mice for a short period of time, corresponding to neonatal life and childhood in humans, with physiological supplements of regular vitamin D_3[75] or one of the less calcemic analogs of $1,25(OH)_2D_3$,[76] is not sufficient to prevent type 1 diabetes. Treated mice did, however, show a clear conservation of their beta cell potential. Not only the diabetic but also the normoglycemic mice had clearly higher insulin levels in their pancreas. These beta cell protective effects of $1,25(OH)_2D_3$ have also been observed *in vitro*, where beta cells are protected against cytokine-induced cell death, and synthesis and secretion of insulin is preserved.[77,78] Additionally, early-life treatment with physiological concentrations of vitamin D_3 or $1,25(OH)_2D_3$ in Biobreeding (BB) rats (another animal model of type 1 diabetes) could not significantly prevent diabetes or alter the time of diabetes onset.[78a]

MOLECULAR PATHWAYS OF $1,25(OH)_2D_3$-MEDIATED IMMUNE REGULATION

In addition to the classical mechanism of action of $1,25(OH)_2D_3$, through binding of the ligand-activated transcription factor VDR to VDREs within the promoter region of $1,25(OH)_2D_3$-responsive genes and thus eventually regulating the transcription rate of these genes, there is abundant evidence that the regulation of the transcription of many cytokine genes by $1,25(OH)_2D_3$ results indirectly from modulation of other intracellular signaling pathways (Table 4.1). In activated T lymphocytes, $1,25(OH)_2D_3$-mediated inhibition of IL-2 transcription occurs through impairment of NF-AT/AP-1 protein complex formation and the subsequent stable association of the ligand-VDR-RXR complex with the NF-AT binding site

Table 4.1 Molecular Pathways of 1,25(OH)$_2$D$_3$-Mediated Immune Regulation

Gene	Effect	Cells and Species	Intracellular Pathway	Mechanism	Ref.
FasL	↓	Murine activated T hybridoma cells	MAPK	Indirect inhibition of c-myc transcriptional activity via interaction of ligand-VDR-RXR with a noncanonical c-myc DNA-binding site	33
GM-CSF	↓	Jurkat T cells	Repressive promoter complex	Binding of ligand-VDR monomers to functional repressive promoter sequence	79,80
IFN-γ	↓	Jurkat T cells	Negative VDRE	Interaction of the ligand-VDR-RXR complex with a negative VDRE and inhibition of an upstream enhancer element	23
IL-2	↓	Jurkat T cells and human tonsilar T lymphocytes	NF-AT/AP-1	Inhibition of NF-AT/AP-1 protein complex formation and direct association of the ligand-VDR-RXR complex with the NF-AT DNA-binding site	24,25
IL-4	↑	Murine naive CD4$^+$ T lymphocytes	GATA-3 and c-*maf*	Increased expression of the Th2-specific transcription factors GATA-3 and c-*maf*	26
	↓	Murine naive CD4$^+$ T lymphocytes	NF-AT	Direct interaction of the ligand bound VDR with a NF-AT DNA-binding site	29
IL-12	↓	Murine activated macrophages and DCs	NF-κB	Inhibition of NF-κB activation and of binding to its DNA-binding site	47
TNF-α	↑	Murine bone marrow macrophages	VDRE + LPS pathway	Direct interaction of the ligand-VDR-RXR complex with a stimulating VDRE and enhanced LPS activity via upregulation of CD14	50

within the IL-2 promoter.[24,25] On the other hand, the inhibition of IFN-γ transcription in activated T lymphocytes occurs directly through interaction of the ligand-VDR-RXR complex with a negative VDRE-like binding sequence in the promoter region of the gene. Moreover, progressive deletion analysis of the IFN-γ promoter revealed that negative regulation by $1,25(OH)_2D_3$ is also exerted at the level of an upstream region containing an enhancer element crucial for activation of the IFN-γ promoter, suggesting a possible interference with normal transcription initiation or progression.[23] Besides the direct effects of $1,25(OH)_2D_3$ on Th1 cytokine expression, Th2 cytokines also stand under direct influence of $1,25(OH)_2D_3$. The reported *in vitro* development of a highly Th2-skewed lymphocyte population after $1,25(OH)_2D_3$ treatment occurs through the increased expression of the Th2-specific transcription factors GATA-3 and *c-maf*, resulting in increased production of IL-4, IL-5, and IL-10.[26] A contradictory study, reporting the inhibition of IL-4 transcription after *in vitro* $1,25(OH)_2D_3$ treatment of naive CD4+ T lymphocytes, demonstrates that the ligand-bound VDR directly interacts with the IL-4 promoter, probably at the level of NF-AT-binding sites.[29] Suppression of the granulocyte/macrophage recruiter GM-CSF by $1,25(OH)_2D_3$ is achieved in yet another, rather exceptional, manner; whereas the ligand-VDR complex typically forms heterodimers with RXR, thus exerting its effects at the DNA level, transcription of GM-CSF is inhibited by the binding of ligand-bound VDR monomers to functional repressive complexes in the promoter region of the cytokine gene.[79,80] In addition, it was recently shown that the expression of Fas ligand (FasL) by activated T lymphocytes could be repressed by $1,25(OH)_2D_3$.[33] Upon binding to its receptor Fas, the Fas/FasL system induces activation-induced cell death of T lymphocytes, a fundamental mechanism for negative selection of immature T lymphocytes in the thymus and for maintenance of peripheral tolerance. This $1,25(OH)_2D_3$-mediated inhibition of FasL transcription occurs through the interaction of the ligand-VDR-RXR complex with a noncanonical c-myc DNA-binding element in close proximity of the transcription start site.

Furthermore, in activated macrophages and DCs, inhibition of IL-12 expression by $1,25(OH)_2D_3$ occurs by targeting the NF-κB pathway.[47] Activation as well as binding of NF-κB to the NF-κB binding site within the promoter of the p40 subunit of IL-12 are downregulated by $1,25(OH)_2D_3$. The study by Hakim et al.[50] demonstrating an enhancement of TNF-α transcription in bone marrow macrophages suggests that $1,25(OH)_2D_3$ may exert its activity via two mechanisms: a direct effect via binding of the ligand-VDR-RXR complex to a VDRE in the promoter of the TNF-α gene and an indirect effect via enhancement of LPS activity through upregulation of the surface expression of CD14 (the membrane receptor for LPS).[50] Also, for the antimycobacterial activity of $1,25(OH)_2D_3$ in human macrophages, a molecular mechanism has been postulated.[37] This antimycobacterial activity would be due to activation of the NADPH oxidase and the production of reactive oxygen intermediates, a process regulated by phosphatidylinositol 3-kinase (PI 3-K). Moreover, it has been suggested that the $1,25(OH)_2D_3$-induced maturation of myeloid cells is dependent on the formation of a VDR-PI 3-K complex.[81]

In contrast with the mechanisms of action of other, more classical immunomodulatory agents that can be linked to minor effects on one or some of the intracellular signaling pathways relevant in activation of lymphocytes and APCs, these multitude of data show that $1,25(OH)_2D_3$ is able to interfere with multiple intracellular pathways orchestrating the differentiation, maturation, and activation events in lymphocyte and APC populations.

IN VIVO IMMUNOMODULATORY POTENTIAL OF $1,25(OH)_2D_3$ AND ITS ANALOGS

The fact that $1,25(OH)_2D_3$ influences the immune system, not merely by suppression but by immune modulation through induction of immune shifts and regulator cells, makes this compound very appealing for clinical use, especially in the treatment and prevention of autoimmune diseases and in the prevention of graft rejection. The earlier-mentioned side effects are, however, a major drawback in its clinical applicability. Various strategies have been developed to overcome these limitations: (1) the use of analogs of $1,25(OH)_2D_3$ with a wider therapeutic window than $1,25(OH)_2D_3$ itself, (2) the use of additive and even synergistic immunomodulatory effects that have been observed between $1,25(OH)_2D_3$ or its analogs and the more classical immunosuppressants, and (3) the use of products specifically counteracting the side effects of $1,25(OH)_2D_3$, and, to a lesser extent, of $1,25(OH)_2D_3$ analogs, without additional effects on its immune properties.

In Vivo Use of $1,25(OH)_2D_3$ Analogs

The less calcemic structural analogs of $1,25(OH)_2D_3$ — analogs with preserved or even enhanced effects on the immune system but attenuated effects on calcium and bone — are those that have been tested most often in *in vivo* models of autoimmunity and transplantation (Table 4.2). Our group has demonstrated that autoimmune type 1 diabetes can be prevented in NOD mice by $1,25(OH)_2D_3$ and its analogs.[57,56] Treatment of NOD mice from weaning until old age not only prevented clinical diabetes, it also downregulated the prevalence of the histological lesion, insulitis.[82] In this model of autoimmune diabetes, upregulation of regulatory immune cells and a shift from Th1 toward Th2 lymphocytes locally in the pancreases of treated mice can be observed. This protective Th2 population is induced not only at the site of the beta cell attack but also in the peripheral immune system.[27] After immunization of $1,25(OH)_2D_3$-treated NOD mice with a diabetes-specific autoantigen (a peptide of GAD65), lymphocytes of the draining lymph nodes showed an increased IL-4 and decreased IFN-γ production *in vitro* and *in vivo*. Strikingly, this immune deviation induced by $1,25(OH)_2D_3$ is limited to pancreatic autoantigens and could not be seen after immunization with the beta cell–irrelevant protein ovalbumin.

Also, other effects on the immune system of NOD mice have been described; the most important was a restoration of the defective apoptosis sensitivity of lymphocytes, leading to a more efficient elimination of potentially dangerous autoimmune effector cells.[58,59] This increased apoptosis induced by

Table 4.2 $1,25(OH)_2D_3$ and Its Analogs in Animal Models of Autoimmune Diseases and Transplantation

Animal Model	Major Effects	Species	Ref.
Autoimmunity			
Arthritis			
Collagen-induced	Prevention of disease, suppression of severity	DA rats	90
		DBA mice	91
Lyme	Prevention of disease, ameliorations of symptoms	C3H mice	91
Diabetes			
Type 1	Prevention of disease, inhibition of insulitis and diabetes	NOD mice	56–58,82,83
Low-dose streptozotocin-induced	Decreased incidence of insulitis and diabetes	CD-1 mice	111
Experimental autoimmune encephalomyelitis	Prevention of disease, attenuation of severity and relapses, delay of onset	SJL mice	89,102–104
		B10.PL mice	88,112,113
		Biozzi mice	30,89,88
Inflammatory bowel disease	Amelioration of symptoms, block of disease progression	C57Bl/6/IL-10-/- mice	93
Systemic lupus erythematosus	Prevention of disease	MLR/l mice	85–87
Nephritis			
Heyman	Reduction of proteinuria and autoantibodies	Lewis rats	92
Lupus	Reduction of proteinuria, prevention of skin lesions	MLR/l mice	85,87
Mercuric chloride-induced	Prevention of proteinuria, downregulation of serum antibody levels	BN rats	114,115
Experimental autoimmune thyroiditis	Reduction of histological lesions and severity	CBA mice	116

Transplantation

Aorta	Reduced chronic allograft rejection	DA to WF rats	100
Bone marrow	Decreased graft-vs.-host disease	Lewis to BN rats	117
Heart	Prolongation of vascularized and nonvascularized allograft survival	B10.A to C57Bl/10 mice and ACI to Lewis rats	94,95
Liver	Decreased severity of acute allograft rejection	ACI to Lewis rats	96
Pancreatic islets	Induction of allotransplantation tolerance	C57Bl/6 to Balb/c	55
	Prevention of autoimmune recurrence after syngeneic transplantation	NOD to NOD mice	84,106
	Prevention of early xenograft failure	BB rats to NOD mice	118
Skin	Prolonged allograft survival	C57Bl/6 to CBA mice	97,98
Kidney	Prolonged allograft survival	ACI to Lewis rats	99
Small bowel	Prolonged allograft survival		95

$1,25(OH)_2D_3$ and its analogs on DCs and T lymphocytes of NOD mice has been described after different apoptosis-inducing signals, such as corticosteroids, and could help to explain why an early short-term treatment with these agents, before onset of autoimmunity, confers long-term protection and promotes tolerance restoration.

Besides preventing their onset, $1,25(OH)_2D_3$ and its analogs are also able to treat ongoing autoimmune diseases. Treatment of NOD mice with analogs of $1,25(OH)_2D_3$ can prevent the progression of an initial beta cell attack (reflected by the presence of insulitis) to clinical overt diabetes.[83] Interestingly, in this model of ongoing autoimmune destruction, no induction of suppressor cells by $1,25(OH)_2D_3$ could be demonstrated. Nevertheless, within the pancreases of protected mice, a shift from Th1 toward Th2 cytokines is again noted. Moreover, analogs of $1,25(OH)_2D_3$ are able to inhibit the recurrence of autoimmune diabetes after syngeneic islet transplantation in NOD mice.[84] These data suggest that by treatment with $1,25(OH)_2D_3$ analogs, a completely established autoimmune disease, in which autoimmune destruction is accomplished and autoimmune memory cells are present, can be overcome and, thus, reinduction of self-tolerance can be achieved. Analysis of the cytokine profile in the surviving islet grafts again demonstrated the induction of the same immune shift from Th1 to Th2.

$1,25(OH)_2D_3$ and its analogs have also been applied in other spontaneous and experimentally induced models of autoimmune diseases. They can prevent systemic lupus erythematosus in *lpr/lpr* mice,[85–87] experimental autoimmune encephalomyelitis,[30,88,89] collagen-induced arthritis,[90,91] Heymann nephritis,[92] and inflammatory bowel disease.[93]

In addition, $1,25(OH)_2D_3$ and its analogs prolong the survival of heart,[94,95] liver,[96] skin,[97,98] small bowel,[95] kidney,[99] and pancreatic islet[55] allografts and inhibit not only acute but also chronic allograft rejection.[100] An additional asset in the use of $1,25(OH)_2D_3$ and its analogs in transplantation therapy is the sustained resistance to opportunistic infections — the most serious side effects encountered with the classical transplantation antirejection drugs.[101]

Synergism with Other Immunosuppressants

An additional way to attenuate the calcemic side effects of $1,25(OH)_2D_3$ and its analogs is to exploit the additive and even synergistic effects observed with more classical immunosuppressants (Table 4.3). The concept of drug synergism reflects the capacity of two or more drugs to promote each other's effects to a higher level than the simple addition of effects. Therefore, each of the synergizing drugs can be used at an individual subtherapeutical level, reducing each compound's adverse side effects. *In vitro*, these synergistic effects between $1,25(OH)_2D_3$ or its analogs and classical immunosuppressants are observed in models of phytohemagglutinin-stimulated T lymphocyte proliferation and in mixed lymphocyte cultures. Varying levels of synergism with $1,25(OH)_2D_3$ can be observed for different immunosuppressants, from cyclosporine A having one of the strongest synergisms with $1,25(OH)_2D_3$ to mycophenolate mofetil belonging to the category of weak synergistic agents.[102]

Moreover, these *in vitro* observed synergisms have been confirmed *in vivo* in different models of autoimmune diseases and in graft rejection. In experimental autoimmune encephalomyelitis, $1,25(OH)_2D_3$ itself[103,104] as well as its analogs[102] are found to act synergistically with the immunosuppressants cyclosporine A, rapamycin, and mycophenolate mofetil, leading to near-total disease protection accompanied by less calcemic side effects compared with monotherapy. Moreover, in syngeneic islet transplantation in NOD mice, combination of analogs of $1,25(OH)_2D_3$ with the classical immunosuppressants cyclosporine A and FK506 resulted in complete prevention of autoimmune recurrence, even after withdrawal of therapy, suggesting a reinduction of self-tolerance.[105] Furthermore, it was found that IFN-β acts synergistically with analogs of $1,25(OH)_2D_3$ in the model of syngeneic islet transplantation.[106] In addition, an arrest in the progression of diabetes development in NOD mice could be achieved by a combined treatment of a $1,25(OH)_2D_3$ analog and mycophenolate mofetil.[60]

Table 4.3 Synergism of $1,25(OH)_2D_3$ and Its Analogs with Other Immunosuppressants

Drugs	Conditions	Species	Models	Ref.
Cyclosporin A	In vitro	Human	Proliferation and IL-2 production of PHA-stimulated PBMC	102,119,120
		Murine	Proliferation and cytokine production in MLR	105
	In vivo	SJL mice	Experimental autoimmune encephalomyelitis	102,103
		NOD mice	Type 1 diabetes	83
		BN rats	Mercuric-chloride induced autoimmunity	115,114
		Rats	Adjuvant arthritis	121
		CBA mice	Thyroiditis	116
		NOD to NOD mice	Transplantation of syngeneic islets	84
		BB rats to NOD mice	Transplantation of xenogeneic islets	118
		ACI to Lewis rats	Transplantation of vascularized renal allografts	99,122
		ACI to Lewis rats	Transplantation of liver allografts	96
		Lewis to BN rats	Bone marrow transplantation	117
		DA to WF rats	Transplantation of aorta allografts	100

Rapamycin	*In vitro*	PVG to WK rats	Transplantation of heart allografts	123
	In vivo	C57Bl/6 to CBA mice	Transplantation of skin allografts	124
	In vitro	Human	Proliferation of PHA-stimulated PBMC	102,104
FK506	*In vivo*	SJL mice	Experimental autoimmune encephalomyelitis	104
	In vitro	Human	Proliferation of PHA-stimulated PBMC	102
		Murine	Proliferation and cytokine production in MLR	105
Mofetil	*In vitro*	Human	Proliferation of PHA-stimulated PBMC	102
	In vivo	SJL mice	Experimental autoimmune encephalomyelitis	102
		C57Bl/6 to Balb/c mice	Transplantation of allogeneic islets	55
Leflunomide	*In vitro*	Human	Proliferation of PHA-stimulated PBMC	102
Dexamethasone	*In vitro*	Human	Proliferation and cytokine production of anti-CD3-stimulated PBMC	125
		Murine	Proliferation, cytokine and chemokine production, and T cell activation of dendritic cells	126
IFN-β	*In vivo*	NOD to NOD mice	Transplantation of syngeneic islets	106

Also, in transplantation models, synergistic effects between analogs of $1,25(OH)_2D_3$ and classical immunosuppressants have been observed. Following combined treatment with a $1,25(OH)_2D_3$ analog and cyclosporine A, chronic rejection of an aortic allograft was inhibited, and vascular injury often associated with chronic rejection was attenuated.[100] In the model of allogeneic pancreatic islet transplantation, combinations of $1,25(OH)_2D_3$ with mycophenolate mofetil induced tolerance to islet allografts not only in treated animals but also in nontreated animals after transfer of protective $CD4^+CD25^+$-regulatory T lymphocytes.[55]

All the immunosuppressants mentioned in the preceding text interact at different levels with the T lymphocyte activation and proliferation cascade (reviewed in Reference 107).[107] Cyclosporine A and FK506, after binding to their binding proteins, interact with calcineurin, inhibiting its dephosphorylating activity and consequently preventing the passage of the transcription factor NF-AT to the nucleus. In this way, activation of T lymphocytes is blocked in a very early phase. Rapamycin blocks a step in the intracellular protein kinase cascade after the binding of IL-2 to its receptor, preventing further autoactivation and proliferation of the T lymphocytes. Mycophenolate mofetil inhibits *de novo* nucleotide synthesis, leading to loss of DNA synthesis and eventually to abrogation of T lymphocyte proliferation. IFN-β, a member of the type 1 IFN family and a promising agent in the treatment of multiple sclerosis, has a broad range of immunomodulatory properties,[108] of which the inhibition of T lymphocyte IFN-γ production (by blocking IL-12 secretion by APCs) and the attenuation of T lymphocyte proliferation (partly through downregulation of IL-12 or upregulation of IL-10 levels) might be largely beneficial in the prevention or treatment of autoimmune diseases and the prevention of graft rejection.

Combination with Bone-Resorption Inhibitors

The synthetic analogs of $1,25(OH)_2D_3$, which are used for their strong immunomodulatory potential, display a decreased hypercalcemic activity compared with the parent compound but, unfortunately, they are still not completely devoid of detrimental side effects such as high-turnover bone demineralization. An additional way to overcome the limitations of the *in vivo* use of $1,25(OH)_2D_3$ and its analogs, other than combined treatment with synergizing immunomodulatory agents at subtherapeutical doses, is to counteract their side effects directly. Bisphosphonates efficiently intervene in blocking and even reverting the loss of bone mass whenever osteolysis exceeds bone accretion.[109] Association of the bisphosphonate pamidronate disodium to a $1,25(OH)_2D_3$ analog treatment in the model of experimental autoimmune encephalomyelitis did not affect analog-induced disease protection, but completely prevented analog-caused acceleration of bone turnover and increased total bone mineral content as well as femoral mineral and calcium content.[110] Combining less-calcemic immune-modulating $1,25(OH)_2D_3$ analogs with bisphosphonates or other potent bone-resorption inhibitors might be a key finding in the *in vivo* applicability of $1,25(OH)_2D_3$ and its analogs.

CONCLUSIONS

The widespread presence of VDR in the immune system and the regulated expression of 1-α-hydroxylase by specific immune signals suggest a paracrine immunomodulatory role for $1,25(OH)_2D_3$. Indeed, $1,25(OH)_2D_3$ has pleiotropic activities in the immune system (summarized in Figure 4.1). Strikingly, $1,25(OH)_2D_3$ uses several different molecular mechanisms to regulate cytokine expression, either directly targeting transcription initiation and regulation or indirectly interfering with other intracellular signaling pathways. Moreover, APCs as well as T lymphocytes can be direct targets of the immunomodulatory effects of $1,25(OH)_2D_3$, leading to the inhibition of pathogenic effector T lymphocytes and the enhanced frequency of regulatory T lymphocytes, largely via the induction of DCs with tolerogenic properties. These strong immunoregulatory properties of $1,25(OH)_2D_3$ justify the many efforts that have been made to reduce its side effects such as hypercalcemia and bone remodeling. The pharmacological

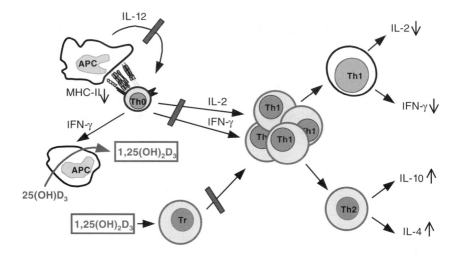

Figure 4.1   Mechanisms of action of $1,25(OH)_2D_3$ and its analogs in the immune system. $1,25(OH)_2D_3$ and its analogs can modulate the immune responses via several mechanisms. They inhibit IL-12 production and down-regulate antigen-presentation and costimulatory molecule expression by APCs, especially DCs, thus indirectly inhibiting the development of T helper lymphocytes along the Th1 pathway and favorizing the induction of Th2 lymphocytes or regulatory T lymphocytes. $1,25(OH)_2D_3$ and its analogs also exert direct effects on T lymphocytes; IL-2 and IFN-γ production by Th1 lymphocytes is inhibited, whereas the production of the Th2 cytokine IL-4 is directly induced, both shifting the Th-balance toward the Th2 phenotype. In addition, antigen-presenting cells can synthesize $1,25(OH)_2D_3$ after stimulation with IFN-γ, which may be a negative feedback signal contributing to the regulation of the immune response.

doses of $1,25(OH)_2D_3$ needed to obtain (locally) a high enough concentration to display its immunological effects results systemically in inadmissible hypercalcemia. Structural $1,25(OH)_2D_3$ analogs display reduced calcemic side effects, combined with equal or stronger immunomodulatory effects compared with the parent molecule. $1,25(OH)_2D_3$ or its analogs have been combined with other immunosuppressants, showing additive or even synergistic immunomodulatory effects, or with bone-resorption inhibitors, reverting the accelerated bone turnover without interfering with the immunomodulatory effects of the VDR ligands. By using the various strategies developed to overcome the dose-limiting side effects of $1,25(OH)_2D_3$, the immunomodulatory potentials of $1,25(OH)_2D_3$ and that of its analogs have been translated into effective *in vivo* immune-intervention in a variety of animal models of autoimmune diseases and graft rejection.

ACKNOWLEDGMENTS

This work was supported by the GOA (Geconcerteerde Ondezoeksactie, grant GOA/2004/10), the IUAP (Interuniversitaire Attractie Polen, grant IUAP-P5-17), the JDRF (Juvenile Diabetes Research Foundation, grant 4-2002-457), and a Fellowship "Fundamenteel Clinisch Narvorser" for C. Mathieu from the Flemisch Research Foundation (FWO, Fonds voor Wetenschappelijk Onderzoek Vlaanderen).

References

1. MF Holick. Photobiology of vitamin D. In: D Feldman, F Glorieux, JW Pike, ed(s). *Vitamin D*. San Diego: Academic Press, 1997, pp 33–39.

2. G Jones, SA Strugnell, HF DeLuca. Current understanding of the molecular actions of vitamin D. *Physiol Rev* 78: 1193–1231, 1998.

3. MR Haussler, GK Whitfield, CA Haussler, JC Hsieh, PD Thompson, SH Selznick, CE Dominguez, PW Jurutka. The nuclear vitamin D receptor: biological and molecular regulatory properties revealed. *J Bone Miner Res* 13: 325–349, 1998.

4. C Carlberg, P Polly. Gene regulation by vitamin D3. *Crit Rev Eukaryot Gene Expr* 8: 19–42, 1998.

5. L Verlinden, A Verstuyf, M Quack, M Van Camp, E Van Etten, P De Clercq, M Vandewalle, C Carlberg, R Bouillon. Interaction of two novel 14-epivitamin D3 analogs with vitamin D3 receptor-retinoid X receptor heterodimers on vitamin D3 responsive elements. *J Bone Miner Res* 16: 625–638, 2001.

6. E Marcinkowska. A run for a membrane vitamin D receptor. *Biol Signals Recept* 10: 341–349, 2001.

7. CM Veldman, MT Cantorna, HF DeLuca. Expression of 1,25-dihydroxyvitamin D3 receptor in the immune system. *Arch Biochem Biophys* 374: 334–338, 2000.

8. C Mathieu, L Adorini. The coming of age of 1,25-dihydroxyvitamin D3 analogs as immunomodulatory agents. *Trends Mol Med* 8: 174–179, 2002.

9. L Overbergh, B Decallonne, D Valckx, A Verstuyf, J Depovere, J Laureys, O Rutgeerts, R Saint-Arnaud, R Bouillon, C Mathieu. Identification and immune regulation of 25-hydroxyvitamin D-1-alpha-hydroxylase in murine macrophages. *Clin Exp Immunol* 120: 139–146, 2000.

10. A Dusso, A Brown, E Slatopolsky. Extrarenal production of calcitriol. *Semin Nephrol* 14: 144–155, 1994.

11. E Van Etten, B Decallonne, L Verlinden, A Verstuyf, R Bouillon, C Mathieu. Analogs of 1alpha,25-dihydroxyvitamin D3 as pluripotent immunomodulators. *J Cell Biochem* 88: 223–226, 2003.

12. R Bouillon, WH Okamura, AW Norman. Structure-function relationships in the vitamin D endocrine system. *Endocr Rev* 16: 200–257, 1995.

13. A Verstuyf, L Verlinden, H Van Baelen, K Sabbe, C D'Hallewyn, P De Clercq, M Vandewalle, R Bouillon. The biological activity of nonsteroidal vitamin D hormone analogs lacking both the C- and D-rings. *J Bone Miner Res* 13: 549–558, 1998.

14. A Verstuyf, L Verlinden, E Van Etten, L Shi, Y Wu, C D'Halleweyn, D Van Haver, GD Zhu, YJ Chen, X Zhou, MR Haussler, P De Clercq, M Vandewalle, H Van Baelen, C Mathieu, R Bouillon. Biological activity of CD-ring modified 1alpha,25-dihydroxyvitamin D analogs: C-ring and five-membered D-ring analogs. *J Bone Miner Res* 15: 237–252, 2000.

15. N Rochel, JM Wurtz, A Mitschler, B Klaholz, D Moras. The crystal structure of the nuclear receptor for vitamin D bound to its natural ligand. *Mol Cell* 5: 173–179, 2000.

16. R Bouillon, K Allewaert, DZ Xiang, BK Tan, H Van Baelen. Vitamin D analogs with low affinity for the vitamin D binding protein: enhanced *in vitro* and decreased *in vivo* activity. *J Bone Miner Res* 6: 1051–1057, 1991.

17. A Brown, E Slatopolsky. Mechanisms for the selective actions of vitamin D analogs. In: D Feldman, F Glorieux, JW Pike, Eds. *Vitamin D*. San Diego: Academic Press, 1997, pp 995–1009.

18. YY Liu, ED Collins, AW Norman, S Peleg. Differential interaction of 1alpha,25-dihydroxyvitamin D3 analogs and their 20-epi homologues with the vitamin D receptor. *J Biol Chem* 272: 3336–3345, 1997.

19. K Takeyama, Y Masuhiro, H Fuse, H Endoh, A Murayama, S Kitanaka, M Suzawa, J Yanagisawa, S Kato. Selective interaction of vitamin D receptor with transcriptional coactivators by a vitamin D analog. *Mol Cell Biol* 19: 1049–1055, 1999.

20. LA Crofts, MS Hancock, NA Morrison, JA Eisman. Multiple promoters direct the tissue-specific expression of novel N-terminal variant human vitamin D receptor gene transcripts. *Proc Natl Acad Sci USA* 95: 10529–10534, 1998.

21. AK Bhalla, EP Amento, B Serog, LH Glimcher. 1,25-Dihydroxyvitamin D3 inhibits antigen-induced T cell activation. *J Immunol* 133: 1748–1754, 1984.

22. WF Rigby, S Denome, MW Fanger. Regulation of lymphokine production and human T lymphocyte activation by 1,25-dihydroxyvitamin D3. Specific inhibition at the level of messenger RNA. *J Clin Invest* 79: 1659–1664, 1987.

23. M Cippitelli, A Santoni. Vitamin D3: a transcriptional modulator of the interferon-gamma gene. *Eur J Immunol* 28: 3017–3030, 1998.

24. I Alroy, TL Towers, LP Freedman. Transcriptional repression of the interleukin-2 gene by vitamin D3: direct inhibition of NFATp/AP-1 complex formation by a nuclear hormone receptor. *Mol Cell Biol* 15: 5789–5799, 1995.

25. A Takeuchi, GS Reddy, T Kobayashi, T Okano, J Park, S Sharma. Nuclear factor of activated T cells (NFAT) as a molecular target for 1alpha,25-dihydroxyvitamin D3-mediated effects. *J Immunol* 160: 209–218, 1998.

26. A Boonstra, FJ Barrat, C Crain, VL Heath, HF Savelkoul, A O'Garra. 1alpha,25-Dihydroxyvitamin D3 has a direct effect on naive CD4(+) T cells to enhance the development of Th2 cells. *J Immunol* 167: 4974–4980, 2001.

27. L Overbergh, B Decallonne, M Waer, O Rutgeerts, D Valckx, KM Casteels, J Laureys, R Bouillon, C Mathieu. 1alpha,25-dihydroxyvitamin D3 induces an autoantigen-specific T-helper 1/T-helper 2 immune shift in NOD mice immunized with GAD65 (p524–543). *Diabetes* 49: 1301–1307, 2000.

28. MT Cantorna, WD Woodward, CE Hayes, HF DeLuca. 1,25-dihydroxyvitamin D3 is a positive regulator for the two anti-encephalitogenic cytokines TGF-beta 1 and IL-4. *J Immunol* 160: 5314–5319, 1998.

29. TP Staeva-Vieira, LP Freedman. 1,25-dihydroxyvitamin D3 inhibits IFN-gamma and IL-4 levels during *in vitro* polarization of primary murine CD4+ T cells. *J Immunol* 168: 1181–1189, 2002.

30. F Mattner, S Smiroldo, F Galbiati, M Muller, P Di Lucia, PL Poliani, G Martino, P Panina-Bordignon, L Adorini. Inhibition of Th1 development and treatment of chronic-relapsing experimental allergic encephalomyelitis by a non-hypercalcemic analog of 1,25-dihydroxyvitamin D3. *Eur J Immunol* 30: 498–508, 2000.

31. JM Lemire, DC Archer, L Beck, HL Spiegelberg. Immunosuppressive actions of 1,25-dihydroxyvitamin D3: preferential inhibition of Th1 functions. *J Nutr* 125: 1704S-1708S, 1995.

32. A Tobler, J Gasson, H Reichel, AW Norman, HP Koeffler. Granulocyte-macrophage colony-stimulating factor. Sensitive and receptor-mediated regulation by 1,25-dihydroxyvitamin D3 in normal human peripheral blood lymphocytes. *J Clin Invest* 79: 1700–1705, 1987.

33. M Cippitelli, C Fionda, D Di Bona, F Di Rosa, A Lupo, M Piccoli, L Frati, A Santoni. Negative regulation of CD95 ligand gene expression by vitamin D3 in T lymphocytes. *J Immunol* 168: 1154–1166, 2002.

34. CD Tsoukas, D Watry, SS Escobar, DM Provvedini, CA Dinarello, FG Hustmyer, SC Manolagas. Inhibition of interleukin-1 production by 1,25-dihydroxyvitamin D3. *J Clin Endocrinol Metab* 69: 127–133, 1989.

35. WF Rigby, M Waugh, RF Graziano. Regulation of human monocyte HLA-DR and CD4 antigen expression, and antigen presentation by 1,25-dihydroxyvitamin D3. *Blood* 76: 189–197, 1990.

36. H Xu, A Soruri, RK Gieseler, JH Peters. 1,25-Dihydroxyvitamin D3 exerts opposing effects to IL-4 on MHC class-II antigen expression, accessory activity, and phagocytosis of human monocytes. *Scand J Immunol* 38: 535–540, 1993.

37. LM Sly, M Lopez, WM Nauseef, NE Reiner. 1alpha,25-Dihydroxyvitamin D3-induced monocyte antimycobacterial activity is regulated by phosphatidylinositol 3-kinase and mediated by the NADPH-dependent phagocyte oxidase. *J Biol Chem* 276: 35482–35493, 2001.

38. AG van Halteren, E Van Etten, EC de Jong, R Bouillon, BO Roep, C Mathieu. Redirection of human autoreactive T-cells upon interaction with dendritic cells modulated by TX527, an analog of 1,25 dihydroxyvitamin D3. *Diabetes* 51: 2119–2125, 2002.

39. G Penna, L Adorini. 1 Alpha,25-dihydroxyvitamin D3 inhibits differentiation, maturation, activation, and survival of dendritic cells leading to impaired alloreactive T cell activation. *J Immunol* 164: 2405–2411, 2000.

40. A Berer, J Stockl, O Majdic, T Wagner, M Kollars, K Lechner, K Geissler, L Oehler. 1,25-Dihydroxyvitamin D3 inhibits dendritic cell differentiation and maturation *in vitro. Exp Hematol* 28: 575–583, 2000.

41. MD Griffin, WH Lutz, VA Phan, LA Bachman, DJ McKean, R Kumar. Potent inhibition of dendritic cell differentiation and maturation by vitamin D analogs. *Biochem Biophys Res Commun* 270: 701–708, 2000.

42. L Piemonti, P Monti, M Sironi, P Fraticelli, BE Leone, E Dal Cin, P Allavena, V Di Carlo, Vitamin D3 affects differentiation, maturation, and function of human monocyte-derived dendritic cells. *J Immunol* 164: 4443–4451, 2000.

43. I Mellman, RM Steinman. Dendritic cells: specialized and regulated antigen processing machines. *Cell* 106: 255–258, 2001.

44. J Banchereau, RM Steinman. Dendritic cells and the control of immunity. *Nature* 19: 245–252, 1998.

45. RM Steinman, S Turley, I Mellman, K Inaba. The induction of tolerance by dendritic cells that have captured apoptotic cells. *J Exp Med* 191: 411–416, 2000.

46. JM Lemire, L Beck, D Faherty, M Gately, HL Spiegelberg. 1,25-dihydroxyvitamin D3 inhibits the production of IL-12 by human monocytes and B cells. In: AW Norman, R Bouillon, M Thomasset, Eds. *Vitamin D, a Pluripotent Steroid Hormone: Structural Studies, Molecular Endocrinology and Clinical Applications*. Berlin: de Gruyter, 1994, pp 531–539.

47. D D'Ambrosio, M Cippitelli, MG Cocciolo, D Mazzeo, P Di Lucia, R Lang, F Sinigaglia, P Panina-Bordignon. Inhibition of IL-12 production by 1,25-dihydroxyvitamin D3. Involvement of NF-kappaB downregulation in transcriptional repression of the p40 gene. *J Clin Invest* 101: 252–262, 1998.

48. R Koren, A Ravid, C Rotem, E Shohami, UA Liberman, A Novogrodsky. 1,25-Dihydroxyvitamin D3 enhances prostaglandin E2 production by monocytes. A mechanism which partially accounts for the antiproliferative effect of 1,25(OH)2D3 on lymphocytes. *FEBS Lett* 205: 113–116, 1986.

49. JL Prehn, DL Fagan, SC Jordan, JS Adams. Potentiation of lipopolysaccharide-induced tumor necrosis factor-alpha expression by 1,25-dihydroxyvitamin D3. *Blood* 80: 2811–2816, 1992.

50. I Hakim, Z Bar-Shavit. Modulation of TNF-alpha expression in bone marrow macrophages: Involvement of vitamin D response element. *J Cell Biochem* 88: 986–998, 2003.

51. L Giovannini, V Panichi, M Migliori, S De Pietro, AA Bertelli, A Fulgenzi, C Filippi, I Sarnico, D Taccola, R Palla, A Bertelli. 1,25-dihydroxyvitamin D3 dose-dependently inhibits LPS-induced cytokines production in PBMC modulating intracellular calcium. *Transplant Proc* 33: 2366–2368, 2001.

52. ML Cohen, A Douvdevani, C Chaimovitz, S Shany. Regulation of TNF-alpha by 1alpha,25-dihydroxyvitamin D3 in human macrophages from CAPD patients. *Kidney Int* 59: 69–75, 2001.

53. K Muller, J Gram, J Bollerslev, M Diamant, T Barington, MB Hansen, K Bendtzen. Down-regulation of monocyte functions by treatment of healthy adults with 1 alpha,25 dihydroxyvitamin D3. *Int J Immunopharmacol* 13: 525–530, 1991.

54. L Chatenoud, B Salomon, JA Bluestone. Suppressor T cells — they're back and critical for regulation of autoimmunity! *Immunol Rev* 182: 149–163, 2001.

55. S Gregori, M Casorati, S Amuchastegui, S Smiroldo, AM Davalli, L Adorini. Regulatory T cells induced by 1 alpha,25-dihydroxyvitamin D3 and mycophenolate mofetil treatment mediate transplantation tolerance. *J Immunol* 167: 1945–1953, 2001.

56. C Mathieu, M Waer, K Casteels, J Laureys, R Bouillon. Prevention of type I diabetes in NOD mice by nonhypercalcemic doses of a new structural analog of 1,25-dihydroxyvitamin D3, KH1060. *Endocrinology* 136: 866–872, 1995.

57. C Mathieu, M Waer, J Laureys, O Rutgeerts, R Bouillon. Prevention of autoimmune diabetes in NOD mice by 1,25 dihydroxyvitamin D3. *Diabetologia* 37: 552–558, 1994.

58. K Casteels, M Waer, R Bouillon, J Depovere, D Valckx, J Laureys, C Mathieu. 1,25-Dihydroxyvitamin D3 restores sensitivity to cyclophosphamide-induced apoptosis in non-obese diabetic (NOD) mice and protects against diabetes. *Clin Exp Immunol* 112: 181–187, 1998.

59. KM Casteels, CA Gysemans, M Waer, R Bouillon, JM Laureys, J Depovere, C Mathieu. Sex difference in resistance to dexamethasone-induced apoptosis in NOD mice: treatment with 1,25(OH)2D3 restores defect. *Diabetes* 47: 1033–1037, 1998.

60. S Gregori, N Giarratana, S Smiroldo, M Uskokovic, L Adorini. A 1alpha,25-dihydroxyvitamin D3 analog enhances regulatory T-cells and arrests autoimmune diabetes in NOD mice. *Diabetes* 51: 1367–1374, 2002.

61. FJ Barrat, DJ Cua, A Boonstra, DF Richards, C Crain, HF Savelkoul, R Waal-Malefyt, RL Coffman, CM Hawrylowicz, A O'Garra. *In vitro* generation of interleukin 10-producing regulatory CD4(+) T cells is induced by immunosuppressive drugs and inhibited by T helper type 1 (Th1)- and Th2-inducing cytokines. *J Exp Med* 195: 603–616, 2002.

62. TK Gray, MS Cohen. Vitamin D, phagocyte differentiation and immune function. *Surv Immunol Res* 4: 200–212, 1985.

63. TY Chan. Vitamin D deficiency and susceptibility to tuberculosis. *Calcif Tissue Int* 66: 476–478, 2000.

64. Z Bar-Shavit, D Noff, S Edelstein, M Meyer, S Shibolet, R Goldman. 1,25-dihydroxyvitamin D3 and the regulation of macrophage function. *Calcif Tissue Int* 33: 673–676, 1981.

65. M Kankova, W Luini, M Pedrazzoni, F Riganti, M Sironi, B Bottazzi, A Mantovani, A Vecchi. Impairment of cytokine production in mice fed a vitamin D3-deficient diet. *Immunology* 73: 466–471, 1991.

66. S Yang, C Smith, JM Prahl, X Luo, HF DeLuca. Vitamin D deficiency suppresses cell-mediated immunity *in vivo*. *Arch Biochem Biophys* 303: 98–106, 1993.

67. MT Cantorna. Vitamin D and autoimmunity: is vitamin D status an environmental factor affecting autoimmune disease prevalence? *Proc Soc Exp Biol Med* 223: 230–233, 2000.

68. RE LaPorte, N Tajima, HK Akerblom, N Berlin, J Brosseau, M Christy, AL Drash, H Fishbein, A Green, R Hamman. Geographic differences in the risk of insulin-dependent diabetes mellitus: the importance of registries. *Diabetes Care* 8: 101–107, 1985.

69. A Green, EA Gale, CC Patterson. Incidence of childhood-onset insulin-dependent diabetes mellitus: the EURODIAB ACE Study. *Lancet* 339: 905–909, 1992.

70. L Nystrom, G Dahlquist, J Ostman, S Wall, H Arnqvist, G Blohme, F Lithner, B Littorin, B Schersten, L Wibell. Risk of developing insulin-dependent diabetes mellitus (IDDM) before 35 years of age: indications of climatological determinants for age at onset. *Int J Epidemiol* 21: 352–358, 1992.

70a. A Giulietti, C Gysemans, K Stoffels, E van Etten, B Decallonne, L Overbergh, R Bouillon, C Mathieu. Vitamin D deficiency in early life accelerates type 1 diabetes in non-obese diabetic mice. *Diabetologia* 47: 451–462, 2004.

71. E Hypponen, E Laara, A Reunanen, MR Jarvelin, SM Virtanen. Intake of vitamin D and risk of type 1 diabetes: a birth-cohort study. *Lancet* 358: 1500–1503, 2001.

72. Vitamin D supplement in early childhood and risk for type I (insulin-dependent) diabetes mellitus. The EURODIAB Substudy 2 Study Group. *Diabetologia* 42: 51–54, 1999.

73. LC Stene, J Ulriksen, P Magnus, G Joner. Use of cod liver oil during pregnancy associated with lower risk of Type I diabetes in the offspring. *Diabetologia* 43: 1093–1098, 2000.

74. S Harris. Can vitamin D supplementation in infancy prevent type 1 diabetes? *Nutr Rev* 60: 118–121, 2002.

75. C Mathieu, E Van Etten, C Gysemans, B Decallonne, R Bouillon. Seasonality of birth in patients with type 1 diabetes. *Lancet* 359: 1248, 2002.

76. E Van Etten, B Decallonne, C Mathieu. 1,25-dihydroxycholecalciferol: endocrinology meets the immune system. *Proc Nutr Soc* 61: 375–380, 2002.

77. R Riachy, B Vandewalle, S Belaich, J Kerr-Conte, V Gmyr, F Zerimech, M d'Herbomez, J Lefebvre, F Pattou. Beneficial effect of 1,25 dihydroxyvitamin D3 on cytokine-treated human pancreatic islets. *J Endocrinol* 169: 161–168, 2001.

78. HJ Hahn, B Kuttler, C Mathieu, R Bouillon. 1,25-Dihydroxyvitamin D3 reduces MHC antigen expression on pancreatic beta-cells *in vitro*. *Transplant Proc* 29: 2156–2157, 1997.

78a. C Mathieu, E van Etten, B Decallone, A Giulietti, C Gysemans, R Bouillon, L Overbergh. Vitamin D and 1,25-dihydroxyvitamin D3 as modulators in the immune system. *J Steroid Biochem Mol Biol* 89–90: 449–452, 2004.

79. TL Towers, LP Freedman. Granulocyte-macrophage colony-stimulating factor gene transcription is directly repressed by the vitamin D3 receptor. Implications for allosteric influences on nuclear receptor structure and function by a DNA element. *J Biol Chem* 273: 10338–10348, 1998.

80. TL Towers, TP Staeva, LP Freedman. A two-hit mechanism for vitamin D3-mediated transcriptional repression of the granulocyte-macrophage colony-stimulating factor gene: vitamin D receptor competes for DNA binding with NFAT1 and stabilizes c-Jun. *Mol Cell Biol* 19: 4191–4199, 1999.

81. Z Hmama, D Nandan, L Sly, KL Knutson, P Herrera-Velit, NE Reiner. 1alpha,25-dihydroxyvitamin D3-induced myeloid cell differentiation is regulated by a vitamin D receptor-phosphatidylinositol 3-kinase signaling complex. *J Exp Med* 190: 1583–1594, 1999.

82. C Mathieu, J Laureys, H Sobis, M Vandeputte, M Waer, R Bouillon. 1,25-Dihydroxyvitamin D3 prevents insulitis in NOD mice. *Diabetes* 41: 1491–1495, 1992.

83. KM Casteels, C Mathieu, M Waer, D Valckx, L Overbergh, JM Laureys, R Bouillon. Prevention of type I diabetes in nonobese diabetic mice by late intervention with nonhypercalcemic analogs of 1,25-dihydroxyvitamin D3 in combination with a short induction course of cyclosporin A. *Endocrinology* 139: 95–102, 1998.

84. K Casteels, M Waer, J Laureys, D Valckx, J Depovere, R Bouillon, C Mathieu. Prevention of autoimmune destruction of syngeneic islet grafts in spontaneously diabetic nonobese diabetic mice by a combination of a vitamin D3 analog and cyclosporine. *Transplantation* 65: 1225–1232, 1998.

85. T Koizumi, Y Nakao, T Matsui, T Nakagawa, S Matsuda, K Komoriya, Y Kanai, T Fujita. Effects of corticosteroid and 1,24R-dihydroxy-vitamin D3 administration on lymphoproliferation and autoimmune disease in MRL/MP-lpr/lpr mice. *Int Arch Allergy Appl Immunol* 77: 396–404, 1985.

86. J Abe, K Nakamura, Y Takita, T Nakano, H Irie, Y Nishii. Prevention of immunological disorders in MRL/l mice by a new synthetic analog of vitamin D3: 22-oxa-1 alpha,25-dihydroxyvitamin D3. *J Nutr Sci Vitaminol* 36: 21–31, 1990.

87. JM Lemire, A Ince, M Takashima. 1,25-Dihydroxyvitamin D3 attenuates the expression of experimental murine lupus of MRL/l mice. *Autoimmunity* 12: 143–148, 1992.

88. MT Cantorna, CE Hayes, HF DeLuca. 1,25-Dihydroxyvitamin D3 reversibly blocks the progression of relapsing encephalomyelitis, a model of multiple sclerosis. *Proc Natl Acad Sci USA* 93: 7861–7864, 1996.

89. JM Lemire, DC Archer. 1,25-dihydroxyvitamin D3 prevents the *in vivo* induction of murine experimental autoimmune encephalomyelitis. *J Clin Invest* 87: 1103–1107, 1991.

90. P Larsson, L Mattsson, L Klareskog, C Johnsson. A vitamin D analog (MC 1288) has immunomodulatory properties and suppresses collagen-induced arthritis (CIA) without causing hypercalcaemia. *Clin Exp Immunol* 114: 277–283, 1998.

91. MT Cantorna, CE Hayes, HF DeLuca. 1,25-Dihydroxychole-calciferol inhibits the progression of arthritis in murine models of human arthritis. *J Nutr* 128: 68–72, 1998.

92. DD Branisteanu, P Leenaerts, B van Damme, R Bouillon. Partial prevention of active Heymann nephritis by 1 alpha, 25 dihydroxyvitamin D3. *Clin Exp Immunol* 94: 412–417, 1993.

93. MT Cantorna, C Munsick, C Bemiss, BD Mahon. 1,25-Dihydroxycholecalciferol prevents and ameliorates symptoms of experimental murine inflammatory bowel disease. *J Nutr* 130: 2648–2652, 2000.

94. DA Hullett, MT Cantorna, C Redaelli, J Humpal-Winter, CE Hayes, HW Sollinger, HF DeLuca. Prolongation of allograft survival by 1,25-dihydroxyvitamin D3. *Transplantation* 66: 824–828, 1998.

95. C Johnsson, G Tufveson. MC 1288 — a vitamin D analog with immunosuppressive effects on heart and small bowel grafts. *Transplant Int* 7: 392–397, 1994.

96. CA Redaelli, M Wagner, YH Tien, L Mazzucchelli, PF Stahel, MK Schilling, JF Dufour. 1 alpha,25-Dihydroxycholecalciferol reduces rejection and improves survival in rat liver allografts. *Hepatology* 34: 926–934, 2001.

97. P Veyron, R Pamphile, L Binderup, JL Touraine. Two novel vitamin D analogs, KH 1060 and CB 966, prolong skin allograft survival in mice. *Transplant Immunol* 1: 72–76, 1993.

98. DL Bertolini, PR Araujo, RN Silva, AJ Duarte, CB Tzanno-Martins. Immunomodulatory effects of vitamin D analog KH1060 on an experimental skin transplantation model. *Transplant Proc* 31: 2998–2999, 1999.

99. CA Redaelli, M Wagner, D Gunter-Duwe, YH Tian, PF Stahel, L Mazzucchelli, RA Schmid, MK Schilling. 1alpha,25-dihydroxyvitamin D3 shows strong and additive immunomodulatory effects with cyclosporine A in rat renal allotransplants. *Kidney Int* 61: 288–296, 2002.

100. AK Raisanen-Sokolowski, IS Pakkala, SP Samila, L Binderup, PJ Hayry, ST Pakkala. A vitamin D analog, MC1288, inhibits adventitial inflammation and suppresses intimal lesions in rat aortic allografts. *Transplantation* 63: 936–941, 1997.

101. MT Cantorna, DA Hullett, C Redaelli, CR Brandt, J Humpal-Winter, HW Sollinger, HF DeLuca. 1,25-Dihydroxyvitamin D3 prolongs graft survival without compromising host resistance to infection or bone mineral density. *Transplantation* 66: 828–831, 1998.

102. E Van Etten, DD Branisteanu, A Verstuyf, M Waer, R Bouillon, C Mathieu. Analogs of 1,25-dihydroxyvitamin D3 as dose-reducing agents for classical immunosuppressants. *Transplantation* 69: 1932–1942, 2000.

103. DD Branisteanu, M Waer, H Sobis, S Marcelis, M Vandeputte, R Bouillon. Prevention of murine experimental allergic encephalomyelitis: cooperative effects of cyclosporine and 1 alpha, 25-(OH)2D3. *J Neuroimmunol* 61: 151–160, 1995.

104. DD Branisteanu, C Mathieu, R Bouillon. Synergism between sirolimus and 1,25-dihydroxyvitamin D3 *in vitro* and *in vivo*. *J Neuroimmunol* 79: 138–147, 1997.

105. C Mathieu, R Bouillon, O Rutgeerts, M Vandeputte, M Waer. Potential role of 1,25(OH)2 vitamin D3 as a dose-reducing agent for cyclosporine and FK 506. *Transplant Proc* 26: 3130, 1994.

106. C Gysemans, E Van Etten, L Overbergh, A Verstuyf, M Waer, R Bouillon, C Mathieu. Treatment of autoimmune diabetes recurrence in non-obese diabetic mice by mouse interferon-beta in combination with an analog of 1alpha,25-dihydroxyvitamin-D3. *Clin Exp Immunol* 128: 213–220, 2002.

107. DA Gerber, CA Bonham, AW Thomson. Immunosuppressive agents: recent developments in molecular action and clinical application. *Transplant Proc* 30: 1573–1579, 1998.

108. CA Biron. Interferons alpha and beta as immune regulators — a new look. *Immunity* 14: 661–664, 2001.

109. AA Licata. Bisphosphonate therapy. *Am J Med Sci* 313: 17–22, 1997.

110. E Van Etten, DD Branisteanu, L Overbergh, R Bouillon, A Verstuyf, C Mathieu. Combination of a 1,25-dihydroxyvitamin D3 analog and a bisphosphonate prevents experimental autoimmune encephalomyelitis and preserves bone. *Bone* 32: 397–404, 2003.

111. M Inaba, Y Nishizawa, K Song, H Tanishita, S Okuno, T Miki, H Morii. Partial protection of 1 alpha-hydroxyvitamin D3 against the development of diabetes induced by multiple low-dose streptozotocin injection in CD-1 mice. *Metabolism* 41: 631–635, 1992.

112. FE Nashold, KA Hoag, J Goverman, CE Hayes. Rag-1-dependent cells are necessary for 1,25-dihydroxyvitamin D3 prevention of experimental autoimmune encephalomyelitis. *J Neuroimmunol* 119: 16–29, 2001.

113. TF Meehan, HF DeLuca. CD8(+) T cells are not necessary for 1 alpha,25-dihydroxyvitamin D3 to suppress experimental autoimmune encephalomyelitis in mice. *Proc Natl Acad Sci USA* 99: 5557–5560, 2002.

114. B Vendeville, D Baran, M Gascon-Barre. Effects of vitamin D3 and cyclosporin A on HgCl2-induced autoimmunity in brown Norway rats. *Nephrol Dial Transplant* 10: 2020–2026, 1995.

115. ST Lillevang, J Rosenkvist, CB Andersen, S Larsen, E Kemp, T Kristensen. Single and combined effects of the vitamin D analog KH1060 and cyclosporin A on mercuric-chloride-induced autoimmune disease in the BN rat. *Clin Exp Immunol* 88: 301–306, 1992.

116. C Fournier, P Gepner, M Sadouk, J Charreire. In vivo beneficial effects of cyclosporin A and 1,25-dihydroxyvitamin D3 on the induction of experimental autoimmune thyroiditis. *Clin Immunol Immunopathol* 54: 53–63, 1990.

117. I Pakkala, E Taskinen, S Pakkala, A Raisanen-Sokolowski. MC1288, a vitamin D analog, prevents acute graft-versus-host disease in rat bone marrow transplantation. *Bone Marrow Transplant* 27: 863–867, 2001.

118. C Gysemans, M Waer, J Laureys, R Bouillon, C Mathieu. A combination of KH1060, a vitamin D3 analog, and cyclosporin prevents early graft failure and prolongs graft survival of xenogeneic islets in nonobese diabetic mice. *Transplant Proc* 33: 2365, 2001.

119. P Gepner, B Amor, C Fournier. 1,25-dihydroxyvitamin D3 potentiates the *in vitro* inhibitory effects of cyclosporin A on T cells from rheumatoid arthritis patients. *Arthritis Rheum* 32: 31–36, 1989.

120. S Gupta, D Fass, M Shimizu, B Vayuvegula. Potentiation of immunosuppressive effects of cyclosporin A by 1 alpha,25-dihydroxyvitamin D3. *Cell Immunol* 121: 290–297, 1989.

121. MC Boissier, G Chiocchia, C Fournier. Combination of cyclosporine A and calcitriol in the treatment of adjuvant arthritis. *J Rheumatol* 19: 754–757, 1992.

122. E Kallio, P Hayry, S Pakkala. MC1288, a vitamin D analog, reduces short- and long-term renal allograft rejection in the rat. *Transplant Proc* 28: 3113, 1996.

123. C Johnsson, L Binderup, G Tufveson. The effects of combined treatment with the novel vitamin D analog MC 1288 and cyclosporine A on cardiac allograft survival. *Transplant Immunol* 3: 245–250, 1995.

124. P Veyron, R Pamphile, L Binderup, JL Touraine. New 20-epi-vitamin D3 analogs: immunosuppressive effects on skin allograft survival. *Transplant Proc* 27: 450, 1995.

125. O Jirapongsananuruk, I Melamed, DY Leung. Additive immunosuppressive effects of 1,25-dihydroxyvitamin D3 and corticosteroids on TH1, but not TH2, responses. *J Allergy Clin Immunol* 106: 981–985, 2000.

126. N Xing, LM ML, LA Bachman, DJ McKean, R Kumar, MD Griffin. Distinctive dendritic cell modulation by vitamin D3 and glucocorticoid pathways. *Biochem Biophys Res Commun* 297: 645–652, 2002.

5

Vitamin D-Regulated Pathways: Impact on Cell Proliferation, Differentiation, and Apoptosis

JOELLEN WELSH

OVERVIEW: BIOLOGY OF VITAMIN D AND THE VITAMIN D RECEPTOR (VDR)

Vitamin D: Sources, Synthesis, and Metabolism

Forms and Functions of Vitamin D

Vitamin D (calciferol) is a fat-soluble vitamin identified as an antirachitic factor in the early 1920s. Rickets and osteomalacia (the adult-onset form of rickets) are bone diseases characterized by impaired mineralization due to insufficient calcium availability. Vitamin D's ability to prevent rickets stems from its function to promote absorption of dietary calcium through regulation of active transport of calcium in enterocytes.[1] More recently, additional roles for vitamin D have been identified, including regulation of cellular proliferation, differentiation, and apoptosis.

The two naturally occurring forms of vitamin D are cholecalciferol (vitamin D_3, from animal sources) and ergocalciferol (vitamin D_2, from plant sources). For clarity, this chapter will focus on vitamin D_3; however, it should be noted that whereas the metabolism and function of ergocalciferol and cholecalciferol are similar, cholecalciferol is somewhat more potent in humans than ergocalciferol. Both forms of vitamin D can be obtained from the diet, although natural foods, with the exception of fish, are relatively low in calciferols. For this reason, milk and other products are fortified with vitamin D in the U.S. and Canada.[2] The actual vitamin D content of fortified milk is highly variable and often less than the stated 400 IU per quart.[3] Vitamin D_3 also can be synthesized from a cholesterol derivative, 7-dehydrocholesterol, in the epidermis, but this process requires ultraviolet radiation and is dependent on sun exposure, which is highly variable.[4]

Vitamin D Deficiency

Despite the fortification of vitamin D in foods and the ability of the body to synthesize the vitamin, nutritional vitamin D deficiency is surprisingly common, especially in populations living in northern climates and in the elderly.[5] In a 1998 sample, more than 50% of patients admitted to Massachusetts General Hospital were vitamin D deficient,[6] and more than 40% of healthy young men in Boston had serum vitamin D levels in the insufficient range at the end of winter.[7] Similar rates of vitamin D insufficiency were reported for middle-aged men in Finland.[8] Factors associated with low vitamin D status include aging, liver or kidney disease, certain medications, poor diet, and limited epidermal synthesis of cholecalciferol (due to infrequent exposure to sunlight, living in geographic areas with low solar radiation, dark pigmentation, and liberal use of sunscreen). Indeed, the recommended daily allowance for vitamin D increases with age (200 IU for those under 50, 400 IU for ages 51 to 70, and 600 IU for ages over 71).

The increasing number of reports of vitamin D insufficiency has prompted a reevaluation of the vitamin D requirement and the appropriate measures of vitamin D status. Although the recommended daily allowance cited above is sufficient for the prevention of rickets, the recommended intake for optimization of more recently identified effects of vitamin D unrelated to calcium homeostasis are not clear.[9] Animal studies have provided evidence that vitamin D status can modify risk for a number of diseases; thus the possibility that chronic, subclinical vitamin D deficiency in humans contributes to chronic disease requires further study. Clarification of the relevant biomarkers of vitamin D status that appropriately reflect newly identified vitamin D actions in target tissues such as colon, prostate, and breast (as discussed in the following text) is also urgently required.

In the majority of cases, overt vitamin D deficiency can be prevented or cured by dietary adjustments or use of a daily multivitamin supplement.[7,9] There are a number of hereditary defects in humans that impair the bioactivation of vitamin D or abolish the function of the VDR, and these are not cured by administration of vitamin D itself. These syndromes (termed vitamin D–resistant rickets) are rare and have been well characterized at the biochemical and molecular levels.[10] Mouse models of these hereditary vitamin D–resistant syndromes have been generated, and have become powerful research tools for identification of new functions of vitamin D.[11,12]

Metabolism of Vitamin D

Metabolism of calciferols is required for biologic activity, and is achieved through a series of enzymatic steps (Figure 5.1), which are the same regardless of the source (dietary or endogenous) or chemical form (cholecalciferol or ergocalciferol). The initial step is hepatic hydroxylation of vitamin D_3 at the 25 position, generating 25-hydroxyvitamin D_3 ($25(OH)D_3$), the major circulating form of the vitamin. This form is also stored in adipose tissue and is an accurate biomarker of the body's overall vitamin D status. Further metabolism of $25(OH)D_3$ generates two major metabolites: 24,25-dihydroxyvitamin D_3 ($24,25(OH)_2D_3$) or $1\alpha,25$-dihydroxyvitamin D_3 ($1,25(OH)_2D_3$). Production of $24,25(OH)_2D_3$ is catalyzed by the vitamin D 24 hydroxylase (also termed CYP24 or P450C24), an enzyme present in the majority of vitamin D target tissues. The

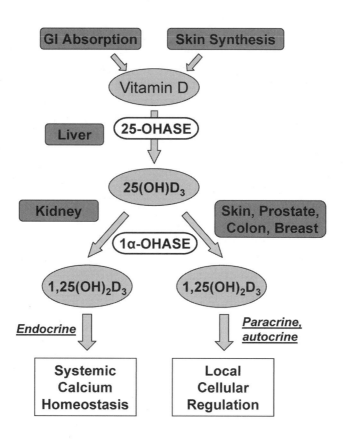

Figure 5.1 Metabolism and activation of vitamin D, emphasizing distinct endocrine and paracrine/autocrine roles of the biologically active compound, $1,25(OH)_2D_3$. Vitamin D in the form of ergocalciferol (D_2) or cholecalciferol (D_3) present in the diet is readily absorbed in the gastrointestinal (GI) tract. Vitamin D_3 can also be synthesized from cholesterol-derived precursors in skin exposed to ultraviolet radiation. Metabolic activation of vitamin D_3 occurs in liver, kidney, and target tissues to produce biologically active $1,25(OH)_2D_3$ which interacts with the vitamin D receptor in either an endocrine or autocrine/paracrine manner. 25-OHASE, 25-hydroxylase; 1α-OHASE, 1α-hydroxylase. (Adapted from J Welsh, JA Wietzke, GM Zinser, B Byrne, K Smith, CJ Narvaez. Vitamin D-3 receptor as a target for breast cancer prevention. *J Nutr* 133: 2425S–2433S, 2003.)

$24,25(OH)_2D_3$ metabolite does not readily bind VDR, and its production has generally been considered the first step in the pathway leading to degradation of $25(OH)D_3$. However, data demonstrating the cellular effects of $24,25(OH)_2D_3$, possibly through interaction with a receptor distinct from the well-characterized nuclear VDR, have accumulated.[13] Production of $1,25(OH)_2D_3$, the biologically active vitamin D metabolite that avidly binds to VDR, is mediated by the vitamin D 1α-hydroxylase (also termed CYP27B1 or P450C1), an enzyme that is highly expressed in renal proximal tubules.[14]

The vitamin D hydroxylases responsible for metabolism of $25(OH)D_3$ are type I (mitochondrial) cytochrome P450 oxidases, which utilize NADPH and molecular oxygen to catalyze the hydroxylation reaction. The activities of the renal vitamin D hydroxylases are tightly regulated to coordinate activation and degradation of vitamin D within the body. Because the kidney 1α-hydroxlase produces $1,25(OH)_2D_3$ (a potent calcium-elevating hormone) for the systemic circulation, its activity reflects the calcium status of the individual. Thus, under conditions of increased demand for calcium, such as growth, pregnancy, and lactation, renal 1α-hydroxlase expression and activity is induced and $1,25(OH)_2D_3$ is generated. The elevated circulating $1,25(OH)_2D_3$ subsequently interacts with VDR in target tissues such as kidney, intestine, and bone to mobilize calcium. Conversely, when calcium demands are low, the activity of the renal 1α-hydroxlase is suppressed and the activity of the 24-hydroxylase is enhanced, leading to formation of $24,25(OH)_2D_3$ and initiation of the catabolic pathway leading to excretion. Both positive (parathyroid hormone, estrogen, and growth hormone) and negative ($1,25(OH)_2D_3$, calcium) regulators of the renal hydroxylases have been identified, and acute regulation of renal $1,25(OH)_2D_3$ production is mediated through the adenylate cyclase and protein kinase C pathways.[15,16] It should be noted that these regulatory concepts of vitamin D hydroxylation have been generated based on data for the renal enzymes, which participate in the endocrine control of extracellular calcium homeostasis, and may not be applicable to regulation of these same enzymes in other tissues. An important objective for the future will be to identify triggers and pathways that regulate $25(OH)D_3$ metabolism in extrarenal tissues.

Extrarenal Production of $1,25(OH)_2D_3$

In recent years, it has become apparent that tissues in addition to the kidney can catalyze the production of $1,25(OH)_2D_3$ from $25(OH)D_3$. Epidermal keratinocytes and activated macrophages, as well as epithelial cells in prostate, breast, and colon[17–21] have been shown to express the vitamin D 1α-hydroxylase. Although the presence of this enzyme suggests that certain extra-renal tissues have the ability to convert $25(OH)D$ to $1,25(OH)_2D_3$, circulating $1,25(OH)_2D_3$ is virtually undetectable in anephric conditions. Thus, $1,25(OH)_2D_3$ produced in extra-renal tissues is apparently not released into the bloodstream, but instead acts locally by binding to VDR present within the same or adjoining cells. Such local actions of $1,25(OH)_2D_3$ might include the effects on regulation of cell proliferation, differentiation, and apoptosis discussed below. The implication of these autocrine or paracrine actions is that cellular production of $1,25(OH)_2D_3$ would likely be regulated in a tissue-specific fashion independently from systemic calcium homeostasis. Similarly, the actions of locally produced $1,25(OH)_2D_3$ would be confined to the immediate cellular environment and would not necessarily impact body calcium homeostasis. The emerging view of systemic vs. cellular pathways of $1,25(OH)_2D_3$ production and action is presented in Figure 5.1.

Mechanisms of Vitamin D Action

The Nuclear Vitamin D Receptor (VDR)

The best characterized mechanism of action of $1,25(OH)_2D_3$ is through the nuclear VDR, a member of the steroid receptor superfamily of proteins that act as ligand-dependent transcription factors to modulate expression of specific genes in a tissue-specific manner.[22,23] These receptors contain ligand-binding domains that selectively bind lipid-soluble ligands, and DNA-binding domains that recognize and bind specific nucleotide sequences in target genes. Along with coactivators, corepressors, and a variety of additional accessory nuclear proteins, the ligand-receptor complexes induce or repress target gene promoters. In the case of the VDR, gene regulation by the liganded receptor requires dimerization, most often heterodimerization with the retinoid X receptor (RXR) family.

Although a variety of structurally distinct vitamin D responsive elements have been identified in the promoter regions of vitamin D regulated genes, the best characterized is a hexanucleotide direct repeat separated by three variable base pairs (DR3) to which VDR:RXR heterodimers bind. However, the recognition that VDR also can function as a homodimer, or as a heterodimer with partners other than RXR, and can bind diverse DNA sequences suggests enormous flexibility to the genomic pathways regulated by vitamin D. In addition, the VDR is subject to posttranslational modifications, including phosphorylation, that affect its transactivation ability.[24]

Cellular sensitivity to vitamin D is in large part dictated by the expression of the VDR, which is subject to transcriptional regulation, ligand-induced stabilization, and proteosomal degradation. The 5′ end of the human VDR gene is highly complex, with multiple promoter regions that generate distinct transcripts.[25-28] Initial characterization of these promoter regions and the resulting transcripts has suggested tissue specific regulatory factors including hormones, dietary components, and growth factors[25,26,29-32] but the relevance of these multiple VDR transcripts in the context of cell signaling is not yet clear.

To add complexity to vitamin D signaling, there is increasing evidence that vitamin D metabolites, including both $1,25(OH)_2D_3$ and $24,25(OH)_2D_3$, can exert rapid, nonenomic effects on signal transduction pathways, leading to biological responses such as calcium uptake or enzyme activation.[13] The mechanism by which rapid, nongenomic effects of vitamin D metabolites are mediated is unclear at present. Some evidence suggests that distinct receptors or binding proteins may exist for $1,25(OH)_2D_3$ and $24,25(OH)_2D_3$, [33,34] whereas other data indicate that the nuclear VDR is required for these nongenomic effects.[35,36] Cloning and characterization of receptors linked to these nongenomic signaling pathways will be necessary to determine their relative importance in mediating cellular actions of vitamin D metabolites.

CELLULAR EFFECTS OF 1,25(OH)$_2$D$_3$

Newly Identified Vitamin D Target Cells

Although originally identified based on its role in calcium and bone homeostasis, it is now recognized that the nuclear VDR and 1,25(OH)$_2$D$_3$ exert effects in almost every tissue in the body. Target tissues for vitamin D signaling not involved in systemic calcium regulation have primarily been identified by the presence of the VDR, and these include the central nervous system, the skin and hair follicles, the immune system, and the endocrine glands. Studies with VDR null mice have identified additional novel vitamin D target tissues based on pathologies associated with disruption of VDR signaling. These include the rennin–angiotensin system and control of blood pressure, ovarian production of reproductive hormones, colonic cell differentiation, and mammary gland morphogenesis.[37–40]

At the cellular level, vitamin D signaling impacts on growth regulatory pathways in a variety of normal and transformed cell types, including leukemic cells, osteoblasts, epidermal keratinocytes, and mammary, colon, and prostate epithelial cells. In the majority of cells, treatment with 1,25(OH)$_2$D$_3$ mediates antiproliferative and proapoptotic effects.[41,42] In some cells, notably keratinocytes and leukemic cell lines, 1,25(OH)$_2$D$_3$ induces differentiation.[42,43] Studies with cells from VDR null mice have indicated that the nuclear VDR is the essential mediator of the growth regulatory effects of 1,25(OH)$_2$D$_3$ in transformed epithelial cells.[44,45] Because VDR expression is retained in the majority of human tumors, this receptor represents a potential therapeutic target for established cancer.[46] Indeed, the primary focus of the work on vitamin D and growth regulation to date has been directed towards the development of synthetic vitamin D based drugs for therapy of hyperproliferative disorders such as psoriasis and cancer.[46]

This chapter will provide a comprehensive analysis of the signaling pathways triggered by $1,25(OH)_2D_3$ that mediate cell cycle arrest, differentiation, and apoptosis, with an emphasis on data from transformed cell lines derived from breast and prostate. A summary of some of the known targets for vitamin D signaling involved in growth regulation is shown in Figure 5.2. These cellular effects (described in the following text) have been identified based on studies with the natural VDR ligand $1,25(OH)_2D_3$, as well as with a variety of structurally distinct synthetic vitamin D analogs. Hundreds of vitamin D analogs have been developed, some of which display enhanced cell regulatory effects with reduced calcemic activity. While the specific mechanisms of action for these vitamin D analogs is not completely understood, most of the widely utilized vitamin D analogs bind VDR. However, it is not entirely clear if their actions are solely mediated via genomic VDR signaling and to what extent these represent physiological effects. Thus, it is important to note that some of the effects discussed here may represent pharmacological consequences of VDR activation, and may not entirely mimic the pathways triggered by $1,25(OH)_2D_3$ itself endogenously.

Antiproliferative Effects of $1,25(OH)_2D_3$

In the majority of transformed cells, including breast, prostate, and leukemic cell lines, $1,25(OH)_2D_3$ induces cell cycle arrest in G_0/G_1.[47–50] In general, cell cycle arrest in response to $1,25(OH)_2D_3$ requires 24 to 48 h and can be demonstrated at nanomolar concentrations. It should be noted that although circulating concentrations of $1,25(OH)_2D_3$ are in the picomolar range, the actual tissue concentrations of $1,25(OH)_2D_3$ are unknown. As noted above, certain cells, including those derived from prostate, breast, and colon, have been demonstrated to express $25(OH)D_3$ 1α-hydroxylase,[19,51,52] suggesting the possibility that circulating $25(OH)D_3$ taken up by cells may be converted to $1,25(OH)_2D_3$. If so, then it is feasible that local tissue concentrations of $1,25(OH)_2D_3$ could indeed reach the nanomolar range required to elicit effects on cell proliferation. In support of this possibility, epithelial cells that

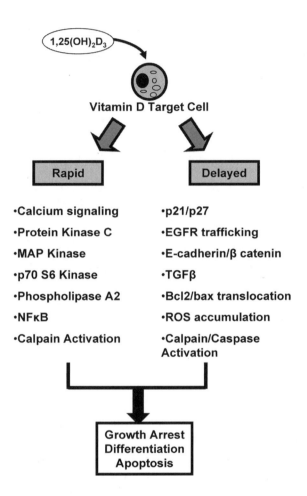

Figure 5.2 VDR targets in regulation of cell cycle, differentiation, and apoptosis. Summary of select vitamin D-modulated proteins, derived from studies conducted with various cell lines. A generic vitamin D target cell is shown; not all responses will be manifest in the same cell. "Rapid" indicates effects mediated within minutes to several hours of $1,25(OH)_2D_3$ exposure; "delayed" indicates effects manifest 24 to 48 h after $1,25(OH)_2D_3$ treatment. The ultimate outcome of $1,25(OH)_2D_3$ treatment (growth arrest, differentiation or apoptosis) will depend on cell type and context. 1,25D, $1,25(OH)_2D_3$; EGFR, epidermal growth factor receptor; ROS, reactive oxygen species.

express functional 1α-hydroxylase have been shown to be growth-inhibited by the precursor metabolite $25(OH)D_3$, presumably due to its bioactivation to $1,25(OH)_2D_3$.[19,53] Concentrations of $25(OH)D_3$ in the nanomolar range, which are within physiological circulating concentrations, can inhibit the growth of normal mammary epithelial cells.[19] These data suggest that circulating $25(OH)D_3$ may be the most appropriate biomarker of vitamin D status in relation to cell regulatory functions.

$1,25(OH)_2D_3$-mediated G_0/G_1 arrest is associated with increases in the cyclin-dependent kinase inhibitors p21 or p27, with resulting inhibition of cyclin-dependent kinases and dephosphorylation of the retinoblastoma protein.[47,54,55] Both p21 and p27 have been identified as targets of the $1,25(OH)_2D_3$–VDR complex.[50,56-58] In the case of p21, a consensus vitamin D response element has been identified in the promoter region.[54] In the case of p27, which lacks well defined vitamin D response elements, the $1,25(OH)_2D_3$–VDR complex cooperates with SP1 and NF-Y transcription factors to upregulate gene expression.[59] An effect of $1,25(OH)_2D_3$ on stabilization of the p27 protein has also been demonstrated.[60] In colon cancer cells, liganded VDR interacts with protein phosphatases PP1c and PP2Ac which inactivate the p70 S6 kinase that is essential for G_1/S phase transition.[61] In leukemic cells, vitamin D-mediated growth arrest is paralleled by activation of Erk, JNK, and p38 mitogen-activated protein kinase (MAPK) pathways.[62] Thus, although the mechanisms vary somewhat in different cell types, the net result of vitamin D signaling is to prevent transition of cells from G_1 into S phase.

Despite the known importance of the p53 tumor-suppressor gene in mediating growth arrest, $1,25(OH)_2D_3$ does not consistently up-regulate p53, and vitamin D compounds can inhibit growth of cells expressing mutant p53 such as T47D breast cancer cells.[30,63] Thus, functional p53 is not required for the antiproliferative effects of $1,25(OH)_2D_3$. This notion is consistent with data indicating that $1,25(OH)_2D_3$-mediated transactivation of the p21[WAF-1/CIP1] gene promoter is p53 independent.[54]

Cells treated with $1,25(OH)_2D_3$ fail to reenter the cell cycle when stimulated with mitogens such as epidermal growth factor (EGF) and insulin like growth factor-1 (IGF-1).

The failure of $1,25(OH)_2D_3$-treated cells to respond to mitogenic stimulation is secondary to the changes in cell cycle machinery discussed above. In addition, effects of $1,25(OH)_2D_3$ on individual signaling pathways have been reported. For example, $1,25(OH)_2D_3$ blocks estrogen-stimulated mitogenesis in breast cancer cells via transcriptional downregulation of the estrogen receptor.[64] Inhibitory effects of $1,25(OH)_2D_3$ on growth-factor-driven mitogenic signaling pathways, including EGF, IGF-1, and keratinocyte growth factor (KGF), have also been demonstrated.[65–67] For example, in A431 epidermoid cells, $1,25(OH)_2D_3$ inhibits EGF receptor phosphorylation and internalization and prevents EGF stimulation of ERK signaling and cyclin D1 transactivation.[68] Thus, the accumulated data suggest that vitamin D signaling decreases cellular sensitivity to stimulation by a variety of mitogens.

Another mechanism that contributes to the antiproliferative effects of $1,25(OH)_2D_3$ is upregulation of negative-growth-factor signaling. The effects of vitamin D signaling on the transforming growth factor beta (TGFβ) growth inhibitory pathway have been studied in a variety of cell types. $1,25(OH)_2D_3$ has been shown to upregulate TGFβ ligands, receptors, and binding proteins, and direct interactions between VDR and SMADS, the intracellular mediators of TGFβ actions have been demonstrated.[69–72] A direct effect of $1,25(OH)_2D_3$ on the TGFβ2 gene is supported by the identification of vitamin D responsive elements in its promoter.[73] Furthermore, the antiproliferative effects of vitamin D compounds can be partially abrogated by neutralizing antibodies to TGFβ, supporting a causal relationship between induction of TGFβ and growth inhibition.[70,74] Collectively, these data suggest a mechanism whereby $1,25(OH)_2D_3$ induces TGFβ secretion which then acts in an autocrine fashion to inhibit cell cycle progression.

Induction of Differentiation by $1,25(OH)_2D_3$

In addition to cell cycle arrest, $1,25(OH)_2D_3$ treatment can be associated with induction of differentiation markers in both normal and transformed cells. These observations have led to the suggestion that a major function of $1,25(OH)_2D_3$ is maintenance of the differentiated phenotype in normal cells. The best-studied example of vitamin D-induced differentiation in normal cells is the keratinocyte.[75] $1,25(OH)_2D_3$ treatment of keratinocytes is associated with induction of involucrin gene expression and formation of the cornified envelope, a marker of epidermal differentiation.[75] In keratinocytes, vitamin D-mediated differentiation is dependent on increases in intracellular calcium and involves genomic induction of the calcium receptor and phospholipase C-gamma 1 (PLCγ 1).[43] Interestingly, PLCγ 1 has a DR6 VDRE (rather than the well-characterized DR3 VDRE) in its promoter, which binds and is activated by VDR-RAR rather than VDR-RXR. In contrast to normal keratinocytes, cells derived from squamous cell carcinomas fail to respond to the prodifferentiating actions of $1,25(OH)_2D_3$, despite normal VDR expression.[76] The failure of $1,25(OH)_2D_3$-induced differentiation in squamous carcinoma cells has been attributed to altered activity of critical VDR cofactors. The ability of vitamin D-compounds to mediate keratinocyte differentiation has become important therapeutically, as vitamin D-based drugs are effective against psoriasis, a hyperproliferative skin disorder. Furthermore, skin from VDR knockout mice exhibits abnormal differentiation[77] and displays enhanced susceptibility to papilloma formation in response to carcinogen treatment, compared with wild-type mice that express VDR.[78]

Cancer cells that retain the ability to differentiate may be programmed to do so in response to $1,25(OH)_2D_3$ treatment. Thus, $1,25(OH)_2D_3$ induces differentiation markers in certain leukemic, skin, colon, and breast cancer cell lines. In breast cancer cell lines, $1,25(OH)_2D_3$ induces intracellular lipid droplets and β-casein, markers of secretory mammary cells.[79,80] In colon cancer cells, $1,25(OH)_2D_3$ modulates the E-cadherin/βcatenin pathway and causes profound changes in the gene expression profile.[81] Expression profiling of vitamin D-responsive squamous carcinoma cells has identified $1,25(OH)_2D_3$-responsive gene clusters involved in regulation of cell cycle, differentiation, cell adhesion, and immune responses, indicating a diverse and broad range of VDR targets potentially involved in epidermal cell differentiation.[82]

In leukemic cell lines such as HL-60, NB4, and U937 cells, $1,25(OH)_2D_3$ induces differentiation along the macrophage/monocyte pathway.[83] Genes identified as targets of $1,25(OH)_2D_3$ in leukemic cells include p21 and the homeobox gene HOXA10.[54,84] Numerous signaling pathways have been implicated in $1,25(OH)_2D_3$-induced monocytic differentiation, including NF-κB, protein kinase C, extracellular signaling-regulating kinases, jun kinase, and p38 mitogen-activated protein kinases.[62] There is evidence that the induction of differentiation by $1,25(OH)_2D_3$ in NB4 leukemic cells is mediated, at least partially, through nongenomic pathways involving protein kinase C, changes in intracellular calcium, calpain activation, and NF-κB signaling.[33] The relative contributions of nongenomic and genomic actions of $1,25(OH)_2D_3$ in mediating growth arrest and differentiation of diverse cell types awaits the molecular identification of additional receptors or novel mechanisms of action for the nuclear VDR, which could mediate rapid effects at the membrane. Regardless of mechanisms, these studies demonstrating induction of differentiation in transformed cells by $1,25(OH)_2D_3$ indicate that vitamin D analogs may be useful therapeutics for early-stage cancers that retain the ability to differentiate. Similar strategies have been successfully used in the development of vitamin A (retinoid) analogs for differentiation therapy of leukemias.[85]

Regulation of Apoptotic Pathways by $1,25(OH)_2D_3$

The ability of $1,25(OH)_2D_3$ to induce morphological and biochemical features of apoptosis (cell shrinkage, chromatin condensation, and DNA fragmentation) in certain cancer cells was first demonstrated in 1995.[74] Cell lines reported to undergo apoptosis in response to $1,25(OH)_2D_3$ or various synthetic analogs include those derived from breast cancer, prostate cancer, squamous carcinoma, myeloma, retinoblastoma, neuroblastoma, and glioma.[86–92] Furthermore, $1,25(OH)_2D_3$ exerts additive or synergistic effects in combination with other triggers of apoptosis, such as radiation and chemotherapeutic agents.[93–95] It is currently not clear whether these synergistic effects result from interactions of $1,25(OH)_2D_3$ with agonist-specific signaling

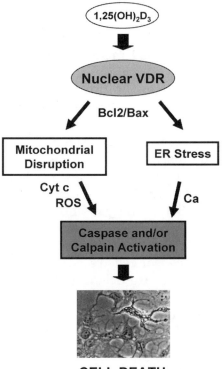

CELL DEATH

Figure 5.3 Pathways leading to $1,25(OH)_2D_3$-mediated apoptosis. Schematic emphasizing the role of the nuclear VDR in regulation of apoptotic regulatory proteins Bcl-2 and Bax. Alteration of Bcl-2 to Bax ratio has been associated with mitochondrial disruption, leading to release of cytochrome c (cyt c) and accumulation of reactive oxygen species (ROS) as well as endoplasmic reticulum (ER) stress, leading to calcium (Ca) depletion. Disruption of these two organelles may activate downstream proteases such as caspases and calpains, facilitating cell death.

pathways or whether $1,25(OH)_2D_3$ impacts on components of a common apoptotic pathway.

The intracellular signaling pathways implicated in $1,25(OH)_2D_3$-mediated apoptosis are depicted in Figure 5.3. Although it is clear that the VDR is required for $1,25(OH)_2D_3$-mediated apoptosis,[44] it is not clear whether activation or repression of gene transcription is required, and if so, what VDR target genes trigger the apoptotic response. As noted above, $1,25(OH)_2D_3$ decreases mitogenic signaling and thereby inhibits numerous intracellular signaling pathways that impact on survival, including MAP kinases and AKT.[96] With respect to specific apoptotic regulatory proteins, several independent studies have reported that sensitivity to $1,25(OH)_2D_3$-mediated apoptosis reflects the relative expression or subcellular localization of the Bcl-2 family of pro- and antiapoptotic proteins, although the specific proteins involved and their modulation vary with cell type. Treatment of MCF-7 breast cancer cells with $1,25(OH)_2D_3$ induces redistribution of the proapoptotic Bcl-2 family member, Bax, from the cytosol to the mitochondria and downregulates the antiapoptotic protein Bcl-2.[55,87,97,98] Furthermore, overexpression of Bcl-2 renders prostate and breast cancer cells resistant to $1,25(OH)_2D_3$-mediated apoptosis.[99–101] Because Bcl-2 and Bax act antagonistically in the regulation of apoptosis, these data suggest that translocation of Bax in conjunction with downregulation of Bcl-2 may be necessary for $1,25(OH)_2D_3$-mediated apoptosis. Vitamin D-mediated Bax translocation triggers reactive oxygen species generation, dissipation of the mitochondrial membrane potential, and release of cytochrome c into the cytosol,[55,98,99] features of the intrinsic (mitochondrial) pathway of apoptosis. These studies, therefore, suggest that vitamin D-mediated apoptosis is triggered through changes in bcl-2 family proteins which trigger mitochondrial death pathways; however, the mechanism by which the $1,25(OH)_2D_3$–VDR complex dictates changes in bcl-2 family proteins has yet to be identified.

Caspases and other proteases have been implicated as both triggers and mediators of apoptosis in response to many different stimuli.[102] The role of proteases in $1,25(OH)_2D_3$-mediated cell death appears to vary with cell type. Cleavage of known substrates such as poly (ADP-ribose) polymerase (PARP) and activation of effector caspases (i.e., caspase 3 and 9) have been implicated during $1,25(OH)_2D_3$-induced apoptosis in several cell lines.[62,89,90,98,99] Furthermore, caspase inhibition can prevent some features of $1,25(OH)_2D_3$-mediated apoptosis, including DNA fragmentation and PARP cleavage.[98] However, caspase inhibitors do not prevent cytochrome c release from mitochondria and are unable to rescue cells from $1,25(OH)_2D_3$-mediated death.[98,99,101,103] Thus, while caspase activation likely facilitates apoptosis, the commitment to $1,25(OH)_2D_3$-mediated cell death appears to be caspase independent. Another pathway recently implicated in vitamin D-mediated apoptosis involves calcium release from the endoplasmic reticulum and activation of μ-calpain.[104] In this study, vitamin D-mediated apoptosis could be prevented by either calpain inhibitors or calcium buffering agents such as calbindin D28K, indicating a requirement for both enzyme activation and calcium signaling.[104] Collectively, these studies indicate that a wide variety of different signaling pathways, apoptotic regulatory proteins, and proteases may contribute to $1,25(OH)_2D_3$-mediated apoptosis, depending on the specific cell type and context.

SUMMARY

Vitamin D is a pleiotropic regulator of proliferation, differentiation and survival for cells of diverse lineages (i.e., hematopoietic, epithelial, mesenchymal). It is clear that vitamin D signaling represents a major negative growth regulatory pathway via inhibition of cell cycle progression, induction of differentiation markers and stimulation of apoptosis in a cell type specific manner. Although the majority of studies reviewed in this chapter have been conducted *in vitro,* animal studies have confirmed the physiological significance of vitamin D regulation of cell turnover. Thus, growth inhibition of tumors *in vivo* in response to vitamin D analog therapy has been attributed to induction of growth arrest and activation of apoptosis.[55,91,105] Furthermore, mice with targeted ablation of the VDR exhibit altered cell turnover in several tissues, including skin and mammary gland.[37,78] A challenge for the future will be to identify the tissue specific molecular targets of the vitamin D signaling pathway involved in cell regulation during normal and pathological processes.

ACKNOWLEDGMENTS

The author is indebted to an excellent research team, consisting of Belinda Byrne, Carly Kemmis, Carmen J. Narvaez, Meggan Valrance, and Glendon Zinser, who contributed significantly to the studies and concepts discussed in this chapter. Work in the author's laboratory is supported by the National Cancer Institute, the Susan G. Komen Foundation, and the Department of Defense Breast Cancer Research Program.

References

1. MF Holick. Vitamin D: A millenium perspective. *J Cell Biochem* 88: 296–307, 2003.

2. V Tangpricha, P Koutkia, SM Rieke, TC Chen, AA Perez, MF Holick. Fortification of orange juice with vitamin D: a novel approach for enhancing vitamin D nutritional health. *Am J Clin Nutr* 77: 1478–1483, 2003.

3. MF Holick, Q Shao, WW Liu, TC Chen. The vitamin D content of fortified milk and infant formula. *N Engl J Med* 326: 1178–1181, 1992.

4. MF Holick. Evolution and function of vitamin D. *Recent Results Cancer Res* 164: 3–28, 2003.

5. P Lips. Vitamin D deficiency and secondary hyperparathyroidism in the elderly: consequences for bone loss and fractures and therapeutic implications. *Endocr Rev* 22: 477–501, 2001.

6. MK Thomas, DM Lloyd_Jones, RI Thadhani, AC Shaw, DJ Deraska, BT Kitch, EC Vamvakas, IM Dick, RL Prince, JS Finkelstein. Hypovitaminosis D in medical inpatients. *N Engl J Med* 338: 777–783, 1998.

7. V Tangpricha, EN Pearce, TC Chen, MF Holick. Vitamin D insufficiency among free-living healthy young adults. *Am J Med* 112: 659–662, 2002.

8. MH Ahonen, L Tenkanen, L Teppo, M Hakama, P Tuohimaa. Prostate cancer risk and prediagnostic serum 25-hydroxyvitamin D levels (Finland). *Cancer Causes Control* 11: 847–852, 2000.

9. A Zittermann. Vitamin D in preventive medicine: are we ignoring the evidence? *Br J Nutr* 89: 552–572, 2003.

10. PJ Malloy, D Feldman. Hereditary 1,25-Dihydroxyvitamin D-resistant rickets. *Endocr Dev* 6: 175–199, 2003.

11. YC Li, AE Pirro, M Amling, G Delling, R Baron, R Bronson, MB Demay. Targeted ablation of the vitamin D receptor: an animal model of vitamin D-dependent rickets type II with alopecia. *Proc Natl Acad Sci USA* 94: 9831–9835, 1997.

12. O Dardenne, J Prud'homme, A Arabian, FH Glorieux, R St. Arnaud. Targeted inactivation of the 25-hydroxyvitamin D3-1(alpha)-hydroxylase gene (CYP27B1) creates an animal model of pseudovitamin D-deficiency rickets. *Endocrinol* 142: 3135–3141, 2001.

13. AW Norman, WH Okamura, JE Bishop, HL Henry. Update on biological actions of 1alpha,25(OH)2-vitamin D3 (rapid effects) and 24R,25(OH)2-vitamin D3. *Mol Cell Endocrinol* 197: 1–13, 2002.

14. WL Miller, AA Portale. Vitamin D biosynthesis and vitamin D 1 alpha-hydroxylase deficiency. *Endocr Dev* 6: 156–174, 2003.

15. J Welsh, V Weaver, M Simboli-Campbell. Regulation of renal 25(OH)D3 1 alpha-hydroxylase: signal transduction pathways. *Biochem Cell Biol* 69: 768–770, 1991.

16. JL Omdahl, HA Morris, BK May. Hydroxylase enzymes of the vitamin D pathway: expression, function, and regulation. *Annu Rev Nutr* 22: 139–166, 2002.

17. HS Cross, E Kallay, H Farhan, T Weiland, T Manhardt. Regulation of extrarenal vitamin D metabolism as a tool for colon and prostate cancer prevention. *Recent Results Cancer Res* 164: 413–425, 2003.

18. M Hewison, V Kantorovich, HR Liker, AJ Van Herle, P Cohan, D Zehnder, JS Adams. Vitamin D-mediated hypercalcemia in lymphoma: evidence for hormone production by tumor-adjacent macrophages. *J Bone Miner Res* 18: 579–582, 2003.

19. J Welsh, JA Wietzke, GM Zinser, B Byrne, K Smith, CJ Narvaez. Vitamin D-3 receptor as a target for breast cancer prevention. *J Nutr* 133: 2425S–2433S, 2003.

20. TC Chen, L Wang, LW Whitlatch, JN Flanagan, MF Holick. Prostatic 25-hydroxyvitamin D-1alpha-hydroxylase and its implication in prostate cancer. *J Cell Biochem* 88: 315–322, 2003.

21. JN Flanagan, L Wang, V Tangpricha, J Reichrath, TC Chen, MF Holick. Regulation of the 25-hydroxyvitamin D-1alpha-hydroxylase gene and its splice variant. *Recent Results Cancer Res* 164: 157–167, 2003.

22. C Carlberg. Current understanding of the function of the nuclear vitamin D receptor in response to its natural and synthetic ligands. *Recent Results Cancer Res* 164: 29–42, 2003.

23. S Christakos, P Dhawan, Y Liu, X Peng, A Porta. New insights into the mechanisms of vitamin D action. *J Cell Biochem* 88: 695–705, 2003.

24. PW Jurutka, PN MacDonald, S Nakajima, JC Hsieh, PD Thompson, GK Whitfield, MA Galligan, CA Haussler, MR Haussler. Isolation of baculovirus-expressed human vitamin D receptor: DNA responsive element interactions and phosphorylation of the purified receptor. *J Cell Biochem* 85: 435–457, 2002.

25. IM Byrne, L Flanagan, MP Tenniswood, J Welsh. Identification of a hormone-responsive promoter immediately upstream of exon 1c in the human vitamin D receptor gene. *Endocrinology* 141: 2829–2836, 2000.

26. K Miyamoto, RA Kesterson, H Yamamoto, Y Taketani, E Nishiwaki, S Tatsumi, Y Inoue, K Morita, E Takeda, JW Pike. Structural organization of the human vitamin D receptor chromosomal gene and its promoter. *Mol Endocrinol* 11: 1165–1179, 1997.

27. LA Crofts, MS Hancock, NA Morrison, JA Eisman. Multiple promoters direct the tissue-specific expression of novel N-terminal variant human vitamin D receptor gene transcripts. *Proc Natl Acad Sci USA* 95: 10529–10534, 1998.

28. KL Sunn, TA Cock, LA Crofts, JA Eisman, EM Gardiner. Novel N-terminal variant of human VDR. *Molecular Endocrinol* 15: 1599–1609, 2001.

29. DL Lazarova, M Bordonaro, AC Sartorelli. Transcriptional regulation of the vitamin D3 receptor gene by ZEB. *Cell Growth Diff* 12: 319–326, 2001.

30. JA Wietzke, J Welsh. Phytoestrogen regulation of a Vitamin D3 receptor promoter and 1,25-dihydroxyvitamin D3 actions in human breast cancer cells. *J Steroid Biochem Mol Biol* 84: 149–157, 2003.

31. TH Lee, J Pelletier. Functional characterization of WT1 binding sites within the human vitamin D receptor gene promoter. *Physiol Genomics* 7: 187–200, 2001.

32. F Jehan, HF DeLuca. The mouse vitamin D receptor is mainly expressed through an Sp1-driven promoter *in vivo*. *Arch Biochem Biophys* 377: 273–283, 2000.

33. AW Norman, JE Bishop, CM Bula, CJ Olivera, MT Mizwicki, LP Zanello, H Ishida, WH Okamura. Molecular tools for study of genomic and rapid signal transduction responses initiated by 1 alpha,25(OH)(2)-vitamin D3. *Steroids* 67: 457–466, 2002.

34. BD Boyan, VL Sylvia, N McKinney, Z Schwartz. Membrane actions of vitamin D metabolites 1alpha,25(OH)2D3 and 24R,25(OH)2D3 are retained in growth plate cartilage cells from vitamin D receptor knockout mice. *J Cell Biochem* 90: 1207–1223, 2003.

35. RG Erben, DW Soegiarto, K Weber, U Zeitz, M Lieberherr, R Gniadecki, G Moller, J Adamski, R Balling. Deletion of deoxyribonucleic acid binding domain of the vitamin D receptor abrogates genomic and nongenomic functions of vitamin D. *Mol Endocrinol* 16: 1524–1537, 2002.

36. LP Zanello, AW Norman. Rapid modulation of osteoblast ion channel responses by 1(alpha),25(OH)2-vitamin D3 requires the presence of a functional vitamin D nuclear receptor. *Proc Natl Acad Sci USA* 101: 1589–1594, 2004.

37. G Zinser, K Packman, J Welsh. Vitamin D3 receptor ablation alters mammary gland morphogenesis. *Development* 129: 3067–3076, 2002.

38. X Li, W Zheng, YC Li. Altered gene expression profile in the kidney of vitamin D receptor knockout mice. *J Cell Biochem* 89: 709–719, 2003.

39. E Kallay, P Bareis, E Bajna, S Kriwanek, E Bonner, S Toyokuni, HS Cross. Vitamin D receptor activity and prevention of colonic hyperproliferation and oxidative stress. *Food Chem Toxicol* 40: 1191–1196, 2002.

40. YC Li, J Kong, M Wei, ZF Chen, SQ Liu, LP Cao. 1,25-Dihydroxyvitamin D3 is a negative endocrine regulator of the renin-angiotensin system. *J Clin Invest* 110: 229–238, 2002.

41. J Welsh, JA Wietzke, GM Zinser, S Smyczek, S Romu, E Tribble, JC Welsh, B Byrne, CJ Narvaez. Impact of the Vitamin D3 receptor on growth-regulatory pathways in mammary gland and breast cancer. *J Steroid Biochem Mol Biol* 83: 85–92, 2002.

42. SY James, MA Williams, AC Newland, KW Colston. Leukemia cell differentiation: cellular and molecular interactions of retinoids and vitamin D. *Gen Pharmacol* 32: 143–154, 1999.

43. DD Bikle, CL Tu, Z Xie, Y Oda. Vitamin D regulated keratinocyte differentiation: role of coactivators. *J Cell Biochem* 88: 290–295, 2003.

44. GM Zinser, K McEleney, J Welsh. Characterization of mammary tumor cell lines from wild type and vitamin D3 receptor knockout mice. *Mol Cell Endocrinol* 200: 67–80, 2003.

45. G Eelen, L Verlinden, M Van Camp, P Van Hummelen, K Marchal, B De Moor, C Mathieu, G Carmeliet, R Bouillon, A Verstuyf. The effects of 1 alpha, 25-dihydroxyvitamin D3 on the expression of DNA replication genes. *J Bone Miner Res* 19: 133–146, 2004.

46. KV Pinette, YK Yee, BY Amegadzie, S Nagpal. Vitamin D receptor as a drug discovery target. *Mini Rev Med Chem* 3: 193–204, 2003.

47. M Simboli-Campbell, CJ Narvaez, K vanWeelden, M Tenniswood, J Welsh. Comparative effects of 1,25(OH)2D3 and EB1089 on cell cycle kinetics and apoptosis in MCF-7 breast cancer cells. *Breast Cancer Res Treat* 42: 31–41, 1997.

48. SH Zhuang, KL Burnstein. Antiproliferative effect of 1alpha,25-dihydroxyvitamin D3 in human prostate cancer cell line LNCaP involves reduction of cyclin-dependent kinase 2 activity and persistent G1 accumulation. *Endocrinol* 139: 1197–1207, 1998.

49. R Munker, A Norman, HP Koeffler. Vitamin D compounds. Effect on clonal proliferation and differentiation of human myeloid cells. *J Clin Invest* 78: 424–430, 1986.

50. WH Park, JG Seol, ES Kim, CW Jung, CC Lee, L Binderup, HP Koeffler, BK Kim, YY Lee. Cell cycle arrest induced by the vitamin D3 analog EB1089 in NCI-H929 myeloma cells is associated with induction of the cyclin-dependent kinase inhibitor p27. *Exp Cell Res* 254: 279–286, 2000.

51. V Tangpricha, JN Flanagan, LW Whitlatch, CC Tseng, TC Chen, PR Holt, MS Lipkin, MF Holick. 25-hydroxyvitamin D-1alpha-hydroxylase in normal and malignant colon tissue. *Lancet* 357: 1673–1674, 2001.

52. GG Schwartz, LW Whitlatch, TC Chen, BL Lokeshwar, MF Holick. Human prostate cells synthesize 1,25-dihydroxyvitamin D3 from 25-hydroxyvitamin D3. *Cancer Epidemiol, Biomarkers Prev* 7: 391–395, 1998.

53. JY Hsu, D Feldman, JE McNeal, DM Peehl. Reduced 1alpha-hydroxylase activity in human prostate cancer cells correlates with decreased susceptibility to 25-hydroxyvitamin D3-induced growth inhibition. *Cancer Res* 61: 2852–2856, 2001.

54. M Liu, MH Lee, M Cohen, M Bommakanti, LP Freedman. Transcriptional activation of the Cdk inhibitor p21 by vitamin D3 leads to the induced differentiation of the myelomonocytic cell line U937. *Genes Dev* 10: 142–153, 1996.

55. L Flanagan, K Packman, B Juba, S O'Neill, M Tenniswood, J Welsh. Efficacy of Vitamin D compounds to modulate estrogen receptor negative breast cancer growth and invasion. *J Steroid Biochem Mol Biol* 84: 181–192, 2003.

56. W Liu, SL Asa, IG Fantus, PG Walfish, S Ezzat. Vitamin D arrests thyroid carcinoma cell growth and induces p27 dephosphorylation and accumulation through PTEN/akt-dependent and -independent pathways. *Am J Pathol* 160: 511–519, 2002.

57. G Hager, M Formanek, C Gedlicka, D Thurnher, B Knerer, J Kornfehl. 1,25(OH)2 vitamin D3 induces elevated expression of the cell cycle-regulating genes P21 and P27 in squamous carcinoma cell lines of the head and neck. *Acta Otolaryngol* 121: 103–109, 2001.

58. BA Scaglione-Ewell, M Bissonnette, S Skarosi, C Abraham, TA Brasitus. A vitamin D3 analog induces a G1-phase arrest in CaCo-2 cells by inhibiting cdk2 and cdk6: roles of cyclin E, p21Waf1, and p27Kip1. *Endocrinol* 141: 3931–3939, 2000.

59. T Inoue, J Kamiyama, T Sakai. Sp1 and NF-Y synergistically mediate the effect of vitamin D3 in the p27(Kip1) gene promoter that lacks vitamin D response elements. *J Biol Chem* 274: 32309–32317, 1999.

60. ES Yang, KL Burnstein. Vitamin D inhibits G1 to S progression in LNCaP prostate cancer cells through p27Kip1 stabilization and Cdk2 mislocalization to the cytoplasm. *J Biol Chem* 278: 46862–46868, 2003.

61. DJ Bettoun, DW Buck, J Lu, B Khalifa, WW Chin, S Nagpal. A vitamin D receptor-Ser/Thr phosphatase-p70 S6 kinase complex and modulation of its enzymatic activities by the ligand. *J Biol Chem* 277: 24847–24850, 2002.

62. C Pepper, A Thomas, T Hoy, D Milligan, P Bentley, C Fegan. The vitamin D3 analog EB1089 induces apoptosis via a p53-independent mechanism involving p38 MAP kinase activation and suppression of ERK activity in B-cell chronic lymphocytic leukemia cells *in vitro*. *Blood* 101: 2454–2460, 2003.

63. JA Eisman, RL Sutherland, ML McMenemy, JC Fragonas, EA Musgrove, GY Pang. Effects of 1,25-dihydroxyvitamin D3 on cell-cycle kinetics of T 47D human breast cancer cells. *J Cell Physiol* 138: 611–616, 1989.

64. A Stoica, M Saceda, A Fakhro, HB Solomon, BD Fenster, MB Martin. Regulation of estrogen receptor-alpha gene expression by 1, 25-dihydroxyvitamin D in MCF-7 cells. *J Cell Biochem* 75: 640–651, 1999.

65. WM Tong, H Hofer, A Ellinger, M Peterlik, HS Cross. Mechanism of antimitogenic action of vitamin D in human colon carcinoma cells: relevance for suppression of epidermal growth factor-stimulated cell growth. *Oncol Res* 11: 77–84, 1999.

66. SP Xie, G Pirianov, KW Colston. Vitamin D analogues suppress IGF-I signaling and promote apoptosis in breast cancer cells. *Eur J Cancer* 35: 1717–1723, 1999.

67. C Crescioli, M Maggie, GB Vannelli, M Luconi, R Salerno, T Barni, M Gulisano, G Forti, M Serio. Effect of a vitamin D3 analogue on keratinocyte growth factor-induced cell proliferation in benign prostate hyperplasia. *J Clin Endocrinol Metab* 85: 2576–2583, 2000.

68. JB Cordero, M Cozzolino, Y Lu, M Vidal, E Slatopolsky, PD Stahl, MA Barbieri, A Dusso. 1,25-Dihydroxyvitamin D downregulates cell membrane growth- and nuclear growth-promoting signals by the epidermal growth factor receptor. *J Biol Chem* 277: 38965–38971, 2002.

69. N Subramaniam, GM Leong, TA Cock, JL Flanagan, C Fong, JA Eisman, AP Kouzmenko. Cross-talk between 1,25-dihydroxyvitamin D3 and transforming growth factor-beta signaling requires binding of VDR and Smad3 proteins to their cognate DNA recognition elements. *J Biol Chem* 276: 15741–15746, 2001.

70. L Yang, J Yang, S Venkateswarlu, T Ko, MG Brattain. Autocrine TGFbeta signaling mediates vitamin D3 analog-induced growth inhibition in breast cells. *J Cell Physiol* 188: 383–393, 2001.

71. CW Jung, ES Kim, JG Seol, WH Park, SJ Lee, BK Kim, YY Lee. Antiproliferative effect of a vitamin D3 analog, EB1089, on HL-60 cells by the induction of TGF-beta receptor. *Leuk Res* 23: 1105–1112, 1999.

72. RG Mehta, RM Moriarty, RR Mehta, R Penmasta, G Lazzaro, A Constantinou, L Guo. Prevention of preneoplastic mammary lesion development by a novel vitamin D analogue, 1alpha-hydroxyvitamin D5. *J Natl Cancer Inst* 89: 212–218, 1997.

73. Y Wu, TA Craig, WH Lutz, R Kumar. Identification of 1 alpha,25-dihydroxyvitamin D3 response elements in the human transforming growth factor beta 2 gene. *Biochemistry* 38: 2654–2660, 1999.

74. J Welsh, M Simboli-Campbell, CJ Narvaez, M Tenniswood. Role of apoptosis in the growth inhibitory effects of vitamin D in MCF-7 cells. *Adv Exp Med Biol* 375: 45–52, 1995.

75. DD Bikle, E Gee, S Pillai. Regulation of keratinocyte growth, differentiation, and vitamin D metabolism by analogs of 1,25-dihydroxyvitamin D. *J Invest Dermatol* 101: 713–718, 1993.

76. DD Bikle, Z Xie, D Ng, CL Tu, Y Oda. Squamous cell carcinomas fail to respond to the prodifferentiating actions of 1,25(OH)2D: why? *Recent Results Cancer Res* 164: 111–122, 2003.

77. KL Xie Z, Yu QC, Elalieh H, Ng DC, Leary C, Chang S, Crumrine D, Yoshizawa T, Kato S, Bikle DD. Lack of the vitamin D receptor is associated with reduced epidermal differentiation and hair follicle growth. *J Invest Dermatol* 118: 11–16, 2002.

78. GM Zinser, JP Sundberg, J Welsh. Vitamin D3 receptor ablation sensitizes skin to chemically induced tumorigenesis. *Carcinogenesis* 23: 2103–2109, 2002.

79. G Lazzaro, A Agadir, W Qing, M Poria, RR Mehta, RM Moriarty, TK Das-Gupta, XK Zhang, RG Mehta. Induction of differentiation by 1alpha-hydroxyvitamin D(5) in T47D human breast cancer cells and its interaction with vitamin D receptors. *Eur J Cancer* 36: 780–786, 2000.

80. Q Wang, D Lee, V Sysounthone, Chandraratna-RAS, S Christakos, R Korah, R Wieder. 1,25-dihydroxyvitamin D3 and retonic acid analogues induce differentiation in breast cancer cells with function- and cell-specific additive effects. *Breast Cancer Res Treat* 67: 157–168, 2001.

81. HG Palmer, JM Gonzalez-Sancho, J Espada, MT Berciano, I Puig, J Baulida, M Quintanilla, A Cano, AG de-Herreros, M Lafarga, A Munoz. Vitamin D3 promotes the differentiation of colon carcinoma cells by the induction of E-cadherin and the inhibition of beta-catenin signaling. *J Cell Biol* 154: 369–387, 2001.

82. R Lin, Y Nagai, R Sladek, Y Bastien, J Ho, K Petrecca, G Sotiropoulou, EP Diamandis, TJ Hudson, JH White. Expression profiling in squamous carcinoma cells reveals pleiotropic effects of vitamin D3 analog EB1089 signaling on cell proliferation, differentiation, and immune system regulation. *Mol Endocrinol* 16: 1243–1256, 2002.

83. E Elstner, M Linker-Israeli, J Le, T Umiel, P Michl, JW Said, L Binderup, JC Reed, HP Koeffler. Synergistic decrease of clonal proliferation, induction of differentiation, and apoptosis of acute promyelocytic leukemia cells after combined treatment with novel 20-epi vitamin D3 analogs and 9-cis retinoic acid. *J Clin Invest* 99: 349–360, 1997.

84. NY Rots, M Liu, EC Anderson, LP Freedman. A differential screen for ligand-regulated genes: identification of HoxA10 as a target of vitamin D3 induction in myeloid leukemic cells. *Mol Cell Biol* 18: 1911–1918, 1998.

85. KH Dragnev, WJ Petty, E Dmitrovsky. Retinoid targets in cancer therapy and chemoprevention. *Cancer Biol Ther* 2: S150–156, 2003

86. J Elias, B Marian, C Edling, B Lachmann, CR Noe, SH Rolf, I Schuster. Induction of apoptosis by vitamin D metabolites and analogs in a glioma cell line. *Recent Results Cancer Res* 164: 319–332, 2003.

87. N Wagner, KD Wagner, G Schley, L Badiali, H Theres, H Scholz. 1,25-dihydroxyvitamin D3-induced apoptosis of retinoblastoma cells is associated with reciprocal changes of Bcl-2 and bax. *Exp Eye Res* 77: 1–9, 2003.

88. I Audo, SR Darjatmoko, CL Schlamp, JM Lokken, MJ Lindstrom, DM Albert, RW Nickells. Vitamin D analogues increase p53, p21, and apoptosis in a xenograft model of human retinoblastoma. *Invest Ophthalmol Vis Sci* 44: 4192–4199, 2003.

89. TF McGuire, DL Trump, CS Johnson. Vitamin D3-induced apoptosis of murine squamous cell carcinoma cells. Selective induction of caspase-dependent MEK cleavage and up-regulation of MEKK-1. *J Biol Chem* 276: 26365–26373, 2001.

90. WH Park, JG Seol, ES Kim, JM Hyun, CW Jung, CC Lee, L Binderup, HP Koeffler, BK Kim, YY Lee. Induction of apoptosis by vitamin D3 analogue EB1089 in NCI-H929 myeloma cells via activation of caspase 3 and p38 MAP kinase. *Br J Haematol* 109: 576–583, 2000.

91. K VanWeelden, L Flanagan, L Binderup, M Tenniswood, J Welsh. Apoptotic regression of MCF-7 xenografts in nude mice treated with the vitamin D3 analog, EB1089. *Endocrinol* 139: 2102–2110, 1998.

92. M Simboli-Campbell, CJ Narvaez, M Tenniswood, J Welsh. 1,25-Dihydroxyvitamin D3 induces morphological and biochemical markers of apoptosis in MCF-7 breast cancer cells. *J Steroid Biochem Mol Biol* 58: 367–376, 1996.

93. M Chaudhry, S Sundaram, C Gennings, H Carter, DA Gewirtz. The vitamin D3 analog, ILX-23-7553, enhances the response to adriamycin and irradiation in MCF-7 breast tumor cells. *Cancer Chemother Pharmacol* 47: 429–436, 2001.

94. GH Posner, KR Crawford, S Peleg, JE Welsh, S Romu, DA Gewirtz, MS Gupta, P Dolan, TW Kensler. A non-calcemic sulfone version of the vitamin D3 analogue seocalcitol (EB 1089): chemical synthesis, biological evaluation and potency enhancement of the anticancer drug adriamycin. *Bioorg Med Chem* 9: 2365–2371, 2001.

95. S Sundaram, A Sea, S Feldman, R Strawbridge, PJ Hoopes, E Demidenko, L Binderup, DA Gewirtz. The Combination of a Potent Vitamin D3 Analog, EB 1089, with Ionizing Radiation Reduces Tumor Growth and Induces Apoptosis of MCF-7 Breast Tumor Xenografts in Nude Mice. *Clin Cancer Res* 9: 2350–2356, 2003.

96. RJ Bernardi, DL Trump, WD Yu, TF McGuire, PA Hershberger, CS Johnson. Combination of 1alpha,25-dihydroxyvitamin D3 with dexamethasone enhances cell cycle arrest and apoptosis: role of nuclear receptor cross-talk and Erk/Akt signaling. *Clin Cancer Res* 7: 4164–4173, 2001.

97. CJ Narvaez, BM Byrne, S Romu, M Valrance, J Welsh. Induction of apoptosis by 1,25-dihydroxyvitamin D3 in MCF-7 Vitamin D3-resistant variant can be sensitized by TPA. *J Steroid Biochem Mol Biol* 84: 199–209, 2003.

98. CJ Narvaez, J Welsh. Role of mitochondria and caspases in vitamin D-mediated apoptosis of MCF-7 breast cancer cells. *J Biol Chem* 276: 9101–9107, 2001.

99. M Guzey, S Kitada, JC Reed. Apoptosis induction by 1alpha,25-dihydroxyvitamin D3 in prostate cancer. *Mol Cancer Ther* 1: 667–677, 2002.

100. SE Blutt, TJ McDonnell, TC Polek, NL Weigel. Calcitriol-induced apoptosis in LNCaP cells is blocked by overexpression of Bcl-2. *Endocrinol* 141: 10–17, 2000.

101. IS Mathiasen, U Lademann, M Jaattela. Apoptosis induced by vitamin D compounds in breast cancer cells is inhibited by Bcl-2 but does not involve known caspases or p53. *Cancer Res* 59: 4848–4856, 1999.

102. A Degterev, M Boyce, J Yuan. A decade of caspases. *Oncogene* 22: 8543–8567, 2003.

103. G Pirianov, KW Colston. Interactions of vitamin D analogue CB1093, TNFalpha and ceramide on breast cancer cell apoptosis. *Mol Cell Endocrinol* 172: 69–78, 2001.

104. IS Mathiasen, IN Sergeev, L Bastholm, F Elling, AW Norman, M Jaattela. Calcium and calpain as key mediators of apoptosis-like death induced by vitamin D compounds in breast cancer cells. *J Biol Chem* 277: 30738–30745, 2002.

6

Effects of Nutrient Deprivation on the Expression of Growth-Arrest Genes

W.D. REES

INTRODUCTION

Mammalian cells require a precise mixture of macro- and micronutrients for growth and development. A diet deficient in one or more essential components limits the growth of cells and leads to metabolic stress. If left unchecked, these stresses can damage important components required for the normal functions of the cell. Environmental stresses such as UV radiation or free radicals also damage the cell. Fortunately, mammalian cells possess mechanisms that respond to stress by suspending progression of the cell cycle and initiating programmed cell death or apoptosis where damage has occurred. The process of apoptosis is also essential in sculpting the developing embryo; for example, Reference 1 describes how the complex process of limb development is dependent on the carefully regulated death of specific cells. It is therefore not surprising to find that common mechanisms regulate the initiation of apoptosis both during differentiation and in response to stress. In this chapter, we will discuss the products of a group of mRNAs that are upregulated as part of the active response to metabolic and environmental stress. Because of their association with growth-arrested cells, the genes coding for these mRNAs are collectively termed the *growth-arrest genes*.

THE IDENTIFICATION OF GROWTH-ARREST GENES

The two seminal studies that initiated the study of growth-arrest genes used subtractive hybridization techniques to identify cDNAs present in growth-arrested cells.[2,2a] This method can be used to isolate nucleotide sequences present in a test sample but absent or present at much lower levels in a reference sample. By removing from growing and non-growing cells the cDNAs that are common to libraries, the subtracted library is enriched with cDNAs that are abundant in quiescent or growth-arrested cells. Using this approach, Schneider and colleagues[3] prepared a subtraction cDNA library from mouse 3T3 cells whose growth was arrested by serum deprivation. The genes isolated from these libraries were called the growth arrest-specific genes (abbreviated to gas). In the second series of experiments, Fornace and colleagues[4,5] utilized a similar procedure to prepare a subtraction library from Chinese hamster ovary cells (CHO) that had been treated with the potent alkylating agent methyl methane sulphonate (MMS) or irradiated with ultraviolet light to induce DNA damage. These experiments identified a number of different cDNAs that were named growth arrest and DNA damage (abbreviated to gadd).

Following their initial identification, a variety of different approaches have been used to identify and understand the function of these mRNAs and their protein products *in vitro* and *in vivo*. It was apparent from an early stage that the gas and gadd genes are highly conserved and widely expressed in many different cell types. They code for a diverse range of products (Table 6.1) that are involved in many different aspects of cell functions. Some mRNAs code for proteins that directly regulate progression through the cell cycle; others code for proteins with more indirect actions such as transcriptional activators or proteins that are part of the interactions

Table 6.1 Growth Arrest Genes and Cell Function

Gene name	Protein Location	Function
Gas1	Membrane	Integrin interactions
Gas2	Cytoskeleton	Caspase substrate, cytoskeletal regulation
Gas3	Integral membrane protein	Unknown
Gas5	Nuclear	rRNA processing
Gas6	Secreted protein	Axl receptor ligand, cell–cell adhesion
Gadd34	Cytosolic	Regulates eIF2 phosphorulation
Gadd45	Nuclear	Inhibits DNA replication, interacts with chromatin
Gadd153/CHOP	Nuclear	Transcriptional activator

between one cell and its neighbors. The common theme from all of these studies has been the involvement of both gas and gadd genes in eventual induction of apoptosis.

Since their initial discovery, many other experiments have added to the catalog of genes associated with the stress response. However, the genes from the gas and gadd families remain representative of the cellular functions that are essential for the stress response. In order to discuss the stress response, this group of mRNAs can be loosely categorized into three different groups: those with mechanisms that are dependent on the activation of p53, RNAs with a role in ribosome biogenesis, and those genes associated with the cell surface.

GROWTH-ARREST GENES ACTIVATED BY THE INHIBITION OF DNA SYNTHESIS AND DEPENDENT ON THE P53 SYSTEM (GAS1 AND GADD45)

When gas1 cDNA is constitutively expressed in cells under the control of the strong SV40 promoter, DNA synthesis is inhibited,[13] but only in cell lines that also possess a functional p53 gene.[14] The p53 protein plays an essential role in the cellular response to genotoxic stress, first slowing the cell cycle and then, if necessary, inducing apoptosis. A failure in this function is known to play crucial roles in tumor development and progression, with the result that mutations of the p53 gene are the most common genetic alterations found in human cancers. Following cellular stress, the p53 protein is phosphorylated, stabilizing the molecule and thereby increasing the steady state concentration in the cell.[6] The p53 protein has four highly conserved functional domains that enable it to interact with proteins and nucleic acids. Each domain has a specific function. The N-terminal domain is required for transcriptional activation,[7] and there is a sequence-specific DNA binding domain,[8] a tetramerization domain,[9] and a C-terminal domain that interacts directly with single-stranded DNA.[10] The interactions of proteins and nucleic acids with the different domains determine whether p53 promotes growth arrest and apoptosis through transcription-dependent or -independent mechanisms.[11] In the case of gas1, studies with specific mutations in each of these domains showed that expression was regulated by a discrete proline-rich domain and did not involve the N-terminal transactivation domain.[15,16] Although p53 is clearly the principal regulator of gas1 expression, p53-mediated expression of gas1 can be repressed in cells expressing the c-myc and v-Src transcriptional activators, which promote cell proliferation.[17,18]

In addition to the p53-dependent stress responses, gas1 also plays an important role during normal development, and like many of the growth arrest genes, it is expressed in the proliferating tissues of the fetus.[19,20] When embryonic limb cells are transfected with gas1, expression of the protein arrests growth and promotes cell death.[21] Studies in the developing embryo have shown that gas1 is expressed in the somite and is induced by the patterning factor *wingless* (Wnt). The gas1 protein is able to bind another patterning factor, *sonic hedgehog* (Shh), to produce an inactive complex. By reducing the availability of Shh, gas1 may play an important part in orchestrating the delicate balance between proliferation and differentiation, which regulates organ development and results in organs of appropriate size.[22] This is illustrated by studies of mice with a knockout mutation of the gas1 gene. At birth, wild-type and gas1 mutant mice have cerebella of similar size; however, once they mature, the cerebella of the mutants are less than half the size of that in the wild-type animals. The reduced cell numbers in the affected areas of the cerebellum correspond to specific and coordinated changes in the expression of gas1.[23] Indeed, there is evidence that gas1 may, under certain circumstances, act as a growth-promoting gene during development.[24]

The gas1 gene codes for a glycoprotein that is attached to the outside of the plasma membrane by a glycosyl-phosphatidylinositol (GPI) anchor.[12] The coding sequence includes a hydrophobic signal peptide at the N-terminal, which targets the nascent peptide to the endoplasmic reticulum (ER), where the signal peptide is removed and the GPI anchor is added.[25] Once processed, the mature protein is exported from the ER to the plasma membrane, where it is randomly distributed over the surface of the cell.[26] However, the GPI anchor is not essential, and experiments with deletion mutants of gas1 showed no loss of activity when the GPI anchor was deleted. A soluble chimeric protein containing the N-terminal half of the mature protein attached to the Fc domain of immunoglobulin was also able to induce growth arrest when it was added to the outside of cells.[25] Although the full mechanism of gas1 is not yet understood, its location on the cell surface does suggest that it has the potential to be a receptor for interactions with adjacent cells. The minimal sequence of gas1 contains motifs that may interact with integrins, and it has been suggested that these enable it to modify the attachment of cells to the extracellular matrix.[14]

Another group of growth-arrest genes regulated by both p53-dependent and -independent pathways are members of the gadd45 family. Following its initial identification, the gadd45 gene was found to be one of three similar genes — gadd45a, gadd45b (also known as MyD118), and gadd45g (also known as CR6/OIG37). The gadd45 family members are highly conserved, with a 60 to 70% amino acid sequence similarity. All three genes are induced with different expression kinetics during the growth arrest associated with terminal differentiation[27] and show distinct expression patterns in a variety of murine tissues.[28] This suggests that the different gene products have specific functions during development; however, it is not clear to what extent each protein is able to overlap with other members of the family.

Early studies of gadd45 suggested at least two distinct signaling pathways, one that responds to damage caused by ionizing radiation and a second that reacts to base-damaging agents such as the alkylating agent methyl methanesulfonate (MMS) or UV irradiation. The induction of gadd45 by ionizing radiation requires p53, which binds to a responsive element within the third intron of the gadd45 gene.[29,30] In contrast, both MMS and medium starvation increase gadd45b expression through a mechanism that is not dependent on p53.[31] Analysis of the promoter sequences suggests the presence of binding sites for the NF-kappaB, the STAT transcriptional activators,[32] and also regions that may allow cell-type specific responses to TGF-beta1.[33] The gadd45 promoter also contains binding motifs for OCT-1 and CAAT transcriptional activators.[34] These appear to be involved in the regulation of gadd45 expression by the BRCA1 breast and ovarian cancer susceptibility gene. Site-directed mutations of these motifs abrogate induction by BRCA1, which is believed to act through protein–protein interactions between BRCA1 and the OCT-1 and NF-YA transcriptional activators.[35]

The protein coded by gadd45a is normally found in very low abundance in exponentially growing cells. The gadd45a protein activates the G2 checkpoint in the cell cycle, suppressing the growth of numerous cell types.[36] Immunohistochemistry has shown that oligomeric complexes of gadd45a, gadd45b, and gadd45g proteins accumulate in discrete foci in the nuclei of growth-arrested cells.[37,38] The protein interacts with nuclear factors associated with cell cycle regulation, including the Cdc2-cyclin B1 kinase complex,[39] the p21 cyclin-dependent kinase inhibitor,[40] and proliferating cell nuclear antigen.[41] In addition, gadd45 proteins activate the p38/Jun N-terminal kinase pathway by binding to MTK1/MEKK4.[42] Gadd45a also associates with core histones and in doing so, it alters the chromatin structure.[43] A 10-amino-acid stretch of the gadd45a sequence is almost identical to another acidic nuclear protein nucleophosmin, suggesting that this sequence may be involved in binding to chromatin. It is possible that gadd45a binding opens the chromatin structure around damaged DNA, facilitating the binding of elements such as topoisomerases, which are involved in relaxing the negative coils in double-stranded DNA.[43] With this variety of different interactions in which gadd45 proteins play a central role in the regulation of cell division, however, there are still many aspects of this fascinating group of proteins that remain to be elucidated.

During normal development, gadd45a is expressed in a wide variety of tissues. Mice with a targeted deletion develop normally apart from a small increase in the frequency of neural tube defects.[44] It is quite probable that the lack of phenotype is due to other members of the same family of proteins being able to substitute for one another during development and in response to stress. Interestingly, embryonic fibroblasts derived from the gadd45a-null mice had a demonstrable increase in genomic instability, with more than twice the normal frequency of aneuploidy and tetraploidy. These cells had lost normal senescence and had a growth advantage compared with wild-type cells.[45,46] However, there was no evidence for an increase in the development of spontaneous tumors in these animals, which again suggests that other systems are able to compensate for the mutation *in vivo*.

ENDOPLASMIC RETICULUM STRESS RESPONSES (GADD153/CHOP AND GADD34)

The transcription factor gadd153 (also known as CHOP-10[47]) is expressed at low levels in all proliferating cells. However, when cells are exposed to genotoxic and metabolic stress, the expression of gadd153/CHOP increases. The mRNA also responds to elevated cellular levels of nitric oxide, reactive oxygen species, hypoxia, and nutrient deprivation (glucose and amino acids). The gadd153/CHOP mRNA level is regulated by a combination of mRNA stabilization,[48] translational control, and transcriptional control. Some of these responses are modulated through amino acid response elements (AARE) in the promoter and by the binding of the transcription factors ATF2 and ATF4.[49]

In addition to the direct signaling through the AARE, gadd153/CHOP expression is also responsive to the accumulation of unfolded proteins in the endoplasmic reticulum (ER).[50–52] The receptor proteins PERK, Ire1, and ATF6 span the ER membrane and are kept in an inactive state by binding to the resident protein chaperone BiP (GRP78). When unfolded proteins accumulate in the ER, there is an increased requirement for BiP, which dissociates from the receptor proteins. The protein kinases, PERK and Ire1, are then free to form active dimers triggering a signaling cascade,[53,54] which results in the phosphorylation of the eukaryotic initiation factor eIF-2. This is one of the key regulators of protein translation (see following discussion). Although phosphorylation of eIF-2 inhibits the translation of most mRNAs, the translation of a few selected transcripts such as those encoding activating transcription factor 4 (ATF4) is increased.[55,56] The increased levels of ATF4 protein can then induce the transcription of downstream genes such as gadd153/CHOP.[55] The expression of gadd153/CHOP is also regulated by activating transcription factor 6 (ATF6), another of the sensor proteins that span the ER membrane. When ATF6 is released from BiP, it is translocated to the Golgi, where its cytosolic domain encoding a transcription factor is cleaved by proteases.[57] The liberated ATF6 then migrates to the nucleus, where it activates the consensus ER stress response element found in the promoters of various genes including gadd153/CHOP.[58]

As a member of the C/EBP family of transcription factors, gadd153/CHOP is able to form dimers with other proteins from the same family through a characteristic leucine-zipper domain.[47] These dimers can then interact with and control the transcription of a variety of target genes. This regulation is both positive and negative, as shown in Figure 6.1. Positive regulation occurs through dimers formed between gadd153/CHOP and other C/EBP factors, which bind to specific regulatory sites in a distinct set of downstream genes (DOC or downstream of CHOP).[59,60] At the same time, the formation of these dimers reduces the availability of other bZip transcription factors, which normally transactivate genes associated with cell proliferation.[47] Inactive

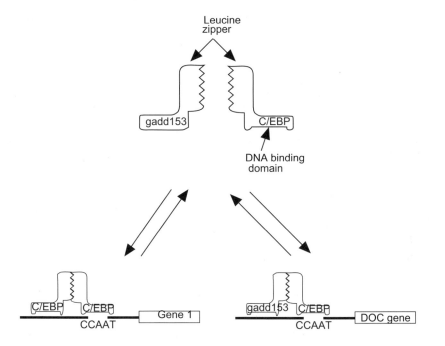

Figure 6.1 Interactions between gadd153/CHOP and other bZip transcriptional activators. An equilibrium exists between C/EBP and other bZip transcriptional activators in the cell. When levels of gadd153/CHOP are low, C/EBP dimers predominate. Increasing levels of gadd153/CHOP compete for dimer formation and are able to bind to new sites, activating genes downstream of CHOP (DOC).

gadd153/CHOP dimers can be formed with members of the C/EBP family and other bZip factors such as the AP1 transcription factors, JunD, c-Fos, and c-Jun.[61] The transcriptional activity of gadd153/CHOP is also regulated by phosphorylation. The stress-activated protein kinase p38-MAP kinase is able to activate the protein,[62] whereas casein kinase 2 inhibits its transcriptional activity.[63]

The expression of gadd153/CHOP is not essential for normal development, and transgenic animals with a targeted deletion in the gene are apparently normal. However, if the adult animals are exposed to agents that produce severe ER stress, there is an abnormal response in the tissues, including a reduction in the initiation of programmed cell death that protects the wild-type animals.[64] The product of the gadd153/CHOP gene may therefore be a protective system that eliminates damaged or defective cells. In tissues where there is a high demand for protein export, such as the pancreas, the gadd153/CHOP system may be particularly important. Reduced beta cell mass is evident in type 2 diabetes, and apoptosis is implicated in this process. The Akita mouse, which carries a mutation in the insulin 2 gene, develops diabetes because of a reduced beta cell mass. Expression of the mutant insulin, in cells secreting insulin, induced gadd153/CHOP expression and led to apoptosis. It was also possible to delay the onset of diabetes by targeted disruption of the gadd153/CHOP gene in Akita mice.[65,66] These data suggest that this system may be particularly important in the pathology of diabetes. A similar system may be important during other stages of development.

Another of the genes associated with the UPR is the growth-arrest and DNA-damage gene gadd34, which has an expression profile paralleling that of gadd153/CHOP.[67] The gadd34 transcript is stabilized under conditions of cellular stress, increasing the steady state levels of its mRNA.[68] The C-terminal domain of the gadd34 protein is highly conserved and homologous to ICP34.5, a virulence factor that allows herpes simplex virus type 1-infected cells to circumvent apoptosis.[69] Both ICP34.5 and mammalian gadd34 bind to protein phosphatase-1, and this complex is able to dephosphorylate the eukaryotic translation elongation factor eIF2.[70] The regulation of eIF-2 provides a critical control point in the regulation of translation as it binds to the initiator methionyl tRNA and brings it to the ribosome. When eIF-2 is phosphorylated by regulatory protein kinases, global protein synthesis is reduced. This is probably a short-term response allowing the cells to clear misfolded proteins from the ER and recover.[71] One possible role for gadd34 is as part of a feedback control system that comes into play if recovery is not possible. By overriding the inhibition of protein synthesis by dephosphorylating eIF2, gadd34 allows the cell to synthesize the proteins required for the later stages of the stress response.[72,73]

During fetal development, gadd34 is widely expressed in many fetal tissues. The expression is not uniform, and in the mouse reaches a peak on embryonic day E12.5. In contrast, almost no expression was detected on embryonic days E10.5, E11.5, and E13.5. Despite the stage-specific expression of gadd34, mice deficient in it had no abnormalities under normal breeding conditions. This suggests that other proteins with similar functions remain to be discovered.[74]

RNA PROCESSING (GAS5), A MEMBER OF THE 5TOP GENE FAMILY

One of the most curious genes identified by subtractive hybridization of growth-arrested cells was gas5. Initial studies suggested that although the mRNA for this gene was abundant in growth-arrested cells, there was no sign of a corresponding protein. This mRNA is also unusual because in some circumstances such as nucleotide depletion, the levels of the gas5 mRNA fall.[75] Analysis of the sequence showed that the gas5 gene is composed of 12 exons that give rise to two alternatively spliced transcripts differing by the presence or absence of one exon and potentially encoding peptides of 20- and 28-amino-acids.[76] However, when synthetic peptides were prepared from these sequences and used to produce antibodies, the resulting antisera did not bind to extracts of cultured cells or tissues. This is despite the ubiquitous expression of the gas5 mRNA in mouse tissues.[76] Comparisons showed that the gas5 exon sequences are poorly conserved between different species. In the rat, the putative coding sequence is interrupted by a stop codon after the first 13 amino acids, further suggesting that this gene does not code for a protein.[77]

This puzzle was finally resolved by the discovery of sequences homologous to the small nucleolar RNAs (snoRNAs) in the introns of both the human and murine gas5 genes. These catalytic RNAs are essential in processing preribosomal RNA during ribosome formation, carrying out a site-specific 2-O-methylation of ribosomal RNA.[78] All of the vertebrate snoRNAs are found in the introns of their host genes, and the functional molecules are produced by cleavage of the pre-RNA. In this respect, gas5 is not unique, and a number of other genes containing short, nonconserved ORFs have also been identified, including U22 host gene (UHG),[79] U17HG, and U19HG.[80,81]

In growing cells, active translation leads to rapid degradation of the spliced gas5 RNA, keeping the steady state level low. However, when cell growth slows and translation is inhibited, the message is stabilized, elevating the steady state level of the gas5 transcript.[82] Sequencing of the gas5 gene revealed that it has a characteristic 5 oligopyrimidine tract (also known as terminal oligo pyrimidine or 5TOP).[83] Other members of the 5TOP gene family include ribosomal proteins and elongation factors. During arrested cell growth, these mRNAs accumulate in mRNP particles, so as much as 15% of the total mRNA in the cell may be from this small family. The gas5 gene and other members of this 5TOP family are important measures of cell growth. As the only member of this group of genes not directly involved in the induction of apoptosis, the gas5 gene may be of great interest in the understanding of nutritional regulation of ribosome biosynthesis and cell growth.

CELL–CELL INTERACTIONS INVOLVING MEMBRANE PROTEINS GAS2, GAS3, AND GAS6

The growth-arrest genes — gas2, gas3, and gas6 — are expressed during the later stages of growth arrest.[84] The protein products are expressed at the cell membrane and have been shown to play an important role in the cell interactions that shape tissue morphology and regulate adhesion and apoptosis.

Microfilaments and Microtubules

The first protein of this group, the product of the gas2 gene, has been shown to associate with the microfilaments and microtubules of mammalian cells.[85,86] During the transition from quiescence to the G_1 phase of cell division, the shape of mammalian cells is rapidly rearranged by the actin cytoskeleton.[87] The gas2 protein plays a role in regulating microfilament function during this transition. Within minutes of the addition of serum to quiescent cells, the protein becomes hyperphosphorylated, relocating the protein from the cell border to the newly formed membrane ruffles.[88] The synthesis of gas2 is then downregulated once the cells are actively dividing.

The gas2 protein is also one of the "death substrates" for a series of cysteine aspartyl proteases (caspases) that form a proteolytic cascade during the induction of apoptosis.[89] A fragment cleaved from the N-terminal of the gas2 protein by the action of caspases causes the microfilament network to collapse around the nucleus, leading to the morphological changes characteristic of apoptotic cells.[90,91] The gas2 protein also enhances the susceptibility of the cells to apoptosis following treatment with UV, etoposide, or MMS.[92] This increase in sensitivity was found to be dependent on the function of the p53 protein. The data suggest that gas2 stabilizes the p53 protein by inhibiting the proteolysis of p53 by calpain, thus increasing levels of p53 in the cell.[92]

This important role in apoptosis is not confined to stress-induced growth arrest; during normal development, gas2 is expressed and cleaved in the hind-limb tissues of mouse and rat embryos between days 13.5 and 15.5.[93,94] *In vitro* studies have also implicated gas2 in the growth and differentiation of keratinocytes[95] and muscle.[96] The caspase-mediated apoptotic pathways play an important role in shaping the developing fetus,[89] suggesting a central role for the gas2 protein in development.

Transmembrane Proteins

The membrane-bound protein product of the gas3 gene was found to be identical to a major component of the peripheral nerves, peripheral myelin protein PMP22.[97] The gene codes for a membrane protein of the tetraspan family.[98] Currently, there are at least six known members in this family of proteins, sharing 30 to 40% amino acid identity.[99,100] The structure of gas3/PMP22 predicts the presence of a 22-kDa glycoprotein with two extracellular loops and a small cytoplasmic tail. When gas3/PMP22 constructs are introduced into cells in culture, the progression from G_0/G_1 to S phase is delayed and proliferation slows.[101,102] The protein appears to affect cell-spreading, and this eventually culminates in the induction of apoptosis in both fibroblasts and Schwann cells.[102-104] Coexpression and inhibitor studies showed that the small GTPase Rho is involved in mediating the effect on cell-spreading, with the result that the GTP status of the cell also influences the signal generated by gas3/PMP22.[105] However, beyond this, little is known about the endogenous functions associated with this family of proteins, and their precise mechanism of action remains to be elucidated.

The expression of gas3/PMP22 can be detected in a variety of different tissues during fetal development.[106,107] However, the highest mRNA levels are detected in the differentiated myelin-forming Schwann cells of the adult peripheral nervous system.[98] Because it accounts for 2 to 5% of the total myelin proteins, mutations of PMP22 are responsible for a number of inherited peripheral neuropathies in mice and humans, including Charcot–Marie tooth disease.[97] The mutations causing the disease behave as dominant negative factors and reduce the apoptotic response when coexpressed with the wild-type gas3/PMP22.[103]

Secreted Proteins

The amino acid sequence of the gas6 gene product codes for a 75-kDa secreted protein with a significant homology to protein S, a negative regulator of the blood coagulation cascade. Both gas6 and protein S share a region near the amino terminus that is rich in gamma-carboxyglutamic acid (known as Gla domains).[108] These domains promote the calcium-dependent binding of the protein to membrane phospholipids (Figure 6.2, panel A). However, critically, the gas6 protein lacks the domain recognized by thrombin, which regulates the coagulation activity of protein S and is therefore not believed to be a part of the coagulation system.

Despite its initial association with growth arrest, the product of the gas6 gene can, under some circumstances, act as a growth factor. The protein is mitogenic when it is over-expressed in serum-starved fibroblasts, inducing the cells to

gas6

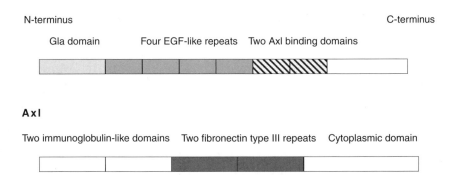

Figure 6.2 Structural motifs of the gas6 and Axl proteins. Sequence information has been used to identify the key domains in the gas6 (panel A) and Axl protein (panel B).

resume division and enter into the S phase.[109] This was shown to be the result of gas6 acting as a ligand for protein tyrosine kinases of the Axl family.[110,111] It was found that gas6 binds, with different affinities, to: Axl (also named Ark, Ufo, or Tyro7), Rse (also named Sky, Brt, Tif, Dtk, or Tyro3), and Mer (Eyk, Nyk, or Tyro12). When quiescent cells are stimulated with growth factors, including gas6, signaling pathways mediated by the enzyme phosphatidylinositol 3-OH kinase (PI3K) are activated.[112] These PI3K-dependent signaling pathways affect downstream targets (such as S6K and Akt) and are important regulators of mitogenic signaling, cytoskeletal interactions, and apoptosis.[116] Addition of recombinant gas6 to the culture medium of serum-starved fibroblasts protects them from cell death by preventing the cleavage of gas2 and blocking the caspase-mediated induction of apoptosis.[109] This antiapoptotic signaling requires functional PI3K.[112]

The action of gas6 is dependent on the cell phenotype; A172 and Wi38 cells are not affected by concentrations of gas6 that are mitogenic in fibroblasts and Schwann cells.[111,113] The extracellular domain of Axl contains both immunoglobulin and fibronectin III repeats, which are similar to those found in cell adhesion molecules (Figure 6.2, panel B). Treating 32D cells that overexpress Axl with recombinant gas6 causes the cells to aggregate. This activity depends on the extracellular domain of the Axl receptor and does not require intracellular signaling. Analysis of the protein structure showed that this binding activity is mediated through the Gla and EGF domains of the gas6 molecule. These are separate from the domain that is responsible for binding and activation of Axl.[114] Therefore, gas6 and Axl appear to have at least two separate functions, one in cell–cell adhesion the other in cell–cell signaling via the Axl family of receptors.[115]

As with other growth-arrest genes, gas6 is expressed extensively in developing tissues. However, it is not essential for development, and transgenic mice with a targeted deletion develop normally, suggesting that there may be other members of this family waiting to be identified. The platelets of adult transgenic mice resemble those of humans with defects in the signaling systems.[117] There are also reports that gas6 is involved in osteoclastic bone reabsorption[118] and in supporting the growth of hematopoietic stem cells.[119] It is therefore likely that gas6 plays an important role in cell–cell signaling during normal development, from early events such as blastocyst formation through to the subsequent development of the organs.[120]

CONCLUDING REMARKS

Over the years since they were first identified, the members of the gas and gadd gene families have been the subjects of intense research. These genes have revealed much about the multiple regulatory mechanisms controlling cell death, either during normal development or in response to environmental and genotoxic stress. Because of their location downstream of the cellular sensors, the products of these genes play a role in downregulating all of the key functions of the proliferating cell (shown schematically in Figure 6.3). Some genes such as gadd153/CHOP and gadd45 are important in the nucleus of the cell; gas5 and gadd34 are important in regulating the ribosome; gas1, gas3, and gas6 affect cell–cell interactions, and gas2, the cytoskeleton.

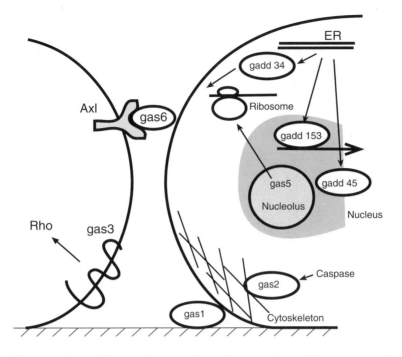

Figure 6.3 The multiple roles of gas and gadd genes in the cell. The gas and gadd genes are involved in all of the cellular functions associated with proliferating cells. The cartoon shows cells attached to a substratum and the different points of interaction between the growth arrest genes and the nucleus, cytoplasm, and cell surface.

Because this group of genes sits at key regulatory points where pathways controlling development and stress responses meet, they also integrate interactions between nutritional status and development. There is increasing evidence that inadequate fetal development represents mechanisms that account for the fetal origins of adult disease. In this role, the gas and gadd genes may be particularly important in maintaining normal development, while at the same time protecting the fetus by responding to nutritional deficiencies and eliminating damaged cells. Despite this apparently central role, most mutants with single-gene deletions develop normally and are quite viable. The loss of one system appears to be compensated for by the action of another. The presence of multiple independent mechanisms that lead to the apoptosis of damaged or inappropriately placed cells reflects the central importance of this protective role.

In addition to the nutritional considerations, the gas and gadd gene families are also frequently encountered in studies of cell transformation. With recent developments in high-throughput methods, there has been great interest in mRNAs overexpressed as part the stress response, and the genes of the gas and gadd families are frequently identified in genome screens. These studies show that the gas and gadd genes can be viewed from different perspectives: that of the nutritionist, who aims to minimize disturbance and achieve optimal cell function or, alternatively, the view of the oncologist, who aims to understand genotoxic damage and disrupt the growth of neoplastic tissues. In both cases, the gas and gadd genes represent important gatekeeper genes, regulating an active response in a wide range of tissues responding to an equally wide range of developmental, nutritional, and environmental stimuli. The gene profiles represent valuable diagnostic tools that can be used to understand the response of cells to these external stresses.

ACKNOWLEDGMENTS

The author's research has been supported by the Scottish Executive Environment and Rural Affairs Department as part of the core funds of the Rowett Research Institute and by the European Union Fifth Framework programme NUTRIX (QLK1-2000-00083)

References

1. X Li, X Cao. BMP signaling and HOX transcription factors in limb development. *Front Biosci* 8: s805–s812, 2003

2. SM Hedrick, DI Cohen, EA Nielsen, MM Davis. Isolation of cDNA clones encoding T cell-specific membrane-associated proteins. *Nature* 308: 149–153, 1984.

2a. T Maniatis, EF Fritsch, J Sambrook. *Molecular Cloning: A Laboratory Manual.* Cold Spring Harbor Lab., Cold Spring Harbor, NY, 1982.

3. C Schneider, RM King, L Philipson. Genes specifically expressed at growth arrest of mammalian cells. *Cell* 54: 787–793, 1988.

4. AJ Fornace, Jr., I Alamo, Jr., MC Hollander. DNA damage-inducible transcripts in mammalian cells. *Proc Natl Acad Sci USA* 85: 8800–8804, 1988.

5. AJ Fornace, Jr., DW Nebert, MC Hollander, JD Luethy, M Papathanasiou, J Fargnoli, NJ Holbrook. Mammalian genes coordinately regulated by growth arrest signals and DNA-damaging agents. *Mol Cell Biol* 9: 4196–4203. 1989.

6. LS Cox, DP Lane. Tumour suppressors, kinases and clamps: how p53 regulates the cell cycle in response to DNA damage. *BioEssays* 17: 501–508, 1995.

7. S Fields, SK Jang. Presence of a potent transcription activating sequence in the p53 protein. *Science* 249: 1046–1049, 1990.

8. Y Cho, S Gorina, PD Jeffrey, NP Pavletich. Crystal structure of a p53 tumor suppressor-DNA complex: understanding tumorigenic mutations. *Science* 265: 346–355, 1994.

9. HW Sturzbecher, R Brain, C Addison, K Rudge, M Remm, M Grimaldi, E Keenan, JR Jenkins. A C-terminal alpha-helix plus basic region motif is the major structural determinant of p53 tetramerization. *Oncogene* 7: 1513–1523, 1992.

10. G Selivanova, V Iotsova, E Kiseleva, M Strom, G Bakalkin, RC Grafstrom, KG Wiman. The single-stranded DNA end binding site of p53 coincides with the C-terminal regulatory region. *Nucleic Acids Res* 24: 3560–3567, 1996.

11. SA Amundson, TG Myers, AJ Fornace Jr. Roles for p53 in growth arrest and apoptosis: putting on the brakes after genotoxic stress. *Oncogene* 17: 3287–3299, 1998.

12. M Stebel, P Vatta, ME Ruaro, G Del Sal, RG Parton, C Schneider. The growth suppressing gas1 product is a GPI-linked protein. *FEBS Lett* 481: 152–158, 2000.

13. G Del Sal, ME Ruaro, L Philipson, C Schneider. The growth arrest-specific gene, gas1, is involved in growth suppression. *Cell* 70: 595–607, 1992.

14. A Evdokiou, PA Cowled. Growth-regulatory activity of the growth arrest-specific gene, GAS1, in NIH3T3 fibroblasts. *Exp Cell Res* 240: 359–367, 1998.

15. G Del Sal, EM Ruaro, R Utrera, CN Cole, AJ Levine, C Schneider. Gas1-induced growth suppression requires a trans-activation-independent p53 function. *Mol Cell Biol* 15: 7152–7160,1995.

16. EM Ruaro, L Collavin, G Del Sal, R Haffner, M Oren, AJ Levine, C Schneider. A proline-rich motif in p53 is required for transactivation-independent growth arrest as induced by Gas1. *Proc Natl Acad Sci USA* 94: 4675–4680, 1997.

17. TC Lee, L Li, L Philipson, EB Ziff. Myc represses transcription of the growth arrest gene gas1. *Proc Natl Acad Sci USA* 94: 12886–12891, 1997.

18. M Grossi, SA La Rocca, G Pierluigi, S Vannucchi, EM Ruaro, C Schneider, F Tato. Role of Gas1 down-regulation in mitogenic stimulation of quiescent NIH3T3 cells by v-Src. *Oncogene* 17: 1629–1638, 1998.

19. CS Lee, CM Fan. Embryonic expression patterns of the mouse and chick Gas1 genes. *Mech Dev* 101: 293–297, 2001.

20. WD Rees, SM Hay, NC Fontanier-Razzaq, C Antipatis, DN Harries. Expression of the growth arrest genes (GAS and GADD) changes during organogenesis in the rat fetus. *J Nutr* 129: 1532–1536, 1999.

21. KK Lee, AK Leung, MK Tang, DQ Cai, C Schneider, C Brancolini, PH Chow. Functions of the growth arrest specific 1 gene in the development of the mouse embryo. *Dev Biol* 234: 188–203, 2001.

22. CS Lee, L Buttitta, CM Fan. Evidence that the WNT-inducible growth arrest-specific gene 1 encodes an antagonist of sonic hedgehog signaling in the somite. *Proc Natl Acad Sci USA* 98: 11347–11352, 2001.

23. Y Liu, NR May, CM Fan. Growth arrest specific gene 1 is a positive growth regulator for the cerebellum. *Dev Biol* 236: 30–45, 2001.

24. JL Mullor, A Altaba. Growth, hedgehog and the price of GAS. *BioEssays* 24: 22–26, 2002.

25. ME Ruaro, M Stebel, P Vatta, S Marzinotto, C Schneider. Analysis of the domain requirement in Gas1 growth-suppressing activity. *FEBS Lett* 481: 159–163, 2000.

26. M Stebel, P Vatta, ME Ruaro, G Del Sal, RG Parton, C Schneider. The growth suppressing gas1 product is a GPI-linked protein. *FEBS Lett* 481: 152–158, 2000.

27. DA Liebermann, B Hoffman. Myeloid differentiation (MyD) primary response genes in hematopoiesis. *Blood Cells Mol Dis* 31: 213–228, 2003.

28. W Zhang, I Bae, K Krishnaraju, N Azam, W Fan, K Smith, B Hoffman, DA Liebermann. CR6: A third member in the MyD118 and Gadd45 gene family which functions in negative growth control. *Oncogene* 18: 4899–4907, 1999.

29. MB Kastan, Q Zhan, WS el Deiry, F Carrier, T Jacks, WV Walsh, BS Plunkett, B Vogelstein, AJ Fornace Jr. A mammalian cell cycle checkpoint pathway utilizing p53 and GADD45 is defective in ataxia-telangiectasia. *Cell* 71: 587–597, 1992.

30. MC Hollander, I Alamo, J Jackman, MG Wang, OW McBride, AJ Fornace Jr. Analysis of the mammalian gadd45 gene and its response to DNA damage. *J Biol Chem* 268: 24385–24393, 1993.

31. S Jin, F Fan, W Fan, H Zhao, T Tong, P Blanck, I Alomo, B Rajasekaran, Q Zhan. Transcription factors Oct-1 and NF-YA regulate the p53-independent induction of the GADD45 following DNA damage. *Oncogene* 20: 2683–2690, 2001.

32. F Chen, Z Zhang, SS Leonard, X Shi. Contrasting roles of NF-kappaB and JNK in arsenite-induced p53-independent expression of GADD45alpha. *Oncogene* 20: 3585–3589, 2001.

33. AG Balliet, MC Hollander, AJ Fornace Jr., B Hoffman, DA Liebermann. Comparative analysis of the genetic structure and chromosomal mapping of the murine Gadd45g/CR6 gene. *DNA Cell Biol* 22: 457–468, 2003.

34. S Takahashi, S Saito, N Ohtani, T Sakai. Involvement of the Oct-1 regulatory element of the gadd45 promoter in the p53-independent response to ultraviolet irradiation. *Cancer Res* 61: 1187–1195, 2001.

35. W Fan, S Jin, T Tong, H Zhao, F Fan, MJ Antinore, B Rajas-ekaran, M Wu, Q Zhan. BRCA1 Regulates GADD45 through Its Interactions with the OCT-1 and CAAT Motifs. *J Biol Chem* 277: 8061–8067, 2002.

36. ML Smith, IT Chen, Q Zhan, I Bae, CY Chen, TM Gilmer, MB Kastan, PM O'Connor, AJ Fornace Jr. Interaction of the p53-regulated protein Gadd45 with proliferating cell nuclear antigen. *Science* 266: 1376–1380, 1994.

37. J Hildesheim, AJ Fornace Jr. Gadd45a: an elusive yet attractive candidate gene in pancreatic cancer. *Clin Cancer Res* 8: 2475–2479, 2002.

38. O Kovalsky, FD Lung, PP Roller, AJ Fornace Jr. Oligomerization of human Gadd45a protein. *J Biol Chem* 276: 39330–39339, 2001.

39. Q Zhan, MJ Antinore, XW Wang, F Carrier, ML Smith, CC Harris, AJ Fornace Jr. Association with Cdc2 and inhibition of Cdc2/Cyclin B1 kinase activity by the p53-regulated protein Gadd45. *Oncogene* 18: 2892–2900, 1999.

40. M Vairapandi, AG Balliet, AJ Fornace Jr., B Hoffman, DA Liebermann. The differentiation primary response gene MyD118, related to GADD45, encodes for a nuclear protein which interacts with PCNA and p21WAF1/CIP1. *Oncogene* 12: 2579–2594,1996.

41. F Carrier, ML Smith, I Bae, KE Kilpatrick, TJ Lansing, CY Chen, M Engelstein, SH Friend, WD Henner, TM Gilmer. Characterization of human Gadd45, a p53-regulated protein. *J Biol Chem* 269: 32672–32677, 1994.

42. M Takekawa, H Saito. A family of stress-inducible GADD45-like proteins mediate activation of the stress-responsive MTK1/MEKK4 MAPKKK. *Cell* 95: 521–530, 1998.

43. F Carrier, PT Georgel, P Pourquier, M Blake, HU Kontny, MJ Antinore, M Gariboldi, TG Myers, JN Weinstein, Y Pommier, AJ Fornace Jr. Gadd45, a p53-responsive stress protein, modifies DNA accessibility on damaged chromatin. *Mol Cell Biol* 19: 1673–1685, 1999.

44. MC Hollander, MS Sheikh, DV Bulavin, K Lundgren, L Augeri-Henmueller, R Shehee, TA Molinaro, KE Kim, E Tolosa, JD Ashwell, MP Rosenberg, Q Zhan, PM Fernandez-Salguero, WF Morgan, CX Deng, AJ Fornace Jr. Genomic instability in Gadd45a-deficient mice. *Nat Genet* 23: 176–184, 1999.

45. MC Hollander, O Kovalsky, JM Salvador, KE Kim, AD Patterson, DC Haines, AJ Fornace Jr. Dimethylbenzanthracene carcinogenesis in Gadd45a-null mice is associated with decreased DNA repair and increased mutation frequency. *Cancer Res* 61: 2487–2491, 2001.

46. MC Hollander, MS Sheikh, DV Bulavin, K Lundgren, L Augeri-Henmueller, R Shehee, TA Molinaro, KE Kim, E Tolosa, JD Ashwell, MP Rosenberg, Q Zhan, PM Fernandez-Salguero, WF Morgan, CX Deng, AJ Fornace Jr. Genomic instability in Gadd45a-deficient mice. *Nat Genet* 23: 176–184, 1999.

47. D Ron, JF Habener. CHOP, a novel developmentally regulated nuclear protein that dimerizes with transcription factors C/EBP and LAP and functions as a dominant-negative inhibitor of gene transcription. *Genes Dev* 6: 439–453, 1992.

48. A Bruhat, C Jousse, XZ Wang, D Ron, M Ferrara, P Fafournoux. Amino acid limitation induces expression of CHOP, a CCAAT/enhancer binding protein-related gene, at both transcriptional and post-transcriptional levels. *J Biol Chem* 272: 17588–17593, 1997.

49. J Averous, A Bruhat, C Jousse, V Carraro, G Thiel, P Fafournoux. Induction of CHOP expression by amino acid limitation requires both ATF4 expression and ATF2 phosphorylation. *J Biol Chem* 279: 5288–5297, 2004.

50. HP Harding, Y Zhang, D Ron. Protein translation and folding are coupled by an endoplasmic-reticulum-resident kinase. *Nature* 397: 271–274, 1999.

51. HP Harding, Y Zhang, A Bertolotti, H Zeng, D Ron. Perk is essential for translational regulation and cell survival during the unfolded protein response. *Mol Cell* 5: 897–904, 2000.

52. K Haze, H Yoshida, H Yanagi, T Yura, K Mori. Mammalian transcription factor ATF6 is synthesized as a transmembrane protein and activated by proteolysis in response to endoplasmic reticulum stress. *Mol Biol Cell* 10: 3787–3799, 1999.

53. A Bertolotti, Y Zhang, LM Hendershot, HP Harding, D Ron. Dynamic interaction of BiP and ER stress transducers in the unfolded-protein response. *Nat Cell Biol* 2: 326–332, 2000.

54. J Shen, X Chen, L Hendershot, R Prywes. ER stress regulation of ATF6 localization by dissociation of BiP/GRP78 binding and unmasking of Golgi localization signals. *Dev Cell* 3: 99–111, 2002.

55. HP Harding, I Novoa, Y Zhang, H Zeng, R Wek, M Schapira, D Ron. Regulated translation initiation controls stress-induced gene expression in mammalian cells. *Mol Cell* 6: 1099–1108, 2000.

56. AG Hinnebusch. Translational regulation of yeast GCN4. A window on factors that control initiator-tRNA binding to the ribosome. *J Biol Chem* 272: 21661–21664, 1997.

57. J Ye, RB Rawson, R Komuro, X Chen, UP Dave, R Prywes, MS Brown, JL Goldstein. ER stress induces cleavage of membrane-bound ATF6 by the same proteases that process SREBPs. *Mol Cell* 6: 1355–1364, 2000.

58. H Yoshida, T Okada, K Haze, H Yanagi, T Yura, M Negishi, K Mori. ATF6 activated by proteolysis binds in the presence of NF-Y (CBF) directly to the cis-acting element responsible for the mammalian unfolded protein response. *Mol Cell Biol* 20: 6755–6767, 2000.

59. J Sok, XZ Wang, N Batchvarova, M Kuroda, H Harding, D Ron. CHOP-dependent stress-inducible expression of a novel form of carbonic anhydrase VI. *Mol Cell Biol* 19: 495–504, 1999.

60. XZ Wang, M Kuroda, J Sok, N Batchvarova, R Kimmel, P Chung, H Zinszner, D Ron. Identification of novel stress-induced genes downstream of chop. *EMBO J* 17: 3619–3630, 1998.

61. M Ubeda, XZ Wang, H Zinszner, I Wu, JF Habener, D Ron. Stress-induced binding of the transcriptional factor CHOP to a novel DNA control element. *Mol Cell Biol* 16: 1479–1489, 1996.

62. XZ Wang, D Ron. Stress-induced phosphorylation and activation of the transcription factor CHOP (GADD153) by p38 MAP Kinase. *Science* 272: 1347–1349, 1996.

63. M Ubeda, JF Habener. CHOP transcription factor phosphorylation by casein kinase 2 inhibits transcriptional activation. *J Biol Chem* 278: 40514–40520, 2003.

64. H Zinszner, M Kuroda, X Wang, N Batchvarova, RT Lightfoot, H Remotti, JL Stevens, D Ron. CHOP is implicated in programmed cell death in response to impaired function of the endoplasmic reticulum. *Genes Dev* 12: 982–995, 1998.

65. C Jousse, S Oyadomari, I Novoa, P Lu, Y Zhang, HP Harding, D Ron. Inhibition of a constitutive translation initiation factor 2alpha phosphatase, CReP, promotes survival of stressed cells. *J Cell Biol* 163: 767–775, 2003.

66. S Oyadomari, E Araki, M Mori. Endoplasmic reticulum stress-mediated apoptosis in pancreatic beta-cells. *Apoptosis* 7: 335–345, 2002.

67. Q Zhan, I Bae, MB Kastan, AJ Fornace Jr. The p53-dependent gamma-ray response of GADD45. *Cancer Res* 54: 2755–2760, 1994.

68. J Jackman, I Alamo Jr., AJ Fornace Jr. Genotoxic stress confers preferential and coordinate messenger RNA stability on the five gadd genes. *Cancer Res* 54: 5656–5662, 1994.

69. WJ Hung, RS Roberson, J Taft, DY Wu. Human BAG-1 proteins bind to the cellular stress response protein GADD34 and interfere with GADD34 functions. *Mol Cell Biol* 23: 3477–3486, 2003.

70. JH Connor, DC Weiser, S Li, JM Hallenbeck, S Shenolikar. Growth arrest and DNA damage-inducible protein GADD34 assembles a novel signaling complex containing protein phosphatase 1 and inhibitor 1. *Mol Cell Biol* 21: 6841–6850, 2001.

71. VM Pain. Initiation of protein synthesis in eukaryotic cells. *Eur J Biochem* 236: 747–771, 1996.

72. I Novoa, H Zeng, HP Harding, D Ron. Feedback inhibition of the unfolded protein response by GADD34-mediated dephosphorylation of eIF2alpha. *J Cell Biol* 153: 1011–1022, 2001.

73. I Novoa, Y Zhang, H Zeng, R Jungreis, HP Harding, D Ron. Stress-induced gene expression requires programmed recovery from translational repression. *EMBO J* 22: 1180–1187, 2003.

74. E Kojima, A Takeuchi, M Haneda, A Yagi, T Hasegawa, K Yamaki, K Takeda, S Akira, K Shimokata, K Isobe. The function of GADD34 is a recovery from a shutoff of protein synthesis induced by ER stress: elucidation by GADD34-deficient mice. *FASEB J* 17: 1573–1575, 2003.

75. N Fontanier-Razzaq, DN Harries, SM Hay, WD Rees. Amino acid deficiency up-regulates specific mRNAs in murine embryonic cells. *J Nutr* 132: 2137–2142, 2002.

76. EM Coccia, C Cicala, A Charlesworth, C Ciccarelli, GB Rossi, L Philipson, V Sorrentino. Regulation and expression of a growth arrest-specific gene (gas5) during growth, differentiation, and development. *Mol Cell Biol* 12: 3514–3521, 1992.

77. AJ Muller, S Chatterjee, A Teresky, AJ Levine. The gas5 gene is disrupted by a frameshift mutation within its longest open reading frame in several inbred mouse strains and maps to murine chromosome 1. *Mamm Genome* 9: 773–774, 1998.

78. CM Smith, JA Steitz. Classification of gas5 as a multi-small-nucleolar-RNA (snoRNA) host gene and a member of the 5′-terminal oligopyrimidine gene family reveals common features of snoRNA host genes. *Mol Cell Biol* 18: 6897–6909, 1998.

79. KT Tycowski, MD Shu, JA Steitz. A mammalian gene with introns instead of exons generating stable RNA products. *Nature* 379: 464–466, 1996.

80. ML Bortolin, T Kiss. Human U19 intron-encoded snoRNA is processed from a long primary transcript that possesses little potential for protein coding. *RNA* 4: 445–454, 1998.

81. P Pelczar, W Filipowicz. The host gene for intronic U17 small nucleolar RNAs in mammals has no protein-coding potential and is a member of the 5′-terminal oligopyrimidine gene family. *Mol Cell Biol* 18: 4509–4518, 1998.

82. C Ciccarelli, L Philipson, V Sorrentino. Regulation of expression of growth arrest-specific genes in mouse fibroblasts. *Mol Cell Biol* 10: 1525–1529, 1990.

83. O Camacho-Vanegas, F Weighardt, C Ghigna, F Amaldi, S Riva, G Biamonti. Growth-dependent and growth-independent translation of messengers for heterogeneous nuclear ribonucleoproteins. *Nucl Acids Res* 25: 3950–3954, 1997.

84. JV Fleming, SM Hay, DN Harries, WD Rees. Effects of nutrient deprivation and differentiation on the expression of growth-arrest genes (gas and gadd) in F9 embryonal carcinoma cells. *Biochem J* 330: 573–579, 1998.

85. C Brancolini, S Bottega, C Schneider. Gas2, a growth arrest-specific protein, is a component of the microfilament network system. *J Cell Biol* 117: 1251–1261, 1992.

86. D Sun, CL Leung, RK Liem. Characterization of the microtubule binding domain of microtubule actin crosslinking factor (MACF): identification of a novel group of microtubule associated proteins. *J Cell Sci* 114: 161–172, 2001.

87. A Aderem. Signal transduction and the actin cytoskeleton: the roles of MARCKS and profilin. *Trends Biochem Sci* 17: 438–443, 1992.

88. C Brancolini, C Schneider. Phosphorylation of the growth arrest-specific protein Gas2 is coupled to actin rearrangements during Go—>G1 transition in NIH 3T3 cells. *J Cell Biol* 124: 743–756, 1994.

89. GS Salvesen. Caspases and apoptosis. *Essays Biochem* 38: 9–19, 2002.

90. C Brancolini, M Benedetti, C Schneider. Microfilament reorganization during apoptosis: the role of Gas2, a possible substrate for ICE-like proteases. *EMBO J* 14: 5179–5190, 1995.

91. A Sgorbissa, R Benetti, S Marzinotto, C Schneider, C Brancolini. Caspase-3 and caspase-7 but not caspase-6 cleave Gas2 *in vitro*: implications for microfilament reorganization during apoptosis. *J Cell Sci* 112: 4475–4482, 1999.

92. R Benetti, G Del Sal, M Monte, G Paroni, C Brancolini, C Schneider. The death substrate Gas2 binds m-calpain and increases susceptibility to p53-dependent apoptosis. *EMBO J* 20: 2702–2714, 2001.

93. KK Lee, MK Tang, DT Yew, PH Chow, SP Yee, C Schneider, C Brancolini. Gas2 is a multifunctional gene involved in the regulation of apoptosis and chondrogenesis in the developing mouse limb. *Dev Biol* 207: 14–25, 1999.

94. JV Fleming, SM Hay, DN Harries, WD Rees. Effects of nutrient deprivation and differentiation on the expression of growth-arrest genes (gas and gadd) in F9 embryonal carcinoma cells. *Biochem J* 330: 573–579, 1998.

95. S Manzow, C Brancolini, F Marks, KH Richter. Expression of growth arrest-specific (Gas) genes in murine keratinocytes: Gas2 is specifically regulated. *Exp Cell Res* 224: 200–203, 1996.

96. PA Cowled, C Ciccarelli, E Coccia, L Philipson, V Sorrentino. Expression of growth arrest-specific (gas) genes in senescent murine cells. *Exp Cell Res* 211: 197–202, 1994.

97. PI Patel, BB Roa, AA Welcher, R Schoener-Scott, BJ Trask, L Pentao, GJ Snipes, CA Garcia, U Francke, EM Shooter. The gene for the peripheral myelin protein PMP-22 is a candidate for Charcot–Marie tooth disease type 1A. *Nat Genet* 1: 159–165, 1992.

98. GJ Snipes, U Suter, AA Welcher, EM Shooter. Characterization of a novel peripheral nervous system myelin protein (PMP-22/SR13). *J Cell Biol* 117: 225–238, 1992.

99. CX Wang, M Wadehra, BC Fisk, L Goodglick, J Braun. Epithelial membrane protein 2, a 4-transmembrane protein that suppresses B-cell lymphoma tumorigenicity. *Blood* 97: 3890–3895, 2001.

100. E Agostoni, S Gobessi, C Brancolini, C Schneider. Identification and characterization of a new member of the gas3/PMP22 gene family in C. elegans. *Gene* 234: 267–274, 1999.

101. G Zoidl, S Blass-Kampmann, D D'Urso, C Schmalenbach, HW Muller. Retroviral-mediated gene transfer of the peripheral myelin protein PMP22 in Schwann cells: modulation of cell growth. *EMBO J* 14: 1122–1128, 1995.

102. G Zoidl, D D'Urso, S Blass-Kampmann, C Schmalenbach, R Kuhn, HW Muller. Influence of elevated expression of rat wild-type PMP22 and its mutant PMP22Trembler on cell growth of NIH3T3 fibroblasts. *Cell Tissue Res* 287: 459–470, 1997.

103. E Fabbretti, P Edomi, C Brancolini, C Schneider. Apoptotic phenotype induced by overexpression of wild-type gas3/PMP22: its relation to the demyelinating peripheral neuropathy CMT1A. *Genes Dev* 9: 1846–1856, 1995.

104. C Baudet, E Perret, B Delpech, M Kaghad, P Brachet, D Wion, D Caput. Differentially expressed genes in C6.9 glioma cells during vitamin D-induced cell death program. *Cell Death Differ* 5: 116–125, 1998.

105. C Brancolini, S Marzinotto, P Edomi, E Agostoni, C Fiorentini, HW Muller, C Schneider. Rho-dependent regulation of cell spreading by the tetraspan membrane protein Gas3/PMP22. *Mol Biol Cell* 10: 2441–2459, 1999.

106. D Baechner, T Liehr, H Hameister, H Altenberger, H Grehl, U Suter, B Rautenstrauss. Widespread expression of the peripheral myelin protein-22 gene (PMP22) in neural and non-neural tissues during murine development. *J Neurosci Res* 42: 733–741, 1995.

107. WD Rees, SM Hay, NC Fontanier-Razzaq, C Antipatis, DN Harries. Expression of the growth arrest genes (GAS and GADD) changes during organogenesis in the rat fetus. *J Nutr* 129: 1532–1536, 1999.

108. G Manfioletti, C Brancolini, G Avanzi, C Schneider. The protein encoded by a growth arrest-specific gene (gas6) is a new member of the vitamin K-dependent proteins related to protein S, a negative coregulator in the blood coagulation cascade. *Mol Cell Biol* 13: 4976–4985, 1993.

109. S Goruppi, E Ruaro, C Schneider. Gas6, the ligand of Axl tyrosine kinase receptor, has mitogenic and survival activities for serum starved NIH3T3 fibroblasts. *Oncogene* 12: 471–480, 1996.

110. TN Stitt, G Conn, M Gore, C Lai, J Bruno, C Radziejewski, K Mattsson, J Fisher, DR Gies, PF Jones. The anticoagulation factor protein S and its relative, Gas6, are ligands for the Tyro 3/Axl family of receptor tyrosine kinases. *Cell* 80: 661–670, 1995.

111. BC Varnum, C Young, G Elliott, A Garcia, TD Bartley, YW Fridell, RW Hunt, G Trail, C Clogston, RJ Toso. Axl receptor tyrosine kinase stimulated by the vitamin K-dependent protein encoded by growth-arrest-specific gene 6. *Nature* 373: 623–626, 1995.

112. S Goruppi, E Ruaro, BC Varnum, C Schneider. Requirement of phosphatidylinositol 3-kinase-dependent pathway and Src for Gas6-Axl mitogenic and survival activities in NIH 3T3 fibroblasts. *Mol Cell Biol* 17: 4442–4453, 1997.

113. R Li, J Chen, G Hammonds, H Phillips, M Armanini, P Wood, R Bunge, PJ Godowski, MX Sliwkowski, JP Mather. Identification of Gas6 as a growth factor for human Schwann cells. *J Neurosci* 16: 2012–2019, 1996.

114. P McCloskey, YW Fridell, E Attar, J Villa, Y Jin, BC Varnum, ET Liu. GAS6 mediates adhesion of cells expressing the receptor tyrosine kinase Axl. *J Biol Chem* 272: 23285–23291, 1997.

115. GC Avanzi, M Gallicchio, G Cavalloni, L Gammaitoni, F Leone, A Rosina, R Boldorini, G Monga, L Pegoraro, B Varnum, M Aglietta. GAS6, the ligand of Axl and Rse receptors, is expressed in hematopoietic tissue but lacks mitogenic activity. *Exp Hematol* 25: 1219–1226, 1997.

116. BM Burgering, PJ Coffer. Protein kinase B (c-Akt) in phosphatidylinositol-3-OH kinase signal transduction. *Nature* 376: 599–602, 1995.

117. A Angelillo-Scherrer, P de Frutos, C Aparicio, E Melis, P Savi, F Lupu, J Arnout, M Dewerchin, M Hoylaerts, J Herbert, D Collen, B Dahlback, P Carmeliet. Deficiency or inhibition of Gas6 causes platelet dysfunction and protects mice against thrombosis. *Nat Med* 7: 215–221, 2001.

118. M Katagiri, Y Hakeda, D Chikazu, T Ogasawara, T Takato, M Kumegawa, K Nakamura, H Kawaguchi. Mechanism of stimulation of osteoclastic bone resorption through Gas6/Tyro 3, a receptor tyrosine kinase signaling, in mouse osteoclasts. *J Biol Chem* 276: 7376–7382, 2001.

119. SP Dormady, XM Zhang, RS Basch. Hematopoietic progenitor cells grow on 3T3 fibroblast monolayers that overexpress growth arrest-specific gene-6 (GAS6). *Proc Natl Acad Sci USA* 97: 12260–12265, 2000.

120. JV Fleming, N Fontanier, DN Harries, WD Rees. The growth arrest genes gas5, gas6, and CHOP-10 (gadd153) are expressed in the mouse preimplantation embryo. *Mol Reprod Dev* 48: 310–316, 1997.

7

Mechanisms for Dietary Regulation of Nitric Oxide Synthesis in Mammals

NICK E. FLYNN, CYNTHIA J. MEININGER,
TONY E. HAYNES, WENJUAN SHI, AND GUOYAO WU

INTRODUCTION

Nitric oxide (NO) is synthesized from L-arginine and O_2 by NO synthase (NOS) in the presence of NADPH, Ca^{2+}, tetrahydrobiopterin (BH_4), FAD, FMN, and calmodulin in virtually all animal cells (Figure 7.1). There are three isoforms of the NOS: neuronal NOS (nNOS, type I), inducible NOS (iNOS, type II), and endothelial NOS (eNOS, type III). Isoforms nNOS and eNOS are collectively termed constitutive NOS (cNOS). NO is a vasodilator, a mediator of immune response, a neurotransmitter, a cytotoxic free radical (at high levels), and a signaling molecule. Studies over the past decade have demonstrated that dietary factors play an important role in regulating NO synthesis by both cNOS and iNOS in a cell- and tissue-specific manner.[1] These factors include protein, amino acids, glucose, fructose, cholesterol, fatty acids, vitamins, minerals, phytoestrogens, ethanol, polyphenols, and other phytochemicals (Tables 7.1 and 7.2). There is growing recognition that changes in arginine availability, intracellular

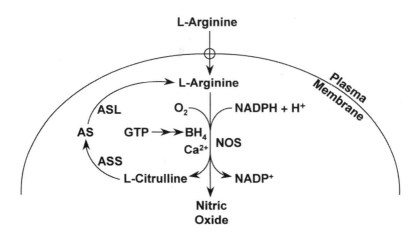

Figure 7.1   Nitric oxide synthesis from L-arginine in animal cells. AS, argininosuccinate; ASL, argininosuccinate lyase; ASS, argininosuccinate synthase; BH_4, tetrahydrobiopterin; NOS, nitric oxide synthase. Calmodulin, FAD, and FMN are also required for enzymatic activity of NOS.

levels of Ca^{2+}, BH_4, NADPH, or NOS activity can modulate the rate of NO production in animal cells.[1] Emerging evidence also shows that dietary factors can influence NOS transcription via nuclear factor-kappa B (NF-κB).[2] The major objective of this article is to discuss cellular and molecular mechanisms whereby dietary factors regulate NO synthesis by cNOS and iNOS in mammalian cells.

ARGININE TRANSPORT AND AVAILABILITY

Arginine transport represents the first step in the regulation of NO synthesis by cells. In a majority of cell types, the most important mechanism for arginine uptake is the system y^+, a high-affinity, Na^+-independent transporter of basic amino acids, including arginine, lysine, histidine, and ornithine.[3] However, other transporters, such as $b^{0,+}$, $B^{0,+}$, and y^{+L}, also transport arginine in a cell-specific manner. There is a positive correlation between plasma and intracellular concentrations of arginine in cells, including endothelial cells, macrophages, and mammary tissue.[4,5] Under normal conditions, arginine transport is the major source of intracellular arginine for NO production.[6] In support of this view, a defect in arginine transport impairs endothelial NO synthesis and results in hypertension in humans.[7] In addition, increasing the extracellular concentration of lysine, which shares the same transport system with arginine, reduces NO synthesis in arteries,[8] vascular smooth muscle cells,[9] and activated macrophages.[10] Further, oxidized low-density lipoprotein (LDL) inhibits arginine transport and NO synthesis in endothelial cells.[11] Interestingly, animal proteins, which contain more lysine than plant proteins, have a greater atherogenic potential in experimental animals.[12] Conversely, nuts, which are rich in arginine, are beneficial for vascular function.[13] This effect of nuts may be mediated by an increased availability of arginine for endothelial NO production.

Another mechanism whereby intracellular arginine availability can be regulated is the provision of its precursors. Metabolically, citrulline (a neutral amino acid) is an effective substrate for arginine synthesis in virtually all cell types.[3] Increasing the extracellular concentration of citrulline increases NO synthesis in activated macrophages, endothelial cells, and neurons.[3,14] Similarly, dietary supplementation of citrulline lowers blood pressure in hypertensive, salt-sensitive rats.[15] Citrulline is synthesized from glutamine or glutamate and proline almost exclusively in enterocytes of the small intestine.[3] Thus, the supply of glutamine or glutamate and proline or branched-chain amino acids (substrates for glutamine synthesis) from the diet may impact the availability of arginine in the whole body. Because citrulline and arginine do not compete for the same amino acid transport systems and the production of arginine from citrulline actually consumes a stoichiometrically equivalent amount of ammonia via aspartate,[3] citrulline may be a preferred dietary treatment for patients with defects in arginine transport.[1]

Cellular arginine availability can also be regulated by arginase, which hydrolyzes arginine into ornithine plus urea. There are two distinct isozymes of mammalian arginase (type

Table 7.1 Effects of Dietary Factors on NO Synthesis by Constitutive NOS

Dietary factor	Effector(s)	Effect	Cell Types
ω-3 PUFA	eNOS, Ca^{2+} & NADPH	↑	EC & VSMC
ω-6 PUFA	eNOS, BH_4 & NADPH	↑	EC & VSMC
ω-9 PUFA	eNOS	→	EC & VSMC
Arginine	eNOS & BH_4	↑	EC, whole body, heart, VSMC
Ca^{2+}	eNOS	↑	EC
Carotenoids	Possibly BH_4 availability	↑	EC
Cholesterol	eNOS & BH_4	→	EC
Citrulline	Arginine availability	↑	EC, neurons, VSMC
Copper	nNOS	→	Glial cells
Ethanol (chronic)	eNOS & BH_4	→	EC
Ethanol (high)	nNOS & eNOS	→	EC and brain
Ethanol (moderate)	eNOS	↑	EC
Fiber (fermentable)	cNOS	→	Colon
Flavonoids	eNOS, BH_4 & NADPH	↑	Vascular tissue
Folic acid	Possibly BH_4 availability	↑	EC
Fructose	eNOS & BH_4	→	EC
Glucosamine	NADPH	→	EC
Glucose (high level)	eNOS, nNOS, BH_4 & glucosamine	→	Cultured EC, VSMC, mesangial cells
Glucose (high level)	nNOS	↑	Cultured neuronal cells
Glutamate	nNOS	↑	Brain
Glutamine	NADPH	→	EC

HDL	eNOS & BH_4	↑	EC
High fat diet	eNOS	↑	Kidney
High fat diet	eNOS & nNOS	→	EC, neuronal tissue
High glucose diet	eNOS	←	EC
High salt diet	eNOS & nNOS	→	EC, kidney
Homocysteine	eNOS	→	EC, platelets, kidney
Iron	nNOS & eNOS	←	Gallbladder, intestine
LDL (oxidized)	eNOS & BH_4	→	EC
Lysine	Arginine transport	→	EC, VSMC
Magnesium	eNOS	←	EC
Manganese	nNOS &NADPH	←	Astrocytes, brain
Melatonin	nNOS	→	Brain
Phytoestrogens	eNOS & NADPH	←	Aorta, vasculature
Polyphenols	eNOS	←	EC
Protein deficiency	iNOS & eNOS	→	Whole body, kidney, brain, muscle
Sphingolipids	eNOS	←	EC
Taurine	eNOS	←	EC
Vitamin A	eNOS & nNOS	←	EC, neuronal cells
Vitamin C	BH_4 stability	←	EC
Vitamin E	eNOS & BH_4	↑	EC

Note: ADMA, asymmetric dimethylarginine; BH_4, tetrahydrobiopterin; EC, endothelial cells; PUFA, polyunsaturated fatty acids; VSMC, vascular smooth muscle cells; ↑, increase; ↓, decrease.

Table 7.2 Effects of Dietary Factors on NO Synthesis by Inducible NOS

Dietary factor	Effector	Effect	Cell Types
ω-3 PUFA	iNOS transcription	→	Macrophages
Arachidonic acid	iNOS transcription	→	Macrophages
Arginine	iNOS translation	←	Astrocytes
Cadmium	iNOS expression	→	Macrophages
Carotenoids	iNOS expression	→	Macrophages
Ceramide	iNOS expression	←	C6 glioma cells, macrophages
Chromium	iNOS expression	→	Macrophages
CLA	iNOS transcription	→	Tumor cells
Cobalt	iNOS expression	←	Macrophages
Copper	iNOS expression	→	Macrophages
Copper	iNOS expression	←	Lung, liver and aorta
Ethanol (high level)	iNOS activity	→	Macrophages
Fructose	iNOS activity	→	Macrophages
Fructose	iNOS activity	←	Kidney
Ginseng	iNOS expression	←	Macrophages
Glucosamine	iNOS expression	→	Macrophages, spleen, lung, liver
Glucose (high level)	iNOS expression	←	Heart, retinal cells
Glucose (high level)	iNOS expression/BH4	→	Macrophages, mesangial cells
Glutamate	iNOS expression	←	Brain
Glutamine	iNOS expression	←	Macrophages
Glutamine	iNOS expression	→	Gut
Glycine	iNOS expression	→	Liver
High fat diet	iNOS activity	←	Islets, liver, colon

Dietary factor	Effect		Cell/tissue types
High salt diet	iNOS expression	→	Heart, aorta, and kidney
Iron	iNOS expression	←	Macrophages, proximal tubules
Iron	iNOS expression	→	Thoracic aorta, kidney
LDL (oxidized)	iNOS activity	←	Macrophages
Lead	iNOS expression	→	Macrophages
Linoleic acid	iNOS expression	←	C6 glioma cells, macrophages
Lysine	Arginine transport	→	Many cell types
Niacin	iNOS expression	→	Lung
Nickel	iNOS expression	←	Macrophages
Phytanic acid	iNOS expression	←	VSMC
Phytoestrogens	iNOS expression	→	Animal cells, tumors
Polyphenols	iNOS expression	→	Macrophages, chondrocytes, colon, leukemic cancers
Polyphenols	iNOS expression	←	Arteries
Protein deficiency	iNOS	→	Whole body, lung, kidney, brain, macrophages
Taurine	iNOS	→	Many cell types
Vitamin A	iNOS expression	←	Macrophages
Vitamin A	iNOS transcription	→	VSMC, EC, myocytes
Vitamin B6	iNOS expression	→	Colon
Vitamin D3	iNOS expression	←	Macrophages
Vitamin D3	iNOS expression	→	Neuronal tissue
Vitamin K2	iNOS expression	→	VSMC
Zinc	iNOS expression	→	Aorta, VSMC, macrophages, gut

Note: CLA, conjugated linoleic acid; EC, endothelial cells; VSMC, vascular smooth muscle cells; ↑, increase; ↓, decrease.

I and type II), which are encoded by separate genes and differ in molecular and immunological properties, tissue distribution, subcellular localization, and regulation of expression.[14] Type I arginase (a cytosolic enzyme) is highly expressed in the liver as a component of the urea cycle, and to a limited extent in a few other tissues. In contrast, type II arginase (a mitochondrial enzyme) is expressed at much lower levels in extrahepatic tissues, including the kidney, brain, small intestine, endothelial cells, mammary gland, and macrophages. Because arginase and NOS use arginine as a common substrate, changes in arginase activity regulate NO synthesis in many cell types (including endothelial cells, neurons, and vascular smooth muscle cells) and in parasites.[16] Functionally, inhibition of arginase II activity improves NO-dependent relaxation of penile smooth muscle and enhances blood flow to sex organs in males and females.[16]

Some dietary factors may play an important role in regulating arginase activity and thus cellular arginine availability. For example, manganese and valine are a cofactor for and an inhibitor of arginase, respectively. Thus, a deficiency of manganese or an increase in valine concentrations may increase arginine availability for NO production in certain cell types. Chitosan, a polymeric derivative of glucosamine, stimulated arginase activity in both resting and activated macrophages.[17] This result suggests that some of the anti-inflammatory effects of chitosan may be mediated through an arginase-mediated reduction in arginine availability for NO synthesis. Interestingly, garlic also increases hepatic arginase activity in rats,[18] but it remains to be determined if this effect is also exhibited in extrahepatic cells.

CALCIUM AVAILABILITY

Changes in intracellular concentrations of Ca^{2+} play an important role in regulating NO synthesis by eNOS and nNOS[3] in animal cells. For example, eicosapentaenoic acid enhances NO generation by endothelial cells and improves endothelium-dependent relaxation through mechanisms including an increase in intracellular Ca^{2+} availability.[1] In some cell types, such as pancreatic β cells, glucose and leucine increase Ca^{2+} levels, which may contribute to an increase in NO production. In addition, vitamin D regulates Ca^{2+} uptake by intestinal cells. Moreover, vitamin A is known to increase the influx of extracellular Ca^{2+} into retinal cells. In iNOS, Ca^{2+} is tightly bound to calmodulin, and thus exogenous Ca^{2+} is not required for its enzymatic activity.[5] However, changes in intracellular Ca^{2+} may indirectly regulate NO synthesis by iNOS in animal cells through alterations in multiple metabolic pathways (including glycolysis, fatty acid oxidation, protein turnover, and amino acid catabolism).

TETRAHYDROBIOPTERIN AVAILABILITY

BH_4 has long been recognized as a cofactor for all isoforms of NO synthase.[3] This biopterin helps stabilize the dimerization of every isoform of NOS and also plays a redox role in its catalytic reaction of NO synthesis. A deficiency of BH_4 results in an uncoupling of NOS for NO generation and subsequent production of superoxide from oxygen.[3] This has important implications for cellular oxidative stress, a major factor responsible for endothelial dysfunction. For example, the aberrant effect of cholesterol or oxidized LDL on NOS activity is partially attributed to decreased BH_4 levels in aorta.[19]

In mammalian cells, BH_4 is synthesized via both *de novo* and salvage pathways. GTP cyclohydrolase I (GTP-CH) is the first and rate-controlling enzyme in *de novo* synthesis of BH_4. Since the initial report that an increase in intracellular BH_4 availability stimulated NO production by endothelial cells, there has been growing interest in the role of BH_4 in cardiovascular function and the therapeutic treatment of endothelial dysfunction.[19] Importantly, administration of BH_4 improves endothelial function in experimental animals and humans with a variety of major cardiovascular risk factors (e.g., hypercholesterolemia, hyperhomocysteinemia, diabetes, smoking, and insulin resistance) or with common cardiovascular disorders (e.g., coronary artery disease and ischemia-reperfusion injury).[20–23] Thus, changes in BH_4 availability can modulate the rate of NO synthesis in cells.

Several dietary factors have been shown to enhance NO synthesis by increasing intracellular concentrations of BH_4. For example, long-term vitamin C treatment increases vascular BH_4 levels and NOS activity in mice, through the stabilization of BH_4.[20] In addition, we recently discovered that increasing extracellular concentration of arginine enhanced BH_4 levels in cultured endothelial cells through an increase in GTP-CH protein expression.[24] Importantly, dietary supplementation of arginine prevented a deficiency of endothelial BH_4 in streptozotocin-induced diabetic rats and augmented endothelial NO production.[25]

Some dietary factors are known to reduce BH_4 availability for NO synthesis in animal cells. For example, the addition of aluminum to drinking water decreased BH_4 levels in the brain, which may explain some of the neurotoxic effects of aluminum on the nervous system.[26] Also, hyperglycemia leads to a BH_4 deficiency and a decrease in NO production in mesangial cells.[27] Most recently, NO-dependent endothelial dysfunction brought about by chronic alcohol consumption was found to result from reduced BH_4 availability.[28] Interestingly, the sphingolipid ceramide decreases NO synthesis in endothelial cells through a BH_4 deficiency owing to oxidative stress.[29] Of note, diets high in fructose reduce hepatic BH_4 levels through the modification of GTP-CH activity.[30] There is also evidence suggesting a deficiency of BH_4 in endothelial cells of fructose-fed rats.[1]

NADPH AVAILABILITY

NADPH is an important cellular reductant providing electrons necessary for NO synthesis by all forms of NOS. Therefore, the regulation of cellular NADPH levels plays an important role in modulating NO production. Glucose is an important dietary component and its metabolism through the pentose phosphate pathway is a major source of NADPH for NO synthesis.[31] Glucose-6-phosphate dehydrogenase (GPD) catalyzes a rate-controlling reaction in the pentose phosphate pathway, and a deficiency of GPD results in impaired NO synthesis and oxidative stress in vascular tissue.[32] Therefore, glucose deprivation decreases NO production in endothelial cells[31] and immunostimulated astrocytes[33] through a depletion of intracellular NADPH. This may help to explain reduced blood flows in fasting subjects and may also have implications for neonates born of mothers with gestational diabetes.

Many dietary factors regulate NADPH levels. For example, high carbohydrate, low fat diets increase NADPH production.[34] In addition, several key enzymes of the pentose phosphate pathway require magnesium for their activity, and therefore this mineral is crucial for NADPH synthesis and NO production by cells. In contrast, glucosamine, a derivative of both glucose and glutamine, inhibits endothelial NO generation by inhibiting GPD activity and thus decreasing cellular NADPH availability.[31] Of note, dietary saturated fat and polyunsaturated fatty acids differentially modulate GPD activity and NADPH concentration in liver and adipose,[35] which may provide a metabolic basis for effects on NO synthesis (Table 7.1). Other mechanisms whereby dietary factors regulate NADPH levels are through the activation or inhibition of NADPH-dependent oxidases. For example, equol, a soy phytoestrogen metabolite, inhibits NADPH-dependent oxidase activity, thereby making more NADPH available for NO synthesis.[36] Additionally, flavonoids have been shown to modify the activity of NADPH-dependent oxidases,[37] thereby increasing the availability of cellular NADPH for NO synthesis.

EXPRESSION AND ACTIVITY OF CNOS

Constitutive NOS is widely expressed in animal cells. Its expression or activity can be modulated by a wide array of dietary factors. These factors include protein, amino acids, glucose, fructose, cholesterol, fatty acids, vitamins, minerals, phytoestrogens, ethanol, and polyphenols.[1]

As building blocks for proteins, amino acids are required for cNOS expression in all cell types. However, little is known about the effects of altering cellular levels of α-amino acids on the expression and activity of eNOS or nNOS in cells. Interestingly, several studies have shown a regulatory role

for the metabolites of some amino acids in cNOS expression
and activity. For example, taurine (a metabolite of methion-
ine/cysteine) restores eNOS expression in endothelial cells of
smokers,[38] which is consistent with its beneficial role in cardiovascular health. In contrast, homocysteine (a metabolite
of methionine)[39] inhibits eNOS expression and activity in
endothelial cells.[1,39] Similarly, melatonin (a derivative of tryptophan) decreases nNOS translation in the brain of older rats,
which may contribute to its beneficial effect on the reversal
of brain aging.[40]

High-glucose diets increase eNOS expression in endothelial cells.[41] Likewise, high-sucrose diets enhance eNOS
expression in the heart.[42] This *in vivo* action of high carbohydrate diets may be partially mediated by insulin. However,
hyperglycemia is associated with an increase in nNOS expression in diabetic renal tubules[43] and medullae,[44] when insulin
action is impaired. This finding suggests that glucose modulates nNOS expression via an insulin-independent mechanism. In support of this view, *in vitro* studies demonstrate
that glucose increases nNOS expression and activity in
human pancreatic β cells.[45] Activities of certain metabolic
pathways may be responsible for the detrimental effects of
hyperglycemia on NO metabolism. For example, inhibition of
glutamine:fructose-6-phosphate transaminase, an enzyme
involved in glucosamine synthesis, reverses the diminishing
effects of hyperglycemia on posttranslational modification of
eNOS protein and activity in endothelial cells.[46]

Because of the recognized role of dietary fat in cardiovascular and neurological disorders, there is growing interest
in its role in eNOS and nNOS expression. A high fat diet
decreases nNOS expression while paradoxically increasing
eNOS expression in kidneys,[47] suggesting differential mechanisms in the dietary regulation of these two NOS isoforms.
Likewise, linoleic acid and n-3 polyunsaturated fatty acids
increase eNOS activity and NO production in endothelial
cells,[48] which may explain enhanced endothelial permeability
in response to these fatty acids. In addition, eicosapentaenoic
acid modulates translocation of eNOS in endothelial cells,
such that the enzyme protein is dissociated with caveolin-1

and its catalytic activity is enhanced.[1] In contrast, glycated, oxidized HDL and LDL decrease eNOS expression in endothelial cells,[49] indicating a deleterious effect of both hyperglycemia and oxidative stress on vascular function. Most recently, the cardioprotective effects of nascent HDL are reinforced by the finding that its binding to scavenger receptors increases eNOS activity in endothelial cells.[50] Additionally, nascent HDL prevents LDL inhibition of eNOS expression, thereby enhancing vasorelaxation in humans.[51]

Vitamins and minerals can also regulate cNOS expression. For example, both vitamin C and vitamin E prevent the decrease in eNOS expression that is normally exhibited in hypercholesterolemic pigs.[52] In addition, vitamin A increases nNOS expression,[53] indicating an important role for this essential nutrient in neurological function. These findings are significant because they are independent of the antioxidant effects of the vitamins. High levels of copper[54] and high salt intake[55] reduce eNOS expression in aorta and endothelial cells. Additionally, high salt intake decreases nNOS expression in brain, renal cortex, and renal medulla.[55] This result may help explain sodium-induced hypertension in animals and humans. In contrast, exposure to elevated levels of manganese and lead increases nNOS expression in the brain,[56] which may explain the neurotoxicity of these two minerals. This effect of manganese is likely independent of its role as a cofactor for arginase.

There is growing interest in the role of phytochemicals in cNOS expression. For example, reservatrol (a polyphenol found in wine)[57] and red wine extracts[58] increase eNOS expression and NO production in endothelial cells. These findings are consistent with the cardioprotective effects of polyphenolic compounds. Although plant carotenoids, including carotenes (e.g., lycopene, α-carotene, and β-carotene), lutein, zeaxanthin, antheraxanthin, violaxanthin, neoxanthin, and β-cryptoxanthin, are natural antioxidants, little is known about their role in eNOS or nNOS expression.

Collectively, there is evidence showing that dietary factors regulate eNOS and nNOS expression at both transcriptional and translational levels. In support of this conclusion, food deprivation leads to decreased expression of nNOS in the gut of pigs.[59] Additionally, reducing dietary intake of energy increases eNOS expression and improves endothelial vasodilation in obese patients with essential hypertension.[60]

EXPRESSION AND ACTIVITY OF INOS

Virtually all cell types express iNOS in response to appropriate stimuli.[3] The molecular regulation of iNOS by dietary factors has important implications for health and disease because this enzyme produces much higher quantities of NO than cNOS. Over the past decade, much attention has been directed towards the role of amino acids, carbohydrates, lipids, vitamins, minerals, polyphenols, and plant components of folk remedies in the regulation of iNOS expression.[1,3,61]

Amino acids are key regulators of iNOS expression in animals. In cytokine-stimulated astrocytes, the translation of iNOS mRNA, but not transcription rate or stability of its mRNA and protein, is enhanced by arginine.[62] Conversely, arginine depletion results in a decrease in the translation rate of iNOS mRNA.[62] Impaired NO production by iNOS may contribute to immunodeficiency in patients with liver transplants and in preterm infants who exhibit hypoargininemia. Glutamine is also required for iNOS expression in activated macrophages, and glutamate mediates an increase in iNOS expression in the ischemic brain.[1] In contrast, glycine inhibits iNOS expression in liver, thereby preventing liver injury in hemorrhagic shock.[63] Likewise, increasing extracellular concentrations of lysine inhibits iNOS expression in keratinocytes,[64] which may be mediated by a reduced availability of intracellular arginine due to competition for the y^+ transporter.

Experimentally induced hyperglycemia, for as short as 2 h, is sufficient to increase cardiac iNOS gene expression.[65] Hyperglycemia also increases iNOS mRNA levels in perfused rat heart and retinal cells.[1] Similarly, fructose induces iNOS expression and increases NO production in the kidney.[66] These findings implicate iNOS expression as an important factor for the pathophysiology of diabetes-induced glaucoma, vision problems, and renal dysfunction. In contrast, cytokine-activated macrophages[67] and mesangial cells[27] exhibit a decrease in iNOS expression in response to hyperglycemia. Thus, different cell types respond differentially to high concentrations of extracellular glucose, which collectively contribute to the dysfunction of multiple organs in diabetic patients.

Because dietary modification of fatty acid content is known to modulate inflammation, it is not surprising that lipids can modulate iNOS expression and activity. A high fat diet increases iNOS activity in animal tissues.[47] Also, increasing the concentration of fatty acids augments iNOS expression in islet cells from normal rats and, to a much larger extent, in islet cells from Zucker obese rats.[68] The *in vivo* effect of high fat diets on iNOS expression may be partially mediated by hormones, but the underlying mechanisms are not known. A recent study suggests that the anticancer effects of conjugated linoleic acid may be due, in part, to inhibition of iNOS transcription.[69] Arachidonic acid and several *n*-3 fatty acids (including docosahexanoic acid, eicosapentanoic acid, and linolenic acid) also inhibit iNOS transcription and NO production in human colon cancer cells.[70] These findings may help explain some of the antiinflammatory effects of polyunsaturated fatty acids. In contrast, the branched-chain fatty acid, phytanic acid, induces iNOS expression in vascular smooth muscle cells[71] while the endogenous fatty acid amide, palmitoylethanolamide, inhibits iNOS expression in response to inflammation.[72]

There is an extensive literature regarding the role of vitamins in iNOS expression. Vitamin A stimulates NO synthesis in stimulated macrophages through an increase of iNOS expression.[73] In other cell types, including endothelial cells, myocytes and mesangial cells,[74–76] vitamin A inhibits iNOS transcription and reduces NO production. Thus, the effects of vitamin A on iNOS expression vary with cell types, which should be taken into consideration in experimental design and data interpretation involving the *in vivo* action of vitamin A. Vitamin B_6 and niacin decrease colonic iNOS expression and NO synthesis in mouse models of colon cancer[77] and fibrosis,[78] respectively. Vitamin K2 also inhibits the expression of iNOS in the vasculature,[79] which may play a role in regulating blood clotting. 1,25-Dihydroxyvitamin D-3, the active form of vitamin D_3, inhibits iNOS expression in experimental allergic encephalomyelitis,[80] but increases iNOS expression in macrophage cell lines.[81] These findings emphasize cell-specific effects of vitamins, like other dietary factors, on iNOS expression.

Minerals play a critical role in iNOS expression. High levels of copper increase iNOS expression in lung, liver, and aorta,[82] thus demonstrating that excessive copper can have detrimental effects on both constitutive and inducible NO synthesis. The combined effects of increased iNOS expression and decreased eNOS activity in the same anatomical location could have profound consequences on inflammatory processes of cells within the cardiovascular bed. Iron either increases iNOS expression in macrophages and proximal tubules[83] or suppresses elevated iNOS protein levels in the heart and kidney.[84] Thus, iron excess or deficiency can impair immunological, cardiovascular, and renal functions. In addition, zinc modulates iNOS expression in the small intestine, thereby preventing cytokine-induced diarrhea.[85] In addition, chromium and lead inhibit, but nickel and cobalt increase, iNOS expression in activated macrophages,[86] suggesting an important role for NO in mediating the cytotoxic effects of environmental contamination by metals.

Most of the published studies demonstrate that polyphenols enhance iNOS expression in aorta but are potent inhibitors of iNOS expression in extracardiac tissues. For example, red wine extracts decrease iNOS mRNA expression in a rat model of colon cancer, suggesting their potential therapeutic role for some forms of cancer.[87] Also, reservatrol and two independent vine shoot extracts reduce iNOS expression in leukemic cells.[88] Quercitin, another polyphenol, inhibits iNOS expression in macrophages.[89] Collectively, these findings suggest an antiinflammatory effect of wine-derived polyphenols on inducible NO production.

Glucosamine[90] and several unidentified plant components of folk remedies[91] regulate iNOS expression in animal cells. Glucosamine inhibits iNOS expression and NO production in activated macrophages in culture.[90] Infusion of glucosamine to lipopolysaccharide-treated rats reduces iNOS expression in many tissues (including liver, kidney, and spleen) and whole body NO synthesis.[90] There is growing interest in dietary supplementation of some phytochemicals for health benefits, although the underlying mechanisms for their action are poorly understood. Indian ginseng has been shown to induce iNOS expression and NO production in macrophages.[91] Thus, Indian ginseng may act as a stimulant of the immune system. Active ingredients of Indian ginseng include steroidal lactones, alkaloids, terpenoids, and a tetracyclic skeleton, but whether these individual components have synergistic effects on iNOS expression is not known. In contrast, phytoestrogen[1] and extracts of *Boerhaavia diffusa* and *Carlowrightia cordifolia*[92] inhibit iNOS expression in macrophages and tumor cells, consistent with their antiinflammatory role. Finally, curcumin, a cancer-preventive polyphenolic phytochemical with antiinflammatory properties, inhibits iNOS expression and NO production in macrophages.[93] The action of these compounds may be mediated by reduced expression of NF-κB.[2,93]

FUTURE PERSPECTIVES

The regulation of NO synthesis by dietary factors is a rich area of research for life scientists. NOS gene expression at the levels of transcription, mRNA stability, translation, and posttranslational modification may be modulated by dietary factors, but little information is available in the literature. Our current knowledge about the transcription factors that modulate iNOS expression is limited to NF-κB expression, but other transcription factors likely play a role in mediating iNOS expression. Except for vitamin C and vitamin E, little is known about the interaction between different dietary factors that affect NO synthesis. This information is critical for dietary recommendation because animal and human diets normally contain a complex mixture of components. Future studies are necessary to define the molecular and cellular mechanisms responsible for dietary regulation of NOS expression and the availability of NOS cofactors including BH_4.[94,95]

Finally, roles of phytochemicals (e.g., extracts of Noni plants[96,97]) or "functional foods" in mammalian NO synthesis deserve intensive investigation as these findings will provide a biochemical basis for explaining their preventive and therapeutic benefits in human health.

ACKNOWLEDGMENTS

Work in our laboratories is supported, in part, by grants from the U.S. Department of Agriculture, the American Heart Association, Juvenile Diabetes Research Foundation, the National Institutes of Health, the Robert A. Welch Foundation (grant #AJ-0029), Texas A&M University and a Faculty Development Grant from Angelo State University. We thank Charlotte Klepac and Frances Mutscher for office support.

References

1. G Wu, CJ Meininger. Regulation of nitric oxide synthesis by dietary factors. *Annu Rev Nutr* 22: 61–86, 2002.

2. M Rodriguez-Porcel, LO Lerman, DR Holmes, D Richardson, C Napoli, A Lerman. Chronic antioxidant supplementation attenuates nuclear factor-κB activation and preserves endothelial function in hypercholesterolemic pigs. *Cardiovasc Res* 53: 1010–1018, 2002.

3. G Wu, SM Morris Jr. Arginine metabolism: nitric oxide and beyond. *Biochem J* 336: 1–17, 1998.

4. NE Flynn, CJ Meininger, TE Haynes, G Wu. The metabolic basis of arginine nutrition and pharmacotherapy. *Biomed Pharmacother* 56: 427–438.

5. G Wu, NE Flynn, SP Flynn, CA Jolly, PK Davis. Dietary protein or arginine deficiency impairs constitutive and inducible nitric oxide synthesis by young rats. *J Nutr* 129: 1347–1354, 1999.

6. G Wu, CJ Meininger. Arginine nutrition and cardiovascular function. *J Nutr* 130: 2626–2629, 2000.

7. Y Kamada, H Nagaretani, S Tamura, T Ohama, T Maruyama, H Hiraoka, S Yamashita, A Yamada, S Kiso, Y Inui, N Ito, Y Kayanoki, S Kawata, Y Matsuzawa. Vascular endothelial dysfunction resulting from L-arginine deficiency in a patient with lysinuric protein intolerance. *J Clin Invest* 108: 717–724, 2001.

8. D Hucks, NM Khan, JPT Ward. Essential role of L-arginine uptake and protein tyrosine kinase activity for NO-dependent vasorelaxation induced by stretch, isometric tension and cyclic AMP in rat pulmonary arteries. *Br J Pharmacol* 131: 1475–1481, 2000.

9. W Durante. Regulation of L-arginine transport and metabolism in vascular smooth muscle cells. *Cell Biochem Biophys* 35: 19–34, 2001.

10. EI Closs, JS Scheld, M Sharafi, U Förstermann. Substrate supply for nitric-oxide synthase in macrophages and endothelial cells: role of cationic amino acid transporters. *Mol Pharmacol* 57: 68–74, 2000.

11. O Hernández-Perera, D Perez-Sala, J Navarro-Antolín, R Sánchez-Pascuala, G Hernández, C Díaz, S Lamas. Effects of the 3-hydroxy-3-methylglutaryl-CoA reductase inhibitors, atorvastatin and simvastatin, on the expression of endothelin-1 and endothelial nitric oxide synthase in vascular endothelial cells. *J Clin Invest* 101: 2711–2719, 1998.

12. LA Cynober. Plasma amino acid levels with a note on membrane transport: characteristics, regulation, and metabolic significance. *Nutrition* 18: 761–766, 2002.

13. EB Feldman. The scientific evidence for a beneficial health relationship between walnuts and coronary heart disease. *J Nutr* 132: 1062S–1101S, 2002.

14. SM Morris Jr. Regulation of arginine availability and its impact on NO synthesis. In: LJ Ignarro, ed. *Nitric Oxide: Biology and Pathobiology.* New York: Academic Press, 2000, pp 187–197.

15. PY Chen, PW Sanders. L-arginine abrogates salt-sensitive hypertension in Dahl/Rapp rats. *J Clin Invest* 88: 1559–1567, 1991.

16. SM Morris Jr. Recent advances in arginine metabolism. *Curr Opin Clin Nutr Metab Care* 7: 45–51, 2004.

17. C Porporatto, ID Bianco, CM Riera, SG Correa. Chitosan induces different L-arginine metabolic pathways in resting and inflammatory macrophages. *Biochem Biophys Res Commun* 304: 266–272, 2003.

18. Y Oi, M Imafuku, C Shishido, Y Kominato, S Nishimura, K Iwai. Garlic supplementation increases testicular testosterone and decreases plasma corticosterone in rats fed a high protein diet. *J Nutr* 131: 2150–2156, 2001.

19. J Vásquez-Vivar, D Duquaine, J Whitsett, B Kalyanaraman, S Rajagopalan. Altered tetrahydrobiopterin metabolism in atherosclerosis: implications for use of oxidized tetrahydrobiopterin analogues and thiol antioxidants. *Arterioscle Thromb Vasc Biol* 22: 1655–1661, 2002.

20. LV d'Uscio, S Milstien, D Richardson, L Smith, ZS Katusic. Long-term vitamin C treatment increases vascular tetrahydrobiopterin levels and nitric oxide synthase activity. *Circ Res* 92: 88–95, 2003.

21. B Dhillon, MV Badiwala, A Maitland, V Rao, SH Li, S Verma. Tetrahydrobiopterin attenuates homocysteine induced endothelial dysfunction. *Mol Cell Biochem* 247: 223–227, 2003.

22. E Stroes, J Kastelein, F Cosentino, W Erkelens, R Wever, H Koomans, T Luscher, T Rabelink. Tetrahydrobiopterin restores endothelial function in hypercholesterolemia. *J Clin Invest* 99: 41–46, 1997.

23. M Ozaki, S Kawashima, T Yamashita, T Hirase, M Namiki, N Inoue, K Hirata, H Yasui, H Sakurai, Y Yoshida, M Masada, M Yokoyama. Overexpression of endothelial nitric oxide synthase accelerates atherosclerotic lesion formation in apoE-deficient mice. *J Clin Invest* 110: 331–340, 2002.

24. G Wu, KA Kelly, K Hatakeyama, CJ Meininger. Regulation of endothelial tetrahydrobiopterin synthesis by L-arginine. In: N Blau, B Thöny, Eds. *Pterins, Folates, and Neurotransmitters in Molecular Medicine*. Heilbronn, Germany: SPS Verlagsgesellschaft mbh, 2004, pp 54–59.

25. R Kohli, CJ Meininger, TE Haynes, W Yan, JT Self, G Wu. Dietary L-arginine supplementation enhances endothelial nitric oxide synthesis in streptozotocin-induced diabetic rats. *J Nutr* 134: 600–608, 2004.

26. RA Amstrong, J Anderson, JD Cowburn, J Cox, JA Blair. Aluminium administered in drinking water but not in the diet influences biopterin metabolism in the rodent. *Biol Chem Hoppe Seyler* 373: 1075–1078, 1992.

27. SS Prabhakar. Tetrahydrobiopterin reverses the inhibition of nitric oxide by high glucose in cultured murine mesangial cells. *Am J Physiol* 281: F179–F188, 2001.

28. H Sun, KP Patel, WG Mayhan. Tetrahydrobiopterin, a cofactor for NOS, improves endothelial dysfunction during chronic alcohol consumption. *Am J Physiol* 281: H1863–H1869, 2001.

29. HG Li, P Junk, A Huwiler, C Burkhardt, T Wallerath, J Pfeilschifter, U Förstermann. Dual effect of ceramide on human endothelial cells: induction of oxidative stress and transcriptional upregulation of endothelial nitric oxide synthase. *Circulation* 106: 2250–2256, 2002.

30. T Guerin, GA Walsh, J Donlon, S Kaufman. Correlation of rat hepatic phenylalanine hydroxylase, with tetrahydrobiopterin and GTP concentrations. *Int J Biochem Cell Biol* 30: 1047–1054, 1998.

31. G Wu, TE Haynes, H Li, W Yan, CJ Meininger. Glutamine metabolism to glucosamine is necessary for glutamine inhibition of endothelial nitric oxide synthesis. *Biochem J* 353: 245–252, 2001.

32. JA Leopold, A Cap, AW Scribner, RC Stanton, J Loscalzo. Glucose-6-phosphate dehydrogenase deficiency promote endothelial oxidant stress and decreases endothelial nitric oxide availability. *FASEB J* 15: 1771–1773, 2001.

33. CY Shin, JW Choi, JR Ryu, KH Ko, JJ Choi, HS Kim, HS Kim, JC Lee, Lee SJ, Kim HC Kim WK. Glucose deprivation decreases nitric oxide production via NADPH depletion in immunostimulated rat primary astrocytes. *Glia* 37: 268–274, 2002.

34. RJ Miksicek, HC Towle. Changes in the rates of synthesis and messenger RNA levels of hepatic glucose-6-phosphate and 6-phosphogluconate dehydrogenases following induction by diet or thyroid hormone. *J Biol Chem* 257: 11829–11835, 1982.

35. SD Clarke, DR Romsos, GA Leveille. Differential effects of dietary methyl esters of long-chain saturated and polyunsaturated fatty acids on rat liver and adipose tissue lipogenesis. *J Nutr* 107: 1170–1181, 1977.

36. J Hwang, J Wang, P Morazzoni, HN Hodis, A Sevanian. The phytoestrogen equol increases nitric oxide availability by inhibiting superoxide production: an antioxidant mechanism for cell-mediated LDL modification. *Free Radical Biol Med* 34: 1271–1282, 2003.

37. B Fuhrman, M Aviram. Flavonoids protect LDL from oxidation and attenuate atherosclerosis. *Curr Opin Lipidol* 12: 41–48, 2001.

38. FM Fennessy, DS Moneley, JH Wang, CJ Kelly, DJ Bouchier-Hayes. Taurine and vitamin C modify monocyte and endothelial dysfunction in young smokers. *Circulation* 107: 410–415, 2003.

39. Z Ungvari, A Csiszar, JG Edwards, PM Kaminski, MS Wolin, G Kaley, A Koller. Increased superoxide production in coronary arteries in hyperhomocysteinemia: role of tumor necrosis factor-alpha, NAD(P)H oxidase, and inducible nitric oxide synthase. *Arterioscler Thromb Vasc Biol* 23: 418–424, 2003.

40. EH Sharman, ND Vaziri, Z Ni, KG Sharman, SC Bondy. Reversal of biochemical and behavioral parameters of brain aging by melatonin and acetyl L-carnitine. *Brain Res* 957: 223–230, 2002.

41. F Cosentino, M Eto, P de Paolis, B van der Loo, M Bachschmid, V Ullrich, A Kouroedov, C Delli-Gatti, H Joch, M Volpe, TF Lüscher. High glucose causes upregulation of cyclooxygenase-2 and alters prostanoid profile in human endothelial cells: role of protein kinase C and reactive oxygen species. *Circulation* 107: 1017–1023, 2003.

42. J Busserolles, W Zimowska, E Rock, Y Rayssiguier, A Mazur. Rats fed a high sucrose diet have altered heart antioxidant enzyme activity and gene expression. *Life Sci* 71: 1303–1312, 2002.

43. A Baines, P Ho. Glucose stimulates O_2 consumption, NOS, and Na/H exchange in diabetic rat proximal tubules. *Am J Physiol* 283: F286–F293, 2002.

44. MC Hsieh, CH Wu, CL Chen, HC Chen, CC Chang, SJ Shin. High blood glucose and osmolality, but not high urinary glucose and osmolality, affect neuronal nitric oxide synthase expression in diabetic rat kidney. *J Lab Clin Med* 141: 200–209, 2003.

45. G Dorff, G Meyer, D Krone, P Pozzilli, H Zühlke. Neuronal NO synthase and its inhibitor PIN are present and influenced by glucose in the human beta-cell line CM and in rat INS-1 cells. *Biol Chem* 383: 1357–1361, 2002.

46. XL Du, D Edelstein, S Dimmeler, Q Ju, C Sui, M Brownlee. Hyperglycemia inhibits endothelial nitric oxide synthase activity by posttranslational modification at the Akt site. *J Clin Invest* 108: 1341–1348, 2001.

47. CK Roberts, ND Vaziri, RK Sindhu, RJ Barnard. A high-fat, refined-carbohydrate diet affects renal NO synthase protein expression and salt sensitivity. *J Appl Physiol* 94: 941–946, 2003.

48. P Meerarani, EJ Smart, M Toborek, GA Boissonneault, B Hennig. Cholesterol attenuates linoleic acid-induced endothelial cell activation. *Metabolism* 52: 493–500, 2003.

49. T Matsunaga, T Nakajima, T Miyazaki, I Koyama, S Hokari, I Inoue, S Kawai, H Shimomura, S Katayama, A Hara, T Komoda. Glycated high-density lipoprotein regulates reactive oxygen species and reactive nitrogen species in endothelial cells. *Metabolism* 52: 42–49, 2003.

50. XA Li, WB Titlow, BA Jackson, N Giltiay, M Nikolova-Kar-akashian, A Uittenbogaard, EJ Smart. High density lipoprotein binding to scavenger receptor, Class B, type I activates endothelial nitric-oxide synthase in a ceramide-dependent manner. *J Biol Chem* 277: 11058–11063, 2002.

51. JT Kuvin, ME Rämet, AR Patel, NG Pandian, ME Mendelsohn, RH Karas. A novel mechanism for the beneficial vascular effects of high-density lipoprotein cholesterol: enhanced vasorelaxation and increased endothelial nitric oxide synthase expression. *Am Heart J* 144: 165–172, 2002.

52. JA Rodríguez, A Grau, E Eguinoa, B Nespereira, M Pérez-Ilzarbe, R Arias, MS Belzunce, JA Páramo, D Martínez-Caro. Dietary supplementation with vitamins C and E prevents downregulation of endothelial NOS expression in hypercholesterolemia in vivo and in vitro. *Atherosclerosis* 165: 33–40, 2002.

53. D Personett, U Fass, K Panickar, M McKinney. Retinoic acid-mediated enhancement of the cholinergic/neuronal nitric oxide synthase phenotype of the medial septal SN56 clone: establishment of a nitric oxide-sensitive proapoptotic state. *J Neurochem* 74: 2412–2424, 2000.

54. A Chiarugi, GM Pitari, R Costa, M Ferrante, L Villari, M Amico-Roxas, T Godfraind, A Bianchi, S Salomone. Effect of prolonged incubation with copper on endothelium-dependent relaxation in rat isolated aorta. *Br J Pharmacol* 136: 1185–1193, 2002.

55. ZM Ni, ND Vaziri. Effect of salt loading on nitric oxide synthase expression in normotensive rats. *Am J Hypertens* 14: 155–163, 2001.

56. CS Chetty, GR Reddy, A Suresh, D Desaiah, SF Ali, WJ Slikker. Effects of manganese on inositol polyphosphate receptors and nitric oxide synthase activity in rat brain. *Int J Toxicol* 20: 275–280, 2001.

57. T Wallerath, G Deckert, T Ternes, H Anderson, H Li, K Witte, U Försterman. Resveratrol, a polyphenolic phytoalexin present in red wine, enhances expression and activity of endothelial nitric oxide synthase. *Circulation* 106: 1652–1658, 2002.

58. JF Leikert, TR Räthel, P Wohlfart, V Cheynier, AM Vollmar, VM Dirsch. Red wine polyphenols enhance endothelial nitric oxide synthase expression and subsequent nitric oxide release from endothelial cells. *Circulation* 106: 1614–1617, 2002.

59. JF Grongnet, JC David. Reciprocal variations of nNOS and HSP90 are associated with fasting in gastrointestinal tract of the piglet. *Dig Dis Sci* 48: 365–372, 2003.

60. S Sasaki, Y Higashi, K Nakagawa, M Kimura, K Noma, S Sasaki, K Hara, H Matsuura, C Goto, T Oshima, K Chayama. A low-calorie diet improves endothelium-dependent vasodilation in obese patients with essential hypertension. *Am J Hypertens* 15: 302–309, 2002.

61. JA Spitzer, MQ Zheng, JK Kolls, CV Stouwe, JJ Spitzer. Ethanol and LPS modulate NF-kappaB activation, inducible NO synthase and COX-2 gene expression in rat liver cells *in vivo*. *Front Biosci* A99–A108, 2002.

62. J Lee, H Ryu, RJ Ferrante, SM Morris Jr, RR Ratan. Translational control of inducible nitric oxide synthase expression by arginine can explain the arginine paradox. *Proc Natl Acad Sci USA* 100: 4843–4848, 2003.

63. JL Mauriz, B Matilla, JM Culebras, P Gonzalez, J Gonzalez-Gallego. Dietary glycine inhibits activation of nuclear factor κB and prevents liver injury in hemorrhagic shock in the rat. *Free Radical Biol Med* 31: 1236–1244, 2001.

64. O Schnorr, CV Suschek, V Kolb-Bachofen. The importance of cationic amino acid transporter expression in human skin. *J Invest Dermatol* 120: 1016–1022, 2003.

65. A Ceriello, L Quagliaro, M D'Amico, C Di Filippo, R Marfella, F Nappo, L Berrino, F Rossi, D Giugliano. Acute hyperglycemia induces nitrotyrosine formation and apoptosis in perfused heart from rat. *Diabetes* 51: 1076–1082, 2002.

66. A Cosenzi, E Bernobich, M Bonavita, F Gris, G Odoni, G Bellini. Role of nitric oxide in the early renal changes induced by high fructose diet in rats. *Kidney Blood Pressure Res* 25: 363–369, 2002.

67. CC Tseng, Y Hattori, K Kasai, N Nakanishi, S Shimoda. Decreased production of nitric oxide by LPS-treated J774 macrophages in high-glucose medium. *Life Sci* 60: PL99–106, 1997.

68. M Shimabukuro, M Ohneda, Y Lee, RH Unger. Role of nitric oxide in obesity-induced beta cell disease. *J Clin Invest* 100: 290–295, 1997.

69. Y Iwakiri, DA Sampson, KG Allen. Suppression of cyclooxygenase-2 and inducible nitric oxide synthase expression by conjugated linoleic acid in murine macrophages. *Prostaglandins Leuk Essential Fatty Acids* 67: 435–443, 2002.

70. BA Narayanan, NK Narayanan, B Simi, BS Reddy. Modulation of inducible nitric oxide synthase and related proinflammatory genes by the omega-3 fatty acid docosahexaenoic acid in human colon cancer cells. *Cancer Res* 63: 972–979, 2003.

71. S Idel, P Ellinghaus, C Wolfrum, JR Nofer, J Gloerich, G Assmann, F Spener, U Seedorf. Branched chain fatty acids induce nitric oxide-dependent apoptosis in vascular smooth muscle cells. *J Biol Chem* 277: 49319–49325, 2002.

72. B Costa, S Conti, G Giagnoni, M Colleoni. Therapeutic effect of the endogenous fatty acid amide, palmitoylethanolamide, in rat acute inflammation: inhibition of nitric oxide and cyclooxygenase systems. *Br J Pharmacol* 137: 413–420, 2002.

73. LM Austenaa, AC Ross. Potentiation of interferon-gamma-stimulated nitric oxide production by retinoic acid in RAW 264.7 cells. *J Leuk Biol* 70: 121–129, 2001.

74. K Hirokawa, KM O'Shaughnessy, P Ramrakha, MR Wilkins. Inhibition of nitric oxide synthesis in vascular smooth muscle by retinoids. *Br J Pharmacol* 113: 1448–1454, 1994.

75. PK Datta, EA Lianos. Retinoic acids inhibit inducible nitric oxide synthase expression in mesangial cells. *Kidney Int* 56: 486–493, 1999.

76. S Grosjean, Y Devaux, C Seguin, C Meistelman, F Zannad, PM Mertes, RA Kelly, D Ungureanu-Longrois. Retinoic acid attenuates inducible nitric oxide synthase (NOS2) activation in cultured rat cardiac myocytes and microvascular endothelial cells. *J Mol Cell Cardiol* 33: 933–945, 2001.

77. S Komatsu, N Yanaka, K Matsubara, N Kato. Antitumor effect of vitamin B6 and its mechanisms. *Biochim Biophys Acta* 1647: 127–130, 2003.

78. G Gurujeyalakshmi, Y Wang, SN Giri. Suppression of bleomycin-induced nitric oxide production in mice by taurine and niacin. *Nitric Oxide* 4: 399–411, 2000.

79. M Sano, H Fujita, I Morita, H Uematsu, S Murota. Vitamin K2 (menatetrenone) induces iNOS in bovine vascular smooth muscle cells: no relationship between nitric oxide production and gamma-carboxylation. *J Nutr Sci Vitaminol* 45: 711–723, 1999.

80. E Garcion, S Nataf, A Berod, F Darcy, P Brachet. 1,25-Dihydroxyvitamin D3 inhibits the expression of inducible nitric oxide synthase in rat central nervous system during experimental allergic encephalomyelitis. *Mol Brain Res* 45: 255–267, 1997.

81. KA Rockett, R Brookes, I Udalova, V Vidal, AV Hill, D Kwiatkowski. 1,25-Dihydroxyvitamin D3 induces nitric oxide synthase and suppresses growth of Mycobacterium tuberculosis in a human macrophage-like cell line. *Infect Immunity* 66: 5314–5321, 1998.

82. S Cuzzocrea, T Persichini, L Dugo, M Colasanti, G Musci. Copper induces type II nitric oxide synthase *in vivo*. *Free Radical Biol Med* 34: 1253–1262, 2003.

83. L Chen, Y Wang, LK Kairaitis, BH Zhang, DC Harris. Molecular mechanisms by which iron induces nitric oxide synthesis in cultured proximal tubule cells. *Exp Nephrol* 9: 198–204, 2001.

84. Z Ni, S Morcos, ND Vaziri. Up-regulation of renal and vascular nitric oxide synthase in iron-deficiency anemia. *Kidney Int* 52: 195–201, 1997.

85. L Cui, Y Takagi, M Wasa, Y Iiboshi, J Khan, R Nezu, A Okada. Induction of nitric oxide synthase in rat intestine by interleukin-1alpha may explain diarrhea associated with zinc deficiency. *J Nutr* 127: 1729–1736, 1997.

86. L Tian, DA Lawrence. Metal-induced modulation of nitric oxide production *in vitro* by murine macrophages: lead, nickel, and cobalt utilize different mechanisms. *Toxicol Appl Pharmacol* 141: 540–547, 1996.

87. C Luceri, G Caderni, A Sanna, P Dolara. Red wine and black tea polyphenols modulate the expression of cycloxygenase-2, inducible nitric oxide synthase and glutathione-related enzymes in azoxymethane-induced f344 rat colon tumors. *J Nutr* 132: 1376–1379, 2002.

88. C Billard, JC Izard, V Roman, C Kern, C Mathiot, F Mentz, JP Kolb. Comparative antiproliferative and apoptotic effects of resveratrol, epsilon-viniferin and vine-shots derived polyphenols (vineatrols) on chronic B lymphocytic leukemia cells and normal human lymphocytes. *Leukemia Lymphoma* 43: 1991–2002, 2002.

89. TL Wadsworth, DR Koop. Effects of the wine polyphenolics quercetin and resveratrol on pro-inflammatory cytokine expression in RAW 264.7 macrophages. *Biochem Pharmacol* 57: 941–949, 1999.

90. CJ Meininger, KA Kelly, H Li, TE Haynes, G Wu. Glucosamine inhibits inducible nitric oxide synthesis. *Biochem Biophys Res Commun* 279: 234–239, 2000.

91. T Iuvone, G Esposito, F Capasso, AA Izzo. Induction of nitric oxide synthase expression by Withania somnifera in macrophages. *Life Sci* 72: 1617–1625, 2003.

92. DE Cruz-Vega, A Aguilar, J Vargas-Villarreala, MJ Verde-Star, MT González-Garza. Leaf extracts of Carlowrightia cordifolia induce macrophage nitric oxide production. *Life Sci* 70: 1279–1284, 2002.

93. MM Chan, HI Huang, MR Fenton, D Fong. In vivo inhibition of nitric oxide synthase gene expression by curcumin, a cancer preventive natural product with antiinflammatory properties. *Biochem Pharmacol* 55: 1955–1962, 1998.

94. Shi W, CJ Meininger, TE Haynes, K Hatakeyama, G Wu. Regulation of tetrahydrobiopterin synthesis and bioavailability in endothelial cells. *Cell Biochem Biophys* 41: 415–433, 2004.

95. Saraswathi V, G Wu, M Toborek, B Hennig. Linoleic acid-induced endothelial activation: role of calcium and peroxynitrite signaling. *J Lipid Res* 45: 794–804, 2004.

96. Wang MY, BJ West, CJ Jensen, D Nowicki, C Su, AK Palu, G Anderson. Morinda citrifolia (Noni): a literature review and recent advances in Noni research. *Acta Pharmacol Sin* 23: 1127–1141, 2002.

97. Ettarh RR, P Emeka. Morinda lucida extract induces endothelium-dependent and independent relaxation of rat aorta. *Fitoterapia* 75: 332–336, 2004.

8

The Roles of Thioredoxin Reductases in Cell Signaling

DAN SU AND VADIM N. GLADYSHEV

INTRODUCTION

Reactive oxygen species (ROS) are generated in all organisms living in aerobic environments. ROS consist of a variety of reactive oxygen compounds, including superoxide anion radical, hydrogen peroxide, hydroxyl radical, and other partially reduced oxygen species.[1] ROS could be produced intracellularly, as an unwanted consequence of aerobic metabolism, or as part of designated pathways of ROS production, but could also be generated by abiotic systems that interact with organisms and cells. Independently of the source, ROS could cause serious damage to DNA, proteins, and lipids, thus posing a constant threat to living organisms. Oxidative stress is implicated in many diseases including atherosclerosis, diabetes, pulmonary fibrosis, and arthritis.[2] Besides the damaging and pathological effects, recent studies show that ROS can also play important roles in numerous physiological functions including transcription, cell signaling, replication, and the cell cycle. The concentration and location of cellular ROS are important factors that determine which role the ROS have.[3]

Numerous defense molecules and systems have evolved to counteract the damaging effects of ROS. Such defense molecules can be small molecules, such as ascorbic acid and glutathione, or proteins, such as thioredoxin (Trx), superoxide dismutase (SOD), catalase (CAT), and glutathione peroxidase (GPx). Some of these biomolecules function by themselves, but most act in a coordinated manner. It was found that cellular defense systems not only protect against the damaging effects of oxidative stress, but also regulate cellular functions of ROS.

There are two key redox regulation systems: the glutathione system and the thioredoxin system. These systems occur in most organisms, and the thioredoxin system has been identified in all species for which completely sequenced genomes are available.[4] Thioredoxin and glutathione systems are composed of distinct enzymes and exhibit different regulatory functions, but both employ NADPH as an ultimate electron donor. Both systems contain an NADPH-dependent enzyme belonging to the pyridine nucleotide disulfide-oxidoreductase family and a small redox-reactive peptide. In the case of the glutathione system, the enzyme is glutathione reductase, which reduces a small tripeptide glutathione, which in turn controls the cellular redox status. Other components of this system are glutaredoxin (Grx), a small protein that can specifically remove glutathione from mixed disulfides formed between glutathione and other molecules, and GPx, which reduces peroxides in a glutathione-dependent manner. In the case of the Trx system, the pyridine nucleotide disulfide oxidoreductase is thioredoxin reductase (TR), which reduces Trx, which in turn transfers its reducing equivalents to numerous substrate proteins such as peroxiredoxin (Prx, TPx), also called thioredoxin peroxidase, and ribonucleotide reductase. The reactive peptide that is involved in the transfer of reducing equivalents from TR to Trx is actually a C-terminal tetrapeptide of thioredoxin reductase (see section titled "Thioredoxin Reductase").

The two redox systems are generally thought to function independently, but some recent findings indicate that they may interact at multiple levels. First, a novel TR containing a Grx domain was found in humans, mice, rats,[5] *Schistosoma mansoni,*[6] *Echinococcus granulosus,*[7] and *Taenia crassiceps.*[8] This TR exhibits Grx activity due to presence of the Grx domain. Second, Trx was recently found to undergo glutathionylation (formation of protein–glutathione mixed disulfides) and a consequent loss of activity.[9] Third, *Drosophila melanogaster* was found to possess a glutathione system, but it was the Trx-TR system that was responsible for recycling oxidized glutathione to its reduced form in this organism.[4] Furthermore, it was recently found that Grx might be a direct substrate of TR. Johansson et al.[10] found that human mitochondrial Grx (Grx2) could be reduced by both TR1 and TR3, although the physiological importance of TR1 having this activity is not clear.

In this chapter, the discussion is focused on the functions and properties of mammalian thioredoxin reductases and their roles in cell signaling.

THIOREDOXIN

Trx and TR constitute the key part of the thioredoxin redox system. Trx is a 12-kDa ubiquitous redox protein containing a conserved WCGPC active site. It functions as disulfide-oxidoreductase by transferring reducing equivalents to numerous proteins and enzymes, such as ribonucleotide reductase,[11] thioredoxin peroxidase (peroxiredoxin),[12,13] and transcription factors,[14–17] whose functions are dependent on the redox status of this protein. Trx has been shown to exhibit additional multiple roles in cells, such as being involved in general regulation of protein redox status[18,19] and cell growth.[20,21]

Multiple Trx isoforms have been found in different organisms, including two such proteins in *E. coli*, three in yeast, and two in humans.[22] Often, Trx isoforms have similar enzyme characteristics and functional roles, but are located in different cellular compartments (e.g., cytosol and mitochondria). However, some thioredoxins with functionally overlapping functions colocalize. In addition, *E. coli* Trx2 is different from other known thioredoxins in that it has two extra CxxC motifs (two cysteines separated by two other amino acids), besides the one that is conserved in the thioredoxin family and responsible for overall Trx function.[23,24] It has been proposed that the additional CxxC motifs are involved in coordination of zinc.[23] Regulation of Trx2 expression and function in *E.coli* is also different from that of the better-studied Trx1.[24]

All thioredoxins share a similar structural fold, exemplified by *E. coli* Trx1,[25] and many other thiol-dependent redox proteins (e.g., glutaredoxin, nucleoredoxin, and protein disulfide isomerase) also adopt this fold,[26] which was designated as the thioredoxin fold. Although thioredoxins from different organisms often exhibit a high degree of homology, there are some important differences between thioredoxins from higher and lower organisms. For example, human Trx1 has three cysteines (C62, C69, and C73) that are located downstream of the catalytic CxxC motif (C32 and C35), whereas *E. coli* and yeast Trx1 do not have these cysteines. Recent findings suggest that the three additional cysteines can be modified and these modifications affect thioredoxin activity and function.[9,27,28] This is consistent with the idea that thioredoxins in higher organisms may be under more complex regulation and may have additional functions compared with those in lower organisms.

Modifications of the three cysteines in Trx are shown in Figure 8.1. It was found that C62 and C69 could form a second disulfide bond (in addition to the active site disulfide) and that the formation of this disulfide inhibits Trx activity.[27] C73 could be glutathionylated, which abolishes activity of the protein, but Trx can also deglutathionylate itself.[9] In addition, C73 is able to form an intermolecular disulfide bond with C73 of another Trx molecule, thus forming a homodimer, which is also inactive.[27] Furthermore, C69 was shown to be S-nitrosylated in endothelial cells, and this modification can boost the Trx activity by about 50%. The S-nitrosylation of Trx at C69 also appeared to stimulate ROS reduction by Trx and was implicated in inhibition of apoptosis.[28] Considering the interplay between these modifications, there appears to be a complex code system for regulation of Trx activity and function. More studies are needed for detailed characterization of this regulation.

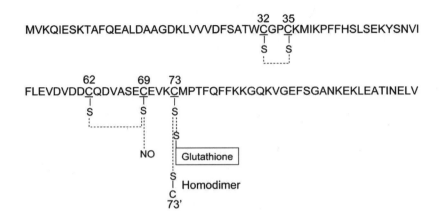

Figure 8.1   The thioredoxin code. Modifications of cysteine residues in mammalian thioredoxins that regulate protein function are shown. Two cysteines that are located in the N-terminal portion of the protein (C32 and C35) constitute the active site of the protein and are conserved within the thioredoxin family. Upon oxidation, they form a disulfide bond as shown in the figure. Three cysteine residues located in the C-terminal region of mammalian thioredoxins (C62, C69, and C73) are regulatory. Their modifications that regulate protein function are illustrated in the figure. Interplay of cysteine modifications provides the basis for the thioredoxin code.

THIOREDOXIN REDUCTASE

TR is the only known enzyme that can reduce oxidized thioredoxins *in vivo*, indicating an important role for this enzyme in all aspects of Trx function. As with Trx, TRs are present in all living organisms. As discussed above, TR is an NADPH-dependent FAD-containing enzyme and a member of the pyridine nucleotide disulfide oxidoreductase family.[29] There are two known TR types that evolved by convergent evolution[30,31]: a homodimer of 35-kDa subunits present in prokaryotes, yeast, and plants,[32] and a homodimer of 55- to 65-kDa subunits that occurs in animals and some lower eukaryotes.[5,33,34] Mammalian TRs are selenoproteins with a C-terminal penultimate selenocysteine (Sec) in its carboxyl-terminal active center.[35,36] TR plays a central role in reduction of disulfides and other oxidized forms of cysteine residues in redox-regulated proteins through its substrate Trx.

Large and small TRs generally do not occur in the same organisms, with the exception of *Chlamydomonas reinhardtii*, which has both enzyme types.[31] Whereas both large and small TRs catalyze Trx reduction, their mechanisms are different.[37] Large TRs evolved from glutathione reductases (GR) by a carboxyl-terminal extension that serves as a substrate for the N-terminal thiol/disulfide active site.[31,34] The catalysis by TRs has been extensively studied.[34,38,39] Consistent with the difference between small and large TRs, the catalytic mechanism of mammalian TRs is similar to that of GR and different from that of *E. coli* TR.

Many organisms, particularly animals, contain several TR genes. To date, three mammalian TRs have been cloned: cytosolic TR (also called TR1 or TrxRα), mitochondrial TR (also called TR3, TrxR, or TrxR2)[40–43] and thioredoxin/glutathione reductase (also called TGR or TR2).[5] TR1 has been studied since 1973,[44] but it was not known to be a selenium-containing protein until 1996.[35,36] TGR and TR3 were discovered several years ago. These enzymes share a high level of homology with TR1 and are also selenium-containing proteins. Although TGR and TR3 show enzyme properties similar to those of TR1, *in vivo* functions of these two enzymes have not been studied in detail. TR3 is a mitochondrial protein, suggesting that it is involved in redox regulation in mitochondria by reducing a mitochondrial Trx. TGR is the only TR that exhibits a tissue-specific expression pattern: it is highly expressed in testis and present at low levels in several other tissues (unpublished data). The tissue-specific expression pattern of TGR suggested that this enzyme might have additional roles, perhaps related to testis function. As discussed earlier, TGR has an N-terminal domain homologous to Grx, which is a component of the glutathione system (Figure 8.2).

EVOLUTION OF THIOREDOXIN REDUCTASE

Small and large TRs evolved by convergent evolution.[30,31] It was suggested that large TRs evolved from GR by the addition

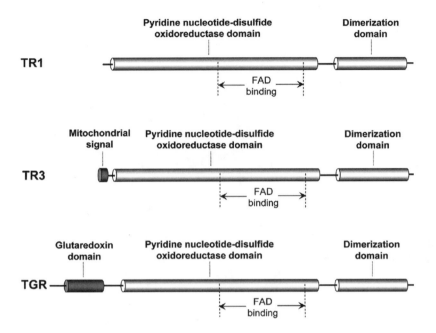

Figure 8.2 Domain organization of mammalian thioredoxin reductases. Mammals contain three TR isozymes. TR1 is a classical thioredoxin reductase. TR3 contains a mitochondrial signal sequence and TGR contains an N-terminal glutaredoxin domain. The catalytic site in all three isozymes is located in the N-terminal portion of the pyridine nucleotide disulfide oxidoreductase domain, and the C-terminal GCUG tetrapeptide is in the extension of the dimerization domain.

of the C-terminal extension containing a GCUG (U represents selenocysteine) tetrapeptide as a second active center. As shown in the alignment of TRs and GRs (Figure 8.3A), both enzymes contain a conserved CVNVGC active site, whereas the *E. coli* and other small TRs do not. GRs employ a small tripeptide glutathione as a substrate, whereas large TRs catalyze the reduction of the C-terminal active center by the N-terminal active center. Thus, the C-terminal peptide in TRs may be viewed as a functional analog of glutathione. Other large TRs are similarly organized, even though some of them

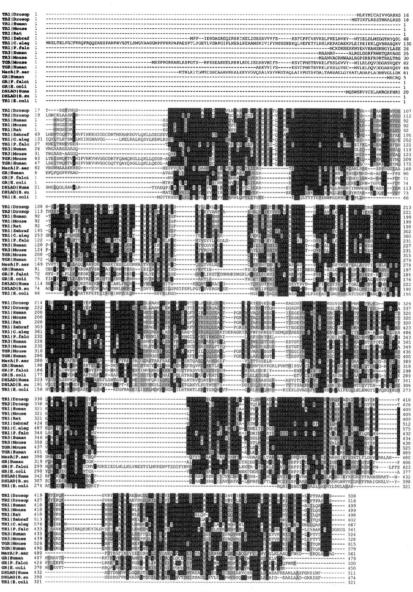

(A)

Figure 8.3 Phylogenetic analysis of thioredoxin reductases and their homologs. (A) Sequence alignment of thioredoxin reductases and their homologs. TR, thioredoxin reductase; GR, glutathione reductase; MerA, mercuric ion reductase; DHLAD, dihydrolipoamide dehydrogenase. (B) phylogenetic tree that is based upon the sequence alignment shown in A.

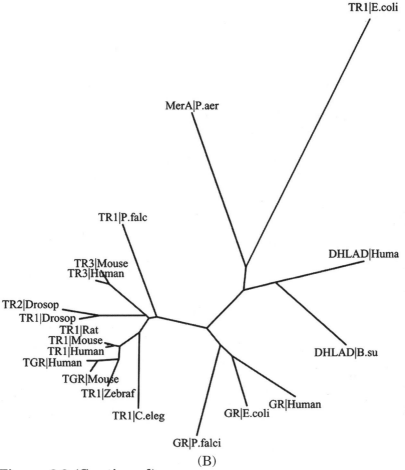

(B)

Figure 8.3 (Continued)

do not possess Sec. For example, *Drosophila melanogaster* TR1 has the same N-terminal CVNVGC active site and a C-terminal SCCS tetrapeptide. It should be noted that this fruit fly enzyme has a k_{cat} value similar to that of mammalian Sec-containing TRs, suggesting that Sec might not necessarily be essential for the function of large TRs.[45] In addition to large TRs, at least one additional enzyme in the pyridine nucleotide disulfide oxidoreductase family, mercuric reductase, has a similar structural organization. It contains an N-terminal

CVNVGC active site and a C-terminal SCCAG active center, which coordinates mercury for reduction by the N-terminal center.

A phylogenetic tree of the TR family (based on primary sequences) is shown in Figure 8.3B. The tree is consistent with the finding that *E. coli* TR1 is evolutionarily more distant from mammalian TRs than mercuric reductase. In addition, mammalian TR1 clusters with TGR rather than with TR3, suggesting that separation of TR1 and TGR is a more recent event than the origin of TR3.

CELLULAR FUNCTIONS OF THIOREDOXIN REDUCTASES

Major functions of TR, such as its roles in cell growth and antioxidant defense, are due to its role in reduction of Trx, which in turn participates in a variety of redox-controlled pathways.[20,46] Trx exerts its regulatory roles only when it is in the reduced form.[47] For example, mutant forms of Trx lacking the active site cysteines failed to stimulate cell proliferation.[48,49] It has been shown that TR activities can be affected by compounds that modulate cell growth. However, sometimes opposing effects of these compounds could be observed. It was found that TR could stimulate[50,51] and inhibit cell growth.[52] Likewise, TR was found to inhibit[20] and induce apoptosis.[53] The conflicting data show that the function of TR in regulation of cell growth is complex. It was suggested that TR and Trx act as a "yin-yang" regulatory system in growth control.[53] Under standard growth conditions, the TR-Trx system maintains the redox status of many redox proteins and controls the redox status of the intracellular environment, thus stimulating growth. On the other hand, when cells are under stress, the system may be involved in removal of damaged cells by slowing cell growth and stimulating apoptosis. For example, the presence of interferon and retinoid in human breast carcinoma cells induces TR expression and cell death, and TR is necessary for the induction of cell death.[53]

In addition to its major role in reducing Trx, mammalian TR can also reduce a variety of cellular proteins and compounds, such as protein disulfide isomerases,[54,55] GPxs,[56] selenite,[57] hydroperoxides, selenodiglutathione,[57] dehydroascorbate,[58] NK-lysin, Sec, vitamin K, alloxan, and 5,5-dithiobis(2-nitrobenzoic acid). These substrates (which are mostly small-molecular-weight compounds) of TRs have various roles in cell growth and redox regulation,[57–60] adding more complexity to the physiological roles of thioredoxin reductases. The broad substrate specificity of TRs was suggested to originate from the highly reactive Sec residue.[61,62] Knockout models could have been very helpful in understanding TR function, but they are not available. Trx knockout in mice is embryonic lethal, and it was suggested that the TR function might also be essential in mammals.[63]

ROLES IN SIGNALING AND REGULATION OF THIOREDOXIN REDUCTASE EXPRESSION

Mammalian TRs are selenium-containing proteins, which adds a further level of complexity to their regulation. Although these proteins can also be regulated at transcriptional and translational levels, the *in vivo* activity of these enzymes depends entirely on levels of selenium in diet (or culture media) due to the essential role of Sec in the catalytic activities of these enzymes.

Selenium is an essential trace element. It occurs in biological systems in the form of selenoproteins and small-molecular-weight compounds. Most selenoproteins are redox proteins and at least one (phospholipid hydroperoxide glutathione peroxidase [PHGPX, GPx4]) was shown to be essential for mammals.[64] All known mammalian selenoproteins contain selenium in the form of Sec, the 21st amino acid in the genetic code. Sec is encoded by TGA, which normally functions as a stop signal. A cellular system, designated the Sec insertion system, evolved to distinguish between stop and Sec insertion functions of UGA codons. A critical component of this system is a cis element present in the 3 untranslated region (UTR) of selenoprotein mRNAs.[65] This cis element is known as selenocysteine insertion sequence (SECIS) element. Many selenoproteins (including TRs) are often misannotated

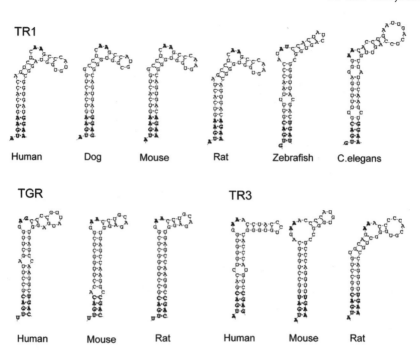

Figure 8.4 Predicted Selenocysteine Insertion Sequence (SECIS) elements in TR genes. SECISearch 2.0 (http://genome.unl.edu/SECI Search.html) was used to predict all SECIS structures, except that in zebra fish the TR1 gene was predicted manually. Nucleotides important for SECIS function are shown in bold.

in genomic databases due to incorrect assignment of termination function to Sec UGA codons. SECIS has been successfully employed in bioinformatics approaches to identify misannotated and novel selenoproteins.[66,67] SECIS elements in TR genes (as predicted by SECISearch, http://genome.unl.edu/SECISearch.html) are shown in Figure 8.4.

Sec is not the only amino acid that is inserted in response to a stop codon. Pyrrolysine, the 22nd amino acid, was recently identified and found to be inserted in response to UAG in certain methanogens.[68,69] Cis elements have not been found for pyrrolysine. The studies on Sec insertion should be of benefit in characterizing the mechanism of pyrrolysine insertion. A possible secondary structure, located immediately downstream of the UAG codon in mRNAs of some pyrrolysine-containing proteins, has been predicted,[70] but the experimental evidence for the role of these structures in pyrrolysine insertion is currently lacking.

Selenoproteins follow a hierarchy in regard to priority for selenium supply *in vivo* because selenoproteins respond differently to dietary changes in selenium levels. This hierarchy allows preserving selenium for more critical selenoproteins when selenium is limited, but its mechanism is not known.[63,71] TR has a higher priority for selenium than most other selenoproteins, and in mild selenium deficiency, its activity is not affected. On the other hand, when selenium supply is sufficient, further increase in selenium does not significantly induce TR activity. Thus, the levels of TR appear to be under a stringent control. Normally, TR is kept at a level that ensures normal function of cellular systems, but under certain conditions (such as stress), its expression can be easily downregulated or upregulated.

TR1 is also under stringent regulation at the transcriptional level. It appears to have a typical housekeeping promoter.[72] AP-1-binding sites and sites for several other transcription factors were identified in the TR promoter region.[73] TR1 is also regulated through its 3'-untranslated region (3'-UTR), which, in addition to the SECIS element, contains several AU-rich elements (AREs) that regulate the stability of TR1 mRNA. The AREs in the 3'-UTR of TR1 mRNA downregulate basal mRNA levels in normal unstimulated cells[63] and are responsive to exogenous agents.[72,74]

It has also been shown that the Sec in TR1 is a direct target of ROS, which likely results in the formation of inactive seleninic acid-containing enzyme.[75,76] Thus, TR is also regulated by ROS at a protein level.

REDOX REGULATION BY ROS

The TR-Trx system is involved in ROS signaling by directly influencing activities of a variety of transcription factors, such as AP-1,[77] NFB,[78] Sp1[79] and Ref-1.[77] By acting on transcription factors, this redox system regulates cell growth, apoptosis, and other cellular processes, as well as the expression of many redox proteins including TR and Trx themselves. It is becoming clear that Trx targets a wide range of substrates, but how the specificity of Trx signaling is achieved is not clear. There is emerging evidence that shows that redox signaling is highly specific,[80] suggesting that specificity should also occur at the level of TR and Trx. The occurrence of a possible mammalian Trx code, discussed above, would allow specificity of Trx signaling and explain why Trx acts on many substrates. In addition, the TR-Trx system could indirectly regulate a variety of transcription factors by regulating the *in vivo* redox status.[16,80] Such regulation could be either specific or nonspecific.

Recent findings indicate that the entire TR-Trx system could function as a switch in response to different ROS levels (Figure 8.5). As discussed above, there are three basic components in the Trx system: TR, Trx, and Trx peroxidase (a member of the peroxiredoxin [Prx] family). Under normal cellular conditions, the ROS level could be controlled by this system at a low level, thus minimizing the detrimental effects of oxidants, yet allowing their signaling functions. However, under stress, ROS levels can be dramatically increased, which may inactivate the Trx system, including TR.[75,76] It is not clear if the inactivated TR could be reactivated or if the TR function is recovered due to elevated expression of the protein. Similarly, Trx might also undergo oxidative inactivation, such as formation of dimers through one or two C-terminal cysteines.[27] The reduction of such dimers is a slow process. Furthermore, Prx has been reported to be inactivated by hydrogen peroxide with the formation of cysteine sulfinic acid.[81,82] Interestingly, the sulfinic acid form of Prx can be reactivated by a newly identified enzyme, sulfiredoxin.[83]

It has been reported that oxidized yeast Trx is a critical signal in upregulation of glutathione synthesis through transcription factors Yap1 and Met4.[16,80] In yeast, Yap1 is activated in response to various stresses, forming disulfide bonds, which stimulate nuclear accumulation of the protein, which result in the expression of stress response genes. In this system,

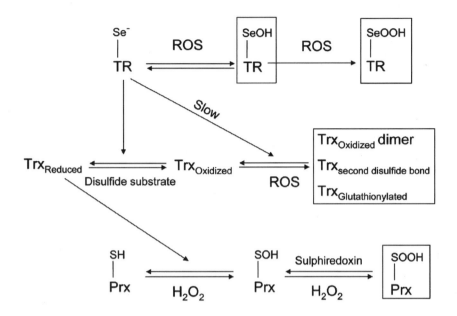

Figure 8.5 Redox regulation by thioredoxin reductase, thioredoxin, and peroxiredoxin. The three major components of the Trx system: TR, Trx, and Tpx or Prx function in concert to regulate cellular redox status. All three proteins could react with ROS, which inactivates the redox system. This regulation may be reversible as shown in the figure, which allows redox regulation of various physiological processes. In contrast, permanent inactivation of the system by excess of ROS results in cell death. In reduced TR, selenocysteine is ionized to selenolate, and ROS could reversibly oxidize it to selenenic acid (SeOH) and irreversibly to seleninic acid (SeOOH). Other details of redox regulation by TR, Trx, and Prx are discussed in the text of the chapter.

reduced Trx serves as Yap1 reductant and the inhibitor of oxidative stress response. This protein also reduces GPx3, a thiol peroxidase that is required for activation of Yap1 by hydrogen peroxide. The role of yeast TR in these processes has not been addressed beyond its function in Trx reduction. Interestingly, transcription of the TR gene is dramatically activated by Yap1, revealing the cyclic nature of redox signaling.

ACKNOWLEDGMENTS

This work is supported by NIH GM065204 (to VNG).

References

1. Thannickal VJ, Fanburg BL; Reactive oxygen species in cell signaling; *Am. J. Physiol. Lung Cell. Mol. Physiol.*; 279: L1005–1028, 2000.

2. Finkel T, Holbrook NJ; Oxidants, oxidative stress and the biology of ageing; *Nature*; 408: 239–247, 2000.

3. Martindale JL, Holbrook NJ; Cellular response to oxidative stress: signaling for suicide and survival; *J. Cell. Physiol.*; 192: 1–15, 2002.

4. Kanzok SM, Fechner A, Bauer H, Ulschmid JK, Muller HM, et al.; Substitution of the thioredoxin system for glutathione reductase in Drosophila melanogaster; *Science*; 291: 643–646, 2001.

5. Sun QA, Kirnarsky L, Sherman S, Gladyshev VN; Selenoprotein oxidoreductase with specificity for thioredoxin and glutathione systems; *Proc. Natl. Acad. Sci. USA*; 98: 3673–3678, 2001.

6. Alger HM, Williams DL; The disulfide redox system of Schistosoma mansoni and the importance of a multifunctional enzyme, thioredoxin glutathione reductase; *Mol. Biochem. Parasitol.*; 121: 129–139, 2002.

7. Agorio A, Chalar C, Cardozo S, Salinas G; Alternative mRNAs arising from trans-splicing code for mitochondrial and cytosolic variants of Echinococcus granulosus thioredoxin glutathione reductase; *J. Biol. Chem.*; 278: 12920–12928, 2003.

8. Rendon JL, del Arenal IP, Guevara-Flores A, Uribe A, Plancarte A, Mendoza-Hernandez G; Purification, characterization and kinetic properties of the multifunctional thioredoxin-glutathione reductase from Taenia crassiceps metacestode (cysticerci); *Mol. Biochem. Parasitol.*; 133:61–69, 2004.

9. Casagrande S, Bonetto V, Fratelli M, Gianazza E, Eberini I, et al.; Glutathionylation of human thioredoxin: a possible crosstalk between the glutathione and thioredoxin systems; *Proc. Natl. Acad. Sci. USA*; 99: 9745–9749, 2002.

10. Johansson C, Lillig CH, Holmgren A; Human mitochondrial glutaredoxin reduces S-glutathionylated proteins with high affinity accepting electrons from either glutathione or thioredoxin reductase; *J. Biol. Chem.*; 279: 7537–7543, 2004.

11. Holmgren A; Thioredoxin; *Annu. Rev. Biochem.*; 54: 237–271, 1985.

12. Kim IH, Kim K, Rhee SG; Induction of an antioxidant protein of Saccharomyces cerevisiae by O2, Fe3+, or 2-mercaptoethanol; *Proc. Natl. Acad. Sci. USA*; 86: 6018–6022, 1989.

13. Kim K, Kim IH, Lee KY, Rhee SG, Stadtman ER; The isolation and purification of a specific "protector" protein which inhibits enzyme inactivation by a thiol/Fe(III)/O2 mixed-function oxidation system; *J. Biol. Chem.*; 263: 4704–4711, 1988.

14. Schenk H, Klein M, Erdbrugger W, Droge W, Schulze-Osthoff K; Distinct effects of thioredoxin and antioxidants on the activation of transcription factors NF-kappa B and AP-1; *Proc. Natl. Acad. Sci. USA*; 91: 1672–1676, 1994.

15. Wang HC, Zentner MD, Deng HT, Kim KJ, Wu R, et al.; Oxidative stress disrupts glucocorticoid hormone-dependent transcription of the amiloride-sensitive epithelial sodium channel alpha-subunit in lung epithelial cells through ERK-dependent and thioredoxin-sensitive pathways; *J. Biol. Chem.*; 275: 8600–8609, 2000.

16. Wheeler GL, Trotter EW, Dawes IW, Grant CM; Coupling of the transcriptional regulation of glutathione biosynthesis to the availability of glutathione and methionine via the Met4 and Yap1 transcription factors; *J. Biol. Chem.*; 278: 49920–49928, 2003.

17. Wei SJ, Botero A, Hirota K, Bradbury CM, Markovina S, et al.; Thioredoxin nuclear translocation and interaction with redox factor-1 activates the activator protein-1 transcription factor in response to ionizing radiation; *Cancer Res.*; 60: 6688–6695, 2000.

18. Stewart EJ, Aslund F, Beckwith J; Disulfide bond formation in the Escherichia coli cytoplasm: an *in vivo* role reversal for the thioredoxins; *EMBO J.*; 17: 5543–5550, 1998.

19. Holmgren A; Enzymatic reduction-oxidation of protein disulfides by thioredoxin; *Methods Enzymol.*; 107: 295–300, 1984.

20. Baker A, Payne CM, Briehl MM, Powis G; Thioredoxin, a gene found overexpressed in human cancer, inhibits apoptosis *in vitro* and *in vivo*; *Cancer Res.*; 57: 5162–5167, 1997.

21. Saitoh M, Nishitoh H, Fujii M, Takeda K, Tobiume K, et al.; Mammalian thioredoxin is a direct inhibitor of apoptosis signal-regulating kinase (ASK) 1; *EMBO J.*; 17: 2596–2606, 1998.

22. Vlamis-Gardikas A, Holmgren A; Thioredoxin and glutaredoxin isoforms; *Methods Enzymol.*; 347: 286–296, 2002.

23. Miranda-Vizuete A, Damdimopoulos AE, Gustafsson J, Spyrou G; Cloning, expression, and characterization of a novel Escherichia coli thioredoxin; *J. Biol. Chem.*; 272: 30841–30847, 1997.

24. Ritz D, Patel H, Doan B, Zheng M, Aslund F, et al.; Thioredoxin 2 is involved in the oxidative stress response in Escherichia coli; *J. Biol. Chem.*; 275: 2505–2512, 2000.

25. Holmgren A, Soderberg BO, Eklund H, Branden CI; Three-dimensional structure of Escherichia coli thioredoxin-S2 to 2.8 A resolution; *Proc. Natl. Acad. Sci. USA*; 72: 2305–2309, 1975.

26. Martin JL; Thioredoxin — a fold for all reasons; *Structure*; 3: 245–250, 1995.

27. Watson WH, Pohl J, Montfort WR, Stuchlik O, Reed MS, et al.; Redox potential of human thioredoxin 1 and identification of a second dithiol/disulfide motif; *J. Biol. Chem.*; 2003.

28. Haendeler J, Hoffmann J, Tischler V, Berk BC, Zeiher AM, Dimmeler S; Redox regulatory and anti-apoptotic functions of thioredoxin depend on S-nitrosylation at cysteine 69; *Nat. Cell Biol.*; 4: 743–749, 2002.

29. Holmgren A; Pyridine nucleotide-disulfide oxidoreductases; *Experientia Suppl*; 36: 149–180, 1980.

30. Kuriyan J, Krishna TS, Wong L, Guenther B, Pahler A, et al.; Convergent evolution of similar function in two structurally divergent enzymes; *Nature*; 352: 172–174, 1991.

31. Novoselov SV, Gladyshev VN; Non-animal origin of animal thioredoxin reductases: Implications for selenocysteine evolution and evolution of protein function through carboxy-terminal extensions; *Protein Sci.*; 12: 372–378, 2003.

32. Williams CH, Jr.; Mechanism and structure of thioredoxin reductase from Escherichia coli; *FASEB J.*; 9:1267–1276, 1995.

33. Mustacich D, Powis G; Thioredoxin reductase; *Biochem. J.*; 346 (Pt 1): 1–8, 2000.

34. Arscott LD, Gromer S, Schirmer RH, Becker K, Williams CH, Jr.; The mechanism of thioredoxin reductase from human placenta is similar to the mechanisms of lipoamide dehydrogenase and glutathione reductase and is distinct from the mechanism of thioredoxin reductase from Escherichia coli; *Proc. Natl. Acad. Sci. USA*; 94: 3621–3626, 1997.

35. Gladyshev VN, Jeang KT, Stadtman TC; Selenocysteine, identified as the penultimate C-terminal residue in human T cell thioredoxin reductase, corresponds to TGA in the human placental gene; *Proc Natl Acad Sci USA*; 93: 6146–6151, 1996.

36. Tamura T, Stadtman TC; A new selenoprotein from human lung adenocarcinoma cells: purification, properties, and thioredoxin reductase activity; *Proc. Natl. Acad. Sci. USA*; 93: 1006–1011, 1996.

37. Lennon BW, Williams CH, Jr., Ludwig ML; Twists in catalysis: alternating conformations of *Escherichia coli* thioredoxin reductase; *Science*; 289: 1190–1194, 2000.

38. Zhong L, Holmgren A; Essential role of selenium in the catalytic activities of mammalian thioredoxin reductase revealed by characterization of recombinant enzymes with selenocysteine mutations; *J. Biol. Chem.*; 275: 18121–18128, 2000.

39. Sandalova T, Zhong L, Lindqvist Y, Holmgren A, Schneider G; Three-dimensional structure of a mammalian thioredoxin reductase: implications for mechanism and evolution of a selenocysteine-dependent enzyme; *Proc. Natl. Acad. Sci. USA*; 98: 9533–9538, 2001.

40. Lee SR, Kim JR, Kwon KS, Yoon HW, Levine RL, et al.; Molecular cloning and characterization of a mitochondrial selenocysteine-containing thioredoxin reductase from rat liver; *J. Biol. Chem.*; 274: 4722–4734, 1999.

41. Miranda-Vizuete A, Damdimopoulos AE, Pedrajas JR, Gustafsson JA, Spyrou G; Human mitochondrial thioredoxin reductase cDNA cloning, expression and genomic organization; *Eur. J. Biochem.*; 261: 405–412, 1999.

42. Miranda-Vizuete A, Damdimopoulos AE, Spyrou G; cDNA cloning, expression and chromosomal localization of the mouse mitochondrial thioredoxin reductase gene(1); *Biochim. Biophys. Acta*; 1447: 113–118, 1999.

43. Gasdaska PY, Berggren MM, Berry MJ, Powis G; Cloning, sequencing and functional expression of a novel human thioredoxin reductase; *FEBS Lett.*; 442: 105–111, 1999.

44. Larsson A; Thioredoxin reductase from rat liver; *Eur. J. Biochem.*; 35: 346–349, 1973.

45. Gromer S, Johansson L, Bauer H, Arscott LD, Rauch S, et al.; Active sites of thioredoxin reductases: why selenoproteins?; *Proc. Natl. Acad. Sci. USA*; 100: 12618–12623, 2003.

46. Gasdaska JR, Berggren M, Powis G; Cell growth stimulation by the redox protein thioredoxin occurs by a novel helper mechanism; *Cell Growth Differ.*; 6: 1643–1650, 1995.

47. Freemerman AJ, Gallegos A, Powis G; Nuclear factor kappaB transactivation is increased but is not involved in the proliferative effects of thioredoxin overexpression in MCF-7 breast cancer cells; *Cancer Res.*; 59: 4090–4094, 1999.

48. Oblong JE, Berggren M, Gasdaska PY, Powis G; Site-directed mutagenesis of active site cysteines in human thioredoxin produces competitive inhibitors of human thioredoxin reductase and elimination of mitogenic properties of thioredoxin; *J. Biol. Chem.*; 269: 11714–11720, 1994.

49. Gasdaska JR, Kirkpatrick DL, Montfort W, Kuperus M, Hill SR, et al.; Oxidative inactivation of thioredoxin as a cellular growth factor and protection by a Cys 73 — >Ser mutation; *Biochem Pharmacol*; 52: 1741–1747, 1996.

50. Gallegos A, Gasdaska JR, Taylor CW, Paine-Murrieta GD, Goodman D, et al.; Transfection with human thioredoxin increases cell proliferation and a dominant-negative mutant thioredoxin reverses the transformed phenotype of human breast cancer cells; *Cancer Res.*; 56: 5765–5770, 1996.

51. Tagaya Y, Maeda Y, Mitsui A, Kondo N, Matsui H, et al.; ATL-derived factor (ADF), an IL-2 receptor/Tac inducer homologous to thioredoxin; possible involvement of dithiol-reduction in the IL-2 receptor induction; *EMBO J.*; 8: 757–764, 1989.

52. Rubartelli A, Bonifaci N, Sitia R; High rates of thioredoxin secretion correlate with growth arrest in hepatoma cells; *Cancer Res.*; 55: 675–680, 1995.

53. Hofmann ER, Boyanapalli M, Lindner DJ, Weihua X, Hassel BA, et al.; Thioredoxin reductase mediates cell death effects of the combination of beta interferon and retinoic acid; *Mol. Cell. Biol.*; 18: 6493–6504, 1998.

54. Lundstrom J, Holmgren A; Protein disulfide-isomerase is a substrate for thioredoxin reductase and has thioredoxin-like activity; *J. Biol. Chem.*; 265: 9114–9120, 1990.

55. Lundstrom-Ljung J, Birnbach U, Rupp K, Soling HD, Holmgren A; Two resident ER-proteins, CaBP1 and CaBP2, with thioredoxin domains, are substrates for thioredoxin reductase: comparison with protein disulfide isomerase; *FEBS Lett.*; 357: 305–308, 1995.

56. Bjornstedt M, Xue J, Huang W, Akesson B, Holmgren A; The thioredoxin and glutaredoxin systems are efficient electron donors to human plasma glutathione peroxidase; *J. Biol. Chem.*; 269: 29382–29384, 1994.

57. Bjornstedt M, Kumar S, Holmgren A; Selenite and selenodiglutathione: reactions with thioredoxin systems; *Methods Enzymol.*; 252: 209–219, 1995.

58. May JM, Mendiratta S, Hill KE, Burk RF; Reduction of dehydroascorbate to ascorbate by the selenoenzyme thioredoxin reductase; *J. Biol. Chem.*; 272: 22607–22610, 1997.

59. Holmgren A, Bjornstedt M; Thioredoxin and thioredoxin reductase; *Methods Enzymol.*; 252: 199–208, 1995.

60. Holmgren A, Lyckeborg C; Enzymatic reduction of alloxan by thioredoxin and NADPH-thioredoxin reductase; *Proc. Natl. Acad. Sci. USA*; 77: 5149–5152, 1980.

61. Gallegos A, Berggren M, Gasdaska JR, Powis G; Mechanisms of the regulation of thioredoxin reductase activity in cancer cells by the chemopreventive agent selenium; *Cancer Res.*; 57: 4965–4970, 1997.

62. Gorlatov SN, Stadtman TC; Human thioredoxin reductase from HeLa cells: selective alkylation of selenocysteine in the protein inhibits enzyme activity and reduction with NADPH influences affinity to heparin; *Proc. Natl. Acad. Sci. USA*; 95: 8520–8525, 1998.

63. Gasdaska JR, Harney JW, Gasdaska PY, Powis G, Berry MJ; Regulation of human thioredoxin reductase expression and activity by 3-untranslated region selenocysteine insertion sequence and mRNA instability elements; *J. Biol. Chem.*; 274:25379–25385, 1999.

64. Yant LJ, Ran Q, Rao L, Van Remmen H, Shibatani T, et al.; The selenoprotein GPX4 is essential for mouse development and protects from radiation and oxidative damage insults; *Free Radic. Biol. Med.*; 34: 496–502, 2003.

65. Berry MJ, Banu L, Chen YY, Mandel SJ, Kieffer JD, et al.; Recognition of UGA as a selenocysteine codon in type I deiodinase requires sequences in the 3 untranslated region; *Nature*; 353: 273–276, 1991.

66. Kryukov GV, Castellano S, Novoselov SV, Lobanov AV, Zehtab O, et al.; Characterization of mammalian selenoproteomes; *Science*; 300: 1439–1443, 2003.

67. Kryukov GV, Kryukov VM, Gladyshev VN; New mammalian selenocysteine-containing proteins identified with an algorithm that searches for selenocysteine insertion sequence elements; *J. Biol. Chem.*; 274: 33888–33897, 1999.

68. Srinivasan G, James CM, Krzycki JA; Pyrrolysine encoded by UAG in Archaea: charging of a UAG-decoding specialized tRNA; *Science*; 296: 1459–1462, 2002.

69. Hao B, Gong W, Ferguson TK, James CM, Krzycki JA, Chan MK; A new UAG-encoded residue in the structure of a methanogen methyltransferase; *Science*; 296: 1462–1466, 2002.

70. Namy O, Rousset JP, Napthine S, Brierley I; Reprogrammed genetic decoding in cellular gene expression; *Mol. Cell*; 13: 157–168, 2004.

71. Bermano G, Arthur JR, Hesketh JE; Role of the 3 untranslated region in the regulation of cytosolic glutathione peroxidase and phospholipid-hydroperoxide glutathione peroxidase gene expression by selenium supply; *Biochem. J.*; 320 (Pt 3): 891–895, 1996.

72. Rundlof AK, Carlsten M, Arner ES; The core promoter of human thioredoxin reductase 1: cloning, transcriptional activity, and Oct-1, Sp1, and Sp3 binding reveal a housekeeping-type promoter for the AU-rich element-regulated gene; *J. Biol. Chem.*; 276: 30542–30551, 2001.

73. Osborne SA, Tonissen KF; Genomic organisation and alternative splicing of mouse and human thioredoxin reductase 1 genes; *BMC Genomics*; 2: 10, 2001.

74. Koishi R, Kawashima I, Yoshimura C, Sugawara M, Serizawa N; Cloning and characterization of a novel oxidoreductase KDRF from a human bone marrow-derived stromal cell line KM-102; *J. Biol. Chem.*; 272: 2570–2577, 1997.

75. Sun QA, Gladyshev VN; Redox regulation of cell signaling by thioredoxin reductases; *Methods Enzymol.*; 347:451–461, 2002.

76. Sun QA, Wu Y, Zappacosta F, Jeang KT, Lee BJ, et al.; Redox regulation of cell signaling by selenocysteine in mammalian thioredoxin reductases; *J. Biol. Chem.*; 274:24522–24530, 1999.

77. Hirota K, Matsui M, Iwata S, Nishiyama A, Mori K, Yodoi J; AP-1 transcriptional activity is regulated by a direct association between thioredoxin and Ref-1; *Proc. Natl. Acad. Sci. USA*; 94: 3633–3638, 1997.

78. Hayashi T, Ueno Y, Okamoto T; Oxidoreductive regulation of nuclear factor kappa B — involvement of a cellular reducing catalyst thioredoxin; *J. Biol. Chem.*; 268: 11380–11388, 1993.

79. Bloomfield KL, Osborne SA, Kennedy DD, Clarke FM, Tonissen KF; Thioredoxin-mediated redox control of the transcription factor Sp1 and regulation of the thioredoxin gene promoter; *Gene*; 319: 107–116, 2003.

80. Delaunay A, Pflieger D, Barrault MB, Vinh J, Toledano MB; A thiol peroxidase is an H2O2 receptor and redox-transducer in gene activation; *Cell*; 111: 471–481, 2002.

81. Wood ZA, Poole LB, Karplus PA; Peroxiredoxin evolution and the regulation of hydrogen peroxide signaling; *Science*; 300: 650–653, 2003.

82. Woo HA, Chae HZ, Hwang SC, Yang KS, Kang SW, et al.; Reversing the inactivation of peroxiredoxins caused by cysteine sulfinic acid formation; *Science*; 300:653–656, 2003.

83. Biteau B, Labarre J, Toledano MB; ATP-dependent reduction of cysteine-sulphinic acid by S. cerevisiae sulphiredoxin; *Nature*; 425: 980–984, 2003.

9

Roles for Biotinylation of Histones in Chromatin Structure

GABRIELA CAMPOREALE, NAGARAMA KOTHAPALLI,
GAUTAM SARATH, AND JANOS ZEMPLENI

INTRODUCTION

The major components of chromatin are (1) DNA, (2) a group of proteins named histones, and (3) various nonhistone proteins. Folding of DNA into chromatin is mediated primarily by histones.[1] Five major classes of histones have been identified in mammals: H1, H2A, H2B, H3, and H4. Histones are small proteins (11 to 22 kDa) consisting of a globular domain and a more flexible and charged amino terminus (histone "tail"). Lysine and arginine residues account for a combined >20% of all amino acid residues in histones, causing the basic properties of these proteins.[1]

DNA and histones form repetitive nucleoprotein units, the nucleosomes.[1] Each nucleosome (nucleosome core particle) consists of 146 basepairs of DNA wrapped around an octamer of core histones (one H3-H3-H4-H4 tetramer and two H2A-H2B dimers). The binding of DNA to histones is of electrostatic nature and is mediated by the association of negatively charged phosphate groups of DNA with positively charged ε-amino groups (lysine moieties) and guanidino groups (arginine moieties) of histones. The amino-terminal tail of histones protrudes from the nucleosomes; covalent modifications of this tail affect the structure of chromatin and form the basis for gene regulation, replication, and DNA repair as described in the following text. The DNA located between nucleosome core particles is called linker DNA and is associated with histone H1.

Histone tails are modified by covalent acetylation,[2-4] methylation,[1] phosphorylation,[1] ubiquitination,[1] and poly(ADP-ribosylation)[5-7] of ε-amino groups (lysine), guanidino groups (arginine), and hydroxyl groups (serine). These modifications of histone tails (histone code) may considerably extend the information potential of the DNA code and gene regulation, given that the enzymes catalyzing these modifications are specific for amino acid residues.[8-10] Multiple signaling pathways converge on histones to mediate covalent modifications of these proteins.[11] For example, binding of nuclear receptors to DNA may recruit histone acetyl transferases to chromatin, leading to acetylation of histone tails.[12] Modifications of histone tails may affect binding of chromatin-associated proteins, triggering cascades of downstream histone modifications. For example, it has been proposed that methylation of arginine-3 in histone H4 recruits the histone acetyl transferase Esa1 to yeast chromatin, leading to acetylation of lysine-5 in histone H4.[10] Histone modifications can influence each other in synergistic or antagonistic ways, mediating gene regulation. For example, phosphorylation of serine-10 inhibits methylation of lysine-9 in histone H3, but is coupled with lysine-9 or lysine-14 acetylation during mitogenic stimulation in mammalian cells.[10] Conversely, deacetylation of lysine-14 in histone H3 is required to facilitate subsequent methylation of lysine-9, leading to transcriptional silencing. Ultimately, modifications of histones impact the access of enzymes such as RNA polymerases and DNA repair enzymes to DNA.

Covalent modifications of histones are reversible, i.e., acetate moieties can be removed by histone deacetylases, and phosphates can be removed by phosphatases. The mechanism leading to removal of methyl groups is uncertain. It has been proposed that methylation marks are removed by proteolytic processing of histones, i.e., the methylated histone tail is clipped off.[10] It is uncertain whether ubiquitination of histones plays a role in this process.

In this chapter we will (1) provide a brief summary of roles for histone modifications in cell biology and (2) review the current state of knowledge for a newly discovered modification of histones: covalent binding of the vitamin biotin.

ROLES FOR HISTONE MODIFICATIONS IN CELL BIOLOGY

Transcriptional Activity of DNA

Roles for acetylation of histones in transcription of DNA have been investigated extensively. These studies have provided evidence that many transcriptional activators have histone acetyltransferase activity[13] and that transcriptional repressors are histone deacetylases.[14] Notwithstanding the important role for acetylation of some lysine moieties in transcriptional activation of chromatin,[15–17] acetylation of lysine-12 in histone H4 might be associated with gene silencing.[10] Ubiquitination and methylation of histones might also play a role in the regulation of transcriptional activity of DNA.[1,10,18]

DNA Repair

Evidence has been provided that poly(ADP-ribosylation) of histones is linked to DNA repair mechanisms[19,20] and apoptosis.[21] Induction of DNA damage in mammalian cells by chemical treatment or UV exposure causes a dramatic increase of poly(ADP-ribosylation) of histones.[22] Poly(ADP-ribosylation) is catalyzed by poly(ADP-ribose) polymerase (E.C. 2.4.2.30).[5,7,23] Among core histones, H2B is poly(ADP-ribosylated) to the highest extent.[7] However, H2A, H3, and H4, as well as linker histone H1, are also poly(ADP-ribosylated).[7] In addition, poly(ADP-ribose) polymerase catalyzes auto-ADP-ribosylation in response to DNA damage.[20] Poly(ADP-ribosylation) of histones mediates a temporary dissociation from and reassociation of histones with DNA.[20] It has been proposed that this mechanism may guide enzymes to sites of DNA repair.[20] Poly(ADP-ribosylation) might also be involved in nucleosomal unfolding of chromatin in DNA excision repair.[20]

Nutritional status might play an important role in DNA repair. The rejoining of DNA strand breaks is prevented by nutritionally depleting cells of NAD, the substrate for poly(ADP-ribosylations).[19] Dietary deficiency of NAD precursors (niacin, tryptophan) decreases poly(ADP-ribose) concentrations in rat liver.[24] Recently, evidence has been provided that acetylation of lysine residues 5, 8, 12, and 16 in histone H4 also plays a role in DNA repair.[25]

Cell Proliferation

Posttranslational modifications of histones are essential to ensure normal cell proliferation. For example, replacement of all four acetylatable lysines in histone H4 of *Saccharomyces cerevisiae* with arginine or glutamine leads to extremely slow growth or a delay in G2/M progression.[26] The pattern of histone modifications changes dramatically during the course of the cell cycle in eukaryotic cells,[1] consistent with a role for these modifications in cell proliferation. For example, core histones are deacetylated during mitosis. Posttranslational modifications of histones during the cell cycle might regulate transcription of genes that are involved in cell cycle progression.[1]

Spermatogenesis

Spermatogenesis requires the packaging of DNA into an inert chromatin structure. During spermatogenesis, histones are heavily acetylated and transiently replaced by "transition proteins."[1] Subsequently, transition proteins are replaced by highly basic protamines, leading to complete condensation of the nucleus. Binding of protamines to DNA is regulated by posttranslational modification of protamines (e.g., by phosphorylation). Condensation makes the DNA in the nucleus of a spermatozoon much more resistant to enzymatic, physical, and chemical degradation. Together, these findings suggest that posttranslational modifications of histones and other nucleic acid–binding proteins play important roles in spermatogenesis.

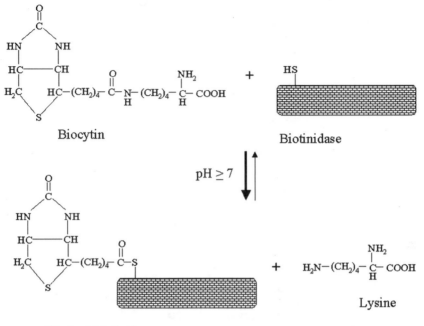

Figure 9.1   Cleavage of biocytin (biotin-ε-lysine) by biotinidase leads to the formation of a biotinyl-thioester intermediate (cysteine-bound biotin; biotinyl biotinidase) at or near the active site of biotinidase. (From J Hymes, B Wolf. Human biotinidase isn't just for recycling biotin. *J Nutr* 129: 485S–489S, 1999.)

ENZYMATIC BIOTINYLATION OF HISTONES

Recently, evidence has been provided that histones are also modified by covalent attachment of the vitamin biotin. Hymes et al. have proposed a reaction mechanism by which cleavage of biocytin (biotin-ε-lysine) by biotinidase (E.C. 3.5.1.12) leads to the formation of a biotinyl-thioester intermediate (cysteine-bound biotin) at or near the active site of biotinidase (Figure 9.1).[27,28] In the next step, the biotinyl moiety is transferred from the thioester to the ε-amino group of lysine in histones (Figure 9.2). The substrate (biocytin) for biotinylation of histones is generated in the breakdown of biotin-dependent

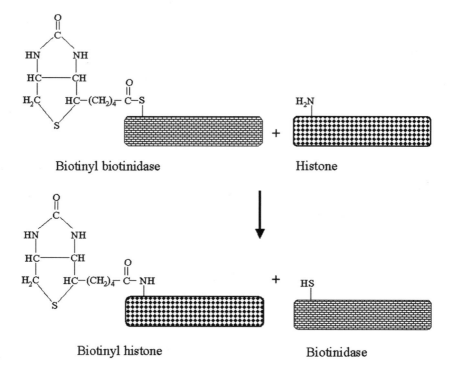

Figure 9.2 The biotinyl moiety from the biotinyl biotinidase intermediate is transferred onto the ε-amino group of lysine (or other
amino groups) in histones. (From J Hymes, B Wolf. Human biotinidase isn't just for recycling biotin. *J Nutr* 129: 485S–489S, 1999.)

carboxylases, which contain biotin linked to the ε-amino group
of a lysine moiety.[29,30]

Biotinidase belongs to the nitrilase superfamily of
enzymes, which consists of 12 families of amidases, *N*-acyl-
transferases, and nitrilases.[31] Some members of the nitrilase
superfamily (vanins-1, -2, and -3) share significant sequence
similarities with biotinidase;[32] it is unknown whether vanins
use histones as acceptor molecules in transferase reactions.
Biotinidase is ubiquitous in mammalian cells, and 26% of the
cellular biotinidase activity is located in the nuclear fraction.[29]
Human biotinidase has been cloned, sequenced, and characterized.[33] Moreover, the structure of the human biotinidase

gene has been determined;[34] the 5-flanking region of exon 1 contains a CCAAT element, three initiator sequences, an octamer sequence, three methylation consensus sites, two GC boxes, and one HNF-5 site, but has no TATA element. The 62-amino-acid region that harbors the active site of biotinidase is highly conserved among various mammals and *Drosophila*.[35]

BIOTINYLATED HISTONES IN HUMAN CELLS

Identification of Biotinylated Histones

We have provided evidence that human cells indeed contain biotinylated histones. Histones were isolated from nuclei of human peripheral blood mononuclear cells (PBMC) by acid extraction.[36] This procedure yielded a pure preparation of the five major classes of histones (H1, H2A, H2B, H3, and H4), as judged by comigration with commercially available histones on polyacrylamide gels and by the absence of quantitatively important nonhistone bands after GelCode Blue staining (Figure 9.3, lanes 1 and 2). Streptavidin-conjugated peroxidase is a well-established probe for biotin and was used to detect biotin in histones from PBMC. Streptavidin-peroxidase bound to histones H1, H3, and H4, suggesting that these proteins were biotinylated (Figure 9.3, lane 3). Streptavidin-peroxidase also bound to histones H2A and H2B, but these histones electrophoresed as one single band. Thus, it remains uncertain whether streptavidin bound to histone H2A, H2B, or both.

The following data suggest that streptavidin specifically bound to biotinylated histones rather than unspecifically to nonbiotinylated proteins: The (nonbiotinylated) proteins aprotinin, α-lactalbumin, β-lactoglobulin, trypsin inhibitor, and trypsinogen have molecular weights similar to the histones; no bands were observed when these proteins were probed with streptavidin-conjugated peroxidase (Figure 9.3, line 4). The synthetic polypeptides poly-L-lysine and poly-L-arginine mimic lysine- and arginine-rich histones; no bands were observed when poly-L-lysine and poly-L-arginine were probed with streptavidin-conjugated peroxidase (Figure 9.3, lanes 5

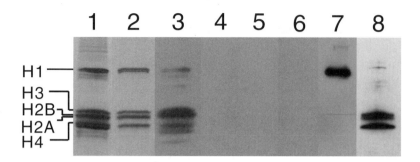

Figure 9.3 Nuclei from human peripheral blood mononuclear cells (PBMC) contain biotinylated histones. Histones were extracted from PBMC nuclei and chromatographed using SDS/gel electrophoresis. Nonhistone proteins, synthetic polypeptides (poly-L-arginine and poly-L-lysine), and commercial histones were used as controls. Samples either were stained with GelCode Blue (lanes 1 and 2), probed with streptavidin-conjugated peroxidase (lanes 3 to 7), or probed with a monoclonal antibody against biotin (lane 8). Lane 1 = histones from human PBMC nuclei; lane 2 = commercially available histones from calf thymus; lane 3 = histones from human PBMC nuclei; lane 4 = mixture of aprotinin, α-lactalbumin, β-lactoglobulin, trypsin inhibitor, and trypsinogen; lane 5 = poly-L-arginine; lane 6 = poly-L-lysine; lane 7 = chemically biotinylated histone H1; lane 8 = histones from human PBMC nuclei. Equal amounts of histones, poly-L-lysine, poly-L-arginine, and nonhistone proteins were used except for the chemically biotinylated histone H1, which was diluted approximately 20,000-fold. (From JS Stanley, JB Griffin, J Zempleni. Biotinylation of histones in human cells: effects of cell proliferation. *Eur J Biochem* 268: 5424–5429, 2001.)

and 6). Next, histone H1 was biotinylated chemically;[37] this synthetic conjugate was used as a positive control.[36] Chemically biotinylated histone H1 produced a strong band if probed with streptavidin-conjugated peroxidase (Figure 9.3, lane 7). Finally, extracts from PBMC nuclei were probed with a monoclonal antibody against biotin (Figure 9.3, lane 8). The antibody bound to histones, providing additional evidence that histones contain biotin.

N-terminus C-terminus

+   −   +   −

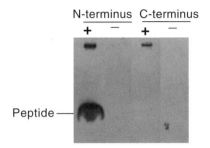

Peptide —

Figure 9.4   Synthetic peptides from the N- and C-termini of histone H4 were incubated with biotinidase and biocytin (denoted "+") to conduct an *in vitro* biotinylation; controls were incubated without biotinidase and biocytin ("−"). Peptides were electrophoresed using SDS-PAGE, and biotin was probed with streptavidin-peroxidase.

Biotinylated histones were also detected in human T cells,[38] human small-cell lung cancer cells,[39] human choriocarcinoma cells,[40] and chicken erythrocytes.[41] In contrast, histones in *Saccharomyces cerevisiae* are not biotinylated (unpublished observation); this may be because these cells do not express biotinidase.

Identification of Biotinylation Sites

We have developed the following strategy to identify amino acid residues that are targets for biotinylation in human histone H4.[42] Peptide fragments of histone H4 are synthesized chemically and are subjected to *in vitro* enzymatic biotinylation using biotinidase and biocytin; peptide-bound biotin is probed with streptavidin-peroxidase. Figure 9.4 depicts a typical example: N-terminal (SGRGKGGKGLGKGGAKRHR) and C-terminal (acetyl-TAMDVVYALKRQGRTLYGFGG) peptide fragments of human histone H4 were biotinylated enzymatically; controls were incubated without biocytin and biotinidase. Biotinidase catalyzed substantial biotinylation of the N-terminal peptide; in contrast, the biotinidase-free controls and the C-terminal peptide were not biotinylated. This is consistent with the

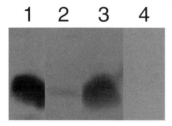

Figure 9.5 Effects of amino acid modifications on the biotinylation of the N-terminal peptide from human histone H4. The following peptides were incubated with biotinidase and biocytin to conduct an *in vitro* biotinylation: lane 1 = native peptide; lane 2 = peptide acetylated at lysine-8; lane 3 = L-lysine in position 12 was substituted with D-lysine; lane 4 = L-lysine in position 8 was substituted with L-glutamate. Peptides were electrophoresed using SDS-PAGE, and biotin was probed with streptavidin-peroxidase.

hypothesis that a biotinylation motif is located in the amino-terminal tail of histone H4. This is also consistent with previous observations that gene regulation is mediated by covalent modifications in the amino terminus in histones.

Next, N-terminal fragments of histone H4 were synthesized that contained various modifications of amino acid residues; these modified peptides were used in enzymatic biotinylations.[42] For example, (1) lysines were acetylated to determine whether acetylation affects biotinylation of neighboring lysines; (2) L-lysine was replaced with D-lysine to study the stereospecificity of biotinylation, and (3) the charge of the peptide was modified (e.g., by lysine-to-glutamate substitutions) to study charge interactions between histones and biotinidase. Figure 9.5 depicts selected examples from these studies, suggesting that (1) acetylation of lysine-8 adversely affects biotinylation of lysine-12 (compare lanes 1 and 2), (2) biotinylation of lysines is not stereospecific (lanes 1 and 3), and (3) charge interactions play an essential role in biotinylation of histones (lanes 1 and 4). Most important, however, these studies provided evidence that lysine-8 and lysine-12 are biotinylation sites in histone H4.[42] Finally, a polyclonal antibody to histone H4 (biotinylated at lysine-12) was developed.[42] The antibody bound to histone H4 from human lymphoid cells, consistent with the hypothesis that lysine-12 is a target site for biotinylation in human cells.

BIOLOGICAL ROLES FOR BIOTINYLATION OF HISTONES

Biotinylation of histones is a relatively new field of research; evidence of biological roles for it is scarce. However, some biological processes have been identified that might be affected by it. These processes are reviewed in the following subsections. The majority of the studies below were conducted before specific biotinylation sites in histones were identified and before biotinylation site-specific antibodies became available. Thus, these studies did not allow pinpointing changes in specific sites; rather, the total biotinylation of histones was quantified by using streptavidin or radiolabeled biotin.

Cell Proliferation

PBMC respond to proliferation with increased biotinylation of histones, as judged by Western blot analysis and by quantifying binding of radioactive biotin to histones.[36] [^{3}H]Biotinylation of histones increases early in the cell cycle (G1 phase) and remains increased during later phases (S, G2, and M phase) compared with quiescent controls (G0); the increase is greater than fourfold (Figure 9.6A). It is uncertain why it is increased throughout the cell cycle rather than at a specific phase of the cycle. Biotinylation increases for all classes of histones rather than for only some classes (Figure 9.6B). Currently, its role in cell proliferation is uncertain. New biotinylation site-specific antibodies will be useful tools to identify changes in the biotinylation pattern of histones in response to cell proliferation.

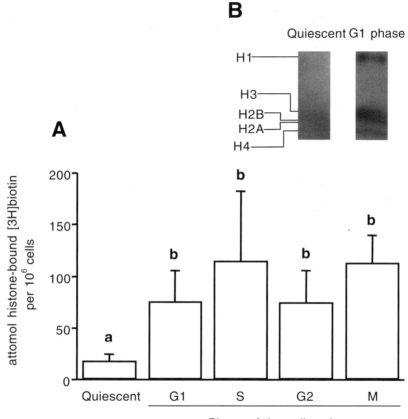

Figure 9.6 Biotinylation of histones in human PBMC at various phases of the cell cycle. (A) [^{3}H]Biotinylation of histones was tracked by liquid scintillation counting; quiescent PBMC were used as controls. Bars not sharing a common letter were significantly different; $p < 0.01$; n = 4. (B) Biotinylation of histones was tracked by SDS-PAGE and streptavidin-peroxidase. PBMC from phases G0 and G1 are depicted; G1-arrested PBMC show a biotinylation pattern that is representative of cells from G1, S, G2, and M phase. (From JS Stanley, JB Griffin, J Zempleni. Biotinylation of histones in human cells: effects of cell proliferation. *Eur J Biochem* 268: 5424–5429, 2001.)

Increased biotinylation of histones in proliferating PBMC is not paralleled by increased expression of the gene encoding biotinidase.[36] Both abundance of mRNA encoding biotinidase and cellular activity of biotinidase are similar in proliferating and quiescent PBMC.[36] Notwithstanding the well-established capability of biotinidase to catalyze biotinylation of histones,[27] it has been proposed that (1) enzymes other than biotinidase are also capable of biotinylating histones, and (2) the level of histone biotinylation is regulated by rates of debiotinylation (see the following text).[36]

Gene Silencing

Studies in chicken erythrocytes have provided circumstantial evidence that biotinylated histones are enriched in transcriptionally silent chromatin.[41]

DNA Damage

Evidence has been provided that UV-induced DNA damage causes (1) increased biotinylation of histones in human lymphoid cells, (2) decreased mitochondrial integrity, and (3) degradation of biotin-dependent carboxylases.[41] Given that degradation of biotinylated carboxylases provides the substrate (biocytin) for biotinylation of histones, it has been proposed that the following mechanistic sequence leads to increased biotinylation of histones in DNA-damaged cells:

1. UV irradiation causes DNA damage and decreased integrity of mitochondria.
2. Biotinylated carboxylases are degraded, perhaps due to mitochondrial disintegration.
3. Breakdown products of carboxylases leak from mitochondria into cytoplasm.
4. Breakdown products of carboxylases diffuse from cytoplasm into the cell nucleus, where they are used as substrate for biotinylation of histones.
5. Histone biotinylation increases in response to UV exposure.

This mechanistic sequence is consistent with previous studies by Wolf and coworkers. They have demonstrated that biotin-dependent carboxylases are degraded by proteases, leading to the release of biotinylated peptides.[30] These peptides are hydrolyzed by biotinidase to release biocytin.[30,33] Biocytin is further hydrolyzed by biotinidase, and the biotinyl moiety is covalently attached to histones.[27,28,43]

Currently, it is uncertain whether increased biotinylation of histones is (1) a mechanism leading to DNA repair, (2) a process leading to apoptosis or necrosis, or (3) a side product of metabolism that does not play a physiologic role in UV-damaged cells. Also, it is uncertain whether biotin nutritional status affects these processes.

DEBIOTINYLATION OF HISTONES

Recent studies are consistent with the hypothesis that biotinidase may catalyze both biotinylation and debiotinylation of histones.[44] How can cells regulate histone biotinylation, given that both biotinylation and debiotinylation are catalyzed by the same enzyme? The following explanations have been offered:

1. Enzymes other than biotinidase might also catalyze biotinylation or debiotinylation of histones. For example, holocarboxylase synthetase (E.C. 6.3.4.10) catalyzes covalent binding of biotin to ε-amino groups of lysine residues in four mammalian carboxylases.[45] The possibility that holocarboxylase synthetase catalyzes binding of biotin to histones is an untested hypothesis.

2. Covalent modification of biotinidase might be a mechanism to favor either biotinylation or debiotinylation of histones. Currently, glycosylation is the only posttranslational modification of biotinidase that has been identified.[33]

3. The presence of cofactors might favor either biotinylation or debiotinylation of histones. For example, high concentrations of the substrate biocytin might increase

the rate of histone biotinylation. The pH in the microenvironment of histones is unlikely to affect the biotinylation equilibrium, given that the pH optimum is similar (pH 8) for both the biotinylating activity[27] and the debiotinylating activity of biotinidase.[44]

4. Other proteins may interact with biotinidase at the chromatin level, favoring either biotinylation or debiotinylation of histones.

EFFECTS OF BIOTIN SUPPLY ON BIOTINYLATION OF HISTONES

Effects of biotin supply on biotinylation of histones have been investigated in various human-derived cell lines.[38–40] In these studies, cell lines were cultured in media containing deficient, physiologic, and pharmacologic concentrations of biotin for several weeks. Biotin concentrations in culture media had only a moderate impact on histone biotinylation; in contrast, biotinylation of carboxylases correlated strongly with biotin concentrations in culture media.[38–40] The reader should note that even small changes in biotinylation of histones might be physiologically meaningful, given that these changes might affect other modifications of histones such as acetylation and methylation.

ACKNOWLEDGMENTS

This research is supported by the National Institutes of Health (DK 60447 and DK 063945) and by a grant from the Nebraska Tobacco Settlement Biomedical Research Enhancement Funds.

References

1. A Wolffe. *Chromatin*. 3rd ed. San Diego, CA: Academic Press, 1998.

2. J Ausio, KE van Holde. Histone hyperacetylation: its effect on nucleosome conformation and stability. *Biochemistry* 25: 1421–1428, 1986.

3. TR Hebbes, AW Thorne, C Crane-Robinson. A direct link between core histone acetylation and transcriptionally active chromatin. *EMBO J* 7: 1395–1402, 1988.

4. DY Lee, JJ Hayes, D Pruss, AP Wolffe. A positive role for histone acetylation in transcription factor access to nucelosomal DNA. *Cell* 72: 73–84, 1993.

5. P Chambon, JD Weill, J Doly, MT Strosser, P Mandel. On the formation of a novel adenylic compound by enzymatic extracts of liver nuclei. *Biochem Biophys Res Commun* 25: 638–643, 1966.

6. T Boulikas. At least 60 ADP-ribosylated variant histones are present in nuclei from dimethylsulfate-treated and untreated cells. *EMBO J* 7: 57–67, 1988.

7. T Boulikas, B Bastin, P Boulikas, G Dupuis. Increase in histone poly(ADP-ribosylation) in mitogen-activated lymphoid cells. *Exp Cell Res* 187: 77–84, 1990.

8. BD Strahl, CD Allis. The language of covalent histone modifications. *Nature* 403: 41–45, 2000.

9. BM Turner. Histone acetylation and epigenetic code. *Bioessays* 22: 836–845, 2000.

10. T Jenuwein, CD Allis. Translating the histone code. *Science* 293: 1074–1080, 2001.

11. P Cheung, CD Allis, P Sassone-Corsi. Signaling to chromatin through histone modifications. *Cell* 103: 263–271, 2000.

12. V Laudet, H Gronemeyer. *The Nuclear Receptor Facts Book.* San Diego, CA: Academic Press, 2002.

13. JE Brownell, J Zhou, T Ranalli, R Kobayashi, DG Edmondson, SY Roth, CD Allis. *Tetrahymena* histone acetyltransferase A: a homolog to yeast Gcn5p linking histone acetylation to gene activation. *Cell* 84: 843–851, 1996.

14. J Taunton, CA Hassig, SL Schreiber. A mammalian histone deacetylase related to a yeast transcriptional regulator Rpd3. *Science* 272: 408–411, 1996.

15. V Allfrey, RM Faulkner, AE Mirsky. Acetylation and methylation of histones and their possible role in the regulation of RNA synthesis. *Proc Soc Natl Acad Sci USA* 51: 786–794, 1964.

16. MA Gorovsky. Macro- and micronuclei of *Tetrahymena pyriformis*: a model system for studying the structure of eukaryotic nuclei. *J Protozool* 20: 19–25, 1973.

17. DJ Mathis, P Oudet, B Waslyk, P Chambon. Effect of histone acetylation on structure and *in vitro* transcription of chromatin. *Nucleic Acids Res* 5: 3523–3547, 1978.

18. A-D Pham, F Sauer. Ubiquitin-activating/conjugating activity of $TAF_{II}250$, a mediator of activation of gene expression in *Drosophila*. *Science* 289: 2357–2360, 2000.

19. BW Durkacz, O Omidiji, DA Gray, S Shall. (ADP-ribose)$_n$ participates in DNA excision repair. *Nature* 283: 593–596, 1980.

20. F Althaus. Poly ADP-ribosylation: a histone shuttle mechanism in DNA excision repair. *J Cell Sci* 102: 663–670, 1992.

21. YS Yoon, JW Kim, KW Kang, YS Kim, KH Choi, CO Joe. Poly(ADP-ribosyl)ation of histone H1 correlates with internucleosomal DNA fragmentation during apoptosis. *J Biol Chem* 271: 9129–9134, 1996.

22. H Juarez-Salinas, JL Sims, MK Jacobson. Poly(ADP-ribose) levels in carcinogen-treated cells. *Nature* 282: 740–741, 1979.

23. Y Nishizuka, K Ueda, O Hayaishi. Adenosine diphosphoribosyltransferase in chromatin. In: DB McCormick, LD Wright, Eds. *Vitamins and Coenzymes.* New York: Academic Press, 1971, pp 230–233.

24. JM Rawling, TM Jackson, ER Driscoll, JB Kirkland. Dietary niacin deficiency lowers tissue poly(ADP-ribose) and NAD^+ concentrations in Fischer-344 rats. *J Nutr* 124: 1597–1603, 1994.

25. AW Bird, DY Yu, MG Pray-Grant, Q Qiu, KE Harmon, PC Megee, PA Grant, MM Smith, MF Christman. Acetylation of histone H4 by Esa1 is required for DNA double-strand break repair. *Nature* 419: 411–415, 2002.

26. PC Megee, BA Morgan, MM Smith. Histone H4 and the maintenance of genome integrity. *Genes Dev* 9: 1716–1727, 1995.

27. J Hymes, K Fleischhauer, B Wolf. Biotinylation of histones by human serum biotinidase: assessment of biotinyl-transferase activity in sera from normal individuals and children with biotinidase deficiency. *Biochem Mol Med* 56: 76–83, 1995.

28. J Hymes, B Wolf. Human biotinidase isn't just for recycling biotin. *J Nutr* 129: 485S–489S, 1999.

29. J Pispa. Animal biotinidase. *Ann Med Exp Biol Fenniae* 43: 4–39, 1965.

30. B Wolf, GS Heard. Biotinidase deficiency. In: L Barness, F Oski, Eds. *Advances in Pediatrics*. Chicago, IL: Medical Book Publishers, 1991, pp 1–21.

31. C Brenner. Catalysis in the nitrilase superfamily. *Curr Opin Struct Biol* 12: 775–782, 2002.

32. B Maras, D Barra, S Dupre, G Pitari. Is pantetheinase the actual identity of mouse and human vanin-1 proteins. *FEBS Lett* 461: 149–152, 1999.

33. H Cole, TR Reynolds, JM Lockyer, GA Buck, T Denson, JE Spence, J Hymes, B Wolf. Human serum biotinidase cDNA cloning, sequence, and characterization. *J Biol Chem* 269: 6566–6570, 1994.

34. H Cole Knight, TR Reynolds, GA Meyers, RJ Pomponio, GA Buck, B Wolf. Structure of the human biotinidase gene. *Mamm Genome* 9: 327–330, 1998.

35. KL Swango, B Wolf. Conservation of biotinidase in mammals and identification of the putative biotinidase gene in *Drosophila melanogaster*. *Mol Genet Metabol* 74: 492–499, 2001.

36. JS Stanley, JB Griffin, J Zempleni. Biotinylation of histones in human cells: effects of cell proliferation. *Eur J Biochem* 268: 5424–5429, 2001.

37. J Zempleni, DM Mock. Chemical synthesis of biotinylated histones and analysis by sodium dodecyl sulfate-polyacrylamide gel electrophoresis/streptavidin-peroxidase. *Arch Biochem Biophys* 371: 83–88, 1999.

38. KC Manthey, JB Griffin, J Zempleni. Biotin supply affects expression of biotin transporters, biotinylation of carboxylases, and metabolism of interleukin-2 in Jurkat cells. *J Nutr* 132: 887–892, 2002.

39. SB Scheerger, J Zempleni. Expression of oncogenes depends on biotin in human small cell lung cancer cells NCI-H69. *Int J Vitam Nutr Res* 73: 461–467, 2003.

40. SERH Crisp, G Camporeale, BR White, CF Toombs, JB Griffin, HM Said, J Zempleni. Biotin supply affects rates of cell proliferation, biotinylation of carboxylases and histones, and expression of the gene encoding the sodium-dependent multivitamin transporter in JAr choriocarcinoma cells. *Eur J Nutr* 43: 23–31, 2004.

41. DM Peters, JB Griffin, JS Stanley, MM Beck, J Zempleni. Exposure to UV light causes increased biotinylation of histones in Jurkat cells. *Am J Physiol Cell Physiol* 283: C878–C884, 2002.

42. G Camporeale, EE Shubert, G Sarath, R Cerny, J Zempleni. K8 and K12 are biotinylated in human histone H4. *Eur J Biochem* 271: 2257–2263, 2004.

43. J Hymes, K Fleischhauer, B Wolf. Biotinylation of biotinidase following incubation with biocytin. *Clin Chim Acta* 233: 39–45, 1995.

44. TD Ballard, J Wolff, JB Griffin, JS Stanley, SV Calcar, J Zempleni. Biotinidase catalyzes debiotinylation of histones. *Eur J Nutr* 41: 78–84, 2002.

45. J Zempleni. Biotin. In: BA Bowman, RM Russell, Eds. *Present Knowledge in Nutrition*. 8th ed. Washington, D.C.: ILSI Press, 2001, pp 241–252.

II

Amino Acids, Lipids, and Glycation

10

Nutrient Signaling to Muscle and Adipose Tissue by Leucine

THOMAS C. VARY AND
CHRISTOPHER J. LYNCH

INTRODUCTION

For the last few years, our laboratories have been examining the regulation of protein synthesis in muscle and adipose tissue by amino acids. Knowledge of this regulation is by no means new. Groups including Buse,[1-7] Morgan,[8,9] Tischler,[10,11] and Goldberg[12] recognized the ability of amino acids to regulate protein synthesis (and their ability to modify insulin effects on protein synthesis) in the mid-1970s and early 1980s. In both muscle and fat, amino acids stimulate protein synthesis, both independently and additively, with growth factors such as insulin and insulin-like growth factors (IGF). The presence of amino acids facilitates the ability of insulin to stimulate protein synthesis, and conversely the presence of insulin appears to facilitate the ability of amino acids to stimulate protein synthesis.

Our research has focused on several questions, among them, what amino acids regulate protein synthesis in muscle and fat? We review the evidence indicating that leucine is

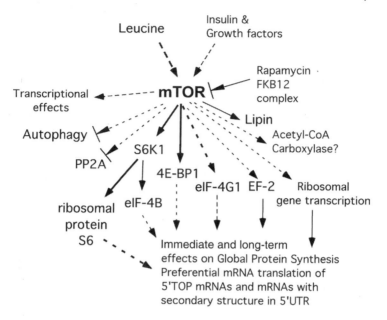

Figure 10.1   Activation of mTOR by amino acids and insulin leads to immediate and delayed effects on protein synthesis by its actions upon multiple targets including translational repressors such as the 4E-BP family, eukaryotic initiation factors, elongation factors, and ribosomal protein S6.

responsible for most of this effect. Research has also focused on the role of the mammalian target of rapamycin (mTOR) in this regulation. As shown schematically in Figure 10.1, mTOR plays a central role both in mediating the additive effects of leucine and metabolic signals from insulin on protein synthesis. In this regard, it is of interest to understand how leucine activates mTOR and how that activation leads to increases in protein synthesis. Regulation of cell processes by mTOR occurs at the levels of transcription of DNA,[13–15] translation of mRNA, and protein degradation (Figure 10.1).[16–22] This chapter will focus on the regulation of mRNA translation. Other questions include whether or not leucine is a *bona fide* nutrient signal, and what are the physiological roles of mTOR and the leucine signal. Review of these areas focuses primarily on results from muscle and adipose tissue.

Muscle makes the largest contribution to whole body protein synthesis in normal lean man. Consequently, it represents an important tissue, particularly in relation to protein turnover disregulation in catabolic diseases such as AIDS wasting, cancer, sepsis, diabetes mellitus, and alcoholism. On the other hand, although muscle is viewed as the most important tissue for whole body protein synthesis, it is difficult to obtain pure muscle preparations or cells for critical mechanistic *in vitro* studies. In contrast to muscle cells, fat cells can be readily and freshly isolated. Studies on the regulation of protein synthesis by amino acids in adipocytes have provided information applicable to muscle. Further, although the importance of muscle is usually agreed upon, it is unclear what percent of the whole body protein synthesis rate that adipose tissue contributes in the growing population of the overweight and obese. This is an underaddressed area of research. Given the growing epidemic of obesity and the evidence that leucine is a nutrient signal to adipose and muscle tissue, as well as the fact that both animal and human forms of obesity are associated with rises in branched-chain amino acids (BCAAs),[23–27] further study of the mechanism and consequences of leucine signaling to adipose tissue seems warranted.

UNIQUE ROLE OF LEUCINE IN AMINO ACID REGULATION OF mTOR AND PROTEIN SYNTHESIS

The synthesis of skeletal muscle proteins is rapidly stimulated after oral intake of nutrients. Of the nutrients provided by a complete meal, amino acids are important for enhancing protein accretion in skeletal muscle and fat. The concept that amino acids *per se* have a stimulatory effect on protein synthesis in muscle is supported by investigations utilizing muscles incubated *in vitro*,[3] as well as the perfused hindlimb.[28–30] In perfused muscles, elevating the amino acids to the concentration of normal rat plasma stimulates protein synthesis.[28–30] Likewise, infusing amino acids *in vivo* enhances muscle protein synthesis, independent of any stimulatory effect induced by changes in plasma anabolic hormone concentrations.[31–36]

Similarly, *in vitro* studies on freshly isolated adipocytes have shown that activation of mTOR and stimulation of protein synthesis can be brought about directly and independently of insulin or other anabolic hormones that rise in response to a meal *in vivo*.[37–41]

Of the BCAAs, leucine was recognized in early studies as being unique in its ability to stimulate protein turnover,[1–12] and much attention has focused subsequently on the use of BCAAs for nutritional support in various catabolic illnesses. The addition of BCAAs to total parenteral nutrition has been reported to improve nitrogen balance and the preservation of plasma protein concentrations. Parenteral administration of BCAA-enriched solutions improves mortality in septic patients.[42] BCAA administration improves nitrogen balance during sepsis[42,43] and results in a faster recovery of muscle protein.[44] BCAAs and leucine in particular intensify protein accretion in skeletal muscle. The ability of BCAA to spare protein is not mimicked by the transamination product of leucine, α-keto-isocaproic acid, indicating that it is the amino acid, rather than the metabolism of leucine, that is important in modulating protein turnover.[45]

Among the amino acids, leucine is best suited as a nutrient signal for several reasons. First, leucine is the most potent of the amino acids with regard to regulating protein metabolism in muscle[46] and activation of mTOR signaling in adipose tissue.[47] Second, leucine is an essential amino acid. Thus, changes in the plasma leucine concentrations must arise from nutrition, tissue breakdown, or changes in the rate of oxidation or loss. Third, leucine is often the most abundant essential amino acid in dietary protein and thus is a crucial nutrient following ingestion of a protein meal. Fourth, once taken into the gut, leucine cannot be directly metabolized by the liver, which lacks the mitochondrial branched-chain amino transferase (BCATm), the enzyme responsible for nitrogen removal. This may facilitate postprandial rises in plasma leucine concentrations, as leucine would not be cleared in the first pass of the liver, where many other amino acids are metabolized. Most other peripheral tissues that respond to leucine (with the exception of liver) have excess BCATm capacity to catalyze this first step in leucine metabolism and thereby begin to terminate any cell signaling stimulated by leucine.

In adipose tissue and muscle, protein synthesis and mTOR activation is stimulated by oral administration of the amino acid leucine. Using adipocytes, mTOR signaling was activated by a mixture of amino acids and by amino acid mixtures from which different amino acids were removed. Only the removal of leucine resulted in a loss of the ability of amino acid mixtures to stimulate 4E-BP1 phosphorylation.[39] The EC_{50} (concentration needed to obtain half maximal effect) for stimulation of 4E-BP1 phosphorylation occurred between 200–400 mM in isolated adipocytes.[39,40] This is within the range of circulating concentrations of leucine found in various nutritional states. This potency was also similar to results of *in vivo* dose–response studies examining S6K1 phosphorylation. *In vivo*, half maximal activation of S6K1 occurred at a mean plasma leucine concentration of 266 mM. The sensitivity of these changes is comparable to plasma leucine concentrations observed after a protein-containing meal. The potency of leucine and the fact that leucine alone can recapitulate the effects of a meal on protein synthesis and mTOR activation provide strong support that leucine is a *bona fide* nutritional signal.[48]

Structure activity studies in adipocytes have revealed that amino acids structurally related to leucine could also activate mTOR signaling *in vitro* but only at unphysiological concentrations.[40,49] The order of potency was Leu > norleucine > threo-L-hydroxyleucine Ile > Met Val. Although Ile, Met, and Val are able to activate mTOR signaling *in vitro*, the concentrations required are much higher than concentrations that these amino acids reach *in vivo*, even after a protein meal. Thus, only leucine appears to be a physiologically relevant activator of mTOR signaling, leading to protein synthesis regulation. The structure activity studies also revealed that a leucine analog, norleucine, was a leucine-like agonist. Norleucine is capable of activating the mTOR signaling pathway and protein synthesis, *in vitro* and *in vivo*. However, in contrast to leucine, norleucine does not stimulate insulin secretion or take part in protein synthesis (i.e., charge tRNA). The lack of effect on insulin secretion is valuable, as it has provided us with a novel tool with which to separate the direct effects of leucine from those that are mediated by the release of insulin (indirect) *in vivo*.

IMPORTANT ROLE OF mTOR

The effects of amino acids on protein synthesis in muscle and adipose tissue appear to be mediated in large part by a cell signaling pathway that centers around mTOR (aka FRAP and RAFT), a serine–threonine protein kinase.[50,51] The kinase is activated by increases in the concentration of amino acids, for example, as might occur after a protein-containing meal.

Figure 10.1 shows that mTOR in adipocytes is regulated by leucine and growth factors such as insulin. Activation of mTOR rapidly affects initiation and elongation steps in protein synthesis.[52–59] These short-term effects can have long-term consequences. For example, two substrates include the ribosomal protein S6 kinase-1 (S6K1) and the translational repressor, 4E-BP1.[54,56,60–68] S6K1-mediated phosphorylation of ribosomal protein S6 is associated with increased translation of mRNAs with a polypyrimidine tract in their 5 untranslated region (5′UTR). Many of these mRNAs code for proteins involved in protein synthesis. Thus, mTOR activation may promote ribosome biogenesis, leading to increased protein synthetic capacity. Similarly, phosphorylation of the translational repressor family of proteins exemplified by 4E-BP1 (aka PHAS-I, first discovered in fat) allows eIF4E to assemble with eIF4G and other factors into the 4F complex. For a rate-limiting step in translation, recognition of the mRNA cap structure 4E is required. This step may be particularly important for the translation of messages with significant secondary structure in their 5′UTR. By leading to the preferential translation of such messages, activation of mTOR may affect cell function.

Figure 10.1 indicates that activation of mTOR by leucine stimulates protein synthesis and inhibits autophagy. Thus, as might be expected from this scheme, infusion of a mixture of

BCAAs improves nitrogen balance following trauma. This effect is not mimicked or mediated by alanine.[69,70] These changes in nitrogen balance could result from a stimulation of protein synthesis, an inhibition of protein degradation, or both. Indeed, several studies have implicated mTOR and leucine in macroautophagy inhibition.[17–20,71–80] Thus, in some tissues, mTOR seems to both stimulate protein synthesis and decrease protein breakdown. With regard to skeletal muscle, however, leucine appears to act through acceleration of protein synthesis because excretion of 3-methylhistidine (a surrogate for the breakdown of myofibrillar proteins) is unaffected by leucine, inferring that degradation of muscle protein is not affected by leucine.[81] Likewise, feeding rats a meal composed of 20 or 25% protein stimulated protein synthesis, whereas protein degradation was unaffected.[82,83] Hence, the effects of amino acids on protein metabolism in skeletal muscle appear mediated mainly through an acceleration of protein synthesis. Although macroautophagy may be important in some peripheral tissues, its contribution to protein degradation may be less important in skeletal muscle.

In adipose tissue, mTOR is mostly membrane- or organelle-associated, with only about a half of that being solubilized by 1% triton-X 100 (i.e., cytoskeletal although Desai et al. has reported a mitochondrial association[22]). The remainder, about 30 to 40%, is found in the cytosolic fraction. Activation does not seem to be associated with a change in the subcellular distribution of mTOR (unpublished studies).

Several phosphorylation sites on mTOR have been reported;[84–86] however, the role of these phosphorylations is uncertain. For some sites, the phosphorylations correlate with mTOR activation; however, little mTOR phosphorylation change can be observed in the first few minutes following addition of amino acids when changes in the phosphorylations of the mTOR substrate 4E-BP1 can be most optimally observed (unpublished data). Thus, the exact mechanism by which leucine and insulin activate mTOR signaling has not been determined. It is hoped that emerging research on mTOR-interacting proteins such as Raptor will provide this answer.[87,88] Because the effects of leucine are independent of

PI3-kinase and Akt, at least in adipocytes,[41,49] and the actions of insulin and leucine are additive, it is thought that they activate mTOR by independent mechanisms.

Mitochondrial metabolism may be required for leucine activation of mTOR. Specifically, it has been proposed that a metabolically linked signal arising from activation of leucine metabolism in the mitochondria results in mTOR activation as described previously.[89] A potential target cited was the branched-chain ketoacid dehydrogenase (BCKD) complex (Figure 10.2). The BCKD multienzyme complex catalyzes the rate-limiting and first irreversible step in leucine oxidation. The reaction converts branched-chain α-keto acids (such as α-KIC) to CoA derivatives (such as isovaleryl-CoA). This conversion is catalyzed by BCKD. There are several compelling arguments in favor of the notion that BCKD kinase activities might provide a link between the leucine signal and the activation of mTOR. The first is that α-KIC is a physiological inhibitor of BCKD kinase. The second is that the order of potency of Leu > Ile > Val and the importance of leucine in the activation of mTOR signaling are similar to that of their respective keto acids' ability to activate the BCKD complex via inhibition of the BCKD kinase. A third observation that supports this potential mechanism is that mTOR appears to be associated with mitochondria membrane *in situ*.[22]

One way to delineate this effect is to simultaneously examine the dose-dependent effects of orally administered leucine on acute activation of S6K1 (an mTOR substrate) and phosphorylation of BCKD.[48] Increasing doses of leucine given orally via gavage directly correlated with elevations in plasma leucine concentration.[48] Phosphorylation of S6K1 (Thr 389, the phosphorylation site leading to activation) in adipose tissue was maximal at a dose of leucine that increased plasma leucine approximately threefold. Changes in BCKD phosphorylation state required higher plasma leucine concentrations. Hence, there is a disconnect between the concentration of leucine required for activation of the S6K1 system and phosphorylation of BCKD. Only at concentrations significantly higher than those required for activation of S6K1 is the phosphorylation state of BCKD lowered, allowing for oxidation of

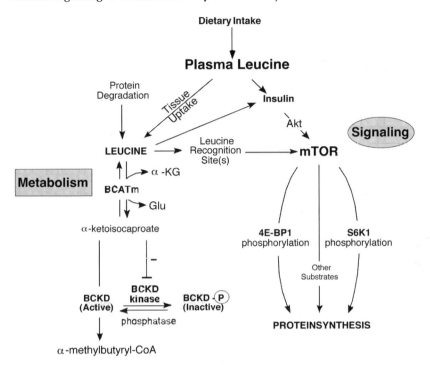

Figure 10.2 Changes in plasma leucine can arise from changes in metabolism, protein turnover, or dietary intake. Leucine activates mTOR directly but also is an insulin secretagogue and contributes to the rise in plasma insulin concentration after a meal. Thus, leucine can affect mTOR, directly and indirectly, by increasing insulin secretion.

leucine to occur. It might be argued, therefore, that leucine metabolism is not involved in mTOR activation. Hence, as the leucine signal becomes too high, there is a stimulation of leucine oxidation in peripheral tissue, thereby lowering the plasma leucine concentration. Under this interpretation as implied by Figure 10.2, leucine metabolism may only be important in regulating the magnitude and duration of the signal for mTOR activation.

It should be noted, however, that there is often a similar disconnect in signaling systems between concentration dependence of the early steps in signaling (e.g., IP3 generation in hepatocytes) and the consequent steps (e.g., calcium mobilization and phosphorylase activation), as shown previously.[90] Given the appeal of the metabolism and mitochondrial signal theory of mTOR activation, further studies are warranted.

Leucine-dependent stimulation of mRNA translation *in vivo* occurs in part, through a rapamycin-sensitive pathway.[50,51] Treatment with rapamycin, an inhibitor of the protein kinase mTOR, reduced the mTOR-dependent 4E-BP1 and S6K1 phosphorylation. However, leucine retained its ability to stimulate protein synthesis. Thus, inhibition of mTOR-mediated signaling was not capable of eliminating the leucine-dependent stimulation of protein synthesis. An additional study examining oral leucine administration in diabetic rats demonstrated enhanced levels of protein synthesis in skeletal muscle without corresponding changes in 4E-BP1 or S6K1.[91] Similarly, rates of skeletal muscle protein synthesis in perfused hindlimbs can be enhanced by elevated leucine concentrations in the absence of phosphorylation changes or 4E-BP1 and S6K1. Collectively, this evidence underscores the ability of leucine to modulate translation initiation and protein synthesis through an mTOR-independent mechanism.

STEPS IN TRANSLATION INITIATION OF PROTEIN SYNTHESIS MODIFIED BY LEUCINE

Two steps in translation initiation have been identified as the major regulatory points in the overall control of protein synthesis (Figure 10.3). The first one is the binding of met-tRNA$_i^{met}$ to the 40S ribosomal subunit to form the 43S preinitiation complex. This reaction is mediated by eukaryotic initiation factor 2 (eIF2) and is regulated by the activity of another eukaryotic initiation factor, eIF2B. The second regulatory step involves the binding of mRNA to the 43S preinitiation complex to form 48S initiation complex, which is mediated by eIF4F (Figure 10.3). A rise in plasma leucine, after a protein-containing meal or oral administration of leucine,

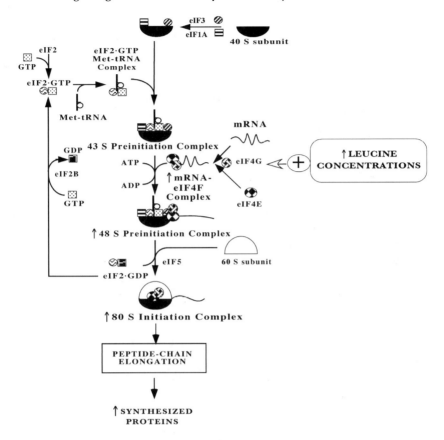

Figure 10.3 Major steps in the initiation of protein synthesis.

mainly affects the latter step. In skeletal muscle and fat, this is occurs by multiple mechanisms.

One of these steps affected by leucine involves the recognition, unwinding and binding of mRNA to the 40S ribosomal subunit (Figure 10.3). This step is catalyzed by the multisubunit complex of eukaryotic factors eIF4F.[92–94] eIF4F is composed of (1) eIF4A (a RNA helicase that functions with eIF4B to unwind secondary structure in 5-untranslated region of mRNA), (2) eIF4E (a protein that binds directly to the m7GTP cap structure present at the 5-end of most eukaryotic mRNAs), and (3) eIF4G (a protein that functions as a

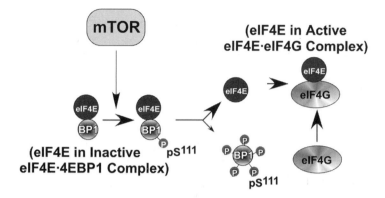

Figure 10.4   4E-BP1 phosphorylation by mTOR leads to increased availability of eukaryotic initiation factor 4E (eIF-4E).

scaffold for eIF4E, eIF4A, the mRNA, and the ribosome). eIF4G appears to be the nucleus around which the initiation complex forms because it has binding sites not only for eIF4E but also for eIF4A and eIF3.[95] eIF4E is the least abundant initiation factor in most cells. eIF4E activity plays a critical role in determining global rates of mRNA translation because essentially all mammalian mRNAs contain the m7GTP cap structure at their 5-ends.

eIF4E is also regulated by reversible phosphorylation.[93,94,96,97] eIF4E is phosphorylated following activation of the MNK1 kinase. Phosphorylation of eIF4E enhances the affinity of the factor for m7GTP cap on mRNA and for eIF4G and eIF4A, and correlates with enhanced rates of protein synthesis in cells in culture stimulated with mitogens, growth factors, or serum,[97,98] or transformed with ras or src oncogenes.[96–98] Conversely, reduced phosphorylation of eIF4E correlates with an inhibition of protein synthesis with serum depletion.[99] No evidence exists to suggest that leucine enhances the phosphorylation of eIF4E in skeletal muscle.[30,50,51,100,101]

Leucine can also stimulate translation initiation through the formation of the eIF4E–eIF4G complex (Figure 10.4). A positive linear relationship between rates of protein synthesis and amount of eIF4G associated with eIF4E in muscle is observed *in vivo*.[102] Although this correlation does not prove cause and effect, the relationship between protein synthesis and the amount of eIF4G associated with eIF4E is consistent with the proposed role of eIF4G–eIF4E complex in the overall regulation of protein synthesis. Reduced amounts of eIF4E associated with eIF4G would be expected to diminish the association of mRNA with the ribosome and hence limit protein synthesis, whereas the exact opposite effect would be expected with increased amounts of eIF4E associated with eIF4G.

The availability of eIF4E for binding to eIF4G is regulated, in part, through the association of eIF4E with a family of translational repressor proteins, 4E-BPs[103–106] (Figure 10.4). In skeletal muscle, 4E-BP1 is the predominant form of the 4E-BPs. When eIF4E is bound to 4E-BP1, eIF4E cannot bind to eIF4G. Consequently, the mRNA cannot bind to the ribosome,[107] and this inhibits cap-dependent translation of mRNA by physically sequestering eIF4E into an inactive 4E-BP1–eIF4E complex. In muscles from control rats, eIF4E binding to 4E-BP1 is reduced following phosphorylation of 4E-BPs by insulin through a PI3-kinase-dependent pathway involving signaling through Akt/PKB and FRAP/mTOR[65,108,109] (Figure 10.2). Leucine also stimulates 4E-BP1 phosphorylation.[18,39,41,46,110–114] Phosphorylation of 4E-BPs releases eIF4E from 4E-BP1–eIF4E complex and allows the eIF4E–mRNA complex to bind to eIF4G and then to the 40S ribosome.[103] The acceleration of protein synthesis with increasing amino acid concentrations is associated with increases in the amount eIF4G bound to eIF4E and a corresponding decrease in the amount of eIF4E in the inactive eIF4E–4E-BP1 complex.[30] The stimulation of protein synthesis is also associated with an increased assembly of active eIF4E–eIF4G complex. Hence, elevation in the plasma leucine concentration accelerates protein synthesis through increased association of eIF4G bound to eIF4E.

In addition to regulation by phosphorylation of 4E-BP1, one potential mechanism for enhanced binding of eIF4E to eIF4G may involve phosphorylation of eIF4G.[98,115] Increased phosphorylation of eIF4G correlates with conditions known to stimulate protein synthesis.[115] Likewise, a decreased eIF4G phosphorylation in presence of mTOR activation is observed in skeletal muscle of septic rats, showing a resistance to stimulation of protein synthesis and formation of active eIF4E–eIF4G complex by insulin.[102] Leucine also stimulates this step.[116]

The effect of an orally administered dose of leucine gavaged alone or in combination with carbohydrate on protein synthesis in gastrocnemius was examined to better understand the role of leucine in stimulating protein synthesis *in vivo*.[101] Rats were either freely fed (F) or food deprived for 18 h. After 18 h, the food-deprived rats were administered either saline (S), carbohydrate (CHO), leucine (L), or a combination of carbohydrate and leucine (CL). CHO and CL meals were isocaloric and provided 15% of the daily energy requirements. L and CL meals each delivered 270 mg of leucine. Muscle protein synthesis was reduced by 35% in S compared with F. Lower rates of protein synthesis in S was associated with a decreased hyperphosphorylation of 4E-BP1, leading to a greater association of 4E-BP1 with eIF4E and a concomitant reduction in eIF4E associated with eIF4G. Oral administration of leucine (L or CL), but not CHO, restored protein synthesis to values observed in F and resulted in a threefold enhanced phosphorylation of 4E-BP1. Consequently, the amount of eIF4E–4E-BP1 complex fell and binding of eIF4E to eIF4G increased. No differences in the phosphorylation state of eIF2α or the activity of eIF2B were observed among the different experimental groups.

Leucine may stimulate protein synthesis *in vivo* by enhancing binding of eIF4E to eIF4G independently of increases in serum insulin. Norleucine, an analog of leucine, was tested for its ability to increase rates of protein synthesis. In these studies, food-deprived (18 h) male Sprague–Dawley rats were orally administered solutions (2.5 ml/100 g body wt) containing normal saline (0.9% NaCl), a carbohydrate mixture (26.2% D-glucose and 26.2% sucrose), leucine (5.4%), or norleucine (5.4%).[110] Both leucine and norleucine administration stimulated protein synthesis in skeletal muscle and adipose tissue. In contrast to leucine, norleucine did not modulate plasma insulin concentrations. The carbohydrate meal had no effect on protein synthesis in any tissue tested but elicited a robust increase in the plasma insulin concentration. These findings also provide support for the role of leucine as a direct-acting nutrient signal for stimulation of protein synthesis in skeletal muscle and fat; its effects are independent of increases in plasma insulin concentrations.

PHYSIOLOGICAL ROLE OF THE LEUCINE SIGNALING TO mTOR

In the case of skeletal muscle, it is presumed that leucine is an anabolic signal to correlate anabolism (protein synthesis) to the availability of substrate for growth. No mTOR knockouts or conditional knockouts have been available to prove this. However, knockouts of some mTOR substrates have reduced body mass consistent with this idea.[117] Critical roles of mTOR and its targets have been implicated in protein synthesis, and in some tissues protein degradation, adipoblast differentiation,[118–121] leptin secretion,[49,114,122] *de novo* lipogenesis,[123–125] regulation of metabolic signals from insulin,[41,126,127] cell cycle progression,[128,129] tissue growth,[130–132] and adipose tissue morphogenesis,[38] lipin phosphorylation,[125] and regulation of acetyl CoA carboxylase activity.[123,124] Consequently, leucine in food has the potential to directly influence some or all of these processes. However, it is recognized that although some of these effects would promote the growth of adipose tissue, others such as the secretion of leptin would decrease adipose tissue size. It seems counterintuitive that leucine would stimulate opposing functions. However, physiologically, this could be viewed as a homeostatic response. Therefore, it seems likely that leucine is a physiological regulator of adipose growth, stimulating the protein synthesis necessary to expand the architecture of the adipocyte in response to nutrient intake but also providing a signal to the brain (leptin) that decreases eating behavior to prevent a feed-forward cycle.

ACKNOWLEDGMENT

This work was supported by the following grants from the National Institutes of Health: GM39277, AA12814 to TCV, and DK53843, DK62880 to CJL.

REFERENCES

1. MG Buse, Reid SS. Leucine. A possible regulator of protein turnover in muscle. *J Clin Invest*, 56: 1250–1261, 1975.

2. MG Buse, Weigand DA. Studies concerning the specificity of the effect of leucine on the turnover of proteins in muscles of control and diabetic rats. *Biochim Biophys Acta*, 475: 81–89, 1977.

3. MP Hedden, Buse MG. Effects of glucose, pyruvate, lactate, and amino acids on muscle protein synthesis. *Am J Physiol*, 242: E184–192, 1982.

4. MG Buse, Atwell R, Mancusi V. *In vitro* effect of branched-chain amino acids on the ribosomal cycle in muscles of fasted rats. *Hormone Metabol Res*, 11: 289–292, 1979.

5. MG Buse. *In vivo* effects of branched-chain amino acids on muscle protein synthesis in fasted rats. *Hormone Metab Res*, 13: 502–505, 1981.

6. ME May, Buse MG. Effects of branched-chain amino acids on protein turnover. *Diabetes Metabol Rev*, 5: 227–245, 1989.

7. MP Hedden, Buse MG. General stimulation of muscle protein synthesis by branched-chain amino acids *in vitro*. *Proc Soc Exp Biol Med*, 160: 410–415, 1979.

8. B Chua, Siehl DL, Morgan HE. Effect of leucine and metabolites of branched-chain amino acids on protein turnover in heart. *J Biol Chem*, 254: 8358–8362, 1979.

9. BH Chua, Siehl DL, Morgan HE. A role for leucine in regulation of protein turnover in working rat hearts. *Am J Physiol*, 239: E510–E514, 1980.

10. M Tischler, Desautels M, Goldberg A. Does leucine, leucyl-tRNA, or some other metabolite of leucine regulate protein synthesis and degradation in skeletal and cardiac muscle? *J Biol Chem*, 257: 1613–1621, 1982.

11. ME Tischler, Ost AH, Spina B, Cook PH, Coffman J. Regulation of protein turnover by glucose, insulin, and amino acids in adipose tissue. *Am J Physiol*, 247: C228–C233, 1984.

12. M Fulks, Li JB, Goldberg AL. Effects of insulin, glucose and amino acids on protein turnover in rat diaphragm. *J Biol Chem*, 250: 290–298, 1975.

13. Y Xia, Wen HY, Young ME, Guthrie PH, Taegtmeyer H, Kellems RE. Mammalian target of rapamycin and protein kinase A signaling mediate the cardiac transcriptional response to glutamine. *J Biol Chem*, 278: 13143–13150, 2003.

14. K Yokogami, Wakisaka S, Avruch J, Reeves SA. Serine phosphorylation and maximal activation of STAT3 during CNTF signaling is mediated by the rapamycin target mTOR. *Curr Biol*, 10: 47–50, 2000.

15. E Erbay, Park IH, Nuzzi PD, Schoenherr CJ, Chen J. IGF-II transcription in skeletal myogenesis is controlled by mTOR and nutrients. *J Cell Biol*, 163: 931–936, 2003.

16. EL Eskelinen, Prescott AR, Cooper J, Brachmann SM, Wang L, Tang X, Backer JM, Lucocq JM. Inhibition of autophagy in mitotic animal cells. *Traffic*, 3: 878–893, 2002.

17. S Mordier, Deval C, Bechet D, Tassa A, Ferrara M. Leucine limitation induces autophagy and activation of lysosome-dependent proteolysis in C2C12 myotubes through a mammalian target of rapamycin-independent signaling pathway. *J Biol Chem*, 275: 29900–29906, 2000.

18. A Beugnet, Tee AR, Taylor PM, Proud CG. Regulation of targets of mTOR (mammalian target of rapamycin) signaling by intracellular amino acid availability. *Biochem J*, 372: 555–566, 2003.

19. T Kanazawa, Taneike I, Akaishi R, Yoshizawa F, Furuya N, Fujimura S, Kadowaki M. Amino acids and insulin control autophagic proteolysis through different signaling pathways in relation to mTOR in isolated rat hepatocytes. *J Biol Chem*, 279: 8452–8459, 2004.

20. K Shigemitsu, Tsujishita Y, Hara K, Nanahoshi M, Avruch J, Yonezawa K. Regulation of translational effectors by amino acid and mammalian target of rapamycin signaling pathways. Possible involvement of autophagy in cultured hepatoma cells. *J Biol Chem*, 274: 1058–1065, 1999.

21. NS Cutler, Pan X, Heitman J, Cardenas ME. The TOR signal transduction cascade controls cellular differentiation in response to nutrients. *Mol Biol Cell*, 12: 4103–4113, 2001.

22. BN Desai, Myers BR, Schreiber SL. FKBP12-rapamycin-associated protein associates with mitochondria and senses osmotic stress via mitochondrial dysfunction. *Proc Natl Acad Sci USA*, 99: 4319–4324, 2002.

23. I Rafecas, Esteve M, Remesar X, Alemany M. Plasma amino acids of lean and obese Zucker rats subjected to a cafeteria diet after weaning. *Biochem Int*, 25: 797–806, 1991.

24. P Felig, Marliss E, Cahill GF, Jr. Plasma amino acid levels and insulin secretion in obesity. *N Eng J Med*, 281: 811–816, 1969.

25. P Felig, Marliss E, Cahill GF, Jr. Are plasma amino acid levels elevated in obesity? *N Eng J Med*, 282: 166, 1970.

26. RA Harris, Powell SM, Paxton R, Gillim SE, Nagae H. Physiological covalent regulation of rat liver branched-chain alpha-ketoacid dehydrogenase. *Arch Biochem Biophys*, 243: 542–555, 1985.

27. LP Opie, Walfish PG. Plasma-free fatty acid concentrations in obesity. *N Eng J Med*, 268: 757–760, 1963.

28. CV Jurasinski, Gray K, Vary TC. Modulation of skeletal muscle protein synthesis by amino acids and insulin during sepsis. *Metabol*, 44: 1130–1138, 1995.

29. JB Li, Jefferson LS. Influence of amino acid availability on protein turnover in perfused skeletal muscle. *Biochem Biophys Acta*, 544: 351–359, 1978.

30. TC Vary, Jefferson LS, Kimball SR. Amino acid-induced stimulation of translation initiation in rat skeletal muscle. *Am J Physiol*, 277: E1077–E1086, 1999.

31. E Svenberg, Moller-Loswick A-C, Mathews DE, Korner U, Anderson M, Lundholm K. Effects of amino acids on synthesis and degradation of skeletal muscle proteins in humans. *Am J Physiol*, 271: E718–E728, 1996.

32. M Giordano, Castellino P, DeFronzo RA. Differential responsiveness of protein synthesis and protein degradation to amino acid availability in humans. *Diabetes*, 45: 393–399, 1996.

33. WM Bennet, Connacher AA, Scrimgeour CM, Smith K, Rennie MJ. Increase in anterior tibialis muscle protein synthesis in healthy man during mixed amino acid infusion: studies of incorporation of [l-^{13}C]leucine. *Clin Sci*, 76: 447–454, 1989.

34. WM Bennet, Connacher AA, Scrimgeour CM, Rennie MJ. The effect of amino acid infusion on leg protein turnover assessed by L-[^{15}N]phenylalanine and L-[l–^{13}C]leucine exchange. *Eur J Clin Invest*, 20: 412–420, 1990.

35. NK Fukagawa, Minaker KL, Young VR. Leucine metabolism in aging humans: effects of insulin and substrate availability. *Am J Physiol*, 256: E288–E294, 1988.

36. P Tessari, Inchiostro S, Biolo G, Trevisan R, Fantin G, Marescotti MC, Iori E, Tiengo A, Crepaldi G. Differential effects hyperinsulinemia and hyperaminoacidemia on leucine-carbon metabolism *in vivo*. Role of substrate availability on estimates of whole-body protein synthesis. *J Clin Invest*, 79: 1062–1069, 1987.

37. S Marshall, Monzon R. Amino acid regulation of insulin action in isolated adipocytes. Selective ability of amino acids to enhance both insulin sensitivity and maximal insulin responsiveness of the protein synthesis system. *J Biol Chem*, 264: 2037–2042, 1989.

38. HL Fox, Kimball SR, Jefferson LS, Lynch CJ. Amino acids stimulate phosphorylation of p70S6k and organization of rat adipocytes into multicellular clusters. *Am J Physiol*, 274: C206–C213, 1998.

39. HL Fox, Pham PT, Kimball SR, Jefferson LS, Lynch CJ. Amino acid effects on translation repressor 4E-BP1 are mediated primarily by L-leucine in isolated adipocytes. *Am J Physiol*, 275: C1232–C1238, 1998.

40. CJ Lynch, Fox HE, Vary TC, Jefferson LS, Kimball SR. Regulation of amino acid-sensitive TOR signaling by leucine analogs in adipocytes. *J Cell Biochem*, 77: 234–251, 2000.

41. PT Pham, Heydrick SJ, Fox HL, Kimball SR, Jefferson LS, Jr., Lynch CJ. Assessment of cell-signaling pathways in the regulation of mammalian target of rapamycin (mTOR) by amino acids in rat adipocytes. *J Cell Biochem*, 79: 427–441, 2000.

42. A Garcia-de-Lorenzo, Ortiz-Leyba C, Plana M, Montejo JC, Nunez R, Ordonez FJ, Aragon C, Jimenez FJ. Parenteral administration of different amounts of branched-chain amino acids in septic patients: clinical and metabolic aspects. *Crit Care Med*, 25: 418–424, 1997.

43. FB Cerra, Hirsh J, Mullen K, Blackburn G, Luther W. The effect of stress level, amino acid formula, and nitrogen dose on nitrogen retention in traumatic and septic stress. *Ann Surg*, 205: 282–287, 1987.

44. FJ Jimenez-Jimenez, Ortiz-Leyba C, Morales-Menedez S, Barros-Perez M, Munoz-Gracia J. Prospective study on the efficacy of branched-chain amino acids in sepsis. *JPEN*, 15: 252–261, 1991.

45. PO Hasselgren, LaFrance R, Pederson P, James JH, Fischer JE. Infusion of branched-chain amino acid-enriched solution and alpha-ketisocaproic acid in septic rats: effects on nitrogen balance and skeletal muscle protein turnover. *JPEN*, 12: 244–249, 1988.

46. TG Anthony, Anthony JC, Yoshizawa F, Kimball SR, Jefferson LS. Oral administration of leucine stimulates ribosomal protein mRNA translation but not global rates of protein synthesis in the liver of rats. *J Nutr*, 131: 1171–1176, 2001.

47. CJ Lynch, Fox HL, Vary TC, Jefferson LS, Kimball SR. Regulation of amino acid-sensitive TOR signaling by leucine analogs in adipocytes. *J Cell Biochem*, 77: 234–251, 2000.

48. CJ Lynch, Halle B, Fujii H, Vary TC, Wallin R, Damuni Z, Hutson SM. Potential role of leuine metabolism in leucine-signaling pathway involving mTOR. *Am J Physiol*, 285: E854–E863, 2003.

49. CJ Lynch. Role of leucine in the regulation of mTOR by amino acids: revelations from structure-activity studies. *J Nutr*, 131: 861S–865S, 2001.

50. JC Anthony, Anthony TG, Kimball SR, Jefferson LS. Signaling pathway involved in translation control of protein synthesis in skeletal muscle by leucine. *J Nutr*, 131: 856S–860S, 2001.

51. JC Anthony, Yoshizawa F, Anthony TG, Vary TC, Jefferson LS, Kimball SR. Leucine stimulates translation initiation in skeletal muscle of postabsorptive rats via a rapamycin-sensitive pathway. *J Nutr*, 130: 2413–2419, 2000.

52. OJ Shah, Anthony JC, Kimball SR, Jefferson LS. 4E-BP1 and S6K1: translational integration sites for nutritional and hormonal information in muscle. *Am J Physiol*, 279: E715–E729, 2000.

53. P Neuhaus, Klupp J, Langrehr JM. mTOR inhibitors: an overview. *Liver Transplant*, 7: 473–484, 2001.

54. O Meyuhas. Synthesis of the translational apparatus is regulated at the translational level. *Eur J Biochem*, 267: 6321–6330, 2000.

55. SR Kimball, Farrell PA, Jefferson LS. Invited Review: Role of insulin in translational control of protein synthesis in skeletal muscle by amino acids or exercise. *J Appl Physiol*, 93: 1168–1180, 2002.

56. B Raught, Gingas A-C, Sonenberg N. The target of rapamycin (TOR) proteins. *Proc Nat Acad Sci USA*, 98: 7037–7044, 2001.

57. AC Gingras, Raught B, Sonenberg N. Control of translation by the target of rapamycin proteins. *Prog Mol Subcell Biol*, 27: 143–174, 2001.

58. AC Gingras, Raught B, Sonenberg N. Regulation of translation initiation by FRAP/mTOR. *Genes Dev*, 15: 807–826, 2001.

59. A Dufner, Andjelkovic M, Burgering BM, Hemmings BA, Thomas G. Protein kinase B localization and activation differentially affect S6 kinase 1 activity and eukaryotic translation initiation factor 4E-binding protein 1 phosphorylation. *Mol Cell Biol*, 19: 4525–4534, 1999.

60. AC Gingras, Kennedy SG, O'Leary MA, Sonenberg N, Hay N. 4E-BP1, a repressor of mRNA translation, is phosphorylated and inactivated by the Akt(PKB) signaling pathway. *Genes Dev*, 12: 502–513, 1998.

61. SR von Manteuffel, Gingras AC, Ming XF, Sonenberg N, Thomas G. 4E-BP1 phosphorylation is mediated by the FRAP-p70s6k pathway and is independent of mitogen-activated protein kinase. *Proc Natl Acad Sci USA*, 93: 4076–4080, 1996.

62. SR von Manteuffel, Dennis PB, Pullen N, Gingras AC, Sonenberg N, Thomas G. The insulin-induced signaling pathway leading to S6 and initiation factor 4E binding protein 1 phosphorylation bifurcates at a rapamycin-sensitive point immediately upstream of p70S6K. *Mol Cell Biol*, 17: 5426–5436, 1997.

63. JC Lawrence, Jr., Fadden P, Haystead TA, Lin TA. PHAS proteins as mediators of the actions of insulin, growth factors and cAMP on protein synthesis and cell proliferation. *Adv Enzyme Regul*, 37: 239–267, 1997.

64. GJ Brunn, Hudson CC, Sekulic A, Williams JM, Hosoi H, Houghton PJ, Lawrence JC, Jr., Abraham RT. Phosphorylation of the translational repressor PHAS-I by the mammalian target of rapamycin. *Science*, 277: 99–101, 1997.

65. S Kimball, Jurasinski C, Lawrence Jr J, Jefferson L. Insulin stimulates protein synthesis in skeletal muscle by enhancing the association of eIF-4E and eIF-4G. *Am J Physiol*, 272 (*Cell Physiol* 41): C754–C759, 1997.

66. JC Lawrence, Jr. mTOR-dependent control of skeletal muscle protein synthesis. *Int J Sport Nutr Exerc Metab*, 11 Suppl: S177–S185, 2001.

67. JC Lawrence, Jr., Brunn GJ. Insulin signaling and the control of PHAS-I phosphorylation. *Prog Mol Subcell Biol*, 26: 1–31, 2001.

68. H Tang, Hornstein E, Stolovich M, Levy G, Livingstone M, Templeton D, Avruch J, Meyuhas O. Amino acid-induced translation of TOP mRNAs is fully dependent on phosphatidylinositol 3-kinase-mediated signaling, is partially inhibited by rapamycin and is independent of S6K1 and rpS6 phosphorylation. *Mol Cell Biol*, 21: 8671–8683, 2001.

69. H Freund, Hoover HC, Atamian S, Fischer JE. Infusion of branched chain amino acids in postoperative patients. Anticatabolic properties. *Ann Surg*, 190: 18–23, 1979.

70. H Freund, Yoshimura N, Fischer JE. The role of alanine in nitrogen conserving quality of the branched-chain amino acids in the post-injury state. *Surg Res*, 29: 23–30, 1980.

71. B Grinde, Seglen PO. Effects of amino acids and amino acid analogs on lysosomal protein degradation in isolated rat hepatocytes. *Acta Biol Med Ger*, 40: 1603–1612, 1981.

72. PO Seglen, Gordon PB, Poli A. Amino acid inhibition of the autophagic/lysosomal pathway of protein degradation in isolated rat hepatocytes. *Biochim Biophys Acta*, 630: 103–118, 1980.

73. B Grinde, Seglen PO. Leucine inhibition of autophagic vacuole formation in isolated rat hepatocytes. *Exp Cell Res*, 134: 33–39, 1981.

74. GE Mortimore, Poso AR. The lysosomal pathway of intracellular proteolysis in liver: regulation by amino acids. *Adv Enzyme Regul*, 25: 257–276, 1986.

75. GE Mortimore, Poso AR. Intracellular protein catabolism and its control during nutrient deprivation and supply. *Annu Rev Nutr*, 7: 539–564, 1987.

76. G Miotto, Venerando R, Marin O, Siliprandi N, Mortimore GE. Inhibition of macroautophagy and proteolysis in the isolated rat hepatocyte by a nontransportable derivative of the multiple antigen peptide Leu8-Lys4-Lys2-Lys-beta Ala. *J Biol Chem*, 269: 25348–25353, 1994.

77. AJ Meijer. Amino acids as regulators and components of nonproteinogenic pathways. *J Nutr*, 133: 2057S–2062S, 2003.

78. EF Blommaart, Luiken JJ, Blommaart PJ, van Woerkom GM, Meijer AJ. Phosphorylation of ribosomal protein S6 is inhibitory for autophagy in isolated rat hepatocytes. *J Biol Chem*, 270: 2320–2326, 1995.

79. XM Leverve, Caro LH, Plomp PJ, Meijer AJ. Control of proteolysis in perfused rat hepatocytes. *FEBS Lett*, 219: 455–458, 1987.

80. LH Caro, Plomp PJ, Leverve XM, Meijer AJ. A combination of intracellular leucine with either glutamate or aspartate inhibits autophagic proteolysis in isolated rat hepatocytes. *Eur J Biochem*, 181: 717–720, 1989.

81. R Jacob, Hu X, Niederstock D, Hasan S, McNulty PH, Sherwin RS, Young LH. IGF-I stimulation of muscle protein synthesis in the awake rat; permissive role of insulin and amino acids. *Am J Physiol*, 270: E60–E66, 1996.

82. M Balage, Sinaud S, Prod'Homme M, Dardevet D, Vary TC, Kimball SR, Jefferson LS, Grizard J. Amino acids and insulin are both required to regulate assembly of eIF4E·eIF4G complex in rat skeletal muscle. *Am J Physiol*, 281: E565–E574, 2001.

83. F Yoshizawa, Kimball SR, Vary TC, Jefferson LS. Effect of dietary protein on translation initiation in rat skeletal muscle and liver. *Am J Physiol*, 275: E814–E820, 1998.

84. V Kumar, Sabatini D, Pandey P, Gingras AC, Majumder PK, Kumar M, Yuan ZM, Carmichael G, Weichselbaum R, Sonenberg N, Kufe D, Kharbanda S. Regulation of the rapamycin and FKBP-target 1/mammalian target of rapamycin and cap-dependent initiation of translation by the c-Abl protein-tyrosine kinase. *J Biol Chem*, 275: 10779–10787, 2000.

85. RT Peterson, Beal PA, Comb MJ, Schreiber SL. FKBP12-rapamycin-associated protein (FRAP) autophosphorylates at serine 2481 under translationally repressive conditions. *J Biol Chem*, 275: 7416–7423, 2000.

86. BT Nave, Ouwens M, Withers DJ, Alessi DR, Shepherd PR. Mammalian target of rapamycin is a direct target for protein kinase B: identification of a convergence point for opposing effects of insulin and amino-acid deficiency on protein translation. *Biochem J*, 344 Pt 2: 427–431, 1999.

87. K Hara, Maruki Y, Long X, Yoshino K, Oshiro N, Hidayat S, Tokuagna C, Avrch J, Yonezawa K. Raptor, binding partner of taget of rapamycin (TOR), mediates TOR action. *Cell*, 110: 177–189, 2002.

88. D-H Kim, Sarbassov DD, Ali SM, King JE, Latek RR, Erdjumnet-Bromage H, Tempest P, Sabatini DM. mTOR interacts with raptor to form a nutrient-sensitive complex that signals to cell growth machinery. *Cell*, 110: 163–175, 2002.

89. M McDaniel, Marshall C, Pappan K, Kwon G. Metabolic and autocrine regulation of the mammalian target of rapamycin by pancreatic β-cells. *Diabetes*, 51: 2877–2885, 2002.

90. CJ Lynch, Blackmore PF, Charest R, Exton JH. The relationships between receptor binding capacity for norepinephrine, angiotensin II, and vasopressin and release of inositol trisphosphate, Ca^{2+} mobilization, and phosphorylase activation in rat liver. *Mol Pharmacol*, 28: 93–99, 1985.

91. JC Anthony, Reiter AK, Anthony TG, Crozier SJ, Lang CH, MacLean DA, Kimball SR, Jefferson LS. Orally administered leucine enhances protein synthesis in muscle of diabetic rats in the absence of increases in 4E-BP1 or S6K1 phopshorylation. *Diabetes*, 51: 928–936, 2002.

92. RE Rhoads, Joshi-Barve S, Minich WB. Participation of initiation factors in recruitment of mRNA to ribosomes. *Biochimie*, 76: 831–838, 1994.

93. RE Rhoads. Regulation of eukaryotic protein synthesis by initiation factors. *J Biol Chem*, 268: 3017–3020, 1993.

94. N Sonenberg. Regulation of translation and cell growth by eIF-4E. *Biochimie*, 76: 839–846, 1994.

95. BJ Lamphear, Kirchweger JR, Skern T, Rhoads RE. Mapping of functional domains in eukaryotic protein synthesis initiation factor 4G (eIF-4G) with pircoviral proteases. Implication for cap-dependent and cap-independent translation initiation. *J Biol Chem*, 270: 21975–21983, 1995.

96. SJ Morley, Traugh JA. Phorbal esters stimulate phosphorylation of eukaryotic initiation factors 3, 4B, and 4F. *J Biol Chem*, 264: 2401–2404, 1989.

97. SJ Morley, Traugh JA. Differential stimulation of phosphorylation of initiation factors eIF-4F, eIF-4B, eIF-3 and ribosomal protein S6 by insulin and phorbol esters. *J Biol Chem*, 265: 10611–10616, 1990.

98. SJ Morley, Traugh JA. Stimulation of translation in 3T3-L1 cells in response to insulin and phorbol ester is directly correlated with increased phosphate labeling of initiation factor (eIF-)4F and ribosomal protein S6. *Biochimie*, 75: 985–989, 1993.

99. R Duncan, Milburn SC, Hershey JWB. Regulated phosphorylation and low abundance of HeLa cell initiation factor eIF-4F suggest a role in translational control. Heat shock effects. *J Biol Chem*, 262: 380–388, 1987.

100. SR Kimball, Hoetsky RL, Jefferson LS. Implication of eIF2B rather than eIF4E in the regulation of global protein synthesis by amino acids in L6 myoblasts. *J Biol Chem*, 273: 30945–30953, 1998.

101. JC Anthony, Gautsch T, Kimball SR, Vary TC, Jefferson LS. Orally administered leucine stimulates protein synthesis in skeletal muscle of postabsorptive rats in association with increased eIF4F formation. *J Nutr*, 130: 139–145, 2000.

102. TC Vary, Jefferson LS, Kimball SR. Insulin fails to stimulate muscle protein synthesis in sepsis despite unimpaired signaling to 4E-BP1 and S6K1. *Am J Physiol*, 281: E1045–E1053, 2001.

103. T Lin, Kong X, Haystead T, Pause A, Belsham G, Sonnenberg N, Lawrence J. PHAS-I as a link between mitogen activated protein kinase and translation initiation. *Science*, 266: 653–656, 1994.

104. T Lin, Kong X, Saltiel AR, Blackshear PJ, Lawrence JC. Control of PHAS-I by insulin in 3T3-L1 adipocytes. Synthesis, degradation, and phosphorylation by a rapamycin-sensitive and mitogen-activated protein kinase dependent pathway. *J Biol Chem*, 270: 18531–18535, 1995.

105. BK Bhandari, Feliers D, Duraisamy S, Stewart JL, Gingas A-C, Abbound HE, Choudhury GG, Sonenberg N, Kasinath BS. Insulin regulation of protein translation repressor 4E-BP1, an eIF4E-binding protein, in renal epithelial cells. *Kidney Int*, 59: 866–875, 2001.

106. A Pause, Belsham GJ, Gingras A-C, Donze O, Lin T-A, Lawrence Jr JC, Sonenberg N. Insulin-dependent stimulation of protein synthesis by phosphorylation of a regulator of 5′-cap function. *Nature*, 371: 762–767, 1994.

107. A Haghihat, Maderr S, Pause A, Sonenberg N. Repression of cap-dependent translation by 4E-binding protein I: competition with p220 for binding to eukaryotic initiation factor-4E. *EMBO J*, 14: 5701–5709, 1995.

108. S Kimball, Jefferson L, Fadden P, Haystead T, Lawrence Jr J. Insulin and diabetes cause reciprocal changes in the association of eIF-4E and PHAS-I in rat skeletal muscle. *Am J Physiol*, 270: C705–C709, 1996.

109. SR Kimball, Horetsky RL, Jefferson LS. Signal transduction pathways involved in the regulation of protein synthesis by insulin in L6 myoblasts. *Am J Physiol*, 274: C221–C228, 1998.

110. CJ Lynch, Patson BJ, Anthony JC, Vaval A, Jefferson LS, Vary TC. Leucine is a direct acting nutrient signal that regulates protein synthesis in adipose tissue. *Am J Physiol*, 283: E824–E835, 2002.

111. SR Kimball, Shantz LM, Horetsky RL, Jefferson LS. Leucine regulates translation of specific mRNAs in L6 myoblasts through mTOR-mediated changes in availability of eIF4E and phosphorylation of ribosomal protein S6. *J Biol Chem*, 274: 11647–11652, 1999.

112. T Peng, Golub TR, Sabatini DM. The immunosuppressant rapamycin mimics a starvation-like signal distinct from amino acid and glucose deprivation. *Mol Cell Biol*, 22: 5575–5584, 2002.

113. CG Proud. Regulation of mammalian translation factors by nutrients. *Eur J Biochem*, 269: 5338–5349, 2002.

114. C Roh, Han J, Tzatsos A, Kandror KV. Nutrient-sensing mTOR-mediated pathway regulates leptin production in isolated rat adipocytes. *Am J Physiol*, 284: E322–330, 2003.

115. B Raught, Gingras A-C, Gygi SP, Imaataka H, Morino S, Gradi A, Aebersold R, Sonenberg N. Serum-stimulated, rapamycin-sensitive phosphorylation sites in eukaryotic translation initiation factor 4GI. *EMBO J*, 19: 434–444, 2000.

116. CH Lang, Frost RA, Deshpande N, Kumar V, Vary TC, Jefferson LS, Kimball SR. Alcohol impairs leucine-mediated phosphorylation of 4E-BP1, S6K1, eIF4G, and mTOR in skeletal muscle. *Am J Physiol*, 285: E1205–1215, 2003.

117. M Miron, Verdu J, Lachance PE, Birnbaum MJ, Lasko PF, Sonenberg N. The translational inhibitor 4E-BP is an effector of PI(3)K/Akt signaling and cell growth in drosophila. *Nat Cell Biol*, 3: 596–601, 2001.

118. A Gagnon, Lau S, Sorisky A. Rapamycin-sensitive phase of 3T3-L1 preadipocyte differentiation after clonal expansion. *J Cell Physiol*, 189: 14–22, 2001.

119. A Bell, Grunder L, Sorisky A. Rapamycin inhibits human adipocyte differentiation in primary culture. *Obes Res*, 8: 249–254, 2000.

120. WC Yeh, Bierer BE, McKnight SL. Rapamycin inhibits clonal expansion and adipogenic differentiation of 3T3-L1 cells. *Proc Natl Acad Sci USA*, 92: 11086–11090, 1995.

121. H Inuzuka, Nanbu-Wakao R, Masuho Y, Muramatsu M, Tojo H, Wakao H. Differential regulation of immediate early gene expression in preadipocyte cells through multiple signaling pathways. *Biochem Biophys Res Commun*, 265: 664–668, 1999.

122. RL Bradley, Cheatham B. Regulation of ob gene expression and leptin secretion by insulin and dexamethasone in rat adipocytes. *Diabetes*, 48: 272–278, 1999.

123. U Krause, Bertrand L, Hue L. Control of p70 ribosomal protein S6 kinase and acetyl-CoA carboxylase by AMP-activated protein kinase and protein phosphatases in isolated hepatocytes. *Eur J Biochem*, 269: 3751–3759, 2002.

124. U Krause, Bertrand L, Maisin L, Rosa M, Hue L. Signalling pathways and combinatory effects of insulin and amino acids in isolated rat hepatocytes. *Eur J Biochem*, 269: 3742–3750, 2002.

125. TA Huffman, Mothe-Satney I, Lawrence JC, Jr. Insulin-stimulated phosphorylation of lipin mediated by the mammalian target of rapamycin. *Proc Natl Acad Sci USA*, 99: 1047–1052, 2002.

126. A Takano, Usui I, Haruta T, Kawahara J, Uno T, Iwata M, Kobayashi M. Mammalian target of rapamycin pathway regulates insulin signaling via subcellular redistribution of insulin receptor substrate 1 and integrates nutritional signals and metabolic signals of insulin. *Mol Cell Biol*, 21: 5050–5062, 2001.

127. ME Patti, Brambilla E, Luzi L, Landaker EJ, Kahn CR. Bidirectional modulation of insulin action by amino acids. *J Clin Invest*, 101: 1519–1529, 1998.

128. M Vilella-Bach, Nuzzi P, Fang Y, Chen J. The FKBP12-rapamycin-binding domain is required for FKBP12-rapamycin-associated protein kinase activity and G1 progression. *J Biol Chem*, 274: 4266–4272, 1999.

129. FJ Dumont, Su Q. Mechanism of action of the immunosuppressant rapamycin. *Life Sci*, 58: 373–395, 1996.

130. A Simm, Schluter K, Diez C, Piper HM, Hoppe J. Activation of p70(S6) kinase by beta-adrenoceptor agonists on adult cardiomyocytes. *J Mol Cell Cardiol*, 30: 2059–2067, 1998.

131. C Rommel, Bodine SC, Clarke BA, Rossman R, Nunez L, Stitt TN, Yancopoulos GD, Glass DJ. Mediation of IGF-1-induced skeletal myotube hypertrophy by PI(3)K/Akt/mTOR and PI(3)K/Akt/GSK3 pathways. *Nat Cell Biol*, 3: 1009–1013, 2001.

132. K Tsukiyama-Kohara, Poulin F, Kohara M, DeMaria CT, Cheng A, Wu Z, Gingras AC, Katsume A, Elchebly M, Spiegelman BM, Harper ME, Tremblay ML, Sonenberg N. Adipose tissue reduction in mice lacking the translational inhibitor 4E-BP1. *Nat Med*, 7: 1128–1132, 2001.

11

Anabolic Effects of Amino Acids and Insulin in the Liver

LUC BERTRAND AND LOUIS HUE

INTRODUCTION

The metabolism of amino acids is well known. Amino acids are not only building blocks for protein synthesis, they also participate in the synthesis of numerous nitrogen-containing substances. In addition, they contribute to glucose homeostasis by being glucose precursors and they directly participate in the transport of ammonia from extra-hepatic tissues to the liver, where they deliver ammonia for ureagenesis.

Amino acids are not only metabolic substrates; they also favor anabolic pathways, such as glycogen synthesis, lipogenesis, and protein synthesis, and inhibit glycogen breakdown, ketogenesis, and protein degradation.[1-6] More recently, the effect of amino acids on protein synthesis has been the focus of intense research, and several laboratories are actively engaged in elucidation of the underlying mechanism.[7-11] It is the purpose of this chapter to briefly describe the various anabolic effects of amino acids and to emphasize the differences in their effects. In a first approximation, regulatory amino acids can be divided into two groups. The first group, represented by glutamine, affects glycogen, lipid, and protein metabolism and induces cell swelling.[7,12,13] The experimental evidence so far accumulated indicates that cell swelling mediates the anabolic effect in this group. The possibility that a common regulatory mechanism is triggered by glutamine and the nature of this mechanism is discussed. The second group is represented by leucine.[14-16] Its main effect is directed towards protein metabolism, i.e., inhibition of protein degradation and stimulation of protein synthesis. Work from our laboratory also indicated that leucine stimulates lipogenesis without affecting glycogen synthesis and cell volume. The underlying mechanism involved in the stimulation of protein synthesis is discussed in the light of recent findings concerning the mammalian target of rapamycin (mTOR), which appears as a central control point. Finally, these effects of amino acids are reminiscent of the anabolic effect of insulin in muscle and adipocytes.[17] Data from our laboratory and others indicate that insulin alone exerts little effect on protein synthesis in hepatocytes, and that leucine exerts permissive effects and allows insulin to stimulate protein synthesis.[8,14,18,19] A possible synergistic mechanism induced by insulin and leucine is reviewed.

STIMULATION OF GLYCOGEN SYNTHESIS AND LIPOGENESIS

Overall Effect of Amino Acids and Insulin

Glycogen synthesis and lipogenesis are controlled by the activity of glycogen synthase (GS) and acetyl CoA carboxylase (ACC), respectively. The activity of both enzymes is controlled by (de)phosphorylation, the active form being dephosphorylated. They possess multiple phosphorylation sites, which follow a hierarchical order. Phosphorylation of the primary sites allows for the phosphorylation of the secondary sites by different protein kinases, eventually leading to inactivation of these enzymes. Conversely, their activation results from dephosphorylation.[20-22]

  In perfused livers and isolated hepatocytes from rats, the synthesis of glycogen and GS activation are stimulated by

amino acids such as glutamine, alanine, asparagine, and proline.[1,23,24] Glutamine, proline, and, to a lesser extent, alanine
also stimulate lipogenesis and inhibit ketogenesis.[2,25] Leucine,
a branched-chain amino acid that is poorly metabolized by
the liver, also activates ACC.[25] However, the effects of leucine
and glutamine clearly differ; first, leucine does not affect
glycogen synthesis and GS; second, leucine does not change
cell volume; third, the time course of ACC activation by leucine differs from the one by glutamine: the maximal effect of
leucine occurs within 10 min, whereas it requires 45 to 60
min with glutamine; fourth, the overall effect of leucine on
ACC activation is smaller than that of glutamine; fifth, leucine antagonizes glutamine regarding GS activation, although
their effect on ACC is additive. In hepatocytes from diabetic
rats, the effects of amino acids on ACC and GS activation
persist, and insulin has no effect by itself or in combination
with glutamine.[19] Taken together, these data indicate that
glutamine and leucine act by different mechanisms and are
independent of insulin.

Incubation of freshly isolated hepatocytes with insulin
alone does not activate GS and ACC, nor does it modify the
effect of glutamine. However, insulin enhances ACC activation
by leucine but not by glutamine.[19] The lack of effect of insulin
alone contrasts with the well-known activation of both
enzymes in skeletal muscle, adipocytes, and hepatocytes in
culture.[17,26,27] It cannot be attributed to a defect in insulin
signaling, which is functional in hepatocytes, from the insulin
receptor down to protein kinase B (PKB or Akt).[19] These data
confirm that the effects of glutamine and leucine differ and
indicate that leucine, but not glutamine, exerts a permissive
effect on insulin action.

Mechanism of Activation of GS and ACC

A stimulation of glycogen synthesis is also observed with
amino-isobutyric acid, a nonmetabolizable amino acid analogue, which is transported in a Na^+-dependent manner similar to glutamine.[24] This led to the hypothesis that stimulation
of glycogen synthesis results from cell swelling and ion

changes following the Na^+-dependent transport of amino acids. In keeping with this hypothesis, a direct relationship between cell swelling and glycogen synthesis is observed in hepatocytes incubated with various amino acids. Moreover, hypoosmotic media, even in the absence of amino acids, also activate GS in relation with cell swelling. Conversely, GS activation by amino acids is blocked by hyperosmotic media, which prevent cell swelling.[24] Similarly, ACC activation by amino acids is also related to changes in cell volume, suggesting the involvement of a common control mechanism resulting from cell swelling.[25] However, this does not apply to leucine, which activates ACC without changing cell volume.[25]

Swollen cells react to this osmotic stress by activating a mechanism known as *regulatory volume decrease*, which tends to restore the initial cell volume.[28] This induces an electrogenic K^+-efflux followed by Cl ions that distribute across the plasma membrane according to its potential. This decrease in intracellular concentrations of Cl ions affects the activity of the protein phosphatases that activate GS and ACC. GS activation is indeed inhibited by the concentrations of Cl ions found in normal cells but not by the decreased concentrations observed in swollen cells.[29] On the other hand, glutamate, the concentration of which increases in cells incubated with glutamine and even further with glutamine and leucine, stimulates ACC activation by a type 2A protein phosphatase, which is called glutamate-activated protein phosphatase (GAPP).[29–32] These data suggest that protein phosphatases are involved in the activation of GS and ACC. The implication of protein phosphatase is also supported by inhibition of the glutamine-induced activation of GS and ACC by inhibitors of type 1 and 2A protein phosphatases.[33]

ACC activity is controlled, among other factors, by its phosphorylation state, and phosphorylation of Ser 79 is known to inactivate ACC. This Ser is phosphorylated by AMPK, a well-conserved protein kinase that senses the energy state of the cell (see the section titled "AMPK").[22,34] GAPP, the glutamate-activated protein phosphatase, dephosphorylates Ser 79 and thus activates ACC.[32] Therefore, a decrease in the phosphorylation state of Ser 79 is expected to occur when ACC is activated in hepatocytes incubated with glutamine. Surprisingly, such a decrease is not found, although ACC activation is clearly mediated by protein phosphatase, as demonstrated by the use of inhibitors of protein phosphatases.[33] Because ACC is known to be multiphosphorylated, we speculate that it contains inactivating sites other than Ser 79 that are dephosphorylated by GAPP.

A stimulation of protein phosphatase is also involved in GS activation by glutamine and it is proposed that the fall in the intracellular concentration of Cl ions stimulates glycogen synthase phosphatase.[29] However, little is known about the actual phosphorylation state of GS in hepatocytes incubated with glutamine. Because GS is also multiphosphorylated, one is tempted to speculate that a common mechanism leading to the dephosphorylation of inactivating sites on both GS and ACC explains their activation by glutamine. Whether GAPP acts on both enzymes is not known.

Little is known about the mechanism of ACC activation by leucine, which is not dependent on cell swelling. However, leucine is able to stimulate liver glutaminase and increase the intracellular concentration of glutamate.[19] The resulting stimulation of GAPP by glutamate could explain the synergism with glutamine in ACC activation.

PROTEIN SYNTHESIS

Control of Protein Synthesis

The control of protein synthesis is complex and involves (de)phosphorylation of various translation factors and ribosomal proteins.[7,18,35–40] Control is exerted on both initiation and elongation. Initiation of translation corresponds to the building up of the ribosomal complex with the correctly positioned mRNA. Elongation represents the actual synthesis of the protein and consumes four ATP equivalents per peptide bond. The eukaryotic initiation factor 2 (eIF2) is a G protein, which recruits the starting methionyl tRNA to the 40S subunit. It is activated by the guanine nucleotide exchange factor (GEF), eIF2B, which is regulated independently of mTOR.[37] Glycogen synthase kinase 3 (GSK3), a protein kinase in the insulin signaling pathway, phosphorylates and inactivates eIF2B. Insulin acts through protein kinase B (PKB or Akt), which is currently believed to mediate most, if not all, short-term metabolic effects of insulin, and which phosphorylates and inactivates GSK3. Because of this double inactivation, insulin action results in eIF2 activation that promotes the formation of the 43S preinitiation complex. The formation of the complex also implies the association of another factor, eIF4E, a process that is prevented by eIF4E binding to 4E-BP1. 4E-BP1 contains several phosphorylation sites and phosphorylation by mTOR releases eIF4E, which then participates in the formation of the preinitiation complex. The importance of these various factors in the control of protein synthesis by amino acids has been extensively reviewed.[36,39] Finally, the eukaryotic elongation factor 2 (eEF2) is phosphorylated and inactivated by a dedicated protein kinase, eEF2-kinase. This kinase is itself inactivated by phosphorylation by the mTOR signaling pathway, which leads to the phosphorylation of 4E-BP1 and activation of p70S6K.[40]

The protein kinase p70S6K controls protein synthesis in response to hormones, mitogens, and nutrients.[8,37,41] It phosphorylates the S6 ribosomal protein present in the 40S ribosomal subunit. S6 is involved in the translation of mRNAs that contain oligopyrimidine sequences upstream of their initiation site and that are members of the terminal oligopyrimidine (TOP) tract family of mRNA. The proteins encoded by these mRNAs are ribosomal proteins and proteins involved in the translation machinery. p70S6K activation results from multiple hierarchical phosphorylations by several protein kinases.[41–43] The most important one is the mammalian target of rapamycin (mTOR) because (1) phosphorylation of p70S6K on Thr 389 by mTOR correlates with p70S6K activation; and (2) rapamycin, the potent inhibitor of mTOR, prevents p70S6K activation in all reported cases of p70S6K activation. It should be added that the phosphorylation of Thr 389 by mTOR is required for the subsequent phosphorylation of Thr 229 by the 3-phosphoinositide-dependent protein kinase (PDK-1), a constitutively active protein kinase.[42]

It is now clear that mTOR, a protein kinase of about 2550 amino acids, plays a central role in the control of protein synthesis by nutrients and energy.[9,44,45] The catalytic domain of mTOR, which is located near the C-terminus, shares common structural features with lipid kinases such as phosphatidylinositol-3-kinase (PtdIns-3-K).[46] Despite this similarity, mTOR acts as a serine–threonine protein kinase.[47] Both mTOR and PtdIns-3-K are inhibited by wortmannin,[48] although mTOR is about ten times less sensitive to wortmannin than PtdIns-3-K. Nevertheless, this sensitivity complicates the study of the implication of PtdIns-3-K in a signaling pathway. However, mTOR is inhibited by rapamycin and PtdIns-3-K is not. The immunosuppressant rapamycin binds to the peptidyl-prolyl *cis-trans* isomerase, FKBP12 (FK506 binding protein), and to the FRB (FKBP12-rapamycin binding) domain of mTOR, which is N-terminally located, relative to the catalytic domain of mTOR.[49]

The affinity of mTOR for ATP is rather poor (K_m about 1 mM in contrast with other protein kinases, K_ms of which for ATP are about tenfold lower). Therefore, mTOR has been proposed to act as an ATP sensor of the cell.[50] However, AMPK seems much better suited for this role, and it is now demonstrated that the inactivation of p70S6K that occurs when ATP content of the cell decreases is probably due to AMPK activation, which inactivates the mTOR signaling pathway[33,51–53] (see section titled "AMPK").

The C-terminal end of mTOR contains a regulatory domain and several residues in this domain are phosphorylated in response to insulin, growth factors, or nutrients.[54–57] Ser 2481 corresponds to an autophosphorylation site, which is believed to reflect mTOR activity. Ser 2448 is phosphorylated in response to insulin through PKB activation. However, it is not clear that Ser 2448 phosphorylation correlates with mTOR activation and is required to phosphorylate the downstream targets of mTOR. In addition, a novel nutrient-regulated phosphorylation site has been identified in mTOR. Thr 2446 is phosphorylated under conditions of nutrient deprivation. This phosphorylation seems to be mediated by AMPK and decreases the subsequent phosphorylation of Ser 2448 in

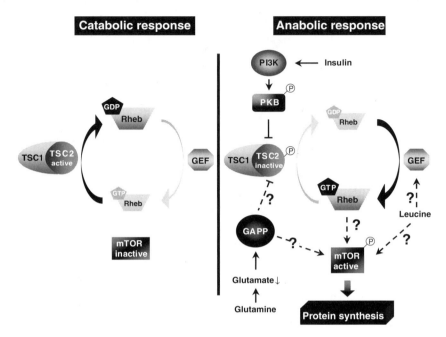

Figure 11.1 Hypothetical mechanisms of activation of mTOR by insulin and amino acids. Under catabolic conditions (left panel), the TSC1-TSC2 complex is active and acts as a GTPase-activating protein, which favors the conversion of the active GTP-bound form of Rheb into its inactive GDP-bound form. Insulin, glutamine, and leucine lead to mTOR phosphorylation by different mechanisms (right panel). Insulin induces the phosphorylation and inactivation of TSC2 by PKB. It is proposed that glutamine activates mTOR through GAPP, a type 2A protein phosphatase. This phosphatase could dephosphorylate the activating site on TSC2 that is phosphorylated by AMPK (see also Figure 11.2). The effect of leucine, which is additive with those of glutamine and insulin, is proposed to be mediated by GEF, the guanine nucleotide exchange factor, which converts Rheb into its active GTP-bound form. See the text for details.

response to insulin.[58] It has been proposed that differential phosphorylation of Thr 2446 and Ser 2448 acts as a switch mechanism to control protein synthesis in response to nutrient availability and insulin or growth factors.

The mechanism of activation of mTOR is complex, and recent evidence points to an indirect mechanism involving a small G protein and a GTPase-activating protein (GAP)[44,45,59] (Figure 11.1). The G protein Rheb (Ras homologue enriched in brain) activates mTOR, although the exact mechanism of this activation is not fully understood. Similar to other G proteins, Rheb acts as a molecular switch. Its inactive, GDP-bound state is favored by a protein complex that stimulates GTPase activity. This protein complex contains two proteins, hamartin (TSC1) and tuberin (TSC2), encoded by the genes that are mutated in a genetic disorder known as tuberous sclerosis, which causes the appearance of benign tumors (hamartomas) in various tissues. The C-terminal domain of TSC2 is homologous to the catalytic domain of the GTPase-activating proteins, and the complex TSC1-TSC2 stimulates the intrinsic GTPase activity of Rheb, thereby promoting its conversion from the GTP-bound active state to the GDP-bound inactive state. Thus, the TSC1-TSC2 complex inhibits mTOR signaling. Moreover, PKB phosphorylates TSC2 on several sites in the C-terminal domain, thereby inactivating the TSC1-TSC2 complex, keeping Rheb in its active form and favoring mTOR activation.[60–63] Interestingly, AMPK also phosphorylates TSC2 but on sites other than those phosphorylated by PKB[64] (Figure 11.2). Phosphorylation of the AMPK sites enhances the GTPase activity of the TSC complex and keeps Rheb in its inactive form. These observations indicate that the effects of TSC2 phosphorylation by PKB and AMPK on mTOR activity depend on the sites phosphorylated in TSC2. On the other hand, activation of Rheb through its conversion from a GDP-bound form to an active GTP-bound form is expected to be favored by a GEF that remains to be discovered.

The activity of mTOR on its downstream targets is also regulated by at least two proteins that form a complex with mTOR and correctly position p70S6K and 4EBP1 to be phosphorylated by mTOR[65–67] (Figure 11.3). These proteins, raptor (regulatory associated protein of mTOR) and the mammalian homologue of the yeast protein LST8 (mLST8), contain numerous WD40 repeats as well as HEAT (Huntington–elongation factor–A subunit of PP2A-TOR) repeats, which are

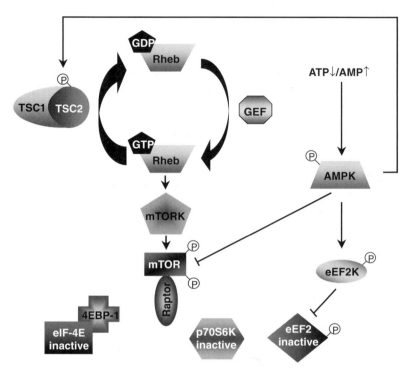

Figure 11.2 Inhibition of protein synthesis by AMPK. AMPK inhibits protein synthesis by phophorylating three targets: (i) TSC2, the site phosphorylated by AMPK, activates TSC2 and thereby inactivates Rheb. This site can be dephosphorylated by GAPP. It differs from the site phosphorylated by PKB, which is activating Rheb. (ii) mTOR, the inactivating site (Ser 2446) phosphorylated by AMPK, is close to the activating site (Ser 2448). (iii) eEF2-kinase, activated by AMPK, induces the phosphorylation and inactivation of eEF2. See the text for details.

also present in mTOR. These repeats could be engaged in multiple protein–protein interactions. A few conserved residues at the N- and C-terminal ends of p70S6K and 4EBP1, respectively, form the so-called TOS (TOR signaling) motif and participate in substrate binding to raptor and in the correct positioning of the substrates to be phosphorylated by mTOR.

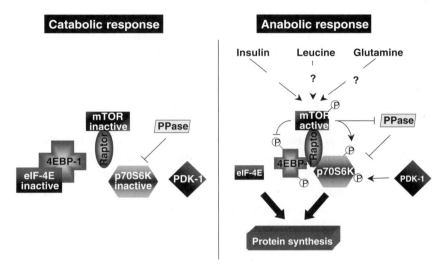

Figure 11.3 Hypothetical mechanisms of activation of the mTOR complex by insulin and amino acids. In the absence of insulin and amino acids, mTOR is inactive (left panel). Under anabolic conditions (right panel), raptor correctly positions 4E-BP1 and p70S6K to be phosphorylated by mTOR; PDK-1 can then phosphorylate and fully activate p70S6K. In addition, mTOR inactivates a protein phosphatase, thereby enhancing the phosphorylation of p70S6K. Activation of mTOR by insulin is not sufficient to activate p70S6K and requires the permissive effect of leucine. This amino acid could act on raptor and so facilitate the phosphorylation and activation of the downstream targets. See the text for details.

Overall Effects of Amino Acids and Insulin

Because amino acids are direct precursors of proteins, their availability controls protein metabolism. Since the pioneering work of Mortimore,[6,68,69] it is now well established that certain amino acids control protein turnover and inhibit autophagy in the liver. Meijer and coworkers[14,70] were the first to demonstrate that certain amino acids stimulate the phosphorylation of the ribosomal protein S6 in hepatocytes. They showed that inhibition of autophagy and S6 phosphorylation are equally sensitive to the same amino acids, namely leucine, tyrosine, and phenylalanine. Rapamycin blocks these effects indicating that the mTOR/p70S6K pathway is involved. Remarkably, insulin alone has no effect unless low concentrations of amino acids are added. These seminal observations[14,70] have later been confirmed in various tissues by many other groups.[8,18,71–75] These studies, including the present one, demonstrate that incubation of hepatocytes with leucine enhances the phosphorylation and activity of p70S6K synergistically with insulin and in a rapamycin-sensitive manner.

The effect of glutamine, alone or in combination with leucine or insulin, was investigated by our group.[19,33,71] It was demonstrated that not only leucine but also glutamine could activate p70S6K. Glutamine was also found to enhance the effect of leucine on p70S6K activation and Ser 2448 phosphorylation in mTOR. However, glutamine did not act in synergy with insulin, whereas leucine did. The effect of leucine on p70S6K activation was slightly more rapid and less pronounced than with glutamine. Maximal activation of p70S6K was obtained with a combination of both amino acids, resulting in one of the highest activity of this protein kinase, i.e., 1 nmol/min/g of cells. It was also observed that in hepatocytes from diabetic rats p70S6K activation by leucine persisted, although the total p70S6K activity was lower than in hepatocytes from control rats. Therefore, the effects of glutamine and leucine do not require insulin and persist in diabetic rats. The liver response to these amino acids, which includes antiketotic and antiproteolytic effects as well as a permissive effect on insulin action, could improve the metabolic state of the diabetic patients.

Amino Acid Signaling

These data, taken together with our results on GS and ACC activation, confirm that leucine signaling to p70S6K activation differs from glutamine signaling, both being different from insulin signaling. Although the upstream signaling pathways may differ, they all converge on mTOR because rapamycin blocked p70S6K activation in all conditions. Thus, mTOR appears as a common target for these amino acids and insulin as well.

AMPK

Glycogen synthesis, lipogenesis, and protein synthesis are energy- and nutrient-sensitive. When the energy supply becomes limited, the fall in ATP concentration turns off these energy-consuming pathways. The mechanism involved is not restricted to ATP limitation but also involves activation of AMPK (Figure 11.2) by the increase in the AMP to ATP ratio.[34,76] AMPK is indeed able to inactivate liver GS and ACC. ACC inactivation results from phosphorylation of Ser 79; the site phosphorylated by AMPK in GS is not known. AMPK activation antagonizes the activation of not only ACC but also p70S6K by amino acids.[33,52] P70S6K is not a direct substrate of AMPK, and its inactivation results from a decreased phosphorylation of Thr 389, the site phosphorylated by mTOR. Recent reports indicate that AMPK could act on two targets upstream of p70S6K: (1) the direct phosphorylation of mTOR on Ser 2446, close to the activating Ser 2448, seems to inactivate mTOR;[58] and (2) AMPK could phosphorylate and activate TSC2, thereby promoting Rheb inactivation.[64] Phosphorylation of TSC2 by PKB has the opposite effect indicating that the sites phosphorylated by PKB and AMPK on TSC2 are different.

Inactivation of mTOR, either directly by phosphorylation of Ser 2446 or indirectly by phosphorylation and activation of TSC2, results in the inactivation of both p70S6K and 4E-BP1, two controlling steps of protein synthesis. Moreover, recent work from our group has demonstrated that AMPK activation results in the phosphorylation and inactivation of eEF2.[77] Here, again, the mechanism of inactivation is indirect. AMPK phosphorylates and activates eEF2-kinase, the upstream protein kinase, which in turn phosphorylates and inactivates eEF2, thereby inhibiting the elongation step in protein synthesis.

Clearly, AMPK activation leads to the inhibition of the stimulation of glycogen synthesis, lipogenesis, and protein synthesis by amino acids. It is, therefore, tempting to speculate that an inhibition of AMPK by amino acids could mediate the stimulation of the anabolic pathways by amino acids.

AMPK would then act as a master switch for nutrients. This is probably not the case. In hepatocytes incubated under normoxic conditions, the activity of AMPK is already barely detectable, and amino acids cannot further decrease this basal activity. In addition, amino acids do not antagonize the dose-dependent activation of AMPK that is induced by incubation of hepatocytes with a known stimulator of AMPK. We can, therefore, conclude that an amino acid-induced inactivation of AMPK is ruled out.

Protein Phosphatases

The activation of ACC and p70S6K by amino acids is blocked by inhibitors of type 1 and 2A protein phosphatases.[33] As discussed in the preceding text, ACC activation does not involve dephosphorylation of the inactivating Ser 79, and it is speculated that GAPP dephosphorylates inactivating sites other than Ser 79 (see the section titled "Mechanism of Activation of GS and ACC").

The active form of p70S6K is phosphorylated and p70S6K activation by amino acids corresponds to an increased phosphorylation of Thr 389 by mTOR. It is expected that Thr 229 phosphorylation by PDK1 could also participate in p70S6K activation by amino acids. In support of the presence of an inhibiting protein phosphatase downstream of mTOR is our observation that p70S6K activation by amino acids is reinforced when inhibitors of protein phosphatase are added after preincubation with amino acids.[33] The mTOR-sensitive protein phosphatase would then be inhibited, thereby reinforcing the direct effect of mTOR on Thr 389 phosphorylation (Figure 11.3). Another protein phosphatase is also involved in p70S6K activation by amino acids,[33] because preincubation of hepatocytes with protein phosphatase inhibitors prevents p70S6K activation by amino acids. Therefore, a protein phosphatase upstream of mTOR is probably involved. We suggest that this phosphatase is GAPP, which could activate mTOR by dephosphorylating either the inactivating Ser 2446 in mTOR or the site phosphorylated by AMPK in TSC2, thereby promoting Rheb activation (Figure 11.1). Taken together these data support the hypothesis that two protein phosphatases are involved, one upstream of mTOR, which could be GAPP leading to ACC and p70S6K activation (Figure 11.1), and the other downstream of mTOR, which inactivates p70S6K (Figure 11.3).

Insulin and Leucine

Insulin alone is unable to activate p70S6K. Because insulin signaling in hepatocytes results in PKB activation, we propose that the lack of p70S6K activation by insulin results from the inhibition of a step downstream of PKB. Our recent work indicates that the limiting control step is at the level, or downstream, of mTOR. Indeed, several studies, including ours (Figure 11.4), indicate that Ser 2448 of mTOR is phosphorylated after incubation of hepatocytes with glutamine, leucine, or even with insulin alone. Therefore, the insulin signaling pathway leads to mTOR phosphorylation, probably through the action of PKB, which inactivates the TSC1-TSC2 complex and keeps Rheb in its active conformation (Figure 11.1). However, phosphorylation of Ser 2481, the autophosphorylation site, which is thought to reflect mTOR activation, is only increased in hepatocytes incubated with amino acids (Figure 11.4). This suggests that the amino acids are able to activate mTOR, and that the synergism between insulin and leucine on p70S6K activation is located downstream of mTOR and could be at the level of raptor mLST8 or another protein recruited on the mTOR complex (Figure 11.3).

Besides its permissive effects on insulin action, leucine alone is able to activate mTOR and p70S6K. On the basis of results reported in the preceding text, we suggest the following mechanisms. Leucine and insulin increase Ser 2448 phosphorylation by different mechanisms. Insulin acts through PKB and TSC2, whereas leucine could act on the GEF that activates Rheb. Therefore, leucine would control p70S6K at two levels, one upstream of mTOR by activating GEF (Figure 11.1), the other at the level of the mTOR complex (Figure 11.3).

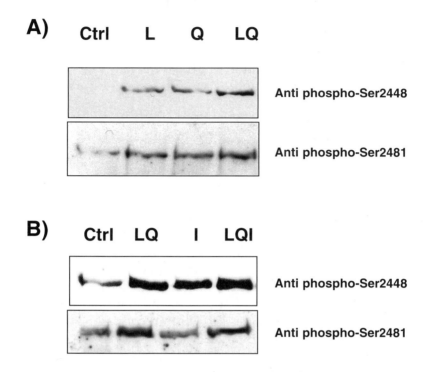

Figure 11.4   Effects of glutamine, leucine, and insulin on the phosphorylation state of Ser 2448 and Ser 2481 of mTOR in hepatocytes. Panel (A): Leucine (L), glutamine (Q), or a combination of both (LQ) increase the phosphorylation state of Ser 2448 and Ser 2481. This results in p70S6K activation. Panel (B): Insulin (I) alone increases the phopshorylation of Ser 2448 without affecting Ser 2481, an autophosphorylation site, which is thought to represent mTOR activity. Accordingly, insulin alone is unable to activate p70S6K. Leucine exerts a permissive effect on insulin action and further activates p70S6K in the presence of insulin.

CONCLUSIONS

In hepatocytes, amino acids and insulin exert anabolic effects that are synergystic and mediated by different but interconnected signaling pathways. Leucine, but not glutamine, exerts a permissive effect on insulin signaling to both p70S6K and ACC. The activation of p70S6K is inhibited by rapamycin indicating that mTOR is crucial in this process. Control of mTOR is distributed between the proteins controlling mTOR activation (i.e., the G protein Rheb, TSC1/TSC2 and GEF) and the proteins forming the mTOR complex (raptor and mLST8), which are involved in the correct positioning of p70S6K and 4EBP1, the substrates of mTOR.

In contrast with leucine, glutamine exerts no permissive effect on insulin action. Glutamine activates GS, ACC, and p70S6K by mechanisms that differ from those of leucine and are mediated by changes in cell volume and in the concentration of intracellular glutamate and Cl ions.

Protein phosphatases are involved in the anabolic effects of these amino acids because the activation of GS, ACC, and p70S6K is prevented by inhibitors of type 1 and type 2A protein phosphatase. We suggest that one of these protein phosphatases is GAPP, which dephosphorylates inactivating sites in GS, ACC, and TSC2. These inactivating sites remain to be identified. Interestingly, AMPK overrules the control by amino acids and inhibits their anabolic effects. This protein kinase acts on the key control steps, namely GS, ACC, and mTOR. By doing so, AMPK conserves cellular energy by switching off ATP-consuming pathways.

Many questions remain unanswered. Is there a specific leucine sensor? Are GEF and the mTOR complex the targets of leucine in the control of p70S6K activation? What are the mechanisms responsible for the insulin resistance that is observed in the liver and for the permissive effect of leucine on insulin action? Clearly, the important progress made recently in our understanding of mTOR signaling are encouraging and augur well for what remains to be found.

ACKNOWLEDGMENTS

The work carried out in the authors' laboratory was supported by grants from the Belgium Fund for Medical Scientific Research, the European Union, and the federal and regional authorities of Belgium.

References

1. A Lavoinne, A Baquet, L Hue. Stimulation of glycogen synthesis and lipogenesis by glutamine in isolated rat hepatocytes. *Biochem J* 248: 429–437, 1987.

2. A Baquet, A Lavoinne, L Hue. Comparison of the effects of various amino acids on glycogen synthesis, lipogenesis and ketogenesis in isolated rat hepatocytes. *Biochem J* 273: 57–62, 1991.

3. D Haussinger, F Lang. Cell volume in the regulation of hepatic function: a mechanism for metabolic control. *Biochim Biophys Acta* 1071: 331–350, 1991.

4. B Stoll, W Gerok, F Lang, D Haussinger. Liver cell volume and protein synthesis. *Biochem J* 287: 217–222, 1992.

5. PO Seglen, PB Gordon, A Poli. Amino acid inhibition of the autophagic/lysosomal pathway of protein degradation in isolated rat hepatocytes. *Biochim Biophys Acta* 630: 103–118, 1980.

6. GE Mortimore, AR Poso. Intracellular protein catabolism and its control during nutrient deprivation and supply. *Annu Rev Nutr* 7: 539–564, 1987.

7. DA van Sluijters, PF Dubbelhuis, EF Blommaart, AJ Meijer. Amino acid–dependent signal transduction. *Biochem J* 351: 545–550, 2000.

8. K Hara, K Yonezawa, QP Weng, MT Kozlowski, C Belham, J Avruch. Amino acid sufficiency and mTOR regulate p70 S6 kinase and eIF-4E BP1 through a common effector mechanism. *J Biol Chem* 273: 14484–14494, 1998.

9. CG Proud. mTOR-mediated regulation of translation factors by amino acids. *Biochem Biophys Res Commun* 313: 429–436, 2004.

10. BD Manning, LC Cantley. United at last: the tuberous sclerosis complex gene products connect the phosphoinositide 3-kinase/Akt pathway to mammalian target of rapamycin (mTOR) signalling. *Biochem Soc Trans* 31: 573–578, 2003.

11. AJ Meijer, PF Dubbelhuis. Amino acid signalling and the integration of metabolism. *Biochem Biophys Res Commun* 313: 397–403, 2004.

12. L Hue. Control of liver carbohydrate and fatty acid metabolism by cell volume. *Biochem Soc Trans* 22: 505–508, 1994.

13. D Haussinger. The role of cellular hydration in the regulation of cell function. *Biochem J* 313: 697–710, 1996.

14. EF Blommaart, JJ Luiken, PJ Blommaart, GM van Woerkom, AJ Meijer. Phosphorylation of ribosomal protein S6 is inhibitory for autophagy in isolated rat hepatocytes. *J Biol Chem* 270: 2320–2326, 1995.

15. G Mortimore, M Kadowaki. Regulation of protein metabolism in Liver, in *Handbook of Physiology*. Eds. Jefferson L S, Cherrington AD. New York: American Physiological Society, Oxford University Press, 2001, pp 553–577.

16. S Vom Dahl, D Haussinger. The role of amino acids in the control of proteolysis in *Metabolic and Therapeutic Aspects of Amino Acids in Clinical Nutrition*. Ed. Cynober LA, Boca Raton, FL: CRC Press, 2004, pp 275–289.

17. AR Saltiel, CR Kahn. Insulin signalling and the regulation of glucose and lipid metabolism. *Nature* 414: 799–806, 2001.

18. LE Campbell, X Wang, CG Proud. Nutrients differentially regulate multiple translation factors and their control by insulin. *Biochem J* 344: 433–441, 1999.

19. U Krause, L Bertrand, L Maisin, M Rosa, L Hue. Signalling pathways and combinatory effects of insulin and amino acids in isolated rat hepatocytes. *Eur J Biochem* 269: 3742–3750, 2002.

20. PJ Roach. Control of glycogen synthase by hierarchal protein phosphorylation. *FASEB J* 4: 2961–2968, 1990.

21. SP Davies, AT Sim, DG Hardie. Location and function of three sites phosphorylated on rat acetyl-CoA carboxylase by the AMP-activated protein kinase. *Eur J Biochem* 187: 183–190, 1990.

22. DG Hardie. Regulation of fatty acid and cholesterol metabolism by the AMP-activated protein kinase. *Biochim Biophys Acta* 1123: 231–238, 1992.

23. J Katz, S Golden, PA Wals. Stimulation of hepatic glycogen synthesis by amino acids. *Proc Natl Acad Sci USA* 73: 3433–3437, 1976.

24. A Baquet, L Hue, AJ Meijer, GM van Woerkom, PJ Plomp. Swelling of rat hepatocytes stimulates glycogen synthesis. *J Biol Chem* 265: 955–959, 1990.

25. A Baquet, L Maisin, L Hue. Swelling of rat hepatocytes activates acetyl-CoA carboxylase in parallel to glycogen synthase. *Biochem J* 278: 887–890, 1991.

26. SK Moule, RM Denton. Multiple signaling pathways involved in the metabolic effects of insulin. *Am J Cardiol* 80: 41A–49A, 1997.

27. M Peak, JJ Rochford, AC Borthwick, SJ Yeaman, L Agius. Signalling pathways involved in the stimulation of glycogen synthesis by insulin in rat hepatocytes. *Diabetologia* 41: 16–25, 1998.

28. EK Hoffmann, LO Simonsen. Membrane mechanisms in volume and pH regulation in vertebrate cells. *Physiol Rev* 69: 315–382, 1989.

29. AJ Meijer, A Baquet, L Gustafson, GM van Woerkom, L Hue. Mechanism of activation of liver glycogen synthase by swelling. *J Biol Chem* 267: 5823–5828, 1992.

30. PJ Plomp, L Boon, LH Caro, GM van Woekom, AJ Meijer. Stimulation of glycogen synthesis in hepatocytes by added amino acids is related to the total intracellular content of amino acids. *Eur J Biochem* 191: 237–243, 1990.

31. A Baquet, V Gaussin, M Bollen, W Stalmans, L Hue. Mechanism of activation of liver acetyl-CoA carboxylase by cell swelling. *Eur J Biochem* 217: 1083–1089, 1993.

32. V Gaussin, L Hue, W Stalmans, M Bollen. Activation of hepatic acetyl-CoA carboxylase by glutamate and Mg2+ is mediated by protein phosphatase-2A. *Biochem J* 316: 217–224, 1996.

33. U Krause, L Bertrand, L Hue. Control of p70 ribosomal protein S6 kinase and acetyl-CoA carboxylase by AMP-activated protein kinase and protein phosphatases in isolated hepatocytes. *Eur J Biochem* 269: 3751–3759, 2002.

34. DG Hardie, D Carling, M Carlson. The AMP-activated/SNF1 protein kinase subfamily: metabolic sensors of the eukaryotic cell? *Annu Rev Biochem* 67: 821–855, 1998.

35. OJ Shah, JC Anthony, SR Kimball, LS Jefferson. 4E-BP1 and S6K1: translational integration sites for nutritional and hormonal information in muscle. *Am J Physiol Endocrinol Metab* 279: E715–729, 2000.

36. AC Gingras, B Raught, N Sonenberg. Regulation of translation initiation by FRAP/mTOR. *Genes Dev* 15: 807–826, 2001.

37. CG Proud. Regulation of mammalian translation factors by nutrients. *Eur J Biochem* 269: 5338–5349, 2002.

38. SR Kimball, LS Jefferson. Control of protein synthesis by amino acid availability. *Curr Opin Clin Nutr Metab Care* 5: 63–67, 2002.

39. AC Gingras, B Raught, N Sonenberg. eIF4 initiation factors: effectors of mRNA recruitment to ribosomes and regulators of translation. *Annu Rev Biochem* 68: 913–963, 1999.

40. GJ Browne, CG Proud. Regulation of peptide-chain elongation in mammalian cells. *Eur J Biochem* 269: 5360–5368, 2002.

41. J Avruch, C Belham, Q Weng, K Hara, K Yonezawa. The p70 S6 kinase integrates nutrient and growth signals to control translational capacity. *Prog Mol Subcell Biol* 26: 115–154, 2001.

42. N Pullen, PB Dennis, M Andjelkovic, A Dufner, SC Kozma, BA Hemmings, G Thomas. Phosphorylation and activation of p70s6k by PDK1. *Science* 279: 707–710, 1998.

43. M Saitoh, N Pullen, P Brennan, D Cantrell, PB Dennis, G Thomas. Regulation of an activated S6 kinase 1 variant reveals a novel mammalian target of rapamycin phosphorylation site. *J Biol Chem* 277: 20104–20112, 2002.

44. E Jacinto, MN Hall. TOR signalling in bugs, brain and brawn. *Nat Rev Mol Cell Biol* 4: 117–126, 2003.

45. TE Harris, JC Lawrence, Jr. TOR signaling. *Sci STKE* 2003 Dec. 9: re15, 2003.

46. SB Helliwell, P Wagner, J Kunz, M Deuter-Reinhard, R Henriquez, MN Hall. TOR1 and TOR2 are structurally and functionally similar but not identical phosphatidylinositol kinase homologues in yeast. *Mol Biol Cell* 5: 105–118, 1994.

47. PE Burnett, RK Barrow, NA Cohen, SH Snyder, DM Sabatini. RAFT1 phosphorylation of the translational regulators p70 S6 kinase and 4E-BP1. *Proc Natl Acad Sci USA* 95: 1432–1437, 1998.

48. GJ Brunn, J Williams, C Sabers, G Wiederrecht, JC Lawrence, Jr., RT Abraham. Direct inhibition of the signaling functions of the mammalian target of rapamycin by the phosphoinositide 3-kinase inhibitors, wortmannin and LY294002. *EMBO J* 15: 5256–5267, 1996.

49. J Choi, J Chen, SL Schreiber, J Clardy. Structure of the FKBP12-rapamycin complex interacting with the binding domain of human FRAP. *Science* 273: 239–242, 1996.

50. PB Dennis, A Jaeschke, M Saitoh, B Fowler, SC Kozma, G Thomas. Mammalian TOR: a homeostatic ATP sensor. *Science* 294: 1102–1105, 2001.

51. HR Samari, PO Seglen. Inhibition of hepatocytic autophagy by adenosine, aminoimidazole-4-carboxamide riboside, and N6-mercaptopurine riboside. Evidence for involvement of amp-activated protein kinase. *J Biol Chem* 273: 23758–23763, 1998.

52. PF Dubbelhuis, AJ Meijer. Hepatic amino acid-dependent signaling is under the control of AMP-dependent protein kinase. *FEBS Lett* 521: 39–42, 2002.

53. DR Bolster, SJ Crozier, SR Kimball, LS Jefferson. AMP-activated protein kinase suppresses protein synthesis in rat skeletal muscle through down-regulated mammalian target of rapamycin (mTOR) signaling. *J Biol Chem* 277: 23977–23980, 2002.

54. PH Scott, GJ Brunn, AD Kohn, RA Roth, JC Lawrence, Jr. Evidence of insulin-stimulated phosphorylation and activation of the mammalian target of rapamycin mediated by a protein kinase B signaling pathway. *Proc Natl Acad Sci USA* 95: 7772–7777, 1998.

55. QP Weng, M Kozlowski, C Belham, A Zhang, MJ Comb, J Avruch. Regulation of the p70 S6 kinase by phosphorylation *in vivo*. Analysis using site-specific anti-phosphopeptide antibodies. *J Biol Chem* 273: 16621–16629, 1998.

56. BT Nave, M Ouwens, DJ Withers, DR Alessi, PR Shepherd. Mammalian target of rapamycin is a direct target for protein kinase B: identification of a convergence point for opposing effects of insulin and amino-acid deficiency on protein translation. *Biochem J* 344: 427–431, 1999.

57. TH Reynolds, SC Bodine, JC Lawrence, Jr. Control of Ser 2448 phosphorylation in the mammalian target of rapamycin by insulin and skeletal muscle load. *J Biol Chem* 277: 17657–17662, 2002.

58. SW Cheng, LG Fryer, D Carling, PR Shepherd. T2446 is a novel mTOR phosphorylation site regulated by nutrient status. *J Biol Chem* in press, 2004.

59. BD Manning, LC Cantley. Rheb fills a GAP between TSC and TOR. *Trends Biochem Sci* 28: 573–576, 2003.

60. K Inoki, Y Li, T Zhu, J Wu, KL Guan. TSC2 is phosphorylated and inhibited by Akt and suppresses mTOR signalling. *Nat Cell Biol* 4: 648–657, 2002.

61. CJ Potter, LG Pedraza, T Xu. Akt regulates growth by directly phosphorylating TSC2. *Nat Cell Biol* 4: 658–665, 2002.

62. X Gao, Y Zhang, P Arrazola, O Hino, T Kobayashi, RS Yeung, B Ru, D Pan. Tsc tumour suppressor proteins antagonize amino-acid-TOR signalling. *Nat Cell Biol* 4: 699–704, 2002.

63. AR Tee, DC Fingar, BD Manning, DJ Kwiatkowski, LC Cantley, J Blenis. Tuberous sclerosis complex-1 and -2 gene products function together to inhibit mammalian target of rapamycin (mTOR)-mediated downstream signaling. *Proc Natl Acad Sci USA* 99: 13571–13576, 2002.

64. K Inoki, T Zhu, KL Guan. TSC2 mediates cellular energy response to control cell growth and survival. *Cell* 115: 577–590, 2003.

65. DH Kim, DD Sarbassov, SM Ali, JE King, RR Latek, H Erdjument-Bromage, P Tempst, DM Sabatini. mTOR interacts with raptor to form a nutrient-sensitive complex that signals to the cell growth machinery. *Cell* 110: 163–175, 2002.

66. K Hara, Y Maruki, X Long, K Yoshino, N Oshiro, S Hidayat, C Tokunaga, J Avruch, K Yonezawa. Raptor, a binding partner of target of rapamycin (TOR), mediates TOR action. *Cell* 110: 177–189, 2002.

67. K Yonezawa, C Tokunaga, N Oshiro, K Yoshino. Raptor, a binding partner of target of rapamycin. *Biochem Biophys Res Commun* 313: 437–441, 2004.

68. GE Mortimore, CM Schworer. Induction of autophagy by amino-acid deprivation in perfused rat liver. *Nature* 270: 174–176, 1977.

69. GE Mortimore, AR Poso. Intracellular protein catabolism and its control during nutrient deprivation and supply. *Annu Rev Nutr* 7: 539–564, 1987.

70. EF Blommaart, JJ Luiken, AJ Meijer. Autophagic proteolysis: control and specificity. *Histochem J* 29: 365–385, 1997.

71. U Krause, MH Rider, L Hue. Protein kinase signaling pathway triggered by cell swelling and involved in the activation of glycogen synthase and acetyl-CoA carboxylase in isolated rat hepatocytes. *J Biol Chem* 271: 16668–16673, 1996.

72. SR Kimball, RL Horetsky, LS Jefferson. Implication of eIF2B rather than eIF4E in the regulation of global protein synthesis by amino acids in L6 myoblasts. *J Biol Chem* 273: 30945–30953, 1998.

73. ME Patti, E Brambilla, L Luzi, EJ Landaker, CR Kahn. Bidirectional modulation of insulin action by amino acids. *J Clin Invest* 101: 1519–1529, 1998.

74. X Wang, LE Campbell, CM Miller, CG Proud. Amino acid availability regulates p70 S6 kinase and multiple translation factors. *Biochem J* 334: 261–267, 1998.

75. G Xu, G Kwon, CA Marshall, TA Lin, JC Lawrence, Jr., ML McDaniel. Branched-chain amino acids are essential in the regulation of PHAS-I and p70 S6 kinase by pancreatic beta-cells. A possible role in protein translation and mitogenic signaling. *J Biol Chem* 273: 28178–28184, 1998.

76. DG Hardie, SA Hawley. AMP-activated protein kinase: the energy charge hypothesis revisited. *BioEssays* 23: 1112–1119, 2001.

77. S Horman, G Browne, U Krause, J Patel, D Vertommen, L Bertrand, A Lavoinne, L Hue, C Proud, M Rider. Activation of AMP-activated protein kinase leads to the phosphorylation of elongation factor 2 and an inhibition of protein synthesis. *Curr Biol* 12: 1419–1423, 2002.

12

Amino Acid-Sensing mTOR Signaling

EBRU ERBAY, JAE EUN KIM, AND JIE CHEN

INTRODUCTION

In multicellular organisms, the need to orchestrate a multitude of responses among numerous tissues, and the consequent emergence of higher modes of regulation such as endocrine and neuronal networks, has overshadowed the regulatory potential of simple nutrients. In recent years, however, various nutritional sources, including glucose, amino acids, inorganic phosphates, lipids, and fatty acids, are being recognized as critical determinants of mammalian cellular behavior. In this chapter we provide an overview of the current understanding of a key amino acid-sensing intracellular signal transduction pathway, namely the TOR pathway, that is defined by its sensitivity to rapamycin. This pathway is conserved in eukaryotes, from yeast to mammals. Our focus will be on the mammalian system.

TARGET OF RAPAMYCIN

Rapamycin, a bacterial macrolide, was isolated in 1975 as an antifungal agent from the soil of Easter Island (called Rapa

Nui by the inhabitants).[1] In the years that followed, rapamycin and its analogs were found to be of high clinical value. As a potent immunosuppressant, rapamycin is used for prevention of graft rejection following organ transplantation.[2] Furthermore, its ability to suppress proliferation has led to the development of rapamycin-eluting stents, which efficiently prevent in-stent restenosis.[3] Most recently, rapamycin analogs such as CCI-779 have emerged as promising antitumor drugs.[4,5]

Two decades after the isolation of rapamycin, the exciting discovery of the cellular target of rapamycin (TOR) came through. TOR is conserved from yeast to mammals. In mammalian cells, the TOR protein (mTOR/FRAP/RAFT1) was identified through its ability to specifically interact with rapamycin in complex with FKBP12, a ubiquitous peptidylisomerase.[6-9] Two TOR proteins are found in *S. cerevisiae*,[10,11] but only a single homolog exists in each of the higher organisms, including *Drosophila, C. elegans, Arabidopsis,* and mammals.[6-9]

The TOR proteins are unusually large serine–threonine protein kinases (280 to 290 kDa, single polypeptide).[12] Due to the homology of their C-terminal kinase domain to the phosphatidylinositol kinases, TOR proteins belong to the family of phosphatidylinositol kinase related-kinases, members of which include ATM, ATR, and DNA-PK.[13] Evidently, a wide range of cellular functions, including regulation of cell growth, proliferation, cell cycle checkpoints, and DNA recombination, has been attributed to this highly diverse family of proteins.[13]

The rapamycin-FKBP12 complex binds to mTOR at a region (FRB domain) immediately N-terminal to the kinase domain[14] (Figure 12.1). A single point mutation in this domain

mTOR (2549 amino acids, 289 kDa)

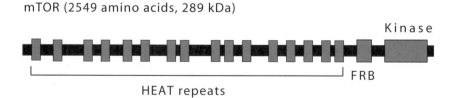

Figure 12.1 Domain structure of mTOR.

at Ser 2035 (to any residue except alanine) confers resistance to rapamycin.[14,15] Interestingly, mTOR's kinase activity remains largely unaffected by rapamycin (reviewed in Reference 16).[16] One potential mechanism for rapamycin action is dislocation of an endogenous ligand normally bound to the FRB region. In fact, a regulatory role for FRB is supported by the observation that microinjection of purified FRB protein results in drastic G1 arrest in a human cancer cell line.[17] The quest for an endogenous ligand for FRB yielded an interesting candidate, the lipid second messenger phosphatidic acid (PA).[18] *In vitro*, FRB was found to selectively bind PA-containing small unilamellar vesicles, in a rapamycin-sensitive manner.[18]

Structural information on mTOR is very limited; the only identified domains, FRB and kinase, both lie within the C-terminus of the protein.[14,19,20] Motif analysis has revealed multiple HEAT repeats throughout the rest of the mTOR sequence[21] (Figure 12.1). This is consistent with the speculation that mTOR serves as a scaffold protein because HEAT repeats are known to mediate protein–protein interactions. In the cell, mTOR is predominantly distributed between the cytosol and intracellular membranes;[22,23] it has been reported to associate with mitochondria,[24] endoplasmic reticulum, Golgi, and granular vesicles.[25] Interestingly, mTOR has also been found to shuttle between the cytoplasm and the nucleus.[26]

NUTRIENT-DEPENDENT TOR FUNCTIONS IN YEAST

The idea that the TOR proteins lie on a nutrient-sensing pathway first arose from observations with rapamycin treatment in yeast.[27] Treatment of *S. cerevisiae* with rapamycin causes cell cycle arrest at the early G1 phase and inhibits translation,[10,28] eliciting a starvation-like phenotype.[28] An identical phenotype was also seen with the loss-of-function TOR mutants.[28] The mechanism whereby TOR proteins signal translation initiation in yeast remains elusive. However, this event seems to be mediated by TAP42, which associates with the catalytic subunits of type 2A or type 2A-related protein phosphatases (SIT4). Active TOR inhibits the phosphatase activity of SIT4 by promoting the association of SIT4 with TAP42, and rapamycin treatment disrupts this complex.[29]

TOR signaling in *S. cerevisiae* also controls nutrient-dependent gene expression at the level of transcription, and the molecular mechanisms are well delineated. Under nutrient-rich growth conditions in yeast, TOR downregulates a small subset of starvation-specific Pol II transcripts, and upregulates Pol II–dependent ribosomal protein synthesis. Upon nutrient limitation, yeast cells enter a stationary phase by reducing protein synthesis as well as inducing starvation-specific transcription in a TOR-dependent manner.[30,31] For example, nitrogen limitation or rapamycin treatment leads to dephosphorylation of the transcription factor GLN3 by SIT4, which results in the dissociation of GLN3 from its cytoplasmic sequester URE2 and subsequent translocation into the nucleus, where it activates target genes such as *MAP2, GAP1, and GLN1*.[30,32] Other nitrogen-responsive transcription factors regulated by TOR include GAT1, RTG1, and RTG3.[30,33] Similarly, with limiting carbon sources, the inhibition of yeast TOR proteins confers nuclear translocation of the carbon source–related transcription factor, MSN.[30] Thus, TOR1 and TOR2 control cell growth and proliferation in yeast by coordinating nutrient availability with transcription and protein synthesis.

mTOR SIGNALING IN CELL GROWTH AND PROLIFERATION

S6K1 and 4EBP1 — Downstream Effectors of mTOR

The best-characterized function of mTOR is the regulation of translation initiation via activation of ribosomal protein S6 kinase 1 (S6K1) and inhibition of eukaryotic translation initiation factor 4E (eIF4E)-binding protein 1 (4EBP1) (Figure 12.2) (reviewed in Reference 16).[16] Growth factors regulate both S6K1 activation and inhibition of 4EBP1 by initiating phosphorylation of multiple sites on these proteins. Upon activation, S6K1 phosphorylates the 40S ribosomal protein S6, allowing the translation of a small subset of mRNAs bearing 5′-terminal oligopyrimidine tracks. These mRNAs

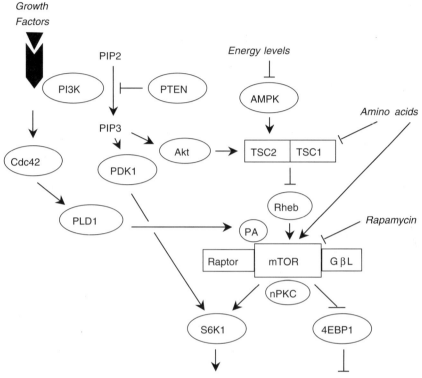

Figure 12.2 The rapamycin-sensitive signaling network regulating cell growth and proliferation.

mostly encode ribosomal proteins and other components of the translation apparatus itself.[34–37] The phosphorylation of 4EBP1 is followed by the release of the cap-binding protein, eIF4E, and the initiation of cap-dependent translation.[16] The multiple phosphorylation of both 4EBP1 and S6K1 is sensitive to rapamycin, and requires the catalytic activity of mTOR.[15,38,39] Although mTOR may directly phosphorylate S6K1 and 4EBP1,[38,40–43] regulation through a phosphatase is also plausible (see discussions in Reference 16).[16] Skeletal muscle differentiation provides an exceptional biological situation, in which mTOR's kinase activity appears dispensable for its function in the initiation of differentiation.[44,45]

mTOR Pathway in Cell Growth and Proliferation

TOR regulates cell cycle progression in organisms ranging from yeast to mammals. In yeast and lymphocytes, rapamycin treatment leads to cell cycle arrest in the G1-phase. In the majority of mammalian cell types, however, rapamycin causes only a delay in cell cycle progression (reviewed in Reference 46).[46] Both S6K1 and 4EBP1 are critically involved in cell cycle progression of mammalian cells. Microinjection of polyclonal antibodies against S6K1 inhibited cell cycle progression,[47] and so did reduction of S6K1 expression by RNAi.[48] In many human tumors eIF4E is overexpressed (reviewed in Reference 49),[49] which leads to transformation of fibroblasts.[50] Consistently, overexpression of dominant-negative 4EBP1 mutants blocks G1 progression.[48] Overexpression of either eIF4E or rapamycin-resistant S6K1 partially rescues cell cycle arrest upon rapamycin treatment, suggesting that both proteins are responsible for mediating mTOR signaling to cell cycle regulation.[48]

Various mutants in the TOR pathway in *Drosophila* have been a valuable resource in illuminating the critical link between this pathway and growth regulation. Loss-of-function mutations of the genes in the TOR pathway, such as dTOR, dPI3K, dPDK1, dAkt, and dS6K, all cause a smaller cell or organism phenotype, whereas loss-of-function mutations of the negative regulators in the TOR pathway, such as dPTEN, dTSC1, and dTSC2, lead to larger cell or organism sizes (reviewed in Reference 51).[51] Consistently, expression of a constitutively active d4EBP mutant results in a smaller cell size.[52] Furthermore, starvation of flies leads to a small-size phenotype, known as microflies, indicating the importance of a nutrient-sensing pathway for cellular and organismic growth (reviewed in Reference 51).[51] These observations in flies have now been extended to mammalian cells. The knockdown of mTOR by RNAi in tissue culture cells results in smaller cells.[53,54] It has been demonstrated that growth-regulating mTOR signaling is mediated by S6K1 and 4EBP1[53,55] and requires the upstream regulator phospholipase D1 (PLD1).[56] It must be noted that the regulation of cell size and cell cycle are separable — but necessary — processes for cellular proliferation. mTOR's role in regulating the mammalian cell cycle is a distinct function independent of its regulation of cell growth.[48,53]

Amino Acid Sensing and Mitogenic Pathways in the mTOR Network

In several different types of mammalian cells, amino acid withdrawal was found to inhibit mitogen-stimulated S6K1 activation and 4EBP1 phosphorylation.[57–60] Readdition of amino acids to the cell medium fully reverses this inhibition, but not in the presence of rapamycin.[58–61] It is thus reasonable to propose that mTOR mediates an amino acid-sensing pathway that serves as a permissive signal for the mitogenic activation of S6K1 and 4EBP1,which also requires phosphatidylinositol-3 kinase (PI3K) (see Reference 62 for a thorough review)[62] (Figure 12.2). One view of this complex signaling network is that two parallel pathways, both essential, converge on the regulation of S6K1/4EBP1 — the amino acid-sensing mTOR pathway and the mitogenic PI3K pathway (Figure 12.3A). This model is

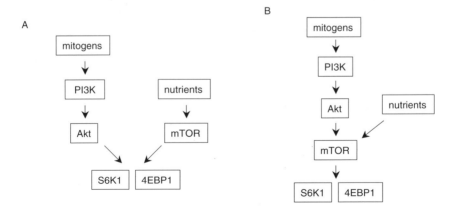

Figure 12.3 Two alternative models for the relationship between the mTOR and PI3K pathways: (A) Parallel and (B) linear.

strongly supported by observations made with a mutant of S6K1 (truncated at both N- and C-termini) that is sensitive to wortmannin (a specific PI3K inhibitor) but resistant to rapamycin.[59,63,64] This mutant S6K1 can be fully stimulated by mitogens, even in the absence of amino acids.[59]

The earlier experiments by Blommaart and coworkers in hepatocytes demonstrated that in the absence of growth factors, amino acids stimulated S6 phosphorylation in a rapamycin-sensitive manner, suggesting that amino acids may activate mTOR signaling, instead of serving solely as a permissive signal, in certain cellular contexts.[65] Similar observations have been made in other types of cells. For instance, in pancreatic cells branched-chain amino acids (leucine, valine, and isoleucine) are potent activators of mTOR downstream signaling independent of insulin, and they modulate insulin secretion through the regulation of pancreatic cell growth and proliferation.[58,66] In L6 skeletal myoblasts, leucine has the unique ability to activate translation initiation and the phosphorylation of 4EBP1 and S6K1, in a rapamycin-sensitive manner.[67] Rapamycin injection into rats prior to leucine administration completely blocks the leucine-induced hyperphosphorylation of 4EBP1 and S6K1 in skeletal muscles.[68] In adipocytes, certain individual amino acids stimulate 4EBP1 phosphorylation with the following order of potency: leucine > norleucine > isoleucine threo-L-hydroxyleucine > methionine valine.[69] A similar potency order was also observed in pancreatic cells and hepatocytes.[70,71] It cannot be ruled out, however, that a basal level of PI3K activity is still required for the amino acids' stimulating effect on S6K1 and 4EBP1 in all the aforementioned cases. Where it was examined, wortmannin or LY294002 did inhibit the stimulation.[58,61,67,72]

Currently, there is an alternative view of the relationship between mTOR and PI3K pathways, in which mTOR is placed downstream of PI3K (Figure 12.3B). This linear mode of signaling has largely been based on the observations that insulin causes a modest increase in mTOR catalytic activity and stimulates the phosphorylation of Ser 2448 on mTOR, a proposed Akt site, in a wortmannin-sensitive manner.[40,73–75] However, the functional relevance of this phosphorylation site on

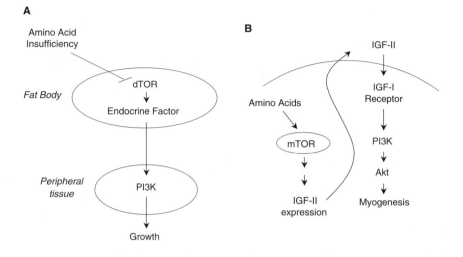

Figure 12.4 (A) In *Drosophila,* dTOR governs nutrient-regulated expression of an endocrine factor, which acts on peripheral tissues to stimulate growth via the PI3K pathway. (B) In mammalian skeletal myoblast, by sensing the availability of amino acids, mTOR controls the expression of IGF-II, which acts in an autocrine fashion to initiate myoblast differentiation.

mTOR has not been demonstrated; mutation of Ser 2448 to alanine does not affect mTOR downstream signaling.[75] Recently, another link between PI3K/Akt and mTOR has been proposed, via the tuberous sclerosis complex proteins TSC1 and TSC2 (Figure 12.2) (see section titled "TSC1-TSC2 Complex and Rheb").

Observations in *Drosophila* have revealed that dTOR can act independently of PI3K as a nutrient sensor in the fat body, a functional equivalent of mammalian liver and pancreas that controls organismic growth through modulating the availability of an endocrine factor and the consequent activation of PI3K signaling in the peripheral tissues[76] (Figure 12.4A). An interesting analogy at the biochemical level is observed in the initiation of skeletal muscle differentiation in mammals, in which mTOR controls IGF-II production independent of PI3K, and IGF-II then signals via the PI3K pathway independent of mTOR[45] (Figure 12.4B). Both cases involve the production of an endocrine or autocrine factor and complete separation of the TOR and PI3K pathways.

A direct link between mTOR and mitogenic signals has recently been revealed to involve phospholipase D (PLD) and its lipid product PA.[18,56,77–79] Mitogenic stimulation results in an increase of intracellular PA levels, which is required for the activation of mTOR downstream signaling.[18] The FRB domain in mTOR selectively binds PA-containing vesicles, and its affinity for PA is correlated with mTOR's signaling capacity.[18] Furthermore, PLD has been shown to be responsible for the PA production upstream of mTOR and cell growth,[56,78,79] and the small G protein Cdc42 is most likely a positive regulator of PLD1 in this pathway.[56] Thus, it appears that mTOR directly integrates nutrient and mitogenic signals (Figure 12.2).

OTHER KEY PLAYERS IN THE AMINO ACID-SENSING mTOR NETWORK

An Amino Acid-Sensing Complex: Raptor and GβL

How mTOR signals to its downstream effectors has long been a mystery. Recent identification of proteins in distinct TOR complexes in yeast and mammals may provide a more detailed picture of these signal transduction events.[54,80–82] One such protein found to interact with mTOR is raptor, a 149-kDa protein widely expressed across tissues in a similar pattern to mTOR.[54,80] Reduction of raptor levels by RNAi leads to decreased mTOR levels and dampened activation of S6K1 and 4EBP1. The interaction between mTOR and raptor appears to be constitutive; mitogenic stimulation has no effect on the association of raptor with mTOR.[54,80] It is found, however, that rapamycin treatment and amino acid withdrawal affect the stability of the complex, suggesting that raptor may mediate nutrient signals.[54]

Multiple regions on mTOR are found responsible for the interaction between raptor and mTOR; stronger interactions occur via the N-terminal half of mTOR, probably through multiple HEAT repeats.[54] Raptor also interacts with 4EBP1 and S6K1 through their TOR signaling motifs (TOS), and this interaction is required for their *in vitro* phosphorylation by mTOR.[54,55,80,83,84] Thus, it appears that raptor serves as a scaffold to bring mTOR kinase to the proximity of its substrates. Consistent with the scaffold idea, overexpression of raptor leads to a dose-dependent inhibition of 4EBP1 and S6K1 phosphorylation *in vivo*.[54]

The subsequent identification of GβL, a third component of the nutrient-sensing mTOR complex, provides further insight into the regulation of mTOR's kinase activity.[82] Formation of the raptor-mTOR complex was found to be highly dependent on GβL, and coexpression of GβL and raptor stimulated mTOR's kinase activity toward S6K1 and 4EBP1 *in vitro*. Thus, raptor and GβL appear to be the minimal components of a nutrient-sensitive mTOR complex (Figure 12.2). As with raptor, depletion of GβL by RNAi abrogated amino acid-dependent and insulin-stimulated S6K1 activation and hindered cell growth.[82]

TSC1-TSC2 Complex and Rheb

The tuberous sclerosis complex TSC1-TSC2 (also known as hamartin–tuberin) have recently emerged as upstream regulators of the TOR pathway (reviewed in References 85 to 87; due to space limitations and the large number of articles published on this topic, readers will be referred to review articles instead of the primary literature, throughout this section).[85–87] Both TSC1 and TSC2 are ubiquitously expressed. TSC1 is a 130-kDa protein with no known sequence motifs, whereas TSC2 is a 198-kDa GTPase-activating protein (GAP) (reviewed in Reference 87).[87] Depletion of dTSC1 or dTSC2 by RNAi induces hyperphosphorylation of dS6K that is resistant to amino acid starvation in *Drosophila* S2 cells. Similarly, in TSC1-null as well as TSC2-null mouse embryonic fibroblasts, S6K1 is constitutively phosphorylated and resistant to amino acid withdrawal. Reintroduction of TSC2 into the TSC2-null cells inhibits S6K1 hyperactivation, and coexpression of recombinant TSC1 and TSC2 in HEK293 cells abrogates mitogenic induction of S6K1 activation. The mutant forms of TSC1, identified from tuberous sclerosis tumors, do not have this negative effect on S6K1. Based on the collective evidence, TSC1 and TSC2 appear to be negative regulators of S6K1 activity in an amino acid-sensing pathway[85–87] (Figure 12.2).

Loss of function of dTSC1 or dTSC2 causes an increase in cell and body size identical to that resulting from loss of dPTEN or overexpression of dPI3K, dS6K, or dAkt. On the other hand, overexpression of dTSC1 or dTSC2 leads to growth inhibition, which can be reversed by simultaneous overexpression of dS6K1. These genetic data in *Drosophila* suggest that dTSC1-dTSC2 is functionally upstream of dS6K1, and downstream of or parallel to PI3K/Akt in the regulation of cell growth[85–87] (Figure 12.2). Evidence for PI3K-Akt being directly upstream of TSC has come from the observation that Akt phosphorylates TSC2 both *in vitro* and *in vivo* (Ser 939 and Thr 1462 on human TSC2, Ser 924 and Thr 1518 on dTSC2). Overexpression of TSC2 mutated at the Akt sites to prevent phosphorylation inhibits activation of S6K1, and the equivalent mutants of dTSC2 can reverse the enlarged cell size phenotype as a result of dAkt overexpression. Exactly how phosphorylation by Akt modifies TSC1-TSC2 heterodimer functioning remains unclear.[85–87]

A better picture of how TSC1-TSC2 functions in the mTOR pathway is now emerging with the discovery of the small GTPase Rheb as the direct target of the GAP activity of TSC1-TSC2 (reviewed in Reference 88)[88] (Figure 12.2). In mammalian cells, overexpression of Rheb leads to S6K1 activation and 4EBP1 hyperphosphorylation in the absence of mitogenic stimulation, which is resistant to amino acid starvation. Insulin-induced activation of S6K1 and 4EBP1 phosphorylation requires Rheb, as demonstrated in siRNA experiments. In *Drosophila,* overexpression of dRheb is associated with increased dS6K activity and overgrowth. Loss-of-function dRheb mutant flies are smaller in size, and the loss of dRheb inhibits overgrowth and increased dS6K activity caused by the loss of dTSC2. It is thus apparent that Rheb mediates TSC1-TSC2 function in the TOR-dependent regulation of cell growth (Figure 12.2). One controversial issue that remains is whether TSC1-TSC2-Rheb lies on a linear pathway upstream of TOR or functions in parallel to TOR. Obviously, future identification of Rheb's effectors will provide important insight into the wiring of this signaling network.

Novel PKCs

Another potential player in the amino acid-sensing mTOR pathway is the novel PKCs. PKCδ and PKCε are found to complex with mTOR constitutively.[89] Mitogen-stimulated phosphorylation and subsequent activation of PKCδ and PKCε are inhibited by rapamycin, and this inhibition is reversed by the introduction of a rapamycin-resistant mTOR. mTOR appears to be responsible for the phosphorylation of Ser 662 on PKCδ, a site analogous to Thr 389 on S6K1 and required for PKCδ activation, and phosphorylation of this site is dependent on amino acid sufficiency.[90] Furthermore, the kinase activity of PKCδ has been shown to be critical for mitogen-induced 4EBP1 phosphorylation.[89] Interestingly, overexpression of PKCδ induces phosphorylation of 4EBP1, which is sensitive to rapamycin but resistant to wortmannin.[89] The collective evidence suggests that the novel PKCs may be a downstream target of mTOR in the amino acid-sensing pathway required for the regulation of 4EBP1 (Figure 12.2), although their role in S6K1 activation has not been examined.

AMINO ACID SENSING BY mTOR IN CELLULAR DIFFERENTIATION

A true master regulator, mTOR is found to play essential roles in various cellular differentiation systems independent of its function in cell growth and proliferation. Some of the signaling mechanisms unraveled in these cellular contexts are unexpected and drastically distinct from those of mTOR signaling in cell growth and proliferation. Although mTOR has been implicated in other differentiation systems, here we will only discuss three systems in which the roles of amino acids have been examined: skeletal myogenesis, adipogenesis, and trophoectoderm differentiation.

Skeletal Myogenesis

Rapamycin inhibits myogenesis in several skeletal muscle cell culture systems, implicating mTOR in the regulation of myogenesis.[44,91–93] As a direct demonstration of mTOR's essential role in muscle differentiation, a rapamycin-resistant mTOR was shown to rescue rapamycin-inhibited differentiation of C2C12 myoblasts.[44,94] Remarkably, the ability of mTOR to initiate differentiation is independent of its wild-type kinase activity, and consequently both S6K1 and 4EBP1 have been ruled out as the relevant effector for mTOR signaling in this process.[44,45] However, we have found that mTOR is also required in a later stage of myogenesis leading to myotube maturation, and this function of mTOR does require its kinase activity and signaling to S6K1 (our unpublished observations). Thus, mTOR appears to regulate different stages of skeletal myogenesis via distinct mechanisms and pathways.

Further investigation into the molecular mechanisms by which mTOR regulates myogenesis has revealed that mTOR primarily controls the autocrine production of insulin-like factor II (IGF-II) in skeletal myoblasts, which is essential for the initiation of differentiation.[45] IGF-II transcription in C2C12 cells is regulated by mTOR, as well as by amino acid sufficiency, through the IGF-II promoter 3 and a downstream enhancer, and mTOR signaling in the initiation of differentiation is indirectly mediated by the PI3K pathway downstream of IGF-II signaling (Figure 12.4B).[45] These observations suggest that the mTOR–IGF axis may be a molecular link between nutritional levels and skeletal muscle development.

Adipogenesis

mTOR was first implicated in adipogenesis by the observation that rapamycin inhibited clonal expansion and differentiation of the 3T3-L1 preadipocytes.[95] Rapamycin also effectively attenuated terminal differentiation when added at day 4 or later, suggesting that a rapamycin-sensitive pathway is involved in the late stages of differentiation independent of the early clonal expansion phase.[96] Contrary to mTOR's kinase-independent roles in myogenesis, its kinase activity is

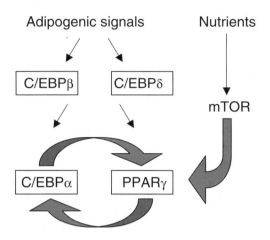

Figure 12.5 mTOR regulates PPARγ activity in the adipogenic transcription network, most likely by sensing amino acid availability.

required for adipocyte differentiation.[96a] Rapamycin treatment of preadipocytes and adipocytes revealed that mTOR may regulate the expression of PPARγ and C/EBPα in the adipogenic transcription network critical for differentiation.[95] Our recent data suggest a novel mechanism whereby mTOR directly regulates PPARγ transactivation activity toward its target genes, most likely by sensing the availability of amino acids[96a] (Figure 12.5). The main function of adipose tissue in animals is maintaining energy homeostasis in response to nutrient availability. In this process, mTOR may serve to coordinate nutrient availability signals with adipocyte differentiation and functions of adipose tissues by directly modulating adipogenic gene transcription.

Trophoectoderm Differentiation

Nutrients are critically involved in the late blastocyst stage epithelial trophoectoderm differentiation. Amino acids provide a permissive signal for the protrusive activity and consequent spreading of trophoectoderm prior to implantation. Addition of rapamycin to an *in vitro* system modeling this specific developmental stage inhibits amino acid sufficiency-dependent S6K1 activation and trophoectoderm differentiation.[97] Thus, mTOR is suggested to mediate nutrient-dependent blastocyst development. The molecular mechanisms of mTOR signaling in this process remain to be explored.

AMINO ACID-SENSING AND ENERGY-SENSING MECHANISMS

The mechanism by which the mTOR pathway senses the amino acid sufficiency signal remains unknown. One proposed model assigns mTOR the sensor role in which it monitors the intracellular levels of uncharged tRNAs. This mode of signaling is analogous to the well-established amino acid-sensing mechanism through the kinase GCN2, where uncharged tRNA binds to a domain on GCN2 that is structurally similar to aminoacyl-tRNA synthetase, and activates the kinase.[98,99] According to this model, under amino acid starvation conditions, increased levels of uncharged tRNAs would be inhibitory to mTOR signaling. Alternatively, mTOR may require the continuous presence of charged tRNAs for constitutive activation. Both models predict that inhibitors of tRNA charging, such as amino acid alcohols, would inhibit mTOR signaling. This has indeed been observed in the Jurkat T cells,[60] but has not been confirmed in other types of cells (see discussions in Reference 100).[100] If the uncharged tRNAs are not the direct signal for the mTOR pathway, then an open question is whether it is the extracellular pool or intracellular pool of amino acids that is being sensed. The existence of a transmembrane amino acid sensor in yeast suggests that sensing of an extracellular pool of amino acids is possible,[101] but such a sensor has not been identified in mammals. Instead, the collective evidence implies that the mTOR pathway may directly sense the intracellular concentrations of certain amino acids.[100,102]

mTOR can also be directly modulated by intracellular ATP levels.[103] This, however, appears to be independent of amino acid-dependent regulation of mTOR signaling. Recent findings have uncovered the AMP kinase (AMPK) as a sensor of cellular ATP deprivation in the mTOR pathway.[104] Activation of AMPK inhibits both mTOR signaling and consequent protein synthesis. Under energy starvation conditions, activated AMPK phosphorylates TSC2 on sites distinct from the Akt sites. This phosphorylation event activates TSC2 and inhibits mTOR signaling, and is critical for cell size control and protection from apoptosis upon energy deprivation.[104] A new branch can thus be added to the TSC–mTOR signaling network (Figure 12.2).

FUTURE DIRECTIONS

It has been an exciting 10 years since the discovery of the target of rapamycin proteins. Our understanding of this nutrient-sensing pathway has advanced exponentially. Identification of the array of regulators and interacting partners of TOR has helped develop a clearer picture of the TOR signaling network. Naturally, this expanding knowledge prompts even more questions. The molecular mechanism of amino acid sensing by the mTOR pathway is currently an open question, needing more detailed models for testing. The much-debated issue of how the mTOR and PI3K pathways relate to one another in this network remains to be resolved. Further characterization of the newly emerged TSC–Rheb link, especially the identification of Rheb's direct effector and assessment of amino acids' role on TSC, may provide some answers to this question. mTOR likely exists in a large protein complex; continued effort to identify the interacting partners will surely cast new light on the regulation and multifunctionality of this unusual kinase. The recognition that mTOR potentially signals via different mechanisms and pathways in different cellular contexts warrants immediate investigation of the functional involvement of all the newly discovered regulators, including TSC1/2, Rheb, Raptor, GβL, PLD, and PA, in various cellular differentiation systems. Unraveling the wiring and regulatory mechanisms of the mTOR signaling network is not only pivotal to a fundamental understanding of nutrient signaling, but has the potential of facilitating the design of precise therapeutic intervention strategies in cancer and other diseases.

ACKNOWLEDGMENT

The work from the authors' laboratory was supported by grants from the National Institutes of Health (GM58064, AR48194), the American Cancer Society, and the American Heart Association.

References

1. Sehgal, S.N., H. Baker, and C. Vezina, Rapamycin (AY-22,989), a new antifungal antibiotic. II. Fermentation, isolation and characterization. *J Antibiot* (Tokyo), 28(10): 727–732, 1975.

2. Abraham, R.T., Mammalian target of rapamycin: immunosuppressive drugs uncover a novel pathway of cytokine receptor signaling. *Curr Opin Immunol*, 10(3): 330–336, 1998.

3. Serruys, P.W., E. Regar, and A.J. Carter, Rapamycin eluting stent: the onset of a new era in interventional cardiology. *Heart*, 87: 305–307, 2002.

4. Hidalgo, M. and E.K. Rowinsky, The rapamycin-sensitive signal transduction pathway as a target for cancer therapy. *Oncogene*, 19(56): 6680–6686, 2000.

5. Vogt, P.K., PI 3-kinase, mTOR, protein synthesis, and cancer. *Trends Mol Med*, 7(11): 482–484, 2001.

6. Brown, E.J., M.W. Albers, T.B. Shin, K. Ichikawa, C.T. Keith, W.S. Lane, and S.L. Schreiber, A mammalian protein targeted by G1-arresting rapamycin-receptor complex. *Nature*, 369(6483): 756–758, 1994.

7. Chiu, M.I., H. Katz, and V. Berlin, RAPT1, a mammalian homolog of yeast Tor, interacts with the FKBP12/rapamycin complex. *Proc Natl Acad Sci USA*, 91(26): 12574–125788, 1994.

8. Sabatini, D.M., H. Erdjument-Bromage, M. Lui, P. Tempst, and S.H. Snyder, RAFT1: a mammalian protein that binds to FKBP12 in a rapamycin-dependent fashion and is homologous to yeast TORs. *Cell*, 78(1): 35–43, 1994.

9. Sabers, C.J., M.M. Martin, G.J. Brunn, J.M. Williams, F.J. Dumont, G. Wiederrecht, and R.T. Abraham, Isolation of a protein target of the FKBP12-rapamycin complex in mammalian cells. *J Biol Chem*, 270(2): 815–822, 1995.

10. Kunz, J., R. Henriquez, U. Schneider, M. Deuter-Reinhard, N.R. Movva, and M.N. Hall, Target of rapamycin in yeast, TOR2, is an essential phosphatidylinositol kinase homolog required for G1 progression. *Cell*, 73(3): 585–596, 1993.

11. Helliwell, S.B., P. Wagner, J. Kunz, M. Deuter-Reinhard, R. Henriquez, and M.N. Hall, TOR1 and TOR2 are structurally and functionally similar but not identical phosphatidylinositol kinase homologues in yeast. *Mol Biol Cell*, 5(1): 105–118, 1994.

12. Thomas, G. and M.N. Hall, TOR signalling and control of cell growth. *Curr Opin Cell Biol*, 9(6): 782–787, 1997.

13. Kuruvilla, F.G. and S.L. Schreiber, The PIK-related kinases intercept conventional signaling pathways. *Chem Biol*, 6(5): R129–136, 1999.

14. Chen, J., X.F. Zheng, E.J. Brown, and S.L. Schreiber, Identification of an 11-kDa FKBP12-rapamycin-binding domain within the 289-kDa FKBP12-rapamycin-associated protein and characterization of a critical serine residue. *Proc Natl Acad Sci USA*, 92(11): 4947–4951, 1995.

15. Brown, E.J., P.A. Beal, C.T. Keith, J. Chen, T.B. Shin, and S.L. Schreiber, Control of p70 s6 kinase by kinase activity of FRAP in vivo. *Nature*, 377(6548): 441–446, 1995.

16. Gingras, A.C., B. Raught, and N. Sonenberg, Regulation of translation initiation by FRAP/mTOR. *Genes Dev*, 15(7): 807–826, 2001.

17. Vilella-Bach, M., P. Nuzzi, Y. Fang, and J. Chen, The FKBP12-rapamycin-binding domain is required for FKBP12-rapamycin-associated protein kinase activity and G1 progression. *J Biol Chem*, 274(7): 4266–4272, 1999.

18. Fang, Y., M. Vilella-Bach, R. Bachmann, A. Flanigan, and J. Chen, Phosphatidic acid-mediated mitogenic activation of mTOR signaling. *Science*, 294(5548): 1942–1945, 2001.

19. Choi, J., J. Chen, S.L. Schreiber, and J. Clardy, Structure of the FKBP12-rapamycin complex interacting with the binding domain of human FRAP. *Science*, 273(5272): 239–242, 1996.

20. Keith, C.T. and S.L. Schreiber, PIK-related kinases: DNA repair, recombination, and cell cycle checkpoints. *Science*, 270(5233): 50–51, 1995.

21. Perry, J. and N. Kleckner, The ATRs, ATMs, and TORs are giant HEAT repeat proteins. *Cell*, 112(2): 151–155, 2003.

22. Sabatini, D.M., R.K. Barrow, S. Blackshaw, P.E. Burnett, M.M. Lai, M.E. Field, B.A. Bahr, J. Kirsch, H. Betz, and S.H. Snyder, Interaction of RAFT1 with gephyrin required for rapamycin-sensitive signaling. *Science*, 284(5417): 1161–1164, 1999.

23. Withers, D.J., D.M. Ouwens, B.T. Nave, G.C. van der Zon, C.M. Alarcon, M.E. Cardenas, J. Heitman, J.A. Maassen, and P.R. Shepherd, Expression, enzyme activity, and subcellular localization of mammalian target of rapamycin in insulin-responsive cells. *Biochem Biophys Res Commun*, 241(3): 704–709, 1997.

24. Desai, B.N., B.R. Myers, and S.L. Schreiber, FKBP12-rapamycin-associated protein associates with mitochondria and senses osmotic stress via mitochondrial dysfunction. *Proc Natl Acad Sci USA*, 99(7): 4319–4324, 2002.

25. Drenan, R.M., X. Liu, P.G. Bertram, and X.F. Zheng, FKBP12-Rapamycin-associated Protein or Mammalian Target of Rapamycin (FRAP/mTOR) Localization in the Endoplasmic Reticulum and the Golgi Apparatus. *J Biol Chem*, 279(1): 772–778, 2004.

26. Kim, J.E. and J. Chen, Cytoplasmic-nuclear shuttling of FKBP12-rapamycin-associated protein is involved in rapamycin-sensitive signaling and translation initiation. *Proc Natl Acad Sci USA*, 97(26): 14340–14345, 2000.

27. Chung, J., C.J. Kuo, G.R. Crabtree, and J. Blenis, Rapamycin-FKBP specifically blocks growth-dependent activation of and signaling by the 70 kd S6 protein kinases. *Cell*, 69(7): 1227–1236, 1992.

28. Barbet, N.C., U. Schneider, S.B. Helliwell, I. Stansfield, M.F. Tuite, and M.N. Hall, TOR controls translation initiation and early G1 progression in yeast. *Mol Biol Cell*, 7(1): 25–42, 1996.

29. Di Como, C.J. and K.T. Arndt, Nutrients, via the Tor proteins, stimulate the association of Tap42 with type 2A phosphatases. *Genes Dev*, 10(15): 1904–1916, 1996.

30. Beck, T. and M.N. Hall, The TOR signalling pathway controls nuclear localization of nutrient-regulated transcription factors. *Nature*, 402(6762): 689–692, 1999.

31. Shamji, A.F., F.G. Kuruvilla, and S.L. Schreiber, Partitioning the transcriptional program induced by rapamycin among the effectors of the Tor proteins. *Curr Biol*, 10(24): 1574–1581, 2000.

32. Crespo, J.L., T. Powers, B. Fowler, and M.N. Hall, The TOR-controlled transcription activators GLN3, RTG1, and RTG3 are regulated in response to intracellular levels of glutamine. *Proc Natl Acad Sci USA*, 99(10): 6784–6789, 2002.

33. Komeili, A., K.P. Wedaman, E.K. O'Shea, and T. Powers, Mechanism of metabolic control. Target of rapamycin signaling links nitrogen quality to the activity of the Rtg1 and Rtg3 transcription factors. *J Cell Biol*, 151(4): 863–878, 2000.

34. Jefferies, H.B., C. Reinhard, S.C. Kozma, and G. Thomas, Rapamycin selectively represses translation of the "polypyrimidine tract" mRNA family. *Proc Natl Acad Sci USA*, 91(10): 4441–4445, 1994.

35. Terada, N., K. Takase, P. Papst, A.C. Nairn, and E.W. Gelfand, Rapamycin inhibits ribosomal protein synthesis and induces G1 prolongation in mitogen-activated T lymphocytes. *J Immunol*, 155(7): 3418–3426, 1995.

36. Peterson, R.T. and S.L. Schreiber, Translation control: connecting mitogens and the ribosome. *Curr Biol*, 8(7): R248–250, 1998.

37. Terada, N., H.R. Patel, K. Takase, K. Kohno, A.C. Nairn, and E.W. Gelfand, Rapamycin selectively inhibits translation of mRNAs encoding elongation factors and ribosomal proteins. *Proc Natl Acad Sci USA*, 91(24): 11477–11481, 1994.

38. Brunn, G.J., C.C. Hudson, A. Sekulic, J.M. Williams, H. Hosoi, P.J. Houghton, J.C. Lawrence, Jr., and R.T. Abraham, Phosphorylation of the translational repressor PHAS-I by the mammalian target of rapamycin. *Science*, 277(5322): 99–101, 1997.

39. Hara, K., K. Yonezawa, M.T. Kozlowski, T. Sugimoto, K. Andrabi, Q.P. Weng, M. Kasuga, I. Nishimoto, and J. Avruch, Regulation of eIF-4E BP1 phosphorylation by mTOR. *J Biol Chem*, 272(42): 26457–26463, 1997.

40. Burnett, P.E., R.K. Barrow, N.A. Cohen, S.H. Snyder, and D.M. Sabatini, RAFT1 phosphorylation of the translational regulators p70 S6 kinase and 4E-BP1. *Proc. Natl. Acad. Sci. USA*, 95: 1432–1437, 1998.

41. Isotani, S., K. Hara, C. Tokunaga, H. Inoue, J. Avruch, and K. Yonezawa, Immunopurified mammalian target of rapamycin phosphorylates and activates p70 S6 kinase alpha in vitro. *J Biol Chem*, 274(48): 34493–34498, 1999.

42. Gingras, A.C., S.P. Gygi, B. Raught, R.D. Polakiewicz, R.T. Abraham, M.F. Hoekstra, R. Aebersold, and N. Sonenberg, Regulation of 4E-BP1 phosphorylation: a novel two-step mechanism. *Genes Dev*, 13(11): 1422–1437, 1999.

43. Saitoh, M., N. Pullen, P. Brennan, D. Cantrell, P.B. Dennis, and G. Thomas, Regulation of an activated S6 kinase 1 variant reveals a novel mammalian target of rapamycin phosphorylation site. *J Biol Chem*, 277(22): 20104–20112, 2002.

44. Erbay, E. and J. Chen, The mammalian target of rapamycin regulates C2C12 myogenesis via a kinase-independent mechanism. *J Biol Chem*, 276: 36079–36082, 2001.

45. Erbay, E., I.H. Park, P.D. Nuzzi, C.J. Schoenherr, and J. Chen, IGF-II transcription in skeletal myogenesis is controlled by mTOR and nutrients. *J Cell Biol*, 163(5): 931–936, 2003.

46. Abraham, R.T. and G.J. Wiederrecht, Immunopharmacology of rapamycin. *Annu Rev Immunol*, 14: 483–510, 1996.

47. Lane, H.A., A. Fernandez, N.J. Lamb, and G. Thomas, p70s6k function is essential for G1 progression. *Nature*, 363(6425): 170–172, 1993.

48. Fingar, D.C., C.J. Richardson, A.R. Tee, L. Cheatham, C. Tsou, and J. Blenis, mTOR controls cell cycle progression through its cell growth effectors S6K1 and 4E-BP1/eukaryotic translation initiation factor 4E. *Mol Cell Biol*, 24(1): 200–216, 2004.

49. Strudwick, S. and K.L. Borden, The emerging roles of translation factor eIF4E in the nucleus. *Differentiation*, 70(1): 10–22, 2002.

50. Lazaris-Karatzas, A., K.S. Montine, and N. Sonenberg, Malignant transformation by a eukaryotic initiation factor subunit that binds to mRNA 5 cap. *Nature*, 345(6275): 544–547, 1990.

51. Jacinto, E. and M.N. Hall, Tor signalling in bugs, brain and brawn. *Nat Rev Mol Cell Biol*, 4(2): 117–126, 2003.

52. Miron, M., J. Verdu, P.E. Lachance, M.J. Birnbaum, P.F. Lasko, and N. Sonenberg, The translational inhibitor 4E-BP is an effector of PI(3)K/Akt signalling and cell growth in Drosophila. *Nat Cell Biol*, 3(6): 596–601, 2001.

53. Fingar, D.C., S. Salama, C. Tsou, E. Harlow, and J. Blenis, Mammalian cell size is controlled by mTOR and its downstream targets S6K1 and 4EBP1/eIF4E. *Genes Dev*, 16(12): 1472–1487, 2002.

54. Kim, D.H., D.D. Sarbassov, S.M. Ali, J.E. King, R.R. Latek, H. Erdjument-Bromage, P. Tempst, and D.M. Sabatini, mTOR interacts with raptor to form a nutrient-sensitive complex that signals to the cell growth machinery. *Cell*, 110(2): 163–175, 2002.

55. Schalm, S.S., D.C. Fingar, D.M. Sabatini, and J. Blenis, TOS motif-mediated raptor binding regulates 4E-BP1 multisite phosphorylation and function. *Curr Biol*, 13(10): 797–806, 2003.

56. Fang, Y., I.H. Park, A.L. Wu, G. Du, P. Huang, M.A. Frohman, S.J. Walker, H.A. Brown, and J. Chen, PLD1 Regulates mTOR Signaling and Mediates Cdc42 Activation of S6K1. *Curr Biol*, 13(23): 2037–2044, 2003.

57. Wang, X., L.E. Campbell, C.M. Miller, and C.G. Proud, Amino acid availability regulates p70 S6 kinase and multiple translation factors. *Biochem J*, 334(Pt 1): 261–267, 1998.

58. Xu, G., G. Kwon, C.A. Marshall, T.A. Lin, J.C. Lawrence, Jr., and M.L. McDaniel, Branched-chain amino acids are essential in the regulation of PHAS-I and p70 S6 kinase by pancreatic beta-cells. A possible role in protein translation and mitogenic signaling. *J Biol Chem*, 273(43): 28178–28184, 1998.

59. Hara, K., K. Yonezawa, Q.-P. Weng, M.T. Kozlowski, C. Belham, and J. Avruch, Amino Acid Sufficiency and mTOR Regulate p70 S6 Kinase and eIF-4E BP1 through a Common Effector Mechanism. *J Biol Chem*, 273: 14484–14494, 1998.

60. Iiboshi, Y., P.J. Papst, H. Kawasome, H. Hosoi, R.T. Abraham, P.J. Houghton, and N. Terada, Amino acid-dependent control of p70(s6k). Involvement of tRNA aminoacylation in the regulation. *J Biol Chem*, 274(2): 1092–1099, 1999.

61. Xu, G., G. Kwon, W.S. Cruz, C.A. Marshall, and M.L. McDaniel, Metabolic regulation by leucine of translation initiation through the mTOR-signaling pathway by pancreatic beta-cells. *Diabetes*, 50(2): 353–360, 2001.

62. Shamji, A.F., P. Nghiem, and S.L. Schreiber, Integration of growth factor and nutrient signaling: implications for cancer biology. *Mol Cell*, 12(2): 271–280, 2003.

63. Cheatham, L., M. Monfar, M.M. Chou, and J. Blenis, Structural and functional analysis of pp70S6k. *Proc Natl Acad Sci USA*, 92(25): 11696–11700, 1995.

64. Dennis, P.B., N. Pullen, S.C. Kozma, and G. Thomas, The principal rapamycin-sensitive p70(s6k) phosphorylation sites, T-229 and T-389, are differentially regulated by rapamycin-insensitive kinase kinases. *Mol Cell Biol*, 16(11): 6242–6251, 1996.

65. Blommaart, E.F., J.J. Luiken, P.J. Blommaart, G.M. van Woerkom, and A.J. Meijer, Phosphorylation of ribosomal protein S6 is inhibitory for autophagy in isolated rat hepatocytes. *J Biol Chem*, 270(5): 2320–2326, 1995.

66. Xu, G., G. Kwon, C.A. Marshall, T.A. Lin, J.C. Lawrence, Jr., and M.L. McDaniel, Branched-chain amino acids are essential in the regulation of PHAS-I and p70 S6 kinase by pancreatic beta-cells. A possible role in protein translation and mitogenic signaling. *J Biol Chem*, 273(43): 28178–28184, 1998.

67. Kimball, S.R., L.M. Shantz, R.L. Horetsky, and L.S. Jefferson, Leucine regulates translation of specific mRNAs in L6 myoblasts through mTOR-mediated changes in availability of eIF4E and phosphorylation of ribosomal protein S6. *J Biol Chem*, 274(17): 11647–11652, 1999.

68. Anthony, J.C., F. Yoshizawa, T.G. Anthony, T.C. Vary, L.S. Jefferson, and S.R. Kimball, Leucine stimulates translation initiation in skeletal muscle of postabsorptive rats via a rapamycin-sensitive pathway. *J Nutr*, 130(10): 2413–2419, 2000.

69. Fox, H.L., P.T. Pham, S.R. Kimball, L.S. Jefferson, and C.J. Lynch, Amino acid effects on translational repressor 4E-BP1 are mediated primarily by L-leucine in isolated adipocytes. *Am J Physiol*, 275(5 Pt 1): C1232–1238, 1998.

70. McDaniel, M.L., C.A. Marshall, K.L. Pappan, and G. Kwon, Metabolic and autocrine regulation of the mammalian target of rapamycin by pancreatic beta-cells. *Diabetes*, 51(10): 2877–2885, 2002.

71. Shigemitsu, K., Y. Tsujishita, H. Miyake, S. Hidayat, N. Tanaka, K. Hara, and K. Yonezawa, Structural requirement of leucine for activation of p70 S6 kinase. *FEBS Lett*, 447(2–3): 303–306, 1999.

72. Fox, H.L., S.R. Kimball, L.S. Jefferson, and C.J. Lynch, Amino acids stimulate phosphorylation of p70S6k and organization of rat adipocytes into multicellular clusters. *Am J Physiol*, 274(1 Pt 1): C206–213, 1998.

73. Nave, B.T., M. Ouwens, D.J. Withers, D.R. Alessi, and P.R. Shepherd, Mammalian target of rapamycin is a direct target for protein kinase B: identification of a convergence point for opposing effects of insulin and amino-acid deficiency on protein translation. *Biochem J*, 344(Pt 2): 427–431, 1999.

74. Scott, P.H., G.J. Brunn, A.D. Kohn, R.A. Roth, and J.C. Lawrence, Jr., Evidence of insulin-stimulated phosphorylation and activation of the mammalian target of rapamycin mediated by a protein kinase B signaling pathway. *Proc Natl Acad Sci USA*, 95(13): 7772–7777, 1998.

75. Sekulic, A., C.C. Hudson, J.L. Homme, P. Yin, D.M. Otterness, L.M. Karnitz, and R.T. Abraham, A direct linkage between the phosphoinositide 3-kinase-AKT signaling pathway and the mammalian target of rapamycin in mitogen-stimulated and transformed cells. *Cancer Res*, 60(13): 3504–3513, 2000.

76. Colombani, J., S. Raisin, S. Pantalacci, T. Radimerski, J. Montagne, and P. Leopold, A nutrient sensor mechanism controls Drosophila growth. *Cell*, 114(6): 739–749, 2003.

77. Chen, J. and Y. Fang, A novel pathway regulating the mammalian target of rapamycin (mTOR) signaling. *Biochem Pharmacol*, 64(7): 1071–1077, 2002.

78. Chen, Y., Y. Zheng, and D.A. Foster, Phospholipase D confers rapamycin resistance in human breast cancer cells. *Oncogene*, 22(25): 3937–3942, 2003.

79. Ballou, L.M., Y.P. Jiang, G. Du, M.A. Frohman, and R.Z. Lin, Ca(2+)- and phospholipase D-dependent and -independent pathways activate mTOR signaling. *FEBS Lett*, 550(1–3): 51–56, 2003.

80. Hara, K., Y. Maruki, X. Long, K. Yoshino, N. Oshiro, S. Hidayat, C. Tokunaga, J. Avruch, and K. Yonezawa, Raptor, a binding partner of target of rapamycin (TOR), mediates TOR action. *Cell*, 110(2): 177–189, 2002.

81. Loewith, R., E. Jacinto, S. Wullschleger, A. Lorberg, J.L. Crespo, D. Bonenfant, W. Oppliger, P. Jenoe, and M.N. Hall, Two TOR complexes, only one of which is rapamycin sensitive, have distinct roles in cell growth control. *Mol Cell*, 10(3): 457–468, 2002.

82. Kim, D.H., D. Sarbassov dos, S.M. Ali, R.R. Latek, K.V. Guntur, H. Erdjument-Bromage, P. Tempst, and D.M. Sabatini, GbetaL, a positive regulator of the rapamycin-sensitive pathway required for the nutrient-sensitive interaction between raptor and mTOR. *Mol Cell*, 11(4): 895–904, 2003.

83. Choi, K.M., L.P. McMahon, and J.C. Lawrence, Jr., Two motifs in the translational repressor PHAS-I required for efficient phosphorylation by mammalian target of rapamycin and for recognition by raptor. *J Biol Chem*, 278(22): 19667–19673, 2003.

84. Nojima, H., C. Tokunaga, S. Eguchi, N. Oshiro, S. Hidayat, K. Yoshino, K. Hara, N. Tanaka, J. Avruch, and K. Yonezawa, The mammalian target of rapamycin (mTOR) partner, raptor, binds the mTOR substrates p70 S6 kinase and 4E-BP1 through their TOR signaling (TOS) motif. *J Biol Chem*, 278(18): 15461–15464, 2003.

85. Marygold, S.J. and S.J. Leevers, Growth Signaling: TSC takes its place. *Curr Biol*, 12(22): R785–787, 2002.

86. McManus, E.J. and D.R. Alessi, TSC1-TSC2: a complex tale of PKB-mediated S6K regulation. *Nat Cell Biol*, 4(9): E214–216, 2002.

87. Manning, B.D. and L.C. Cantley, United at last: the tuberous sclerosis complex gene products connect the phosphoinositide 3-kinase/Akt pathway to mammalian target of rapamycin (mTOR) signalling. *Biochem Soc Trans*, 31(Pt 3): 573–578, 2003.

88. Manning, B.D. and L.C. Cantley, Rheb fills a GAP between TSC and TOR. *Trends Biochem Sci*, 28(11): 573–576, 2003.

89. Kumar, V., P. Pandey, D. Sabatini, M. Kumar, P.K. Majumder, A. Bharti, G. Carmichael, D. Kufe, and S. Kharbanda, Functional interaction between RAFT1/FRAP/mTOR and protein kinase cdelta in the regulation of cap-dependent initiation of translation. *EMBO J*, 19(5): 1087–1097, 2000.

90. Parekh, D., W. Ziegler, K. Yonezawa, K. Hara, and P.J. Parker, Mammalian TOR controls one of two kinase pathways acting upon nPKCdelta and nPKCepsilon. *J Biol Chem*, 274(49): 34758–34764, 1999.

91. Cuenda, A. and P. Cohen, Stress-activated protein kinase-2/p38 and a rapamycin-sensitive pathway are required for C2C12 myogenesis. *J Biol Chem*, 274(7): 4341–4346, 1999.

92. Conejo, R., A.M. Valverde, M. Benito, and M. Lorenzo, Insulin produces myogenesis in C2C12 myoblasts by induction of NF-kappaB and downregulation of AP-1 activities. *J Cell Physiol*, 186(1): 82–94, 2001.

93. Coolican, S.A., D.S. Samuel, D.Z. Ewton, F.J. McWade, and J.R. Florini, The mitogenic and myogenic actions of insulin-like growth factors utilize distinct signaling pathways. *J Biol Chem*, 272(10): 6653–6662, 1997.

94. Shu, L., X. Zhang, and P.J. Houghton, Myogenic differentiation is dependent on both the kinase function and the N-terminal sequence of mammalian target of rapamycin. *J Biol Chem*, 277(19): 16726–16732, 2002.

95. Yeh, W.C., B.E. Bierer, and S.L. McKnight, Rapamycin inhibits clonal expansion and adipogenic differentiation of 3T3-L1 cells. *Proc Natl Acad Sci USA*, 92(24): 11086–11090, 1995.

96. Gagnon, A., S. Lau, and A. Sorisky, Rapamycin-sensitive phase of 3T3-L1 preadipocyte differentiation after clonal expansion. *J Cell Physiol*, 189(1): 14–22, 2001.

96a. JE Kim, Chen, J. Regulation of PPARgamma activity by mammalian target of rapamycin and amino acids in adipogenesis. *Diabetes* 53: 2748–2756, 2004.

97. Martin, P.M. and A.E. Sutherland, Exogenous amino acids regulate trophectoderm differentiation in the mouse blastocyst through an mTOR-dependent pathway. *Dev Biol*, 240(1): 182–193, 2001.

98. Hinnebusch, A.G., Translational regulation of yeast GCN4. A window on factors that control initiator-trna binding to the ribosome. *J Biol Chem*, 272(35): 21661–21664, 1997.

99. Dong, J., H. Qiu, M. Garcia-Barrio, J. Anderson, and A.G. Hinnebusch, Uncharged tRNA activates GCN2 by displacing the protein kinase moiety from a bipartite tRNA-binding domain. *Mol Cell*, 6(2): 269–279, 2000.

100. Proud, C.G., Regulation of mammalian translation factors by nutrients. *Eur J Biochem*, 269(22): 5338–5349, 2002.

101. Iraqui, I., S. Vissers, F. Bernard, J.O. de Craene, E. Boles, A. Urrestarazu, and B. Andre, Amino acid signaling in Saccharomyces cerevisiae: a permease-like sensor of external amino acids and F-Box protein Grr1p are required for transcriptional induction of the AGP1 gene, which encodes a broad-specificity amino acid permease. *Mol Cell Biol*, 19(2): 989–1001, 1999.

102. Beugnet, A., A.R. Tee, P.M. Taylor, and C.G. Proud, Regulation of targets of mTOR (mammalian target of rapamycin) signalling by intracellular amino acid availability. *Biochem J*, 372(Pt 2): 555–566, 2003.

103. Dennis, P.B., A. Jaeschke, M. Saitoh, B. Fowler, S.C. Kozma, and G. Thomas, Mammalian TOR: a homeostatic ATP sensor. *Science*, 294(5544): 1102–1105, 2001.

104. Inoki, K., T. Zhu, and K.L. Guan, TSC2 mediates cellular energy response to control cell growth and survival. *Cell*, 115(5): 577–590, 2003.

13

Alterations in Glutamate Trafficking and Signal Transduction Pathways in Hyperammonemia

HELEN CHAN AND ROGER F. BUTTERWORTH

INTRODUCTION

Hyperammonemia is a consistent feature of a large number of diseases including liver disease, Reye's syndrome, inborn errors of the urea cycle, and organic acidemias.[1] Regardless of the etiology of these disorders, there is overwhelming evidence that demonstrates that elevated levels of ammonia are toxic and have adverse effects on brain function. Excess levels of ammonia affect bioenergetics, neurotransmission, electrophysiology, and, in particular, glutamate signaling pathways involving neurons and astrocytes.

Glutamate is the major excitatory neurotransmitter in the mammalian central nervous system (CNS). It is present at high concentrations in the CNS and is released in a calcium-dependent manner upon electrical stimulation *in vitro*.[2] In addition to its role in synaptic transmission, glutamate is involved in a myriad of other cell functions. It is incorporated into proteins, is involved in fatty acid synthesis, contributes to nitrogen homeostasis, is involved in the control of osmotic and anion balance, and serves as a precursor for gamma-aminobutyric acid (GABA) as well as various tricarboxylic acid cycle intermediates. Glutamate's concentration in the CNS is six times that of the principal inhibitory neurotransmitter GABA, and its transmitter pool constitutes approximately 30% of the total glutamate concentration in the brain.[3]

Glutamate is compartmentalized into separate pools in the brain that subserve different metabolic pathways. This compartmentation is an essential factor in the separate regulation of the various functions of glutamate, such as neurotransmission. As a result of more sophisticated technology, recent studies demonstrate that glutamate metabolism in the CNS is intricately compartmentalized in neurons and astrocytes, hence reflecting pools that have different turnover rates and sizes, which synchronize to result in normal brain functioning.[4] ^{13}C-Nuclear magnetic resonance studies demonstrate that there are at least two glutamate pools, small and large, associated with two kinetically distinct tricarboxylic acid cycles localized in astrocytes and neurons, respectively.[5] Neuronal and glial cycles interact closely, utilizing common substrates such as glucose and oxygen and exchanging a variety of metabolites including glutamate. Any dysfunction at any point in any of the interlinked pools can result in deleterious consequences for CNS function.

Imbalance of glutamate pools in hyperammonemic disorders is frequently observed. Of particular importance in this respect is the function of the glutamate–glutamine cycle because it comprises not only an important aspect of glutamate cycling but also compartmentalizes ammonia metabolism in the brain. The following is an overview of the current knowledge of ammonia's effects on the glutamate–glutamine cycle.

THE GLUTAMATE–GLUTAMINE CYCLE

The glutamate–glutamine cycle in the brain is a classical example of the functional symbiotic relationship that exists between neurons and astrocytes and is shown in a simplified

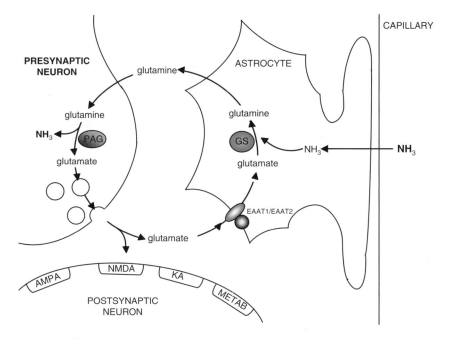

Figure 13.1   The glutamate–glutamine cycle. This simple schematic diagram depicts the major steps involved in the glutamate–glutamine cycle. Glutamate is released into the extracellular space where it can bind to glutamate-specific receptors including the *N*-methyl-D-aspartate (NMDA) receptor, kainate (KA) receptor, AMPA receptor, and metabotropic glutamate (METAB) receptor. High-affinity glutamate transporters on perineuronal astrocytes (EAAT1 and EAAT2) clear glutamate from the extracellular space where a portion is converted to glutamine by glutamine synthetase (GS). Glutamine is released from the astrocyte into the extracellular space and taken up by the surrounding neurons for conversion to glutamate by action of glutaminase (PAG). This glutamate is stored as part of the synaptic (vesicular) pool of releasable glutamate.

schematic form in Figure 13.1. Briefly, upon synaptic activation, neurons release glutamate into the synaptic cleft in order to stimulate glutamate receptors on postsynaptic neurons. Perineuronal astrocytes take up the released glutamate via the glutamate transporters EAAT1 and EAAT2. Subsequently, glutamate is used as a substrate for the enzyme glutamine synthetase (GS) to catalyze the production of glutamine. Glutamine is then either stored in the astrocyte, released to the cerebrospinal fluid, or released into the extracellular space and taken up by the surrounding neurons. By the action of the neuronal enzyme glutaminase (PAG), a considerable fraction of this glutamine is converted back to glutamate, which is consequently used in part to replenish the releasable pool of glutamate.

Ammonia is a significant player in the operation of the glutamate–glutamine cycle. Glutamine synthetase acts to "detoxify" ammonia by conversion to glutamine. Glutamine synthesis by the astrocyte is the only major route for ammonia detoxification by the brain. Furthermore, operation of the glutamate–glutamine cycle results in the production of ammonia by PAG, an event that occurs principally in neurons. Thus, a single turn of the cycle leads to ammonia production in the neuron, which is unequipped to remove it; ammonia must diffuse to the astrocyte in order to be removed by glutamine synthetase. The following sections will outline the current knowledge of the various steps of the glutamate–glutamine cycle and the effects of hyperammonemia on its functioning.

AMMONIA AND GLUTAMATE RELEASE

Glutamate is the principal excitatory neurotransmitter in the CNS, and its controlled release from neurons is a critical factor in excitatory neurotransmission. It is synthesized and stored in synaptic vesicles in the presynaptic terminal and is released upon synaptic transmission via a Ca^{2+}-dependent mechanism. There have been many studies examining the effects of ammonia on excitatory synaptic transmission and the release of glutamate. Excitatory transmission was observed to be reduced in rat hippocampal slices perfused with ammonia,[6] and ammonium ions depress excitatory neurotransmission in neurons.[7] Release of glutamate as determined *in vivo* by microdialysis[8] and the cortical cup method[9] is increased in experimental chronic liver failure, a condition associated with hyperammonemia. *In vitro* studies have further supported the latter findings. For instance, Hamberger et al.[10] demonstrated an elevation in the spontaneous release of glutamate from hippocampal slices exposed to pathophysiological concentrations of ammonium ions, whereas cerebrocortical minislices from rats with thioacetamide-induced liver failure were observed to release higher amounts of glutamate in comparison to similar preparations from control animals.[11] Furthermore, increased electrically stimulated Ca^{2+}-dependent release of glutamate from superfused hippocampal slices has been observed in rats with experimental liver failure.[12]

AMMONIA AND GLUTAMATE UPTAKE

The clearance of glutamate from the extracellular space is an essential step in the termination of excitatory synaptic transmission in the CNS. Glutamate released into the synaptic cleft is cleared by the action of high-affinity glutamate transporters localized in the forebrain as in most brain structures, mainly on perineuronal astrocytes. Hence, a perturbation in the glutamate uptake system could potentially lead to increased extracellular glutamate levels with potentially deleterious consequences.

A myriad of studies have demonstrated that the glutamate uptake system functions abnormally under hyperammonemic conditions (see Table 13.1). Schmidt et al. (1990)[13] observed a dose-dependent inhibition of D-aspartate (a nonmetabolizable analog of glutamate) uptake in rat hippocampal slices exposed to blood extracts from patients with liver failure. Despite the fact that these blood extracts could have contained various other toxins associated with liver failure (such as manganese, short-chain fatty acids, and mercaptans), the relative potency of this inhibition was best correlated with elevations of ammonia levels. *In vitro* studies in neural cell culture systems also contribute to the understanding of ammonia's effects on high-affinity glutamate uptake. Cultured astrocytes exposed to pathophysiological concentrations of ammonia exhibited significant reductions in radiolabeled glutamate/aspartate uptake.[14] Analyses of uptake kinetics

Table 13.1 Effects of Ammonia on High-Affinity Glutamate Uptake by Brain Preparations: A Review of the Literature

	Model	Reference
Decreased ³H-D-aspartate uptake	Rat hippocampal slices exposed to serum extracts from hyperammonemic patients	Schmidt et al.[13]
	Cultured astrocytes exposed to ammonia	Bender and Norenberg[14]
	Cultured neurons exposed to ammonia	Chan et al.[22]
	Crude rat brain synaptosomes exposed to ammonia	Mena and Cotman[24]
Decreased EAAT1 expression	Cultured astrocytes exposed to ammonia	Chan et al.[17] Zhou and Norenberg[18]
Decreased EAAT2 expression	Acute liver failure model (rat)	Knecht et al.[19]
	Thioacetamide-induced acute liver failure model (rat)	Norenberg et al.[20]
Unchanged EAAT3 expression	Cultured neurons exposed to ammonia	Chan et al.[22]

Note: Effects shown are a composite of the results of ammonia exposure on glutamate uptake in various astrocyte/neuron preparations and animal models.

revealed a significant reduction in the V_{max} of glutamate uptake into these cultured cells, suggesting that the reduction of uptake was a consequence of a loss of transporter protein or an inhibition of uptake capacity.

The high-affinity glutamate transporters were cloned in the early 1990s. The sodium-dependent astrocytic glutamate transporters EAAT1[15] and EAAT2[16] were characterized and shown to be the high-affinity glutamate transporters responsible for the majority of glutamate clearance in the mammalian forebrain. The identification of these astrocytic glutamate transporters as well as their specific agonists and antagonists have facilitated the analysis and characterization of astrocytic glutamate uptake. Several studies have focused on ammonia's

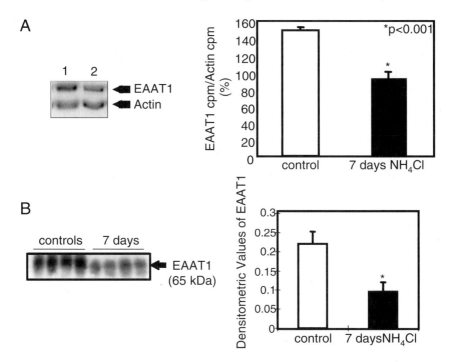

Figure 13.2 Effects of 5 mM ammonium chloride on EAAT1 mRNA (A) in which controls (1) (media alone) are compared with treated samples (2). Protein (B) expression levels are subsequently measured by Western blotting. Representative blots are accompanied by histograms displaying radioactive counts (A) and image densitometric analysis (B) *p < 0.05. Adapted from H Chan, AS Hazell, P Desjardins, RF Butterworth. Effects of ammonia on glutamate transporter (GLAST) protein and mRNA in cultured rat cortical astrocytes. *Neurochem Int* 37: 243–248, 2000.

effects on glutamate transporters. Chan et al.[17] demonstrated that prolonged exposure of cultured astrocytes to pathophysiological concentrations of ammonia results in significant reductions in both mRNA and protein levels of EAAT1 (Figure 13.2), which were associated with a loss in high-affinity radiolabeled D-aspartate uptake. Similarly, Northern blot analyses showed dose-dependent reductions in *EAAT1* mRNA levels in differentiated cultured astrocytes.[18] Evidence for abnormal glutamate transporter expression has also been demonstrated *in vivo*. Significant reductions in both mRNA and protein expression of the EAAT2 glutamate transporter were observed in the frontal cortices of rats with ischemic liver failure.[19] Other investigators have reported a loss of *EAAT2* mRNA by Northern blotting in rats with thioacetamide-induced liver failure,[20] and immunohistochemical studies in cerebella of rats with experimental chronic liver failure showed significant losses in EAAT1 and EAAT2 immunostaining.[21] Clearly, the above evidence demonstrates that the suppression of glutamate uptake observed in *in vitro* and *in vivo* models of hyperammonemia is a consequence (in part) of reduced glutamate transporter expression.

Although high-affinity glutamate uptake and clearance of extracellular glutamate have classically been the task of astrocytes, increasing evidence suggests that ammonia may also contribute to dysfunction of neuronal glutamate uptake. Chan et al. (2003)[22] demonstrated a significant loss in high-affinity glutamate uptake by cultured neurons exposed to pathophysiological concentrations of ammonia. Kinetic analysis revealed associated reductions in V_{max} and K_m values, implying that the depression in high-affinity glutamate uptake is the result of a loss of glutamate transporter activity or a loss of glutamate transporter protein levels. The reduction in K_m is an indication of increased sensitivity to the uptake substrate, possibly resulting in compensation to a loss of glutamate uptake activity. Examination of protein levels of the neuronal glutamate transporter EAAT3 revealed no significant alterations in the presence of ammonia, indicating that ammonia's lowering effect on neuronal glutamate uptake occurs at the level of transporter function.

Earlier studies focused on ammonia's effects on glutamate uptake in synaptosomal (nerve terminal) preparations. A significant loss of synaptosomal glutamate uptake was observed in preparations from rats with thioacetamide-induced liver failure,[23] whereas exogenous application of ammonia to rat synaptosomal preparations resulted in reductions in glutamate uptake.[24] In contrast, Raghavendra Rao and Murthy (1991)[25] reported elevated high-affinity glutamate uptake by synaptosomes isolated from brains of animals with acute or subacute hyperammonemia, whereas others observed no alterations in preparations from rats with liver failure due to bile-duct and portal-vein ligation.[26] The contrasting reports of ammonia's effects on glutamate uptake in synaptosomal preparations could result from several factors. Synaptosomal preparations often contain varying amounts of glial contaminants (gliosomes) and hence any variations in purity would be reflected as glial glutamate uptake as opposed to uptake originating from nerve terminals. Furthermore, there is no currently known neuronal glutamate transporter localized to the synaptic region in the forebrain, and until one is discovered, it is possible that any synaptosomal glutamate uptake activity may result from glial contamination in these preparations. Other possible causes of inconsistent findings in neuronal preparations include differing levels of ammonia and duration of exposure used in the various studies.

The consistent ammonia-lowering effect on high-affinity glutamate uptake in the brain is further strengthened and supported by reports of increased extracellular concentrations of glutamate in hyperammonemic conditions. Brain dialysate concentrations of glutamate are increased two- to four fold in brains of rats with ischemic liver failure.[27] Similar findings have been reported by several other groups in a range of hyperammonemic models.[28–30]

AMMONIA AND GLUTAMINE SYNTHETASE

Glutamine synthetase (GS) is a critical component in the regulation of brain nitrogen homeostasis because its product acts not only as a building block to proteins but delivers nitrogen atoms to enzymes that build nitrogen-rich molecules, such as deoxyribonucleic acid bases and amino acids. In particular, GS is principally an astroglial protein[31] and is a key cytoplasmically localized enzyme in the glutamate–glutamine cycle that catalyzes the amidation of glutamate to glutamine. Under normal physiological conditions, glutamate synthesized *in situ* together with extracellular glutamate transported by high-affinity transporters is converted by the action of GS into glutamine.

Table 13.2 Effects of Ammonia on Brain Glutamine Levels and
GS Activity: A Review of the Literature

	Model	Reference
Increased glutamine levels	Brains of cirrhotic patients who died in coma	Weiser et al.[34]
	Rats with experimental chronic liver failure	Butterworth and Giguère[32]
	Rats with thioacetamide-induced acute liver failure	Hilgier and Olsen[33]
	Patients with acute liver failure	Strauss et al.[35]
	Rats infused with ammonium acetate	Takahashi et al.[36a]
Decreased GS activity	Autopsied brain tissue from patients with chronic liver failure	Lavoie et al.[40]
	Rats with experimental chronic liver failure	Butterworth and Giguère[32]
Unchanged GS activity	Rats with experimental chronic liver failure	Colombo et al.[37]
	Rats with experimental chronic liver failure	Cooper et al.[39]
	Rats with experimental chronic liver failure	Desjardins et al.[44]

Note: Effects shown are a composite of the results of ammonia exposure on glutamine
levels and glutamine synthetase (GS) activity in various animal models and patient
studies.

Numerous studies have examined GS under hyperammonemic conditions. Glutamine levels in the brain are elevated in both chronic and acute hyperammonemic conditions
(see Table 13.2). Butterworth and Giguère,[32] using a sensitive
double-isotope dansyl microtechnique, demonstrated
increased glutamine levels in rats with chronic liver failure,
whereas rats with thioacetamide-induced acute liver failure
showed similar elevations in cerebral glutamine levels.[33] In
addition, brains of cirrhotic patients who died in hepatic
coma[34] and patients with fulminant hepatic failure[35] were also
found to contain elevated levels of glutamine. Glutamine synthesis is enhanced in cultured astrocytes exposed to patho-
physiologically relevant concentrations of ammonia.[36]

However, despite the consistent reports of elevation of glutamine levels in both human and experimental hyperammonemia, most studies have failed to show increased GS activity.[37–39] Some studies have even reported reduced GS activities in certain brain regions in chronic hyperammonemic states.[40,41]

Studies evaluating the protein and gene expression of GS under hyperammonemic conditions must also be considered. Quantitative RNA blot hybridization analysis in brains of rats with thioacetamide-induced acute liver failure demonstrated elevated levels of glutamine synthetase RNA expression,[42] whereas GS immunoreactivity is selectively increased in regions of high glutamatergic activity in rats made hyperammonemic by portacaval shunts.[43] In contrast, others reported no alterations in either GS mRNA or protein levels in brain regions of rats with experimental chronic liver failure.[44] The conflicting nature of the data reported for GS gene and protein expression may be a consequence of a number of factors. It is known that the half-life of GS is relatively short and that the enzyme is highly regulated by a number of factors.[45] In addition, GS has a region-specific localization in fine astrocytic processes surrounding excitatory synapses.[46] Such factors make investigation of GS expression difficult because its short half-life, variable modulation, and region-specific localization may act to mask any alterations in expression. More intensive examinations of GS transcriptional and translational rates in cultured astrocytes exposed to ammonia may yield better insight into ammonia's effects of GS expression.

AMMONIA AND GLUTAMINE RELEASE AND UPTAKE

Glutamine release and uptake are mediated by glutamine-specific transporters in neurons and astrocytes. Recent studies have demonstrated that neurons and astrocytes express different glutamine transporter systems. Astrocytes express System N transporters (SN1), which are bidirectional and electroneutral (transport of glutamine is coupled to the cotransport of a sodium ion and the efflux of a proton),[47] whereas neurons possess System A transporters (SA1 and SA2), which are electrogenic and not so readily able to mediate flux reversal.[48] SN1 transporters are pH sensitive and the uptake action of glutamine through this system results in an intracellular alkalinization[49] in astrocytes. Although no studies have directly assessed ammonia's effects on SN1 transporters, it has been demonstrated that ammonia induces rapid intracellular alkalinization followed by sustained acidification.[50] Such fluctuations in astrocytic pH could presumably affect the uptake or release of glutamine from these cells. Further studies are clearly required in order to assess this issue.

AMMONIA AND GLUTAMINASE

Following its synthesis by GS, glutamine is released into the extracellular space where adjacent neurons are able to transport it for cellular requirements. A portion of this glutamine is converted to glutamate by the action of the hydrolase-class enzyme glutaminase. Glutaminase is a ubiquitously expressed protein that is responsible for the breakdown of glutamate to produce glutamine with the subsequent release of ammonia. It is phosphate activated (often referred to as phosphate-activated glutaminase or PAG) and bears a mitochondrial localization in neural cells.[51] Extensive studies have demonstrated that glutamate produced by glutaminase-dependent hydrolysis of glutamine on the outer side of the inner mitochondrial membrane is transported across the inner membrane in exchange for aspartate. It is subsequently transaminated in the mitochondrial matrix to α-ketoglutarate and, with the aid of the ketodicarboxylate carrier, is transferred to the cytoplasm[52,53] in which it is retransaminated to glutamate. Cytoplasmic glutamate is accumulated in vesicles and kept in storage for release in a Ca^{2+}-dependent manner.[54,55]

Glutaminase has been a subject of study with respect to ammonia. Neuronal glutaminase is subject to inhibition by its products glutamate and ammonia.[56–58] Wallace and Dawson (1992)[59] reported dose-dependent reductions of PAG activity in rat brain synaptosomes exposed to ammonia, and a reduced rate of conversion of glutamine to glutamate was observed in the brain under hyperammonemic conditions.[60]

As previously mentioned, the conversion of glutamine to glutamate via glutaminase involves a circuitous pathway in the mitochondria. Glutamate produced via glutaminase in the mitochondrial intermembrane space traverses the inner mitochondrial membrane in exchange with aspartate. It is subsequently transaminated to α-ketoglutarate via aspartate aminotransferase. α-Ketoglutarate crosses the inner mitochondrial membrane in exchange with malate through a ketodicarboxylate carrier, and cytosolic α-ketoglutarate is then transaminated by cytosolic aspartate aminotransferase into glutamate.[61] Studies investigating ammonia's effects on the elements of this pathway reveal that activity of the aspartate aminotransferases (cytosolic and mitochondrial) is significantly inhibited by it.[62,63]

The data outlining ammonia's effects on glutaminase suggest a serious dysfunction in the neuronal portion of the glutamate–glutamine cycle. The reduction in glutaminase activity due to substrate inhibition results in a loss in glutamine deamidation and a consequent increase in neuronal glutamine content. It is not known, however, whether the inhibition of glutaminase activity is simply a reflection of inhibition by glutamate and ammonia or a loss in enzyme protein levels as well. Thomas et al. (1988)[42] demonstrated an increase of glutaminase mRNA levels in brains of rats with thioacetamide-induced acute liver failure. Whether or not this change in gene expression is reflected in altered protein levels of glutaminase is yet to be reported. In the meantime, one has to entertain the probability that loss of glutaminase activity likely contributes to the increased levels of brain glutamine consistently observed in hyperammonemic disorders.

AMMONIA AND GLUTAMATE UPTAKE INTO SYNAPTIC VESICLES

Synaptic vesicular uptake and release of glutamate into the extracellular space are important components in the glutamate–glutamine cycle. Glutamate accumulation into synaptic vesicles is an energy-dependent process that involves glutamate-specific vesicular transport proteins. The transport of glutamate into synaptic vesicles is driven by an electrochemical proton gradient generated by a vacuolar-type H^+-ATPase,[64] which establishes an electrochemical and pH gradient that couples ATP hydrolysis and glutamate uptake. To date, only two vesicular glutamate transporters have been identified. These are VGLUT1[65] and VGLUT2.[66] Both transporters show similar functional characteristics including ATP dependence, substrate affinity, substrate specificity, and mode of energization. They differ, however, in localization. VGLUT1 appears localized to glutamatergic neurons of the cerebral cortex, hippocampus, and cerebellum, whereas VGLUT2 is expressed in cerebral cortex, hippocampus, and thalamus.[66]

Recent evidence has demonstrated that glutamate accumulation in synaptic vesicles is affected by hyperammonemia. Albrecht et al.[67] demonstrated that exogenous addition of 3 mM ammonia to crude rat synaptic vesicle preparations stimulated glutamate uptake and H^+-ATPase activity. In contrast, synaptic vesicles isolated from rats with moderate hyperammonemia (with ammonia levels of approximately 0.6 mM) showed no alterations in vesicular transport of glutamate or H^+-ATPase activity. In addition to the differences in ammonia concentration in the models, the two preparations also show another important distinction. Synaptic vesicles isolated from normal rats were treated with exogenously applied ammonia during the uptake study, whereas those isolated from rats with moderate hyperammonemia were not. It has been postulated that ammonia in its gaseous form can cross the vesicular membrane and gain a proton to produce ammonium, to which the membrane is impermeable.[68,69] This event would consequently result in intracellular alkalinization and a disruption of the vesicular pH gradient. Hence, the increase in H^+-ATPase activity may be a compensatory mechanism for this alteration in the pH gradient, which would in turn drive the uptake of glutamate. However, studying the vesicular uptake of glutamate in the absence of ammonia (as was the case with vesicles isolated from rats with moderate hyperammonemia) would not have the same effect. The importance of studying these two different preparations is that the results indicate that a change in vesicular glutamate transporter expression is not a factor in ammonia-induced stimulation of vesicular transport.

Ammonia's stimulatory effect on vesicular glutamate transport is in accordance with the previously mentioned reports of ammonia-induced enhanced glutamate release. Whether or not the enhanced release of glutamate is a consequence of increased vesicular concentrations of glutamate or increased calcium-dependent quantal release of glutamate into the synaptic cleft remains unknown. Preliminary data from Rose (2002)[70] indicate that ammonia can elevate intracellular calcium ion levels in cultured astrocytes. Therefore, it may be hypothesized that an increase in intracellular calcium levels may be a causal factor in elevation of calcium-dependent glutamate release by neurons. Further studies would be required to address this issue.

CONSEQUENCES OF AMMONIA-INDUCED DYSFUNCTION OF THE GLUTAMATE–GLUTAMINE CYCLE

From the above discussion, it is clear that hyperammonemia has adverse effects on CNS function. Implications of ammonia-induced dysfunction of the glutamate–glutamine cycle are numerous.

Reduced astrocytic uptake and increased neuronal release of glutamate in the CNS under hyperammonemic conditions result in elevations in glutamate in the extracellular space. Normally, glutamate binds to and stimulates specific receptors on postsynaptic neurons to elicit events that would culminate in propagation of excitatory synaptic transmission. However, in the event of loss of astrocytic clearance of glutamate and increased neuronal release of glutamate, altered glutamate signaling results. Recent evidence demonstrates that this occurs in hyperammonemia in the brain. For example, activation of NMDA receptors by glutamate results in an elevation of intracellular Ca^{2+}, which binds to the regulatory protein calmodulin and activates a variety of enzymes, including neuronal nitric oxide synthase (nNOS) (Figure

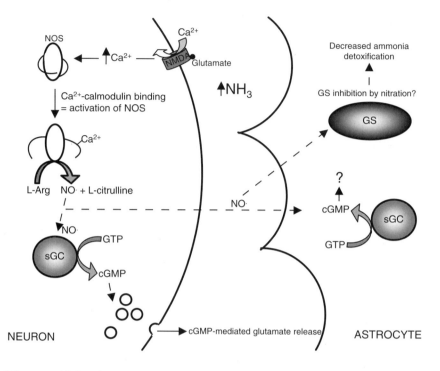

Figure 13.3 Ammonia-induced NMDA receptor activation leads to increased glutamate release and inhibition of GS. This simple schematic diagram outlines the proposed signal transduction pathway following activation of NMDA receptors by ammonia. Briefly, ammonia releases the Mg^{2+} ion block on NMDA receptors and renders them more susceptible to activation. NMDA activation leads to a Ca^{2+} influx into the neuron. Elevated levels of Ca^{2+} bind to calmodulin and activate nitric oxide synthase (NOS), which leads to the formation of nitric oxide (NO). NO is able to freely diffuse across membranes and stimulate activity of soluble guanylate cyclase (sGC) to produce cyclic GMP (cGMP), resulting in enhanced cGMP-mediated glutamate release from neurons. Furthermore, NO may also contribute to the nitration (and consequently inhibition of GS), which would result in a reduction of the ammonia detoxification capacity of the astrocyte.

13.3). Activation of nNOS leads to formation of nitric oxide (NO), which enhances the action of soluble guanylate cyclase and consequently increases the formation of cyclic GMP.

Hermenegildo et al.[71] reported that acute ammonia intoxication induces an increase in cGMP levels in rat brain. This increase was blocked by administration of the NMDA antagonist MK-801. Interestingly, elevation of extracellular glutamate levels was found to occur later than activation of NMDA receptors and could also be prevented by inhibition with MK-801. These findings suggest that the ammonia-induced increase in extracellular levels of glutamate is (at least initially) a consequence of (as opposed to being the cause of) NMDA receptor activation. A probable explanation for this is that ammonia-induced depolarization of neurons[72] leads to a release of the voltage-dependent Mg^{2+} block in the NMDA receptor channel, thus rendering NMDA receptors more sensitive to activation by normal levels of glutamate. In chronic hyperammonemia, in contrast to acute ammonia toxicity, extracellular levels of glutamate are unchanged and NMDA receptor expression is reduced,[73,74] indicating that the activation of the NMDA receptor–cGMP pathway may not occur.

In addition to the elevation of extracellular glutamate levels, the activation of the NMDA receptor–cGMP pathway by acute ammonia intoxication has been shown to manifest more adverse effects on neuronal and astrocytic functions. Kosenko et al. (1994)[75] demonstrated that acute ammonia intoxication leads to a significant reduction in ATP content in rat brain, which could be blocked by pretreatment with MK-801, suggesting that the depletion of ATP is a consequence of NMDA receptor activation. Moreover, the same group reported a significant increase in Na^+/K^+-ATPase activity in brains of rats with acute ammonia intoxication that could also be blocked by MK-801 administration. It has been estimated that approximately 40% of brain energy utilization is attributable to Na^+/K^+-ATPase activity,[76] suggesting that NMDA activation may lead to significant elevations in Na^+/K^+-ATPase activity and, ultimately, a depletion of cellular energy.

It has also been reported that inhibition of NMDA receptor activity in acute ammonia intoxication leads to elevation of levels of glutamine and GS activity in rat brain.[75] Blocking NMDA receptor activation with MK-801 afforded partial amelioration of brain glutamine levels. These reports (along with similar data mentioned in previous sections concerning GS activity) suggest that ammonia-induced NMDA receptor activation may act as the initial event that triggers a cascade of events resulting in inhibition of GS activity. A possible signal for this is NO because studies have demonstrated that incubation of cultured astrocytes with the NO-generating agent SNAP inhibits the activity of GS.[77] Furthermore, Schliess et al.[78] observed ammonia-induced nitration of GS associated with reduced GS activity in cultured astrocytes exposed to ammonia. These studies suggest that, in acute ammonia intoxication, activation of NMDA receptors results in an increase in NO formation, which in turn leads to inhibition of GS activity (Figure 13.3).

THERAPEUTIC IMPLICATIONS OF AMMONIA-INDUCED DYSFUNCTION OF THE GLUTAMATE–GLUTAMINE CYCLE

Based on the above discussion, elevation of extracellular glutamate levels in hyperammonemia appears to be a culmination of both increased release of neuronal glutamate and reduced neuronal and glial uptake of glutamate. The results of these aberrations may contribute to the pathogenesis of clinical symptoms in hyperammonemic syndromes. For example, although the exact pathophysiological mechanisms are unclear, there is evidence to suggest that elevations of extracellular glutamate may be a contributing factor in seizure activity. Seizures are not uncommon in both acute liver failure[79] and congenital hyperammonemic syndromes resulting from urea cycle enzymopathies.[80] Dogs with liver failure associated with congenital portal-systemic shunts were observed to have seizures and concomitant elevations in levels of cerebrospinal fluid glutamate.[81] Furthermore, another study demonstrated that EAAT1 glutamate transporter knockout mice were more susceptible to amygdala-kindled seizures displaying a more severe type of seizure accompanied by a longer duration.[82]

Brain edema is also a clinical symptom of hyperammonemias associated with acute liver failure and congenital urea-cycle enzymopathies. For example, rats made hyperammonemic with ischemic liver failure were observed to have brain edema along with increased brain levels of both glutamine and glutamate.[83] Studies demonstrate that ammonia-induced cellular swelling is localized to astrocytes, and astrocytic swelling has been observed in cerebral cortices of primates following ammonia infusions[84] and in cultured astrocytes exposed to millimolar levels of ammonia.[85] These observations are further strengthened by other studies using hypothermia as a preventative measure. Hypothermia was observed to reduce brain edema, brain levels of ammonia, and extracellular brain glutamate levels in rats with acute liver failure.[86] Moreover, *in vitro* studies have demonstrated that excessive levels of glutamate can lead to cellular swelling.[87] Furthermore, administration of memantine (a noncompetitive NMDA antagonist) to rats with complete liver ischemia resulted in a significant reduction in brain edema and cerebrospinal fluid levels of glutamate.[88] These studies provide strong evidence that elevation of extracellular brain glutamate levels due to impairment of the glutamate–glutamine cycle may be significant factors causing brain edema in hyperammonemic syndromes.

SUMMARY

The studies reviewed in this article provide convincing evidence that ammonia exerts deleterious effects on the glutamate–glutamine cycle, which results in the loss of glutamate homeostasis and lead to abnormal glutamate-stimulated pathways. Excess ammonia exerts effects on most (if not all) of the steps involved in this cycle, resulting in increased glutamate release from neurons, reduced perineuronal astrocytic glutamate uptake, altered GS activity, reduced glutaminase activity, enhanced vesicular glutamate uptake in neurons, and, very possibly, altered glutamate–glutamine release by astrocytes. The end result of ammonia's effects on the various steps of the glutamate–glutamine cycle is increased levels of extracellular glutamate, which lead to activation of the NMDA–NO–cGMP signal transduction pathway. The resulting elevation of NO and cGMP formation further aggravates ammonia's effects by increasing extracellular glutamate levels (cGMP-mediated glutamate release) and limiting the astrocyte's capability to detoxify ammonia (by inhibition of GS), thus resulting in a "vicious cycle." The activation of NMDA receptor-mediated NO–cGMP signal transduction pathway could be responsible for the neurological symptoms observed in inherited and acquired hyperammonemic disorders.

ACKNOWLEDGMENTS

Studies from the authors' research unit were funded by the Canadian Institute for Health Research (CIHR). Helen Chan is a recipient of the CIHR Doctoral Research Award.

References

1. AJL Cooper, F Plum. Biochemistry and physiology of brain ammonia. *Physiol Rev* 67: 440–519, 1987.

2. A Hamberger, JT Cummins, E Keller, CW Cotman. Glutamate secretion and NAD(P)H levels during calcium-dependent depolarization of slices of the dentate gyrus. *Brain Res* 156(2): 253–264, 1978.

3. R Dingledine, CJ McBain. Excitatory amino acid transmitters. In: GJ Siegel, Ed. *Basic Neurochemistry*, 5th edition. New York: Raven Press, 1994, pp. 367–387.

4. HS Waagepetersen, U Sonnewald, A Schousboe. Compartmentation of glutamine, glutamate, and GABA metabolism in neurons and astrocytes: functional implications. *Neuroscientist* 9(5): 398–403, 2003.

5. F Cruz, S Cerdan. Quantitative 13C NMR studies of metabolic compartmentation in the adult mammalian brain. *NMR Biomed* 12(7): 451–462, 1999.

6. Y Theoret, MF Davies, B Esplin, R Capek. Effects of ammonium chloride on synaptic transmission in the rat hippocampal slice. *Neuroscience* 14: 798–806, 1985.

7. JC Szerb, RF Butterworth. Effect of ammonium ions on synaptic transmission in the mammalian nervous system. *Prog Neurobiol* 39: 135–153, 1992.

8. U Tossman, A Delin, LS Eriksson, T Ungerstedt. Brain cortical amino acids measured by intracerebral dialysis in portacaval shunted rats. *Neurochem Res* 12: 265–269, 1987.

9. F Moroni, G Lombardi, G Moneti, C Cortesini. The release and neosynthesis of glutamic acid are increased in experimental models of hepatic encephalopathy. *J Neurochem* 40: 850–854, 1983.

10. A Hamberger, P Lindroth, B Nyström. Regulation of glutamate biosynthesis and release *in vitro* by low levels of ammonium ions. *Brain Res* 237: 339–350, 1982.

11. M Zieliska, W Hilgier, RO Law, P Gorynski, J Albrecht. Effects of ammonia and hepatic failure on the net efflux of endogenous glutamate, aspartate and taurine from rat cerebrocortical slices: modulation by elevated K⁺ concentrations. *Neurochem Int* 41: 87–93, 2002.

12. RF Butterworth, O Le, J Lavoie, JC Szerb. Effect of portacaval anastomosis on electrically-stimulated release of glutamate from rat hippocampal slices. *J Neurochem* 56: 1481–1841, 1991.

13. W Schmidt, G Wolf, K Grungreiff, M Meier, T Rheum. Hepatic encephalopathy influences high affinity uptake of transmitter glutamate and aspartate into the hippocampal formation. *Metab Brain Dis* 5: 19–32, 1990.

14. AS Bender, MD Norenberg. Effects of ammonia on L-glutamate uptake in cultured astrocytes. *Neurochem Res* 21: 567–573, 1996.

15. T Storck, S Schulte, K Hofmann, W Stoffel. Structure, expression and functional analysis of Na⁺-dependent glutamate/aspartate transporter from rat brain. *Proc Natl Acad Sci USA* 89: 10955–10959.

16. NC Danbolt, G Pines, BI Kanner. Purification and reconstitution of the sodium and potassium-coupled glutamate transport glycoprotein from rat brain. *Biochemistry* 29: 6734–6740, 1990.

17. H Chan, AS Hazell, P Desjardins, RF Butterworth. Effects of ammonia on glutamate transporter (GLAST) protein and mRNA in cultured rat cortical astrocytes. *Neurochem Int* 37: 243–248, 2000.

18. BG Zhou, and MD Norenberg. Ammonia downregulates GLAST mRNA glutamate transporter in rat astrocyte cultures. *Neurosci Lett* 276: 145–148, 1999.

19. K Knecht, A Michalak, C Rose, JD Rothstein, RF Butterworth. Decreased glutamate transporter (GLT-1) expression in frontal cortex of rats with acute liver failure. *Neurosci Lett* 229: 201–203, 1996.

20. MD Norenberg, Z Huo, JT Neary, A Roig-Cantesano. The glial glutamate transporter in hyperammonemia and hepatic encephalopathy: relation to energy metabolism and glutamatergic neurotransmission. *Glia* 21: 124–133, 1997.

21. I Suarez, G Bodega, B Fernandez. Modulation of glutamate transporters (GLAST, GLT-1 and EAAC1) in the rat cerebellum following portocaval anastomosis. *Brain Res* 859: 293–302, 2000.

22. H Chan, C Zwingmann, M Pannunzio, and RF Butterworth. Effects of ammonia on high affinity glutamate uptake and glutamate transporter EAAT3 expression in cultured rat cerebellar granule cells. *Neurochem Int* 43: 137–146, 2003.

23. KNW Oppong, K Bartlett, CO Record, AH Mardini. Synaptosomal glutamate transport in thioacetamide-induced hepatic encephalopathy in the rat. *Hepatology* 22: 553–558, 1995.

24. EE Mena, CW Cotman. Pathological concentrations of ammonium ions block L-glutamate uptake. *Exp Neurol* 89: 259–263, 1985.

25. VL Raghavendra Rao, CRK Murthy. Hyperammonemic alterations in the uptake and release of glutamate and aspartate by rat cerebellar preparations. *Neurosci Lett* 130: 49–52, 1991.

26. JE Maddison, C Mickelthwaite, WE Watson, GA Johnson. Synaptosome and brain slice cerebrocortical [^{3}H]-L-glutamate uptake in a rat model of chronic hepatic encephalopathy. *Neurochem Int* 28: 89–93, 1996.

27. A Michalak, C Rose, J Butterworth, RF Butterworth. Neuroactive amino acids and glutamate (NMDA) receptors in frontal cortex of rats with experimental acute liver failure. *Hepatology* 24: 908–914, 1996.

28. DK Bosman, NEP Deutz, MAW Mass, HMJ van Eijk, JJH Smit, JG de Hann, RA Chamuleau. Amino acid release from cerebral cortex in experimental acute liver failure, studied by *in vivo* cerebral cortex microdialysis. *J Neurochem* 59: 591–599.

29. RJ de Knegt, SW Schalm, CCD van der Rijt, D Fekkes, E Dalm, I Hekking-Weyma. Extracellular brain glutamate during acute liver failure and during acute hyperammonemia simulating acute liver failure: an experimental study based on *in vivo* brain dialysis. *J Hepatol* 20: 19–26.

30. W Hilgier, M Zieliska, HD Borkowska, R Gadamski, M Walski, SS Oja, P Saransaari, J Albrecht. Changes in the extracellular profiles of neuroactive amino acids in the rat striatum at the asymptomatic stage of hepatic failure. *J Neurosci Res* 56: 76–84, 1999.

31. MD Norenberg, A Martinez-Hernandez. Fine structural localization of glutamine synthetase in astrocytes of rat brain. *Brain Res* 161: 303–310, 1979.

32. RF Butterworth, JF Giguère. Cerebral aminoacids in portal-systemic encephalopathy: lack of evidence for altered gamma-aminobutyric acid (GABA) function. *Metab Brain Dis* 1: 221–228, 1986.

33. W Hilgier, JE Olsen. Brain ion and amino acid contents during edema development in hepatic encephalopathy. *J Neurochem* 62(1): 197–204, 1994.

34. M Weiser, P Riederer, G Kleinberger. Human cerebral free amino acids in hepatic coma. *J Neural Transm Suppl* (14): 95–102, 1978.

35. GI Strauss, GM Knudsen, J Kondrup, K Moller, FS Larsen. Cerebral metabolism of ammonia and amino acids in patients with fulminant hepatic failure. *Gastroenterology* 121(5): 1109–1119, 2001.

36. ACH Yu, A Schousboe, L Hertz. Influence of pathological concentrations of ammonia on the metabolic fate of ^{14}C-labeled glutamate in astrocytes in primary culture. *J Neurochem* 42: 594–597, 1984.

36a. H Takahashi, RC Koehler, T Hirata, SW Brusilow, RJ Traystman. Restoration of cerebrovascular CO_2 responsivity by glutamine synthesis inhibition in hyperammonemic rats. *Circ Res* 71(5): 1220-1230, 1992.

37. JP Colombo, C Bachman, E Peheim, J Beruter. Enzymes of ammonia detoxification after portacaval shunt in the rat. II. Enzymes of glutmate metabolism. *Enzyme* 22: 399–406, 1977.

38. B Sadasivu, CRK Murthy, GN Rao, M Swamy. Studies on acetylcholinesterase and gamma-glutamyl transpeptidase in mouse brain in ammonia toxicity. *J Neurosci Res* 9: 127–134, 1983.

39. AJL Cooper, SN Mora, NF Cruz, AS Gelbard. Cerebral ammonia metabolism in hyperammonemic rats. *J Neurochem* 44: 1716–1723, 1985.

40. J Lavoie, JF Giguère, G Pomier-Layrargues, RF Butterworth. Activities of neuronal and astrocytic marker enzymes in autopsied brain tissue from patients with hepatic encephalopathy. *Metab Brain Dis* 2: 283–290, 1987.

41. JF Giguère, RF Butterworth. Amino acid changes in regions of the CNS in relation to function in experimental portal-systemic encephalopathy. *Neurochem Res* 9: 1309–1321, 1984.

42. JW Thomas, C Banner, J Whitman, KD Mullen, E Freese. Changes in glutamate-cycle enzyme mRNA levels in a rat model of hepatic encephalopathy. *Metab Brain Dis* 3(2): 81–90, 1988.

43. I Suarez, G Bodega, E Arilla, B Fernandez. Long-term changes in glial fibrillary acidic protein and glutamine synthetase immunoreactivities in the supraoptic nucleus of portacaval shunted rats. *Metab Brain Dis* 11(4): 369–379, 1996.

44. P Desjardins, KV Rao, A Michalak, C Rose, RF Butterworth. Effect of portacaval anastomosis on glutamine synthetase protein and gene expression in brain, liver and skeletal muscle. *Metab Brain Dis* 14(4): 273–280, 1999.

45. J De Vellis, D Wu, S Kumar. Enzyme induction and regulation of protein synthesis. In: S Federoff, A Vernadakis, Eds. Astrocyte, Vol 2. New York: Alan Liss, 1986, pp. 209–248.

46. T Miyake, T Kitamura. Glutamine synthetase immunoreactivity in two types of mouse brain glial cells. *Brain Res* 586: 53–60, 1992.

47. A Broer, A Albers, I Setiawan, RH Edwards, FA Chaudhry, F Lang, CA Wagner, S Broer. Regulation of the glutamine transporter SN1 by extracellular pH and intracellular sodium ions. *J Physiol* 539: 3–14, 2002.

48. FA Chaudhry, D Schmitz, RJ Reimer, P Larsson, AT Gray, R Nicoll, M Kavanaugh, RH Edwards. Glutamine uptake by neurons: interaction of protons with system A transporters. *J Neurosci* 22(1): 62–72, 2002.

49. T Nakanishi, R Kekuda, YJ Fei, T Hatanaka, M Sugawara, RG Martindale, FH Leibach, PD Prasad, V Ganapathy. Cloning and functional characterization of a new subtype of the amino acid transport system N. *Am J Physiol Cell Physiol* 281(6): C1757–C1768, 2001.

50. TN Nagaraja, N Brookes. Intracellular acidification induced by passive and active transport of ammonium ions in astrocytes. *Am J Physiol* 274: C883–C891, 1998.

51. G Dienel, E Ryder, O Greengard. Distribution of mitochondrial enzymes between perikaryal and synaptic fractions of immature and adult rat brain. *Biochem Biophys Acta* 496: 484–494, 1977.

52. G Palaiologos, L Hertz, A Schousboe. Role of aspartate aminotransferase and mitochondrial dicarboxylate transport for release of endogenously and exogenously supplied neurotransmitter in glutamatergic neurons. *Neurochem Res* 14: 359–366, 1989.

53. R Svarna, A Georgopoulos, G Palaiologos. Effectors of D-[^{3}H]-aspartate release from rat cerebellum. *Neurochem Res* 21: 603–608, 1996.

54. F Fonnum, EM Fykse, S Roseth. Uptake of glutamate into synaptic vesicles. *Prog Brain Res* 116: 87–101, 1998.

55. ED Ozkan, T Ueda. Glutamate transport and storage in synaptic vesicles. *Jap J Pharmacol* 77: 1–10, 1998.

56. E Kvamme, and K Lenda. Evidence for compartmentalization of glutamate in rat brain synaptosomes using the glutamate sensitivity of phosphate-activated glutaminase as a functional test. *Neurosci Lett* 25: 193–198, 1982.

57. E Kvamme, K Lenda. Regulation of glutaminase by exogenous glutamate, ammonia and 2-oxoglutarate in synaptosomal enriched preparation from rat brain. *Neurochem Res* 7: 667–678, 1982.

58. AM Benjamin. Control of glutaminase activity in rat brain cortex *in vitro*: influence of glutamate, phosphate, ammonium, calcium and hydrogen ions. *Brain Res* 208: 363–377, 1981.

59. DR Wallace, RJ Dawson. Ammonia regulation of phosphate-activated glutaminase displays regional variation and impairment in the brain of aged rats. *Neurochem Res* 17(11): 1113–11122, 1992.

60. DF Matheson, CJ van den Berg. Ammonia and brain glutamine: inhibition of glutamine degradation by ammonia. *Biochem Soc Trans* 3(4): 525–528, 1975.

61. L Hertz, ACH Yu, G Kala, A Schousboe. Neuronal-astrocytic and cytosolic-mitochondrial metabolite trafficking during brain activation, hyperammonemia and energy deprivation. *Neurochem Int* 37: 83–102, 2000.

62. L Ratnakumari, CRK Murthy. Activities of pyruvate dehydrogenase, enzymes of citric acid cycle, and aminotransferases in the subcellular fractions of cerebral cortex in normal and hyperammonemic rats. *Neurochem Res* 14: 221–228, 1989.

63. L Faff-Michalak, J Albrecht. Aspartate aminotransferase, malate dehydrogenase, and pyruvate carboxylase activities in rat cerebral synaptic and nonsynaptic mitochondria: effects of *in vitro* treatment with ammonia, hyperammonemia and hepatic encephalopathy. *Met Brain Dis* 6: 187–196, 1991.

64. S Naito, T Ueda. Characterization of L-glutamate uptake into synaptic vesicles. *J Neurochem* 44: 99–109, 1985.

65. S Takamori, JS Rhee, C Rosenmund, R Jahn. Identification of a vesicular glutamate transporter that defines a glutamatergic phenotype in neurons. *Nature* 407(6801): 189–194, 2000.

66. L Bai, H Xu, JF Collins, FK Ghishan. Molecular and functional analysis of a novel neuronal vesicular glutamate transporter. *J Biol Chem* 276(39): 36764–36769, 2001.

67. J Albrecht, W Hilgier, M Walski. Ammonia added *in vitro*, but not moderate hyperammonemia *in vivo*, stimulates glutamate uptake and H(+)-ATPase activity in synaptic vesicles of the rat brain. *Metab Brain Dis* 9(3): 257–266, 1994.

68. DC Anderson, SC King, SM Parsons. Proton gradient linkage in active uptake of [^{3}H]-acetylcholine by Torpedo electric organ synaptic vesicles. *Biochemistry* 21: 3037–3043, 1982.

69. M Erecinska, A Pastuzko, DF Wilson, D Nelson. Ammonia-induced release of neurotransmitters from rat brain synaptosomes: differences between the effects of amines and amino acids. *J Neurochem* 49: 1258–1265, 1987.

70. C Rose. Increased extracellular brain glutamate in acute liver failure: decreased uptake or increased release? *Metab Brain Dis* 17: 251–261, 2002.

71. C Hermenegildo, P Monfort, V Felipo. Activation of NMDA receptors in rat brain *in vivo* following acute ammonia intoxication: characterization by *in vivo* brain microdialysis. *Hepatology* 31: 709–715, 2000.

72. W Raabe. Effects of NH_4^+ on the function of the CNS. *Adv Exp Med Biol* 272: 99–120, 1990.

73. L Ratnakumari, IA Qureshi, RF Butterworth. Loss of [^{3}H]-MK-801 binding sites in brain in congenital ornithine transcarbamylase deficiency. *Metab Brain Dis* 10: 249–255, 1995.

74. C Peterson, JF Giguère, CW Cotman, RF Butterworth. Selective loss of N-methyl-D-aspartate-sensitive L-[^{3}H]-glutamate binding sites in rat brain following portacaval anastomosis. *J Neurochem* 55: 386–390, 1990.

75. E Kosenko, Y Kaminsky, E Grau, MD Miñana, G Marcaida, S Grisolía, V Felipo. Brain ATP depletion induced by acute ammonia intoxication in rats is mediated by activation of NMDA receptor and Na$^+$,K$^+$-ATPase. *J Neurochem* 63: 2172–2178, 1994.

76. RW Albers, GJ Siegel, WL Stahl. Membrane transport. In: GJ Siegel, Ed. *Basic Neurochemistry*, 5th edition. New York: Raven Press, 1994, pp. 49–73.

77. MD Miñana, E Kosenko, G Marcaida, C Hermenegildo, C Montoliu, S Grisolía, V Felipo. Modulation of glutamine synthesis in cultured astrocytes by nitric oxide. *Cell Mol Neurobiol* 17: 433–445, 1997.

78. F Schliess, B Gorg, R Fischer, P Desjardins, HJ Bidmon, A Herrmann, RF Butterworth, K Zilles, D Haussinger. Ammonia induces MK-801-sensitive nitration and phosphorylation of protein tyrosine residues in rat astrocytes. *FASEB J* 16(7): 739–41, 2002.

79. SW Brown, MA Clarke, PA Tomlin. Fatal liver failure following generalized tonic-clonic seizures. *Seizure* 1: 75–77, 1992.

80. DJ Lacey, PK Duffner, ME Cohen, L Mosovich. Unusual biochemical and clinical features in a girl with ornithine transcarbamylase deficiency. *Pediatr Neurol* 2(1): 51–53, 1986.

81. J Butterworth, CR Gregory, LR Aronson. Selective alterations of cerebrospinal fluid amino acids in dogs with congenital portosystemic shunts. *Metab Brain Dis* 12: 299–306, 1997.

82. N Tsuro, Y Ueda, T Doi. Amygdaloid kindling in glutamate transporter (GLAST) knock-out mice. *Epilepsia* 43: 805–811, 2002.

83. M Swain, RF Butterworth, AT Blei. Ammonia and related amino acids in the pathogenesis of brain edema in acute ischemic liver failure in rats. *Hepatology* 15(3): 449–453, 1992.

84. TM Voorhies, ME Ehrlich, TE Duffy, CK Petito, F Plum. Acute hyperammonemia in the young primate: physiologic and neuropathologic correlates. *Pediatr Res* 17: 970–975, 1983.

85. MD Norenberg, L Baker, LO Norenberg, J Blicharska, JH Bruce-Gregorios, JT Neary. Ammonia-induced astrocyte swelling in primary culture. *Neurochem Res* 16: 833–836, 1991.

86. C Rose, A Michalak, M Pannunzio, N Chatauret, A Rambaldi, RF Butterworth. Mild hypothermia delays the onset of coma and prevents brain edema and extracellular brain glutamate accumulation in rats with acute liver failure. *Hepatology* 31(4): 872–877, 2000.

87. GH Schneider, A Baethmann, O Kempski. Mechanisms of glial swelling induced by glutamate. *Can J Physiol Pharmacol* 70 Suppl: S334–S343, 1992.

88. BA Vogels, MA Maas, J Daalhuisen, G Quack, RA Chamuleau. Memantine, a noncompetitive NMDA receptor antagonist improves hyperammonemia-induced encephalopathy and acute hepatic encephalopathy in rats. *Hepatology* 25(4): 820–827, 1997.

14

The Influence of Lipids on Nuclear Protein Import, Cell Growth, and Gene Expression

RANDOLPH S. FAUSTINO,
MELANIE N. LANDRY, NICOLE T. GAVEL,
MICHAEL P. CZUBRYT, AND GRANT N. PIERCE

NUCLEAR TRANSPORT

The movement of molecules between the nuclear and cytosolic compartments is essential in promoting proper cell functioning. It is an elaborate process requiring the interplay of both cytosolic and membrane-bound components. Nucleocytoplasmic trafficking is both signal mediated and energy dependent and is regulated throughout the transport pathway. The movement of molecules into the nucleus is referred to as nuclear import. Movement in the opposite direction is referred to as nuclear export, and the mechanism of transport is similar in both directions.

Nuclear import occurs when a protein exposes its nuclear localization signal (NLS) and becomes competent for import. The NLS itself is a polybasic amino acid sequence that can occur in a mono- or bipartite form.[1] Other classes of NLSs exist that are distinct from the "classical" signal sequence.[2]

Exposure of the NLS can occur by phosphorylation,[3-5] dephosphorylation,[6] or controlled proteolysis.[7] In addition to unmasking the nuclear localization signal, such modifications may also increase or decrease the affinity of the NLS for its

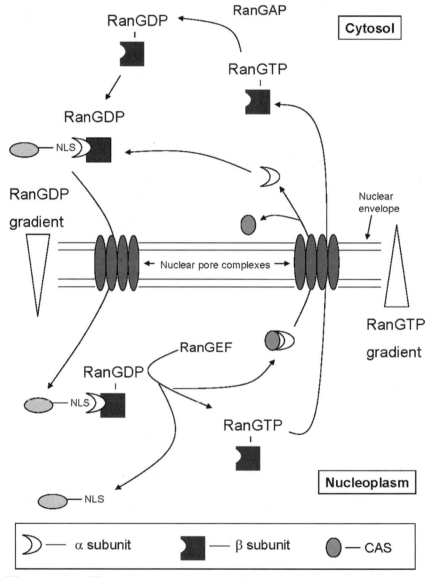

Figure 14.1 (See caption on opposite page)

cognate receptor, called a karyopherin or importin.[8,9] Karyopherins are a class of heterodimeric molecules that recognize the NLS. They consist of an alpha subunit that recognizes the NLS motif, and a beta subunit that is responsible for mediating the interaction between the receptor-NLS complex and the nuclear pore complex (NPC). Movement of the protein substrate-karyopherin complex to the NPC is an energy independent process, and the NLS-receptor complex interacts with additional accessory proteins *en route* to the nuclear envelope.[10]

Translocation through the pore complex into the nuclear interior is energy dependent and requires a gradient of RanGTP across the nuclear envelope.[11] High concentrations of RanGDP are present outside the nucleus and a high concentration of RanGTP is within the nucleus. These levels are maintained by the presence of RanGAP and RanGEF in the cytoplasm and nucleus, respectively.[12] Once inside the nucleus, the import complex is dissociated by the high concentration of RanGTP,[12] and the NLS-bearing protein is released into the nuclear interior whereas the karyopherin subunits are transported back through the NPC into the cytoplasm for another round of import[12] (Figure 14.1).

Figure 14.1 (Opposite page) Overview of nuclear import. Proteins bearing a nuclear localization sequence (NLS) are recognized by the NLS receptor. The receptor, referred to as a karyopherin or an importin, is a heterodimeric complex consisting of an alpha and a beta subunit. The NLS binds to importin alpha upon unmasking. Importin alpha subsequently binds to the beta subunit and this protein-receptor complex is ready to be imported into the nucleus. High concentrations of RanGDP outside the nucleus are maintained by active RanGAP, a Ran GTPase-activating enzyme constitutive found in the cytoplasm. RanGDP associates with the NLS-receptor complex and promotes its import. Once inside the nucleus, RanGDP is exchanged for RanGTP by the nuclear enzyme RanGEF. RanGTP bound to the beta subunit allosterically displaces the alpha subunit, ultimately dissociating the entire import complex. The RanGTP-bound beta subunit is then recycled to the cytosol, while the alpha subunit is exported to the cytosol by its endogenous export receptor, CAS, permitting another round of nuclear import to occur.

The NPC is the central gateway through which molecules transiting the nuclear envelope must pass. It is a large 125-MDa multiproteinaceous structure composed of roughly 40 to 50 different nuclear pore proteins, or nucleoporins, as they are collectively referred to. Molecules of 40 to 60 kDa or larger must be actively transported through the NPC.[10] In this manner, the NPC imparts its own selectivity on the nuclear transport process.

The NPC spans both membranes of the nuclear envelope. The central, barrel-like region is composed of eight subunits arranged in an annular configuration, leaving an aqueous pore through which substances are transported.[13,14] The pore itself is a dynamic structure that can adjust its diameter depending on the metabolic state of the cell.[15,16]

Adjacent to the barrel-like core region are rings on both the nuclear and cytoplasmic faces. The cytoplasmic ring possesses filaments that extend for about 50 nm[12] into the cytoplasm and possess sites to which the protein–NLS-receptor complex docks.[17] The nuclear ring also possesses filaments that extend 50 to 120 nm into the nucleoplasm[12] and are connected at their distal ends to another ring, forming a basket-like structure. Much like its cytoplasmic counterpart, sites on the nuclear ring and its components are believed to be the first sites that interact with molecules being exported out of the nucleus.[18–21]

A report by Goldberg and Allen indicates that the distal ring of the nuclear basket is associated with the nuclear envelope lattice,[22] a fibrous network distinct from the nuclear lamina. The nuclear ring has been reported to interact with the nuclear lamina and anchors the NPC within the nuclear envelope.[23] It has been hypothesized that this interaction also allows for contact between the nuclear envelope and interphase chromatin,[24] suggesting a role in nonrandom organization of the genetic machinery as well as gene regulation.

At both the cytosolic and membrane-bound levels, nuclear transport remains a complex and tightly controlled eukaryotic process that is dynamically regulated according to the needs of the cell. For example, proliferating smooth muscle cells exhibit an increase in nuclear protein import in comparison to quiescent cells.[25] Furthermore, previous work by Maul et al. demonstrated a relationship between nuclear pore density and metabolic activity of the cell.[26] Perez-Terzic et al. have reported a structural adaptation in NPCs obtained from proliferative, stem cell-derived cardiomyocytes as opposed to terminally differentiated, postmitotic cardiac cells.[27] The same group has also shown that regulators of nuclear import cause dynamic changes in the NPC of adult cardiomyocytes.[16] While proliferative signals typically affect the movement of individual transcription factors,[6,28] it is becoming apparent that more than one level of the nuclear transport machinery can be involved[29] regardless of the stimulus.

It is well recognized that cells within the vasculature are in an advanced proliferative state in atherosclerotic conditions.[30] A variety of factors may stimulate cell proliferation within the vascular wall during atherosclerosis. Lipids are an interesting class of molecules that appear to have important pro- and antiproliferative effects in vascular cells. For example, various free fatty acids,[31,32] cholesterol,[33-35] and sphingolipids such as ceramide[36-38] and sphingosine-1-phosphate[39-41] have a variety of proliferative, apoptotic, and growth-arresting effects in different cell types. The mechanism by which these effects occur remains relatively unclear. The purpose of this chapter is to discuss these lipid species and their role as potential regulators of nuclear protein import, gene expression, and cell proliferation.

FATTY ACIDS

Fatty acids are an integral part of the human diet because they are required for proper cell function and maintaining membrane flexibility and fluidity. They are essential in reproductive functioning, for growth and development, and for cholesterol metabolism.[42,43] The basic structure of fatty acids consists of a carbon chain with a carboxylic acid moiety on one end and a methyl group on the other. The carboxylic acid and methyl ends are known as the delta (Δ) and omega (ω) end, respectively. Fatty acids can be divided into two groups — saturated and unsaturated fatty acids, based on the

Saturated Fatty Acid – NO double bonds

$$CH_3-CH_2-CH_2-CH_2-CH_2-CH_2-CH_2-CH_2-CH_2-CH_2-CH_2-COOH$$
(Lauric Acid)

Monounsaturated Fatty Acid – ONE double bond

$$CH_3-(CH_2)_7-\mathbf{CH=CH}-(CH_2)_7-COOH$$
(Oleic Acid – Omega-9)

Polyunsaturated Fatty Acid – MORE THAN ONE double bond

$$CH_3-(CH_2)_4-\mathbf{CH=CH}-CH_2-\mathbf{CH=CH}-(CH_2)_7-COOH$$
(Linoleic Acid - Omega-6)

$$CH_3-CH_2-\mathbf{CH=\text{-}CH}-CH_2-\mathbf{CH=CH}-CH_2-\mathbf{CH=CH}\text{-}-(CH_2)_7-COOH$$
(Alpha-Linolenic Acid - Omega-3)

Figure 14.2 Structure of fatty acids. Illustrated are examples of typical fatty acids. Unsaturated fatty acids contain double bonds and can be further subclassified into polyunsaturated and monounsaturated fatty acids, possessing more than one or only one double bond, respectively.

absence or presence of double bonds.[44] Saturated fatty acids (SFA) do not contain any double bonds and are identified by the number of carbons that make up the chain.[44] Dodecanoic acid (lauric acid) is illustrated in Figure 14.2 as an example. Conversely, unsaturated fatty acids contain at least one double bond and are further subclassified based on the number of double bonds present. Monounsaturated fatty acids (MUFAs) contain only one double bond and polyunsaturated fatty acids (PUFAs) contain more than one (Figure 14.2). By convention, the carbon atoms of a chain are numbered for identification purposes, starting with the methyl or omega (ω) end of the molecule. When one or more double bonds are present in the structure, the location of the double bond is given by the number of the first carbon of the double bond.

Two isomers can be formed when a double bond is present in a molecule, known as *cis-* and *trans-* isomers. The two hydrogen atoms that remain at the site of the double bond can be in the *cis*- configuration (on the same side of the double bond) or in the *trans*- configuration (on opposite sides of the double bond). Natural oils exist in the *cis*- configuration, which are known to be beneficial for the maintenance of membrane fluidity. In contrast, animal fats, which are solid at room temperature, typically exist in the *trans*- configuration.[43,45] *Trans*- fatty acids, due to their configuration, allow for PUFAs present in cell membranes to pack tightly together, and can thus have an impact on the fluidity of the membranes. Consequently, a change in membrane fluidity may negatively affect cell function.

Polyunsaturated fatty acids can be further subdivided into nonessential and essential fatty acids. Nonessential fatty acids are those that can be produced by the body from other lipid precursors, and the majority of fatty acids fall into this category. Essential fatty acids, alternatively, must be obtained from external sources because the human body cannot synthesize them.[46] The human body lacks the enzymes necessary to convert single bonds to double bonds at a position further than nine carbons from the delta end of the molecule.[45] The only fats that are essential are alpha-linolenic acid (omega-3) and linoleic acid (omega-6) PUFAs (Figure 14.3). These essential fatty acids are required for optimal functioning of the cell, the immune and organ systems, as well as for overall healthy growth and development.[42,45] Once these essential fatty acids are ingested, they are metabolized through a series of desaturation and elongation reactions that produce eicosapentaenoic acid (EPA) and docosahexaenoic acid (DHA), which are in turn involved in eicosanoid metabolism.[45] The enzymes involved in the metabolism of both linoleic acid (LA) and alpha-linolenic acid (ALA) are elongase, $\Delta 4$-, $\Delta 5$-, and $\Delta 6$-desaturase.[45] The same enzymes are involved in each set of reactions. As a result, competition exists between the two pathways.[43,47]

Polyunsaturated fatty acids and their metabolites can affect vascular smooth-muscle cell (VSMC) migration and proliferation. The development of neointimal hyperplasia and the subsequent generation of an atherogenic lesion occurs when

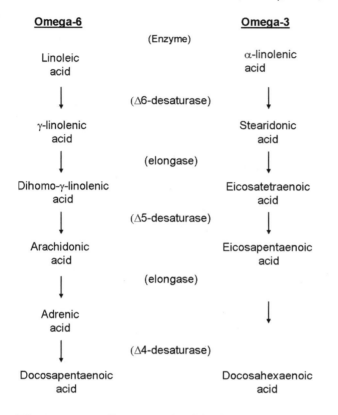

Figure 14.3   Steps involved in the synthesis of n-3 and n-6 PUFAs. Also known as omega-3 and omega-6 fatty acids, these PUFAs are essential to the human diet. Both pathways share enzymes and therefore competition between the two is common.

smooth-muscle cells proliferate. Epidemiological and clinical evidence suggests that omega-3 polyunsaturated fatty acids may reduce VSMC proliferation[32] and excitability,[48] but the precise cellular mechanisms by which these fatty acids modulate VSMC growth are not well characterized. However, evidence does exist in support of the capability of PUFAs to modulate VSMC growth. For example, serotonin-induced smooth-muscle cell proliferation can be blocked by EPA and DHA.[32] Serotonin mediates its proliferative effects through the $5\text{-}HT_2$ receptor in VSMC.[49] In their study, Pakala et al.[32]

demonstrated that incubation of smooth-muscle cells with EPA and DHA resulted in downregulation of 5-HT_2 mRNA expression resulting in blockade of the serotonin-signaling cascade. Omega-3 PUFAs have also been demonstrated to inhibit platelet-derived growth factor (PDGF)-induced mitogenesis in mesangial cells.[50]

Studies using smooth muscle cells[51] and T-cells[52] have reported an alteration in NFκB signaling due to arachidonic acid-, EPA- and DHA- dependent regulation. Other studies have demonstrated PUFA-mediated lipid raft remodeling, resulting in displacement of resident proteins within the raft leading to subsequent suppression of T-cell activation.[53–55] In this manner, direct effects of fatty acids on nuclear signaling and nuclear transport of NFκB in particular have been demonstrated. The effects of MUFAs and PUFAs on this type of signaling modality, however, remain to be determined.

Although the significance of fatty acids to the human diet and cellular metabolism is well appreciated, other cardiovascular lipid molecules are important too. Cholesterol and sphingolipids, for example, play important roles not only as membrane structural elements but function as important signaling molecules as well.

CHOLESTEROL

Cholesterol is an important component of all animal cell membranes and is synthesized predominantly in the endoplasmic reticulum of most cell types.[56] It is a steroid lipid molecule possessing a four-ringed carbon backbone structure. The smaller portion of the cholesterol molecule consists of a hydrophilic hydroxyl group that is found embedded in the lipid bilayer facing the membrane surface. Functionally, cholesterol increases the fluidity of membranes by interfering with the tight packing of the fatty acid tails of the phospholipid membrane,[56] which consequently decreases the permeability of smaller water-soluble molecules. Additionally, cholesterol prevents membrane phase shifts and the crystallization of hydrocarbons within the membrane.[56]

Recent studies suggest that cholesterol exhibits a direct role in the regulation of cell growth and membrane biosynthesis. Abnormal cell growth and proliferation can ultimately lead to the development and progression of cancer[34] and the development of atherosclerosis.[57] Changes in cholesterol metabolism in the cell may be associated with these changes in cellular proliferation. However, the way in which cholesterol can influence cell growth is poorly understood at present. Several potential mechanisms explaining the effects of cholesterol on cell growth and proliferation have been proposed.

Cholesterol has been implicated in tightly regulating the activity of the sterol regulatory element-binding protein (SREBP) within the nuclear membrane.[58] SREBP is a membrane-bound transcription factor that exhibits a direct role in maintaining cholesterol homeostasis.[59] Excessive cholesterol may inhibit the entry of SREBP into the nucleus, thus preventing transcriptional activation. Maxwell et al.[60] identified the role of high cholesterol-supplemented diets on changes in gene expression. These changes were found to be the cause of alterations in the activity of transcription factors such as SREBPs.[60] Another example of cholesterol-induced changes in gene expression involved the peripheral-type benzodiazepine receptor (PBR) protein. Hardwick et al. revealed that the PBR protein functions to regulate the transport of cholesterol and cell proliferation in aggressive breast tumor cells. Their data suggests that nuclear PBR regulates the transport of cholesterol into the nucleus, which ultimately induces an excessive proliferative response.[61]

The subcellular and molecular mechanisms responsible for these changes in gene expression and cell proliferation are not always entirely clear. It may involve alterations at the level of the nucleus itself. The nuclear membrane is unusually rich in cholesterol.[62] Cholesterol greatly increases the fluidity of membranes, which directly affects the role of membrane proteins or enzymatic activity within the membrane.[63] A number of studies have indicated that nuclear membrane cholesterol significantly stimulates NTPase activity,[62,64] and it has been proposed that the NTPase enzyme found in the nuclear membrane regulates the function of the nuclear pore.[62] This would be predicted, therefore, to have a significant and direct impact on nucleocytoplasmic transport, which would consequently have an impact on gene expression and cell growth because of the close relationship of nuclear protein import with gene expression (see the preceding text for detailed discussion).

The location of cholesterol may also affect gene expression directly. The JCR:LA-cp corpulent rat is a model of obesity and glucose intolerance, and the livers from these rats typically exhibit large, frequent lipid deposits throughout the tissue (Figure 14.4). We have also identified (in this tissue) an unusual deposition of lipid within the nucleus itself (Figure 14.5). It is difficult to imagine that such a large physical intrusion into the genetic machinery within the nucleus would not pose transcriptional problems to the cell.

Preliminary data have demonstrated that cholesterol may also have a direct impact on nuclear pore density. Nuclear pores are clearly visible when isolated rabbit hepatic nuclei are stained with antibodies specific for p62, an integral nuclear pore protein (data not shown). A significant decrease in fluorescence intensity and punctate staining in nuclei obtained from livers from rabbits fed a cholesterol-enriched diet for 8 weeks was observed when compared with control hepatic nuclei (data not shown). This indicates a decrease in the density of the nuclear pore complexes found on the surface of the nuclear membrane. Furthermore, western blots performed in parallel also demonstrated a decrease in the expression of p62 in hepatic nuclei isolated from cholesterol-fed rabbits (Figure 14.6). These data would strongly suggest that the expression of NPC channels within the nucleus is influenced by dietary cholesterol and its deposition within the liver.

Modifications in the structure of cholesterol can also affect its function and its effects on cell growth. For example, free radicals can oxidize membrane lipids like cholesterol, which can then alter cell growth. In a study conducted by Ramjiawan et al., oxidation of cell cholesterol induced significant changes in nuclear NTPase activity.[65] This was suggested to have an impact on nuclear trafficking and subsequently cell growth and proliferation.

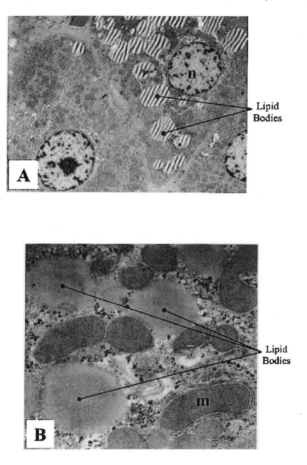

Figure 14.4   Electron micrograph of lipid bodies in JCR:LA-cp liver tissue. Liver tissue samples from a 6-month-old corpulent female JCR:LA-cp rat were prepared for electron microscopy. Numerous large lipid-laden bodies are indicated. (A) magnification ~3300x. (B) magnification ~19000x. Abbreviations: n — nucleus, m — mitochondria.

In summary, cholesterol is most often implicated in having an important but mostly passive role to play in gene expression and cell growth. It is becoming increasingly clear that this is not the case. Disturbances in cholesterol metabolism are closely associated with proliferative diseases such as cancer and atherosclerosis. Recent evidence now more strongly implicates cholesterol as having both direct and

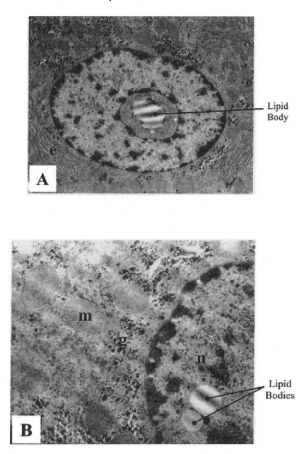

Lipid
Body

Lipid
Bodies

Figure 14.5 Electron micrographs of lipid body inclusions in the nuclei of JCR:LA-cp liver tissue samples. Samples of liver tissue from a 6-month-old corpulent female JCR:LA-cp rat were prepared for electron microscopy. (A) Hepatocyte revealing a large nuclear inclusion containing a lipid body (magnification ~7700x). (B) Example of multiple lipid bodies in the nucleus of another hepatocyte (magnification ~18500x). Abbreviations: e — endoplasmic reticulum, m — mitochondria, n — nucleus, g — glycogen deposits.

indirect actions that can alter gene expression and play an important role in determining cell growth. The effects that cholesterol has on the nucleus itself may play a significant role in these actions.

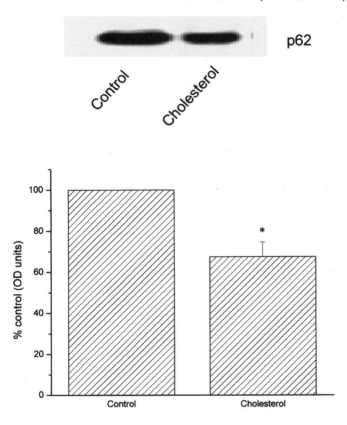

Figure 14.6 Specific nuclear pore proteins (NPC) show decreased expression in cholesterol-fed animals. NPC proteins (nucleoporins) were isolated from the hepatic tissue of both control and cholesterol fed rabbits and analyzed by immunoblotting with mAb414. A decreased expression of p62 was observed in cholesterol-fed animals vs. animals fed with a control diet. Densitometric analysis confirms a significant 30% decrease in levels of p62. *p < .05, n = 3.

SPHINGOLIPIDS

Ceramides are a class of sphingolipids synthesized *de novo* from serine and palmitoyl-CoA precursors.[66] Alternatively, they may be made by the conversion of sphingomyelin by the enzyme sphingomyelinase.[67,68] *De novo* synthesis of ceramide begins when serine and palmitoyl-CoA form 3-ketosphinganine

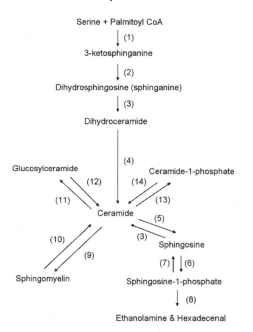

Figure 14.7 Basic pathway of sphingolipid metabolism. Once ceramide has been synthesized *de novo* from serine and palmitoyl-CoA precursors or generated from sphingomyelin, it can easily convert to and from its subsequent metabolic products, thus serving as an essential building block for a variety of sphingolipids. Shown are steps leading to the formation of ceramide as well as the products arising from its conversion. Enzymes involved: (1) serine palmitoyl-transferase, (2) 3-ketosphinganine reductase, (3) ceramide synthase, (4) dihydroceramide synthase, (5) ceramidase (acidic, neutral, or basic), (6) sphingosine kinase (1 and 2), (7) S1P phosphatase, (8) S1P lyase, (9) sphingomyelin synthase, (10) sphingomyelinase (acidic or neutral), (11) glucosylceramide synthase, (12) glucosyl-ceramidase, (13) ceramide kinase, (14) ceramide-1-phosphate phosphatase.

via serine palmitoyl-transferase (Figure 14.7). Then, 3-keto-sphinganine undergoes a reductive step to form dihydrosphin-gosine (sphinganine). Ceramide synthase converts sphinganine to dihydroceramide, which is subsequently acylated to form ceramide by dihydroceramide desaturase. Ceramide can also be synthesized by the removal of a phosphocholine group from sphingomyelin, catalyzed by sphingomyelinase. Ceramides serve as the building block for other sphingolipids, namely ceramide-1-phosphate, glucosylceramide, and sphingosine-1-phosphate (S1P), in addition to sphingomyelin.[67] Whereas ceramide is primarily made via *de novo* synthesis and sphingomyelin breakdown, interconversion between ceramide and its three other metabolic products occurs quite easily. Of these molecules, ceramide and S1P have been the best characterized.

Ceramide has the capacity to induce growth arrest and apoptosis in a variety of cell types.[69–74] In a variety of nonvascular cells, levels of ceramide were observed to increase following UVA/B irradiation, subsequently leading to apoptosis.[37,75] In an elaborate study demonstrating the involvement of ceramide in the apoptotic signaling pathway, one group incubated ceramine, a nonhydrolyzable analogue of ceramide, with mouse fibroblast cells and found that levels of apoptosis increased by approximately sevenfold when compared with cells incubated with hydrolyzable ceramide.[76] The antiproliferative and apoptotic effects of ceramide have been linked to activation of stress-activated MAP kinases, most notably p38 and JNK.[36,69,77,78] In addition to this, a variety of new substrates for ceramide have been identified including ceramide-activated protein kinases[79] and ceramide-activated protein phosphatases.[80] Both of these protein classes may affect cellular proliferation.

In addition to serving as a second messenger that activates apoptotic signaling cascades, ceramide has demonstrated membrane-altering properties that may play a part in causing irreversible apoptosis. Siskind et al. first demonstrated that ceramide could form large channels in artificial planar membranes[81] that were large enough to permit the release of proapoptotic factors.[82] The extremely high ceramide concentrations needed to produce these effects may limit the physiological relevance. Despite this limitation, it is clear from these studies and others that ceramide has a strong proapoptotic activity.

These apoptotic effects of ceramide have great pathological relevance for the vasculature. Apoptosis is involved in vascular development and is suggested to be involved in atherosclerosis.[83] Therefore, ceramide-induced apoptosis may be involved in the atherosclerotic process. Ceramide signaling is important in cell growth in other pathological states as well. For example, ceramide signaling was impaired in genetically hypertensive animals, and it was hypothesized that this deficiency could be responsible for the accelerated smooth-muscle cell proliferation observed in the hypertensive vessel.[73]

Ceramide has other actions and targets within the cell that may explain its proapoptotic and antiproliferative effects. For example, the addition of ceramide globally downregulates nuclear import in VSMCs, both *in vitro* and *in situ* (Faustino et al., in preparation). An overall reduction in nuclear import such as that observed with ceramide-treated VSMCs, therefore, may affect the movement of antiapoptotic and mitogenic transcription factors into the nucleus, resulting in apoptosis and growth arrest, respectively.

A large body of evidence also exists to implicate S1P as having a role in determining cell fate. S1P is a mitogenic factor involved in proliferative signaling[40,84] in a variety of cell types.[39,41,76,85,86] S1P activates ERK-1 and ERK-2, both of which are known proliferative MAP kinases.[41] In addition to possessing mitogenic properties, S1P also participates in cell survival and has been shown to protect against apoptotic signaling in mouse fibroblasts.[76] Thus, the signaling properties of both ceramide and S1P are directly opposed. However, they are also part of the same system, as these two molecules are directly interconvertible. Therefore, it has been proposed that the balance between cellular levels of ceramide and S1P is a cellular "rheostat" responsible for balancing the signals from both the mitogenic, S1P-dependent side and the apoptogenic, antiproliferative, ceramide-dependent side.[87] Evidence for this rheostat model is provided by a study in which it was observed that addition of exogenous S1P rescued ceramide-mediated apoptosis.[76] Additionally, "classic" proliferative signals from growth factors and cytokines can upregulate the activities of ceramidase and sphingosine kinase, two enzymes essential to the production of S1P. Concomitant with this upregulation, the same signals downregulated the activity of enzymes involved in ceramide synthesis.[88] Other studies have also demonstrated a similar differential regulation of the ceramide and S1P synthetic pathways by the same stimulus, whether that stimulus is proliferative or apoptotic. These data argue in support of the ceramide–S1P rheostat model.

While numerous evidence exists demonstrating the role of both proliferative and apoptotic enzymes involved in sphingolipid signaling, the exact mechanisms by which this occurs still need to be identified. The effects of ceramide and S1P on membrane topology, membrane potential, and nuclear transport are only a few of the other ways in which these molecules may exert their effects. The properties of sphingolipids in cellular metabolism, therefore, remain a complex area that requires further study to clarify their role in determining cell fate.

CONCLUSIONS

Nuclear protein import is an important regulatory process in cell growth. Protein entry through the pore channels regulates protein transcription and gene expression. This channel also regulates the exit of RNA from the nucleus to begin protein translation. It is, therefore, a central pathway through which cell growth and proliferation may be modulated. Lipids have proven to be important modulators of gene expression, cell growth, and cell proliferation. Several classes of lipids can achieve these effects. Fatty acids, cholesterol, and sphingolipids can all influence cell growth. These lipids may affect a variety of processes within the cytoplasm of the cell, or, alternatively, they may alter cell growth through direct and indirect actions on the nucleus itself, which may involve effects on nuclear protein import. This is, however, a novel area of study, and further work is clearly warranted because of the importance of the process to cell proliferation.

REFERENCES

1. J Robbins, SM Dilworth, RA Laskey, C Dingwall. Two interdependent basic domains in nucleoplasmin nuclear targeting sequence: identification of a class of bipartite nuclear targeting sequence. *Cell* 64: 615–623, 1991.

2. S Nakielny, MC Siomi, H Siomi, WM Michael, V Pollard, G Dreyfuss. Transportin: nuclear transport receptor of a novel nuclear protein import pathway. *Exp Cell Res* 229: 261–266, 1996.

3. EK Heist, M Srinivasan, H Schulman. Phosphorylation at the nuclear localization signal of Ca^{2+}/calmodulin-dependent protein kinase II blocks its nuclear targeting. *J Biol Chem* 273: 19763–19771, 1998.

4. DA Jans, S Hübner. Regulation of protein transport to the nucleus: central role of phosphorylation. *Physiol Rev* 76: 651–685, 1996.

5. K Mishra, VK Parnaik. Essential role of phosphorylation in nuclear transport. *Exp Cell Res* 216: 124–134, 1995.

6. H Okamura, J Aramburu, C Garcia-Rodriguez, JP Viola, A Raghavan, M Tahiliani, X Zhang, J Qin, PG Hogan, A Rao. Concerted dephosphorylation of the transcription factor NFAT1 induces a conformational switch that regulates transcriptional activity. *Mol Cell* 6: 539–550, 2000.

7. RT Hay, L Vuillard, JM Desterro, MS Rodriguez. Control of NF-kappa B transcriptional activation by signal induced proteolysis of I kappa B alpha. *Philos Trans R Soc Lond B Biol Sci* 354: 1601–1609, 1999.

8. D Görlich, RA Laskey. Roles of importin in nuclear protein import. Cold Spring Harbor Symposium on Quantitative Biology LX: 695–699, 1995.

9. MA Powers, DJ Forbes. Cytosolic factors in nuclear transport: what's importin? *Cell* 79: 931–934, 1994.

10. SA Adam. The nuclear pore complex. *Genome Biol* 2: 1–6, 2001.

11. TR Kau, PA Silver. Nuclear transport as a target for cell growth. *Drug Discov Today* 8: 78–85, 2003.

12. SK Lyman, T Guan, J Bednenko, H Wodrich, L Gerace. Influence of cargo size on Ran and energy requirements for nuclear protein import. *J Cell Biol* 159: 55–67, 2002.

13. KJ Ryan, SR Wente. The nuclear pore complex: a protein machine bridging the nucleus and cytoplasm. *Curr Opin Cell Biol* 12: 361–371, 2000.

14. MP Rout, JD Aitchison, A Suprapto, K Hjertaas, Y Zhao, BT Chait. The yeast nuclear pore complex: composition, architecture, and transport mechanism. *J Cell Biol* 148: 635–651, 2000.

15. H Wang, DE Clapham. Conformational changes of the *in situ* nuclear pore complex. *Biophys J* 77: 241–247, 1999.

16. C Perez-Terzic, AM Gacy, R Bortolon, PP Dzeja, M Puceat, M Jaconi, FG Prendergast, A Terzic. Structural plasticity of the cardiac nuclear pore complex in response to regulators of nuclear import. *Circ Res* 84: 1292–1301, 1999.

17. C Delphin, T Guan, F Melchior, L Gerace. RanGTP targets p97 to RanBP2, a filamentous protein localized at the cytoplasmic periphery of the nuclear pore complex. *Mol Biol Cell* 8: 2379–2390, 1997.

18. KS Ullman, S Shah, MA Powers, DJ Forbes. The nucleoporin nup153 plays a critical role in multiple types of nuclear export. *Mol Biol Cell* 10: 649–664, 1999.

19. S Nakielny, S Shaikh, B Burke, G Dreyfuss. Nup153 is an M9 containing mobile nucleoporin with a novel Ran-binding domain. *EMBO J* 18: 1982–1995, 1999.

20. S Shah, S Tugendreich, D Forbes. Major binding sites for the nuclear import receptor are the internal nuclcoporin nup153 and the adjacent nuclear filament protein Tpr. *J Cell Biol* 141: 31–49, 1998.

21. J Moroianu, G Blobel, A Radu. RanGTP-mediated nuclear export of karyopherin α involves its interaction with the nucleoporin Nup153. *Proc Natl Acad Sci USA* 94: 9699–9704, 1997.

22. MW Goldberg, TD Allen. High resolution scanning electron microscopy of the nuclear envelope: demonstration of a new, regular, fibrous lattice attached to the baskets of the nucleoplasmic face of the nuclear pores. *J Cell Biol* 119: 1429–1440, 1992.

23. N Daigle, J Beaudouin, L Hartnell, G Imreh, E Hallberg, J Lippincott-Schwartz, J Ellenberg. Nuclear pore complexes form immobile networks and have a very low turnover in live mammalian cells. *J Cell Biol* 154: 71–84, 2001.

24. WF Marshall. Order and disorder in the nucleus. *Curr Biol* 12: R185–R192, 2002.

25. CM Feldherr, D Akin. Regulation of nuclear transport in proliferating and quiescent cells. *Exp Cell Res* 205: 179–186, 1993.

26. GG Maul, LL Deaven, JJ Freed, GL Campbell, W Beçak. Investigation of the determinants of nuclear pore number. *Cytogenet Cell Genet* 26: 175–190, 1980.

27. C Perez-Terzic, A Behfar, A Méry, JMA vanDeursen, A Terzic, M Pucéat. Structural adaptation of the nuclear pore complex in stem cell-derived cardiomyocytes. *Circ Res* 92: 444–452, 2003.

28. RS Polizotto, MS Cycrt. Calcineurin-dependent nuclear import of the transcription factor Crz1p requires Nmd5p. *J Cell Biol* 154: 951–960, 2001.

29. RH Kehlenbach, L Gerace. Phosphorylation of the nuclear transport machinery down-regulates nuclear protein import *in vitro*. *J Biol Chem* 275: 17848–17856, 2000.

30. RA Hegele. The pathogenesis of atherosclerosis. *Clin Chim Acta* 246: 21–38, 1996.

31. KH Baumann, F Hessel, I Larass, T Muller, P Angerer, R Kiefl, C von Schacky. Dietary omega-3, omega-6, and omega-9 unsaturated fatty acids and growth factor and cytokine gene expression in unstimulated and stimulated monocytes: A randomized volunteer study. *Arterioscler Thromb Vasc Biol* 19: 59–66, 1999.

32. R Pakala, WL Sheng, CR Benedict. Eicosapentaenoic acid and docosahexaenoic acid block serotonin-induced smooth muscle cell proliferation. *Arterioscler Thromb Vasc Biol* 19: 2316–2322, 1999.

33. B Batetta, RR Bonatesta, F Sanna, M Putzolu, MF Mulas, M Collu, S Dessi. Cell growth and cholesterol metabolism in human glucose-6-phosphate dehydrogenase deficient lympho-mononuclear cells. *Cell Prolif* 35: 143–154, 2002.

34. KN Rao. The significance of the cholesterol biosynthetic pathway in cell growth and carcinogenesis. *Anticancer Res* 15: 309–314, 1995.

35. B Batetta, MF Mulas, F Sanna, M Putzolu, RR Bonatesta, A Gasperi-Campani, L Roncuzzi, D Baiocchi, S Dessi. Role of cholesterol ester pathway in the control of cell cycle in human aortic smooth muscle cells. *FASEB J* 17: 746–748, 2003.

36. S Willaime-Morawek, K Brami-Cherrier, J Mariani, J Caboche, B Brugg. C-Jun N-terminal kinases/c-Jun and p38 pathways cooperate in ceramide-induced apoptosis. *Neuroscience* 119: 387–397, 2003.

37. Y Uchida, AD Nardo, V Collins, PM Elias, WM Holleran. *De novo* ceramide synthesis participates in the ultraviolet B irradiation-induced apoptosis in undifferentiated cultured human keratinocytes. *J Invest Dermatol* 120: 662–669, 2003.

38. R Kolesnick, Z Fuks. Radiation and ceramide-induced apoptosis. *Oncogene* 22: 5897–5906, 2003.

39. V Thamilselvan, W Li, BE Sumpio, MD Basson. Sphingosine-1-phosphate stimulates human Caco-2 intestinal epithelial proliferation via p38 activation and activates ERK by an independent mechanism *in vitro*. *Cell Dev Biol Anim* 38: 246–253, 2002.

40. T Hla. Signaling and biological actions of sphingosine-1-phosphate. *Pharmacol Res* 47: 401–407, 2003.

41. D-S Kim, E-S Hwang, J-E Lee, S-Y Kim, K-C Park. Sphingosine-1-phosphate promotes mouse melanocyte survival via ERK and Akt activation. *Cell Signal* 15: 919–926, 2003.

42. AP Simopoulos. Essential fatty acids in health and chronic disease. *Am J Clin Nutr* 70 (suppl): 560S–569S, 1999.

43. LA Horrocks, YK Yeo. Health benefits of docosahexaenoic acid (DHA). *Pharmacol Res* 40: 211–225, 1999.

44. EW Hardman. Omega-3 fatty acids to augment cancer therapy. *J Nutr* 132: 3508S–3512S, 2002.

45. TM Devlin. Textbook of Biochemistry with Clinical Correlations. 3rd ed. New York: Wiley-Liss, 1992.

46. H Grimm, K Mayer, P Mayser, E Eigenbrodt. Regulatory potential of n-3 fatty acids in immunological and inflammatory processes. *Br J Nutr* 87: S59–S67, 2002.

47. L Demaison, D Moreau. Dietary n-3 polyunsaturated fatty acids and coronary heart disease-related mortality: a possible mechanism of action. *Cell Mol Life Sci* 59: 463–477, 2002.

48. M Hirafuji, T Machida, N Hamaue, M Minami. Cardiovascular protective effects of n-3 polyunsaturated fatty acids with special emphasis on docosahexaenoic acid. *J Pharmacol Sci* 92: 308–316, 2003.

49. M Corson, R Alexander, B Berk. 5-HT$_2$ receptor mRNA is overexpressed in cultured rat aortic smooth muscle cells relative to normal aorta. *Am J Physiol* 262: C309–C315, 1992.

50. M Hida, H Fujita, K Ishikura, S Omori, M Hoshiya, M Awazu. Eicosapentaenoic acid inhibits PDGF-induced mitogenesis and cyclin D1 expression via TGF-β in mesangial cells. *J Cell Physiol* 196: 293–300, 2003.

51. S Bousserouel, A Brouillet, G Béréziat, M Raymondjean, M Andréani. Different effects of n-6 and n-3 polyunsaturated fatty acids on the activation of rat smooth muscle cells by interleukin-1B. *J Lipid Res* 44: 601–611, 2003.

52. M Zeyda, AB Szekeres, MD Säemann, R Geyeregger, H Stockinger, GJ Zlabinger, W Waldhäusl, TM Stulnig. Suppression of T cell signaling by polyunsaturated fatty acids: selectivity in inhibition of mitogen-activated protein kinase and nuclear factor activation. *J Immunol* 170: 6033–6039, 2003.

53. Y Fan, D McMurray, L Ly, R Chapkin. Dietary (n-3) polyunsaturated fatty acids remodel mouse T-cell lipid rafts. *J Nutr* 133: 1913–1920, 2003.

54. M Zeyda, G Staffler, V Horejší, W Waldhäusl, TM Stulnig. LAT displacement from lipid rafts as a molecular mechanism for the inhibition of T-cell signaling by polyunsaturated fatty acids. *J Biol Chem* 277: 28418–28423, 2002.

55. TM Stulnig, J Huber, N Leitinger, E-M Imre, P Angelisová, P Nowotny, W Waldhäusl. Polyunsaturated eicosapentaenoic acid displaces proteins from membrane rafts by altering raft lipid composition. *J Biol Chem* 276: 37335–37340, 2001.

56. PL Yeagle. Cholesterol and the cell membrane. *Biochim Biophys Acta* 822: 267–287, 1985.

57. ME Zettler, GN Pierce. Growth-promoting effects of oxidized low density lipoprotein. *Can J Cardiol* 17: 73–79, 2001.

58. BL Knight, DD Patel, SM Humphreys, D Wiggins, GF Gibbons. Inhibition of cholesterol absorption associated with a PPAR-alpha -dependent increase in ABC transporter A1 in mice. *J Lipid Res* 44:2049–2058, 2003.

59. JD Horton, I Shimomura, S Ikemoto, Y Bashmakov, RE Hammer. Overexpression of sterol regulatory element-binding protein-1a in mouse adipose tissue produces adipocyte hypertrophy, increased fatty acid secretion, and fatty liver. *J Biol Chem* 278: 36652–36660, 2003.

60. KN Maxwell, RE Soccio, EM Duncan, E Sehayek, JL Breslow. Novel putative SREBP and LXR target genes identified by microarray analysis in liver of cholesterol-fed mice. *J Lipid Res* 44:2109–2119, 2003.

61. M Hardwick, D Fertikh, M Culty, H Li, B Vidic, V Papadopoulos. Peripheral-type benzodiazepine receptor (PBR) in human breast cancer: correlation of breast cancer cell aggressive phenotype with PBR expression, nuclear localization, and PBR-mediated cell proliferation and nuclear transport of cholesterol. *Cancer Res* 59: 831–842, 1999.

62. MP Czubryt, JC Russell, J Sarantopoulos, GN Pierce. Nuclear cholesterol content and nucleoside triphosphatase activity are altered in the JCR: LA-cp corpulent rat. *J Cell Biochem* 63: 349–357, 1996.

63. A Sachinidis, M Carniel, S Seewald, C Seul, I Gouni-Berthold, Y Ko, H Vetter. Lipid-induced changes in vascular smooth muscle cell membrane fluidity are associated with DNA synthesis. *Cell Prolif* 32: 101–105, 1999.

64. ML Tomassoni, D Amori, MV Magni. Changes of nuclear membrane lipid composition affect RNA nucleocytoplasmic transport. *Biochem Biophys Res Commun* 258: 476–481, 1999.

65. B Ramjiawan, MP Czubryt, JS Gilchrist, GN Pierce. Nuclear membrane cholesterol can modulate nuclear nucleoside triphosphatase activity. *J Cell Biochem* 63: 442–452, 1996.

66. AH Merrill. *De novo* sphingolipid biosynthesis: a necessary but dangerous pathway. *J Biol Chem* 277: 25843–25846, 2002.

67. T Levade, N Augé, RJ Veldman, O Cuvillier, A Nègre-Salvayre, R Salvayre. Sphingolipid mediators in cardiovascular cell biology and pathology. *Circ Res* 89: 957–968, 2001.

68. YA Hannun, C Luberto, KM Argraves. Enzymes of sphingolipid metabolism: from modular to integrative signaling. *Biochemistry* 40: 4893–4903, 2001.

69. S Willaime, P Vanhoutte, J Caboche, Y Lemaigre-Dubreuil, J Mariani, B Brugg. Ceramide-induced apoptosis in cortical neurons is mediated by an increase in p38 phosphorylation and not by the decrease in ERK phosphorylation. *Eur J Neurosci* 13: 2037–2046, 2001.

70. A Erdreich-Epstein, LB Tran, NN Bowman, H Wang, MC Cabot, DL Durden, J Vlckova, CP Reynolds, MF Stins, S Groshen, M Millard. Ceramide signaling in fenretinide-induced endothelial cell apoptosis. *J Biol Chem* 277: 49531–49537, 2002.

71. CV Haefen, T Wieder, B Gillissen, L Stärck, V Graupner, B Dörken, PT Daniel. Ceramide induces mitochondrial activation and apoptosis via a Bax-dependent pathway in human carcinoma cells. *Oncogene* 21: 4009–4019, 2002.

72. EPD Chaves, M Bussiere, B MacInnis, DE Vance, RB Campenot, JE Vance. Ceramide inhibits axonal growth and nerve growth factor uptake without compromising the viability of sympathetic neurons. *J Biol Chem* 276: 36207–36214, 2001.

73. DG Johns, RC Webb, JR Charpie. Impaired ceramide signaling in spontaneously hypertensive rat vascular smooth muscle: a possible mechanism for augmented cell proliferation. *J Hypertens* 19: 63–70, 2001.

74. RK Malik, BA Thornhill, AY Chuang, SC Kiley, RL Chevalier. Apoptosis parallels ceramide content in the developing rat kidney. *Dev Biol* 15: 188–191, 2000.

75. C Mazière, M-A Conte, L Leborgne, T Levade, W Hornebeck, R Santus, J-C Mazière. UVA radiation stimulates ceramide production: relationship to oxidative stress and potential role in ERK, JNK, and p38 activation. *Biochem Biophys Res Commun* 281: 289–294, 2001.

76. SS Castillo, D Teegarden. Ceramide conversion to sphingosine-1-phosphate is essential for survival in C3H10T½ cells. *J Nutr* 131: 2826–2830, 2001.

77. SP Li, MR Junttila, J Han, VM Kahari, J Westermarck. p38 Mitogen-activated protein kinase pathway suppresses cell survival by inducing dephosphorylation of mitogen-activated protein/extracellular signal-regulated kinase kinase 1,2. *Cancer Res* 63: 3473–3477, 2003.

78. B Brenner, U Koppenhoefer, C Weinstock, O Linderkamp, F Lang, E Gulbin. Fas- or ceramide-induced apoptosis is mediated by a Rac1-regulated activation of Jun N-terminal kinase/p38 kinases and GADD153. *J Biol Chem* 272: 22173–22181, 1997.

79. S Mathias, KA Dressler, RN Kolesnick. Characterization of a ceramide-activated protein kinase: stimulation by tumor necrosis factor α. *Proc Natl Acad Sci USA* 88: 10009–10013, 1991.

80. S Galadari, A Hago, M Patel. Effects of cations on ceramide-activated protein phosphatase 2A. *Exp Mol Med* 33: 240–244, 2001.

81. LJ Siskind, M Colombini. The lipids C_2- and C_{16}-ceramide form large stable channels. *J Biol Chem* 275: 38640–38644, 2000.

82. LJ Siskind, RN Kolesnick, M Colombini. Ceramide channels increase the permeability of the mitochondrial outer membrane to small proteins. *J Biol Chem* 277: 26796–26803, 2002.

83. W Martinet, DM Schrivers, GR De Meyer, J Thielemans, MW Knaapen, AG Herman, MM Kockx. Gene expression profiling of apoptosis-related genes in human atherosclerosis: upregulation of death-associated protein kinase. *Arterioscler Thromb Vasc Biol* 22: 2023–2029, 2002.

84. H Zhang, NN Desai, A Olivera, T Seki, G Booker, S Spiegel. Sphingosine-1-phosphate, a novel lipid, involved in cellular proliferation. *J Cell Biol* 114: 155–167, 1991.

85. N Augé, M Nikolova-Karakashian, S Carpentier, S Parthasarathy, A Nègre-Salvayre, R Salvayre, AH Merrill Jr., T Levade. Role of sphingosine-1-phosphate in the mitogenesis induced by oxidized low density lipoprotein in smooth muscle cells via activation of sphingomyelinase, ceramidase, and sphingosine kinase. *J Biol Chem* 274: 21533–21538, 1999.

86. Y Rikitake, K Hirata, S Kawashima, M Ozaki, T Takahashi, W Ogawa, N Inoue, M Yokoyama. Involvement of endothelial nitric oxide in sphingosine-1-phosphate-induced angiogenesis. *Arterioscler Thromb Vasc Biol* 22: 108–114, 2001.

87. S Spiegel, S Milstien. Sphingosine-1-phosphate, a key cell signaling molecule. *J Biol Chem* 277: 25851–25854, 2002.

88. S Pyne, NJ Pyne. Sphingosine 1-phosphate signalling in mammalian cells. *Biochem J* 349: 385–402, 2000.

15

Isoprenoid Regulation of the Expression of Ras and Ras-Related Proteins

SARAH A. HOLSTEIN AND RAYMOND J. HOHL

INTRODUCTION

Isoprenoids

The isoprenoid biosynthetic pathway (Figure 15.1) is the source of an amazingly diverse group of compounds. More than 23,000 naturally occurring isoprenoids have been identified thus far, with new structures continuing to be reported.[1] As natural substances, these isoprenoids are ubiquitous in animal diets and thus might be considered nutrients. Products and derivatives of the isoprenoid biosynthetic pathway play key roles in all aspects of life. For example, isoprenoids serve as components of signal transduction pathways, visual pigments, vitamins, constituents of membranes, mating pheromones, reproductive hormones, regulators of gene expression, antimicrobial agents, and constituents of electron transport and photosynthetic machinery.

Biosynthesis

The isoprenoid pathway and its products are shown in Figure 15.1. β-Hydroxymethlyglutaryl coenzyme A (HMG-CoA), ultimately derived from acetyl CoA, is converted to mevalonate

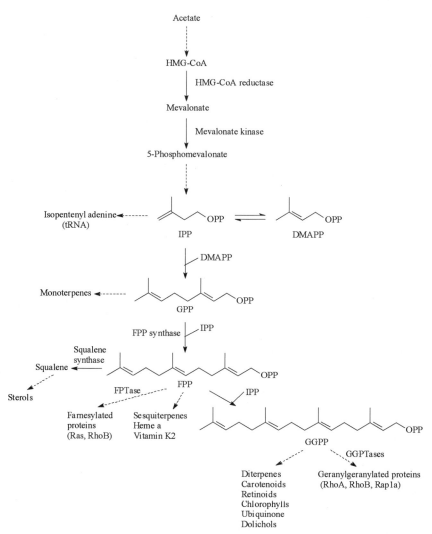

Figure 15.1   The isoprenoid biosynthetic pathway. IPP, isopentenyl pyrophosphate; DMAPP, dimethylallyl pyrophosphate; GPP, geranyl pyrophosphate; FPP, farnesyl pyrophosphate; GGPP, geranylgeranyl pyrophosphate.

via the enzyme HMG-CoA reductase in the rate-limiting step of the pathway. Mevalonate is subsequently phosphorylated by mevalonate kinase to yield 5-phosphomevalonate. The five-carbon compounds isopentenyl pyrophosphate (IPP) and dimethylallyl pyrophosphate (DMAPP) are condensed to form the 10-carbon geranyl pyrophosphate (GPP). The addition of another IPP unit to GPP yields the 15-carbon farnesyl pyrophosphate (FPP). The enzyme FPP synthase catalyzes the synthesis of both GPP and FPP. Geranylgeranyl pyrophosphate (GGPP) synthase catalyzes the addition of IPP to FPP to form GGPP. The enzyme squalene synthase catalyzes the head-to-head condensation of two FPP molecules to form the sterol precursor squalene. It should be noted that in plants a newly discovered deoxy-D-xylulose 5-phosphate (DOXP) pathway exists, through which pyruvate and D-glyceraldehyde-3-phosphate are converted to IPP.[2] Genomic analyses have indicated that the mevalonate-dependent pathway is the ancestral pathway in archaebacteria, whereas the DOXP pathway is the ancestral source of IPP in eubacteria.[3]

Monoterpenes

Monoterpenes and monoterpenoids, derived from GPP, are synthesized within secretory cells in the oil glands of plants. It has been estimated that there are nearly 1000 monoterpenes.[4] While the function of monoterpenes in plants is not completely understood, some of these compounds have been implicated in playing roles in attracting seed-dispersing animals or pollinating insects, repelling browsing animals or insect pests, resisting microbial attack, and inhibiting the growth of competitors.[5] Although monoterpenes were once viewed as inert waste products, the finding that they are rapidly synthesized and catabolized suggests critical metabolic functions.

Monoterpenes, as components of essential oils, herbs, and spices, were early items of commerce and of interest to alchemists. In modern times, monoterpenes can be found as ingredients in a wide variety of products including food flavorings, cosmetics, and cleaning products. Monoterpenes such as limonene and perillyl alcohol have been the focus of investigation because of their chemopreventative and chemotherapeutic activities. The development of spontaneous neoplasms as well as chemically induced tumors in rodents is inhibited by limonene.[6] In rats, dietary limonene inhibits the development of *ras* oncogene-induced mammary tumors.[7] Carvone, carveol, menthol, geraniol, and perillyl alcohol have also been reported to have chemopreventative activities.[6] Limonene and perillyl alcohol have been shown to cause the complete regression of chemically induced rat mammary tumors.[8,9] Perillyl alcohol also inhibits the growth rate and induces complete regression of transplanted hamster pancreatic tumors.[10] Phase I and II clinical trials for limonene and perillyl alcohol are under way.

Isoprenylated Proteins

It has been estimated that at least 300 proteins undergo the posttranslational modification of isoprenylation involving the addition of a 15-carbon farnesyl or a 20-carbon geranylgeranyl chain to a cysteine residue at the carboxyl terminus, and that this accounts for approximately 1% of the total proteins in a cell.[11] These proteins include members of the Ras small GTPase superfamily (e.g., Ras, Rap, Rho, Rab), heterotrimeric G protein γ subunits, nuclear lamins (prelamin A, lamin B), centromere-associated proteins (CENP-E, CENP-F), heat-shock proteins (HDJ2), and protein tyrosine phosphatases (PTP(CAAXI) and PTP(CAAX2)). Approximately 70 to 80% of isoprenylated proteins are geranylgeranylated. A γ subunit of heterotrimeric G proteins was the first geranylgeranylated protein to be identified.[12] Many members of the Ras superfamily have been shown to be geranylgeranylated, including Rap1a,[13] RhoA,[14] and RhoB, although the latter protein can also be farnesylated.[15] Members of the Rab family of small GTPases are modified by two geranylgeranyl chains.[16] These isoprenylation reactions are catalyzed by the enzymes farnesyl transferase, geranylgeranyl transferase I, and geranylgeranyl transferase II (Rab proteins). Isoprenylated proteins share a consensus sequence termed the "CAAX" box at the carboxyl terminus. The specific CAAX sequence, where C is cysteine, A is an aliphatic amino acid, and X is the C-terminal residue, dictates the type of modification.

Ras

Ras is a regulator of eukaryotic cell growth and plays a key role in signal transduction pathways initiated by a variety of extracellular stimuli including hormones, cytokines, and peptide growth factors. Ras genes were first identified as the transforming principle of the Harvey and Kirsten strains of rat sarcoma viruses.[17] Ras genes have since been identified in mammals, birds, insects, plants, fungi, and yeasts,[17] and have been highly conserved throughout evolution.[18]

Ras, like other G proteins, cycles between an active GTP-bound state and an inactive GDP-bound state. Guanine nucleotide exchange factors (GEFs) facilitate the release of GDP, allowing GTP to bind. Ras has intrinsic GTPase activity that is enhanced by GTPase-activating proteins, thus restoring Ras to its inactive form.[19] The best-studied Ras pathway is the Ras-MAPK pathway. Ras, situated at the plasma membrane and activated by upstream receptor tyrosine kinases, interacts with Raf and initiates a kinase cascade consisting of Raf, MEK, and ERK1/2 with subsequent activation of transcription factors and cell proliferation.[20]

Mutations that lead to impaired Ras GTPase activity or interfere with the ability of Ras to interact with GAP proteins result in constitutively active Ras capable of transforming cells.[21] These mutations have been found in 90% of pancreatic adenocarcinomas, 50% of colon cancers, 30% of lung cancers, and 30% of myeloid leukemias.[22] The overall incidence of mutated Ras in cancer is estimated to be 30%.[22] Although K-Ras is mutated most frequently, other Ras isoforms have also been found to be mutated in a tissue-specific manner.[22] The contribution of the Ras pathway in human cancers may be underestimated because upregulation of the pathway can occur in the absence of mutated Ras.[20]

Rap1

Rap1, also known as Krev-1, has two closely related isoforms Rap1a and Rap1b and shares 50% sequence identity with Ras.[23] In particular, the GTP–GDP-binding, effector-binding, and membrane-attachment domains of Ras and Rap1 are conserved.[23] Rap1 was first identified from studies screening proteins for their ability to suppress the transformation of fibroblasts by mutated *K-ras*.[24] It has been suggested that Rap1 antagonizes Ras by competing for the regulatory and target proteins of Ras. In particular, Rap1 may inhibit Ras-mediated signaling by binding and preventing the activation of Raf-1.[25] More recent work has indicated that Rap1 may play roles in regulating diverse cellular processes, including integrin-mediated cell adhesion,[26] cAMP-induced neurite out-growth,[27] exocytosis,[28] as well as growth and differentiation.[29] Rap1 is activated following the stimulation of several types of transmembrane receptors, including receptor tyrosine kinases, heterotrimeric G-protein-coupled receptors, cytokine receptors, and cell-adhesion molecules.[29] Although Rap1, like Ras, contains a CAAX motif, Rap1 is geranylgeranylated.[13] Geranylgeranylation and subsequent membrane association are critical for Rap1 function.

Rho Proteins

The Rho proteins are a subfamily of the Ras superfamily. The Rho subfamily is further divided into subgroups of Rho, Rac, and Cdc42 proteins. These proteins were initially identified because of their roles as regulators of actin cytoskeletal organization.[30] The Rho subgroup has received the most study and includes RhoA, RhoB, and RhoC. Despite significant structural similarities (90% identity), RhoA and RhoB have been shown to have unique properties. RhoA appears to be located primarily in the cytoplasm, with some association with the plasma membrane.[31] Initial studies showed RhoB localized to early endosomes and prelysosomes.[32] More recent studies have suggested that RhoB is localized predominantly to the plasma membrane.[31]

Despite numerous studies in transformed cells and in cells treated with isoprenyltransferase inhibitors, the physiological function of RhoB remains largely unknown. The use of RhoB knockout mice has revealed that loss of RhoB does not affect mouse development, fertility, or wound healing.[33] However, the knockout mice did display an increased susceptibility to chemical carcinogenesis, and embryo fibroblasts have altered actin and proliferative responses to transforming growth factor (TGF) β.[33]

RESULTS

Mevalonate Depletion Upregulates Ras, Rap1a, RhoA, and RhoB

With the exception of RhoB, remarkably little work has been done with regards to the transcriptional and posttranscriptional regulatory determinants of Ras and Ras-related proteins. The extent to which these proteins are regulated by their isoprenylation status or by specific isoprenoid species has only recently been elucidated. The discovery of the HMG-CoA reductase inhibitors[34] enabled understanding of the importance of the roles of isoprene moieties in posttranslational modification of many signaling proteins. HMG-CoA reductase inhibitors, through their ability to deplete cells of the mevalonate-derived products FPP and GGPP (Figure 15.1), have long been known to result in the inhibition of isoprenylation.[35,36] An observation that was made several years ago by our laboratory was that treatment of cells with lovastatin, an HMG-CoA reductase inhibitor, in addition to inhibiting Ras farnesylation, also appeared to increase the total amount of Ras protein.[37]

More recently, we have investigated the effects of mevalonate depletion on the expression of Ras and Ras-related proteins. Figure 15.2a shows a time-dependent increase in Ras, Rap1a, RhoA, and RhoB protein levels in K562 cells incubated with lovastatin. Interestingly, temporal differences in this upregulation were seen, suggesting that the isoprenylated proteins are not uniformly regulated in response to mevalonate depletion. Further studies demonstrated that there are diverse mechanisms underlying this upregulation. Cycloheximide and [^{35}S]-methionine pulse and pulse-chase experiments revealed that mevalonate depletion increases the *de novo* synthesis of Ras and RhoA and decreases the degradation of existing Ras and RhoA protein.[38] Pretreatment with actinomycin D was found to completely prevent the induced upregulation of RhoB and to only partially prevent the upregulation of Ras, Rap1a, and RhoA.[38] Although depletion of mevalonate does not alter steady-state levels of Ras or RhoA mRNA, a marked increase in RhoB mRNA levels was observed (Figure 15.2b). The induced increase in RhoB mRNA was completely prevented by pretreatment with actinomycin D, suggesting that the increased levels are due to increased

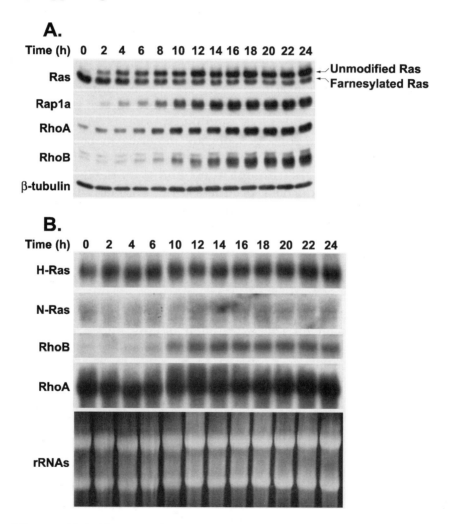

Figure 15.2 (See caption on opposite page)

transcription and not to changes in message stability. Thus these studies demonstrate that mevalonate depletion results in the upregulation of Ras and Ras-related proteins by discrete mechanisms including modulation of transcriptional, translational, and posttranslational processes.

Isoprenoids and Regulation of Ras-Related Proteins

The observed upregulation of the Ras-related proteins in response to mevalonate depletion could be a result of global inhibition of isoprenylation or a response to depletion of key regulatory isoprenoid species. If the upregulation merely was due to inhibition of isoprenylation, then we reasoned that specific inhibitors of the isoprenyl transferases would yield similar results. In experiments utilizing the peptidomimetic farnesyl transferase inhibitor FTI-277 and geranylgeranyl transferase inhibitor GGTI-286, we demonstrated that inhibition of farnesylation and geranylgeranylation does not mimic lovastatin's ability to increase Ras and RhoA synthesis, decrease Ras and RhoA degradation, increase RhoB mRNA, or increase total levels of Ras, Rap1a, RhoA, and RhoB protein.[39] These results suggested that inhibition of isoprenylation was not the signal required for upregulation, but instead that there

---

Figure 15.2 (Opposite page) Mevalonate depletion induces upregulation of Ras and Ras-related proteins. (A) K562 cells were incubated with 10 μM lovastatin for up to 24 h, with cells collected every 2 h. Whole cell lysate was prepared and Western blot analysis was subsequently performed. Each lane contains an equivalent amount of protein from cell lysate. (B) K562 cells were incubated with 10 μM lovastatin for up to 24 h. Total RNA was isolated and Northern blots were performed using H-Ras-, N-Ras-, RhoA-, or RhoB-specific riboprobes. The lower panel depicts the photograph of the ethidium bromide–stained gel. (From SA Holstein, CL Wohlford-Lenane, RJ Hohl. Consequences of mevalonate depletion. Differential transcriptional, translational, and post-translational upregulation of Ras, Rap1a, RhoA, and RhoB. *J Biol Chem* 277: 10678–10682, 2002. With permission, conveyed through the Copyright Clearance Center.)

might be key regulatory isoprenoid species. In order to address this hypothesis, a series of experiments were performed in which cells were incubated with lovastatin and isoprenoid pathway intermediates (mevalonate, IPP, DMAPP, GPP, FPP, GGPP, and squalene). The results of these experiments, as well as additional studies utilizing specific FPP synthase and squalene synthase inhibitors, revealed that the nonsterol species FPP and GGPP were the critical regulatory

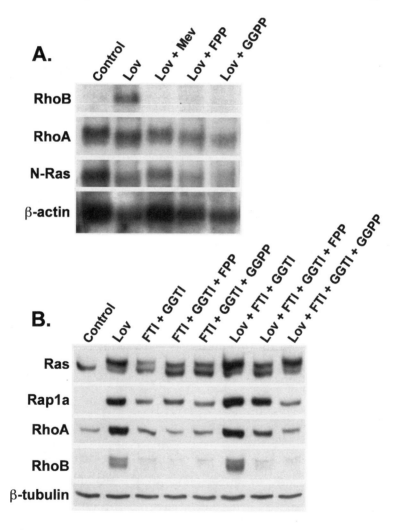

Figure 15.3 (See caption on opposite page)

isoprenoids.[39] Specifically, FPP prevents the upregulation of Ras and RhoB, whereas GGPP prevents the upregulation of Rap1a, RhoA, and RhoB.[39]

Furthermore, when the effects of the isoprenoids on mRNA levels were examined, it was found that either FPP or GGPP completely prevents lovastatin-induced upregulation of RhoB mRNA (Figure 15.3a). Finally, although we had shown that inhibition of isoprenyl transferases does not result in upregulation of Ras-related proteins and that FPP and GGPP can prevent lovastatin-induced upregulation, the possibility remained that FPP or GGPP restores the isoprenylation of a protein that is responsible for regulating the levels of the Ras-related proteins. Therefore, experiments were performed to examine the ability of FPP and GGPP to prevent lovastatin-induced upregulation of RhoA and RhoB in the presence of isoprenyl transferase inhibitors. As shown in Figure 15.3b, the combination of FTI and GGTI resulted in the appearance of unmodified Ras and Rap1a. Treatment with FPP and GGPP did not restore Ras or Rap1a processing in the presence of the peptidomimetic inhibitors. Levels of unmodified Ras and Rap1a were further increased by the addition of lovastatin to FTI–GGTI. Coincubation with FPP and GGPP in lovastatin–FTI–GGTI-treated cells restored levels of unmodified Ras and Rap1a to those observed in cells treated only with FTI–GGTI. This demonstrated the ability

---

Figure 15.3 (Opposite page) Regulatory effects of FPP and GGPP. (A) K562 cells were incubated for 24 h in the presence of 10 μM lovastatin (Lov), 5 mM mevalonate (Mev), 10 μM FPP, and 10 μM GGPP. Total RNA was isolated, and Northern blots were performed using RhoB-, RhoA-, or N-Ras-specific riboprobes. β-actin levels are shown as a control. (B) Cells were incubated with 10 μM lovastatin, 10 μM FPP, 10 μM GGPP, 100 nM FTI, and 2 μM GGTI for 24 h and Western blot analysis was performed. Each lane contains an equivalent amount of protein from cell lysate, and β-tubulin levels are shown as a control. (From SA Holstein, CL Wohlford-Lenane, RJ Hohl. Isoprenoids influence the expression of Ras and Ras-related proteins. *Biochemistry* 41: 13698–13704, 2002. With permission, Copyright © 2002 American Chemical Society.)

A.

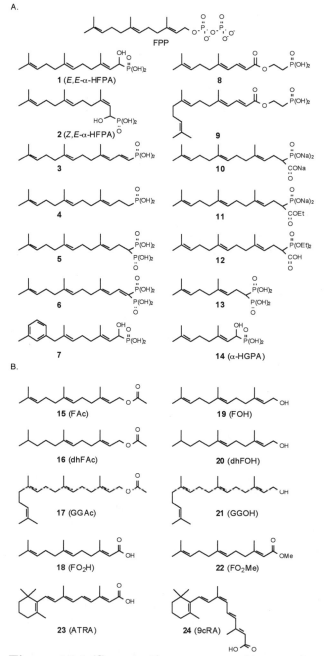

B.

Figure 15.4 (See caption on opposite page)

of FPP and GGPP to prevent upregulation of Ras and Rap1a in the presence of continued inhibition of isoprenylation. Furthermore, the upregulation of RhoA and RhoB induced by lovastatin was also prevented by coincubation with either FPP or GGPP in the presence of FTI and GGTI. These findings provided additional evidence that the upregulation of the Ras-related proteins induced by mevalonate depletion is a consequence of isoprenoid, rather than isoprenylated protein, depletion.

Isoprenoid Analogues and Regulation of Ras-Related Proteins

To further investigate the regulatory roles of the isoprenoid pyrophosphates, the effects of isoprenoid analogues on Ras-related protein expression were examined. A variety of analogues were employed including phosphonates, alcohols, acetates, and even the related retinoid species (Figure 15.4). The results of these experiments are summarized in Table 15.1. Intriguingly, in contrast to the isoprenoid pyrophosphates that prevent the upregulation of these small GTPase proteins induced by mevalonate depletion, a number of the farnesyl phosphonic acids were shown to potentiate the upregulation of these proteins. The potentiation of RhoB upregulation is at both the mRNA and protein level.[40] In this context, these analogues are functional antagonists for FPP and GGPP. The

---

Figure 15.4 Structures of isoprenoid analogues. The abbreviation or compound number are shown below the corresponding structure. E,E-α-HFPA, E,E-α-hydroxyfarnesyl phosphonic acid; Z,E-α-HFPA, Z,E-α-hydroxyfarnesyl phosphonic acid; α-HGPA, -α-hydroxygeranyl phosphonic acid; FAc, farnesyl acetate; dhFAc, dihydrofarnesyl acetate; GGAc, geranylgeranyl acetate; FO_2H, farnesoic acid; FOH, farnesol; dhFOH, dihydrofarnesol; GGOH, geranylgeraniol; FO_2Me, methyl farnesoate; ATRA, all-*trans*-retinoic acid; 9cRA, 9-*cis*-retinoic acid. (From SA Holstein, CL Wohlford-Lenane, DF Wiemer, RJ Hohl. Isoprenoid pyrophosphate analogues regulate expression of Ras-related proteins. *Biochemistry* 42: 4384–4391, 2003. With permission, Copyright © 2003 American Chemical Society)

Table 15.1 Summary of the Effects of Isoprenoid Analogues (from Figure 15.4) on Ras-Related Protein Levels in the Absence (lovastatin) or Presence (+lovastatin) of Mevalonate Depletion

Compound	Ras		Rap1a		RhoA		RhoB	
	lov[a]	+lov	lov	+lov	lov	+lov	lov	+lov
1, 3	↑[b]	↑[c]	—[d]	↑	—	↑	—	↑
2, 4, 11	—	—	—	↑	—	↑	—	↑
5, 6, 10	—	—	↑	↑	↑	↑	—	↑
7–9, 12–14	—	—	—	—	—	—	—	—
15, 16, 18, 19, 20, 22	—	—	—	—	—	—	—	↓[e]
17, 21	—	—	—	—	—	↓	—	↓
23, 24	—	↓	—	↓	—	↓	—	↓

[a] Lovastatin.
[b] Increased protein expression compared with untreated control (↑).
[c] Increased protein expression compared with lovastatin-treated control (↑).
[d] No effect observed (—).
[e] Decreased protein expression compared with lovastatin-treated control (↓).

Source: From SA Holstein, CL Wohlford-Lenane, DF Wiemer, RJ Hohl. Isoprenoid pyrophosphate analogues regulate expression of Ras-related proteins. *Biochemistry* 42: 4384–4391, 2003. (With permission, Copyright © 2003 American Chemical Society.)

ability of these analogues to serve as functional antagonists of the isoprenoid pyrophosphates was found to be dependent on the nature of the functional group at the head of the molecule, the charge of the molecule, and the length of the isoprenoid chain. Importantly, the ability of these compounds to alter the expression of the Ras-related proteins were found to be independent of their effects on FPTase, GGPTase I, FPP synthase, or GGPP synthase activities.[40] Metabolites and additional analogues of isoprenoid pyrophosphates were found to possess agonist properties relative to FPP and GGPP; that is, they attenuated the mevalonate depletion-induced upregulation of the Ras-related proteins (Table 15.1). Interestingly, the structurally related retinoids all-*trans*-retinoic acid and 9-*cis*-retinoic acid were found to have slight agonist properties. The identification of isoprenoids with either functional agonist or antagonist properties with respect to the endogenous isoprenoid pyrophosphates suggest the existence of specific isoprenoid binding factors that are involved in the regulation of Ras-related protein expression. Studies are ongoing to further delineate the nature of these isoprenoid-binding factors.

Monoterpenes and Regulation of Ras-Related Proteins

While the majority of our work up to this point had been focused on C15 and C20 length isoprenoids, we were also interested in the family of C10 isoprenoids known as the monoterpenes. Prior studies have suggested an interaction between monoterpenes and isoprenylated proteins. In particular, incorporation of radiolabeled mevalonate into small GTPase proteins was shown to be decreased in the presence of limonene metabolites.[41] Although these data were interpreted to result from monoterpene inhibition of isoprenyl transferases,[41] we clarified that the decrease in farnesylated Ras levels by perillyl alcohol was a consequence of decreased *de novo* synthesis of Ras protein.[37] We therefore hypothesized that the effect of perillyl alcohol on Ras might be generalizable to other naturally occurring monoterpenes and to other small GTPase proteins. To this end, we investigated the effects of a number of monoterpenes (Figure 15.5) on Ras and Ras-related protein expression, in the absence and presence of mevalonate depletion. The results of these studies are displayed in Table 15.2. Our studies showed that select monoterpenes inhibit upregulation of Ras and the Ras-related proteins. Some of the more active monoterpenes are dietary constituents and flavoring agents (e.g., perillyl alcohol and menthol). A structure-activity relationship model for these effects was defined, in which the activity of the monoterpene was found to be dependent on the flexibility of the menthane skeleton, the nature of the oxygenated species at specific substituent positions, and even the specific stereochemistry of compounds with multiple stereocenters.[42] The ability of monoterpenes to regulate the expression of the Ras-related proteins was found to be independent of effects on cell proliferation or total cellular protein synthesis and degradation.[42] These studies advance the understanding of the relationship between Ras-related proteins and isoprenoids to reveal that plant-derived isoprenoids

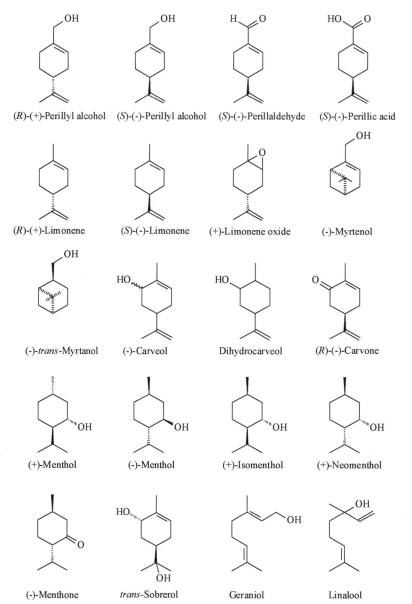

(R)-(+)-Perillyl alcohol   (S)-(-)-Perillyl alcohol   (S)-(-)-Perillaldehyde   (S)-(-)-Perillic acid

(R)-(+)-Limonene   (S)-(-)-Limonene   (+)-Limonene oxide   (-)-Myrtenol

(-)-*trans*-Myrtanol   (-)-Carveol   Dihydrocarveol   (R)-(-)-Carvone

(+)-Menthol   (-)-Menthol   (+)-Isomenthol   (+)-Neomenthol

(-)-Menthone   *trans*-Sobrerol   Geraniol   Linalool

Figure 15.5 Structures of monoterpenes. (From SA Holstein, RJ Hohl. Monoterpene regulation of Ras and Ras-related protein expression. *J Lipid Res* 44: 1209–1215, 2003. With permission, conveyed through the Copyright Clearance Center.)

Table 15.2 Effects of Monoterpenes on Ras, Rap1a, RhoA, and RhoB Protein Levels in the Presence or Absence of Mevalonate Depletion

	Lovastatin[a]				+Lovastatin[b]			
	Ras	Rap1a	RhoA	RhoB	Ras	Rap1a	RhoA	RhoB
(*R*)-perillyl alcohol	↓↓	—	↓	—	↓↓↓	↓↓	↓↓↓	↓↓↓
(*S*)-perillyl alcohol	↓↓↓	—	↓	—	↓↓↓	↓↓	↓	↓↓↓
(*S*)-perillaldehyde	—	—	—	—	↓↓	↓↓	↓↓	↓↓
(*S*)-perillic acid	—	—	—	—	—	—	↓	↓
(*R*)-limonene	—	—	—	—	—	—	—	—
(*S*)-limonene	—	—	—	—	—	—	—	—
Limonene oxide	—	—	—	—	—	—	—	—
Myrtenol	—	—	—	—	↓	↓	↓	↓
Myrtanol	—	—	—	—	↓↓	↓	↓↓	↓
(+)-Menthol	—	—	—	—	—	—	—	—
(−)-Menthol	—	—	—	—	↓↓	↓	—	↓
Isomenthol	—	—	—	—	↓	↓	↓	↓↓
Neomenthol	—	—	—	—	↓	—	↓	↓↓↓
(−) Menthone	—	—	—	—	—	—	—	—
Carveol	—	—	—	—	↓↓	—	—	↓
Dihydrocarveol	—	—	—	—	↓	—	↓	↓
Carvone	—	—	—	—	↓	—	↓	↓↓↓
Sobrerol	—	—	—	—	—	—	—	—
Geraniol	—	—	—	—	—	—	—	—
Linalool	—	—	—	—	—	—	—	—

Note: K562 cells were incubated for 24 h with monoterpenes (0.5 m*M*) in the absence or presence of mevalonate depletion (10 μM lovastatin). Western blots with subsequent densitometric analyses were performed.

[a] Decrease of <20% in protein levels compared with untreated control (—), 20–39% decrease in protein level compared with untreated control (↓), 40–59% decrease in protein levels compared with untreated control (↓↓), 60–79% decrease in protein levels compared with untreated control (↓↓↓).
[b] Decrease of <20% in protein levels compared with lovastatin-treated control (—), 20–39% decrease in protein level compared with lovastatin-treated control (↓), 40–59% decrease in protein levels compared with lovastatin-treated control (↓↓), 60–79% decrease in protein levels compared with lovastatin-treated control (↓↓↓).

Source: From SA Holstein, RJ Hohl. Monoterpene regulation of Ras and Ras-related protein expression. *J Lipid Res* 44: 1209–1215, 2003. (With permission, conveyed through the Copyright Clearance Center.)

influence expression of mammalian Ras superfamily proteins. Further studies will clarify the consequences of these induced changes in the expression of Ras proteins as related to the ascribed biological effects of the monoterpenes.

CONCLUSIONS AND FUTURE DIRECTIONS

Given the multifaceted functions of the isoprenoids, including their regulatory properties with respect to the Ras-related proteins, it is perhaps not surprising that the isoprenoid bio-synthetic pathway is so highly conserved. The only recognized enzyme deficiency for nonsterol isoprenoid synthesis is that of mevalonate kinase (Figure 15.1).[43] The phenotype of this deficiency, however, appears to be related more to excessive levels of mevalonic acid and resultant mevalonic aciduria and psychomotor retardation than to disruption of synthesis of nonsterol and sterol products. In fact, radiolabeled acetate incorporation into cholesterol in the fibroblasts from patients with mevalonate kinase deficiency has been shown to be com-parable to controls.[44] This observation and the lack of any other recognized abnormalities in the production of the non-sterol isoprenoids would suggest that marked congenital dis-turbances of isoprenoid biosynthesis are not compatible with life. This is in contrast to the widespread utility of the HMG-CoA reductase inhibitors for manipulating serum lipoprotein levels and perhaps altering the clinical course of diseases ranging from cancers to Alzheimer's disease. The extent to which the isoprenoids and their regulatory effects on Ras-related proteins reviewed in this chapter impact these diverse disease processes remains to be established.

REFERENCES

1. JC Sacchettini, CD Poulter. Creating isoprenoid diversity. *Science* 277: 1788–1789, 1997.

2. W Eisenreich, M Schwarz, A Cartayrade, D Arigoni, MH Zenk, A Bacher. The deoxyxylulose phosphate pathway of terpenoid biosynthesis in plants and microorganisms. *Chem Biol* 5: R221–233, 1998.

3. BM Lange, T Rujan, W Martin, R Croteau. Isoprenoid biosynthesis: the evolution of two ancient and distinct pathways across genomes. *Proc Nat Acad Sci USA* 97: 13172–13177, 2000.

4. SS Mahmoud, RB Croteau. Strategies for transgenic manipulation of monoterpene biosynthesis in plants. *Trends Plant Sci* 7: 366–373, 2002.

5. JW Porter, SL Spurgeon. *Biosynthesis of Isoprenoid Compounds*. New York: John Wiley & Sons, 1981.

6. PL Crowell. Prevention and therapy of cancer by dietary monoterpenes. *J Nutr* 129: 775S–778S, 1999.

7. MN Gould, CJ Moore, R Zhang, B Wang, WS Kennan, JD Haag. Limonene chemoprevention of mammary carcinoma induction following direct in situ transfer of v-Ha-ras. *Cancer Res* 54: 3540–3543, 1994.

8. JA Elegbede, CE Elson, MA Tanner, A Qureshi, MN Gould. Regression of rat primary mammary tumors following dietary d-limonene. *J Natl Cancer Inst* 76: 323–325, 1986.

9. JD Haag, MN Gould. Mammary carcinoma regression induced by perillyl alcohol, a hydroxylated analog of limonene. *Cancer Chemother Pharmacol* 34: 477–483, 1994.

10. MJ Stark, YD Burke, JH McKinzie, AS Ayoubi, PL Crowell. Chemotherapy of pancreatic cancer with the monoterpene perillyl alcohol. *Cancer Lett* 96: 15–21, 1995.

11. RA Gibbs. Farnesyltransferase inhibitors: novel anticancer mechanisms and new therapeutic applications. *Curr Opin Drug Discov Devel* 3: 585–596, 2000.

12. SM Mumby, PJ Casey, AG Gilman, S Gutowski, PC Sternweis. G protein gamma subunits contain a 20-carbon isoprenoid. *Proc Nat Acad Sci USA* 87: 5873–5877, 1990.

13. JE Buss, LA Quilliam, K Kato, PJ Casey, PA Solski, G Wong, R Clark, F McCormick, GM Bokoch, CJ Der. The COOH-terminal domain of the Rap1A (Krev-1) protein is isoprenylated and supports transformation by an H-Ras: Rap1A chimeric protein. *Mol Cell Biol* 11: 1523–1530, 1991.

14. M Katayama, M Kawata, Y Yoshida, H Horiuchi, T Yamamoto, Y Matsuura, Y Takai. The posttranslationally modified C-terminal structure of bovine aortic smooth muscle rhoA p21. *J Biol Chem* 266: 12639–12645, 1991.

15. P Adamson, CJ Marshall, A Hall, PA Tilbrook. Post-translational modifications of p21rho proteins. *J Biol Chem* 267: 20033–20038, 1992.

16. MC Seabra, JL Goldstein, TC Sudhof, MS Brown. Rab geranylgeranyl transferase. A multisubunit enzyme that prenylates GTP-binding proteins terminating in Cys-X-Cys or Cys-Cys. *J Biol Chem* 267: 14497–14503, 1992.

17. M Barbacid. Ras genes. *Annu Rev Biochem* 56: 779–827, 1987.

18. BZ Shilo, RA Weinberg. DNA sequences homologous to vertebrate oncogenes are conserved in Drosophila melanogaster. *Proc Nat Acad Sci USA* 78: 6789–6792, 1981.

19. MS Boguski, F McCormick. Proteins regulating Ras and its relatives. *Nature* 366: 643–654, 1993.

20. AB Vojtek, CJ Der. Increasing complexity of the Ras signaling pathway. *J Biol Chem* 273: 19925–19928, 1998.

21. DR Lowy, BM Willumsen. Function and regulation of ras. *Annu Rev Biochem* 62: 851–891, 1993.

22. JL Bos. Ras oncogenes in human cancer: a review. *Cancer Res* 49: 4682–4689, 1989.

23. M Kawata, Y Matsui, J Kondo, T Hishida, Y Teranishi, Y Takai. A novel small molecular weight GTP-binding protein with the same putative effector domain as the ras proteins in bovine brain membranes. Purification, determination of primary structure, and characterization. *J Biol Chem* 263: 18965–18971, 1988.

24. H Kitayama, Y Sugimoto, T Matsuzaki, Y Ikawa, M Noda. A ras-related gene with transformation suppressor activity. *Cell* 56: 77–84, 1989.

25. CD Hu, K Kariya, G Kotani, M Shirouzu, S Yokoyama, T Kataoka. Coassociation of Rap1A and Ha-Ras with Raf-1 N-terminal region interferes with ras-dependent activation of Raf-1. *J Biol Chem* 272: 11702–11705, 1997.

26. KA Reedquist, E Ross, EA Koop, RM Wolthuis, FJ Zwartkruis, Y van Kooyk, M Salmon, CD Buckley, JL Bos. The small GTPase, Rap1, mediates CD31-induced integrin adhesion. *J Cell Biol* 148: 1151–1158, 2000.

27. RD York, H Yao, T Dillon, CL Ellig, SP Eckert, EW McCleskey, PJ Stork. Rap1 mediates sustained MAP kinase activation induced by nerve growth factor. *Nature* 392: 622–626, 1998.

28. N Ozaki, T Shibasaki, Y Kashima, T Miki, K Takahashi, H Ueno, Y Sunaga, H Yano, Y Matsuura, T Iwanaga, Y Takai, S Seino. cAMP-GEFII is a direct target of cAMP in regulated exocytosis. *Nat Cell Biol* 2: 805–811, 2000.

29. JL Bos, J de Rooij, KA Reedquist. Rap1 signalling: adhering to new models. *Nat Rev Mol Cell Biol* 2: 369–377, 2001.

30. GC Prendergast. Actin' up: RhoB in cancer and apoptosis. *Nature Rev Cancer* 1: 162–168, 2001.

31. D Michaelson, J Silletti, G Murphy, P D'Eustachio, M Rush, MR Philips. Differential localization of Rho GTPases in live cells: regulation by hypervariable regions and RhoGDI binding. *J Cell Biol* 152: 111–126, 2001.

32. P Adamson, HF Paterson, A Hall. Intracellular localization of the P21rho proteins. *J Cell Biol* 119: 617–627, 1992.

33. AX Liu, N Rane, JP Liu, GC Prendergast. RhoB is dispensable for mouse development, but it modifies susceptibility to tumor formation as well as cell adhesion and growth factor signaling in transformed cells. *Mol Cell Biol* 21: 6906–6912, 2001.

34. A Endo, M Kuroda, Y Tsujita. ML-236A, ML-236B, and ML-236C, new inhibitors of cholesterogenesis produced by Penicillium citrinium. *J Antibiotics* 29: 1346–1348, 1976.

35. WR Schafer, R Kim, R Sterne, J Thorner, SH Kim, J Rine. Genetic and pharmacological suppression of oncogenic mutations in ras genes of yeast and humans. *Science* 245: 379–385, 1989.

36. S Leonard, L Beck, M Sinensky. Inhibition of isoprenoid biosynthesis and the post-translational modification of pro-p21. *J Biol Chem* 265: 5157–5160, 1990.

37. RJ Hohl, K Lewis. Differential effects of monoterpenes and lovastatin on RAS processing. *J Biol Chem* 270: 17508–17512, 1995.

38. SA Holstein, CL Wohlford-Lenane, RJ Hohl. Consequences of mevalonate depletion. Differential transcriptional, translational, and post-translational up-regulation of Ras, Rap1a, RhoA, and RhoB. *J Biol Chem* 277: 10678–10682, 2002.

39. SA Holstein, CL Wohlford-Lenane, RJ Hohl. Isoprenoids influ-
     ence the expression of Ras and Ras-related proteins. *Biochem-
     istry* 41: 13698–13704, 2002.

40. SA Holstein, CL Wohlford-Lenane, DF Wiemer, RJ Hohl. Iso-
     prenoid pyrophosphate analogues regulate expression of Ras-
     related proteins. *Biochemistry* 42: 4384–4391, 2003.

41. PL Crowell, RR Chang, ZB Ren, CE Elson, MN Gould. Selec-
     tive inhibition of isoprenylation of 21-26-kDa proteins by the
     anticarcinogen d-limonene and its metabolites. *J Biol Chem*
     266: 17679–17685, 1991.

42. SA Holstein, RJ Hohl. Monoterpene regulation of Ras and
     Ras-related protein expression. *J Lipid Res* 44: 1209–1215,
     2003.

43. G Hoffmann, KM Gibson, IK Brandt, PI Bader, RS Wappner,
     L Sweetman. Mevalonic aciduria — an inborn error of choles-
     terol and nonsterol isoprene biosynthesis. *N Engl J Med* 314:
     1610–1614, 1986.

44. KM Gibson, G Hoffmann, A Schwall, RL Broock, S Aramaki,
     L Sweetman, WL Nyhan, IK Brandt, RS Wappner, W Lehnert
     et al. 3-Hydroxy-3-methylglutaryl coenzyme A reductase
     activity in cultured fibroblasts from patients with mevalonate
     kinase deficiency: differential response to lipid supplied by
     fetal bovine serum in tissue culture medium. *J Lipid Res* 31:
     515–521, 1990.

16

Amino Acid Signaling and the Control of Autophagy

ALFRED J. MEIJER, PETER F. DUBBELHUIS,
EDWARD F.C. BLOMMAART, AND
PATRICE CODOGNO

INTRODUCTION

In eukaryotic cells, autophagy is an essential cell biological process that serves several functions. The process is involved in the production of molecules that are required for cell survival when nutrients fall short, the elimination of functionally redundant or defective cell organelles, the remodeling of cells during differentiation, and in apoptosis. In many cell types, autophagy contributes significantly to the degradation of protein and thus to the control of cell growth. In some cases, defective autophagy promotes tumor cell growth.

Amino acids are the classical (product) inhibitors of autophagy. In search of a mechanism responsible for this inhibition, we discovered, in the past, an amino acid-stimulated signaling pathway that was unknown at the time and that inhibits autophagy and simultaneously stimulates protein synthesis. This pathway shares components of signaling pathways stimulated by insulin and other growth factors.

In this chapter we confine ourselves to a discussion of the regulation of autophagy by amino acid signaling.

AUTOPHAGY

Introduction

In the turnover of cell components, (macro)autophagy is essential. During sequestration, cytoplasmic material to be degraded is surrounded by a double membrane to form a closed vesicle, the so-called autophagosome. The autophagosome acquires hydrolytic enzymes by fusion with lysosomes to form an autophagolysosome. After the fusion step, degradation of the sequestered material occurs. It is also possible that the autophagosome, prior to its fusion with lysosomes, first fuses with an endosome to form an amphisome.[1] It is the sequestration step that largely controls autophagic flux.[1-3]

The origin of the sequestering membrane is still unknown. A number of possibilities have been proposed: the ribosome-free regions of the ER, post-Golgi and Golgi membranes, or a preexisting membrane structure, the so-called phagophore.[1-3] A problem in studies on the origin of the isolation membrane has been the lack of good marker molecules caused by the low number of transmembrane and peripheral-surface membrane proteins, as demonstrated by freeze-fracture electron microscopy.[4] Studies on yeast have suggested *de novo* synthesis of a preautophagosomal structure.[5]

A large amount of information on the molecular mechanism of autophagy has been obtained from genetic studies with the yeast *Saccharomyces cerevisiae*. At least 16 genes involved in autophagosome formation were identified, the *APG* (autophagy) and *AUT* (autophagocytosis) genes (see Reference 6 for a review).[6] A new unifying nomenclature for these *APG* and *AUT* genes has been introduced recently.[7] They are now termed *ATG* (autophagy-related) genes. In this chapter, this new nomenclature will be used. Autophagy-deficient mutants are unable to survive under nutrient-deficient conditions. Among the proteins encoded by these genes are components of two ubiquitin-like systems that conjugate some of the components of the autophagic machinery: conjugation of Atg12 to Atg5 and conjugation of Atg8 (mammalian homologue: LC3) to phosphatidylethanolamine.[6,8] Mammalian counterparts of most of the *ATG* genes have also been found.[6,8]

As might be expected for this complicated membrane flow-dependent process, flux through the system is ATP-dependent, ATP being required at all three steps of the pathway, sequestration, fusion, and degradation.[3] In the last step, ATP is required for operation of the lysosomal proton pump. In addition, ATP is required for the ubiquitin-like protein-conjugation system that is part of the autophagic protein machinery (see previous paragraph). Although autophagic flux is sensitive to relatively small changes in the intracellular concentration of ATP,[3] recent evidence suggests that changes in the activity of the AMP-activated protein kinase resulting from changes in ATP levels may also contribute to the control of flux.[9]

It has long been assumed that autophagy is a nonspecific process in which cytoplasmic structures and macromolecules are randomly sequestered[10] in order to generate molecules (e.g., amino acids) that are essential for cell survival when nutrients fall short. We now know that autophagy can also be very specific under some conditions. For example, the process may be involved in the removal of damaged mitochondria[11] or in the selective removal of organelles that are functionally redundant (e.g., peroxisomes[12,13]). Autophagy also participates in complicated cell biological processes such as differentiation[6,14,15] and cell death.[16]

In this chapter, we confine ourselves to a discussion of signaling mechanisms that contribute to the control of non-specific autophagy that is induced by starvation. The signaling mechanism controlling autophagy is similar to that involved in the (opposite) control of protein synthesis. For a discussion of the role of signaling in the regulation of protein synthesis, the reader is referred to other chapters in this book.

Inhibition of Autophagy by Amino Acids

Because amino acids are the products of autophagic proteolysis, they are the classical feedback inhibitors of autophagic sequestration. In addition, the process is hormonally controlled: insulin inhibits and (in the liver) glucagon stimulates autophagy.[1–3] The hormones only affect autophagy at intermediate amino acid concentrations, but not at either very low or very high amino acid concentrations when autophagic flux is maximal or minimal, respectively.[2] As we will discuss later, this closely parallels the effect of the two hormones and amino acids on signal transduction.

Our interest in the regulation of autophagy by amino acids was triggered by experiments we performed with isolated perfused rat hepatocytes under true steady-state conditions. To our surprise, we found that a simple combination of leucine and alanine, at near-physiological concentrations, was equally effective in inhibiting autophagic proteolysis as a complete mixture of all amino acids; alanine or leucine alone had no effect.[17] This observation was confirmed by Mortimore and colleagues in the perfused liver.[18] Combinations of low concentrations of leucine and either proline, glutamine, or asparagine were also effective as inhibitors of autophagy.[19] With leucine present, the inhibition by alanine, but not by proline, glutamine, or asparagine, was sensitive to inhibition by the transaminase inhibitor aminooxyacetate.[17,19] This indicated that metabolism of alanine, but not that of leucine, was required for its inhibitory effect on autophagy. As conversion of proline and glutamine to glutamate and of asparagine to aspartate does not require transamination reactions, we suspected that a combination of leucine with intracellular glutamate or aspartate (which do not readily leave hepatocytes) was sufficient to inhibit autophagy maximally. Indeed, in the presence of leucine and either alanine, proline, glutamine, or asparagine there was a clear inverse relationship between the rate of autophagic proteolysis and the intracellular concentration of glutamate plus aspartate, with maximal inhibition of proteolysis at an intracellular concentration of about 6 mM.[19]

These and other studies[20] led us to conclude that leucine (and perhaps also phenylalanine and tyrosine) in combination with a few other amino acids (glutamine, alanine, asparagine, and proline) are sufficient to inhibit autophagy in hepatocytes.[3]

The same amino acids that inhibit autophagic proteolysis in liver also inhibit proteolysis in other tissues. In skeletal muscle, leucine and glutamine are potent inhibitors of proteolysis. In the kidney, leucine and phenylalanine are most important, whereas in the heart only leucine is involved (see Reference 3 for the literature).[3] In C2C12 myotubes, autophagic proteolysis was induced by leucine limitation and was sensitive to the autophagy inhibitor 3-methyladenine.[21] Rat muscle proteolysis, measured as the production of 3-methylhistidine from myofibrillar proteins, was inhibited after oral feeding with proteins or leucine alone.[22] The assumption that muscle proteolysis mainly occurs via an extralysosomal pathway, e.g., the ubiquitin–proteasome pathway that is not controlled by amino acids,[23] therefore needs to be reconsidered.[21,24,25] The presence of autophagosomes and the accumulation of these organelles in certain muscle diseases with defects in the autophagic pathway indicate that in muscle this process may be more important than hitherto assumed.[26]

A new factor in the control of autophagic protein degradation was introduced with the observation, initially made in liver but later confirmed for other tissues, that an increase in cell volume is able to mimic most of the anabolic properties of insulin, including inhibition of proteolysis and stimulation of protein synthesis.[27] Cell swelling can occur under hypoosmotic conditions or be caused by Na^+-dependent concentrative transport of amino acids across the plasma membrane. Intracellular accumulation of impermeant catabolites in the course of amino acid degradation (glutamate, aspartate) helps to increase intracellular osmolarity and contributes to cell swelling. In response to an increase in cell volume, cells undergo "regulatory volume decrease" in which they try to restore, at least in part, the original volume by releasing KCl.[27] In freshly isolated hepatocytes, a hypoosmotically induced increase in cell volume potentiates the ability of a complete mixture of amino acids, or of a mixture of leucine, phenylalanine, and tyrosine, to inhibit autophagic proteolysis. The effect of cell swelling thus strongly resembles the effect of insulin.[28–30] In the isolated flow-through perfused liver, and also in cultured hepatocytes, hypoosmotically induced cell swelling *per se* can inhibit autophagy.[27,31]

In summary, leucine appears to be the most important amino acid in the control of autophagy. In addition, its anti-proteolytic effect is potentiated by insulin and cell swelling.

AMINO ACID SIGNALING AND AUTOPHAGY

Signaling, Autophagy, and Amino Acid Specificity

An important step forward in understanding the mechanism by which amino acids inhibit autophagy (and stimulate protein synthesis) was obtained when we discovered that the same amino acid mixtures that inhibited autophagy in hepatocytes greatly stimulated the phosphorylation of ribosomal protein S6 (up to fivefold).[29,30] Furthermore, similar to their ability to inhibit autophagy, we observed synergy between insulin (or cell swelling) and low concentrations of a complete mixture of amino acids (or a mixture of leucine, phenylalanine, and tyrosine) with regard to their ability to promote S6 phosphorylation. Addition of insulin alone, in the absence of added amino acids, did not stimulate S6 phosphorylation. The relationship between the inhibition of autophagic proteolysis (measured in the presence of cycloheximide to inhibit simultaneous protein synthesis) and the degree of phosphorylation of S6 was perfectly linear.[30] Amino acid-induced S6 phosphorylation was completely prevented by rapamycin, indicating that mTOR and p70S6 kinase were components of the signaling pathway. Of great significance was the fact that rapamycin could partly (but not completely) reverse the inhibition of autophagy by amino acids. Interestingly, in the absence of cycloheximide, rapamycin also did not completely inhibit protein synthesis. We concluded that (global) protein synthesis and (autophagic) protein degradation are under the opposite control of the same signaling pathway, which is efficient from the point of view of metabolic regulation.[30] The observation that rapamycin increases autophagy has been confirmed in other cell types,[21,32,33] including yeast.[34]

Amino acid stimulation of S6 phosphorylation was inhibited by the PI 3-kinase inhibitors wortmannin or LY294002 even in the absence of insulin, but did not increase autophagy in the presence of amino acids. By contrast, these two compounds inhibited autophagy in the absence of amino acids.[35] This was quite unexpected because interruption of signaling with rapamycin increased autophagy. The solution to this apparent anomaly was found in experiments with HT-29 cells, a human colon cancer cell line, showing that the product of PI 3-kinase class III, PI(3)P, is required for autophagy, but that the products of PI 3-kinase class I, PI(3,4)P2 and PI(3,4,5)P3, are inhibitory.[36] Indeed, overexpression of PTEN (phosphatase and tensin homologue deleted on chromosome 10), which removes the phosphate from the 3-position of PI(3,4)P2 and PI(3,4,5)P3, increases autophagy.[37] The PI 3-kinase inhibitors do not distinguish between the enzymes, and also inhibit PI(3)P formation and thus autophagy. Intriguingly, 3-methyladenine, a generally used specific inhibitor of autophagy, turned out to be an inhibitor of PI 3-kinase, thus explaining its inhibitory effect on autophagy[35,36] and S6 phosphorylation.[35]

The requirement of PI 3-kinase class III (and its adaptor p150) for autophagy was also found in yeast. This organism contains Vps34, a homologue of PI 3-kinase class III, which is part of a complex that contains Vps15 (the homologue of p150 in mammalian cells), Vps30, and Atg14.[38] This complex is required to recruit the Atg12–Atg5 conjugate to the preautophagosomal structure,[39] possibly by interaction with the protein domain FYVE.[40] Interestingly, Beclin1, the mammalian homologue of Atg6, is also required for autophagy[41] and is found in a complex with PI 3-kinase class III and p150.[42]

The ability of amino acids to stimulate signaling and the synergy of amino acids with insulin were confirmed in hepatocytes and in other insulin-sensitive cell types including muscle cells, adipocytes, hepatoma cells, CHO cells, and pancreatic beta cells (see References 43 to 47 for the literature).[43–47] In addition to S6, other downstream targets of mTOR, such as p70S6 kinase, 4E-BP1, eIF2α kinase (the equivalent of Gcn2, general control nondepressible 2, in yeast), and eEF2 kinase were found to be phosphorylated in response to amino acid addition in a wortmannin-sensitive or LY294002-sensitive manner (see References 43 to 48; see also other chapters in this book).[43-48] Direct evidence that amino acids can increase mTOR phosphorylation and activity was also obtained (compare reference 51).[49,50]

However, amino acids do not directly affect protein kinase B activity and probably also not PI 3-kinase class I, which are two signaling proteins that are stimulated by insulin (see References 44 and 47 for the literature).[44,47] Presumably, PI 3-kinase class I is on a pathway parallel to that of the amino acid signaling pathway so that the activation of both PI 3-kinase class I (by insulin) and mTOR (by amino acids) is required for full activation of mTOR downstream targets (see Reference 48 for the literature).[48] This scenario would explain the synergy between insulin and amino acids. In order to account for the ability of high concentrations of amino acids to activate p70S6 kinase or S6 in the absence of added insulin by a mechanism that is sensitive to the PI 3-kinase inhibitors, one has to assume that either basal activity of PI 3-kinase class I or only a very slight stimulation of PI 3-kinase class I by amino acids is sufficient to stimulate p70S6 kinase phosphorylation.[47,52]

Most studies agree that leucine (but not the other branched-chain amino acids[53,54]), independent of the cell type, is the most effective amino acid in stimulating signaling but that, in addition, some other amino acids are required. In analogy with the inhibition of autophagy and the stimulation of S6 phosphorylation by amino acids (see the preceding discussion), we previously proposed that leucine (and perhaps phenylalanine plus tyrosine), in combination with amino acid–induced cell swelling, would be sufficient to stimulate signalling.[44,47] In line with this, Krause et al.[55] showed synergy between glutamine, a potent amino acid in promoting cell swelling, and leucine with regard to p70S6 kinase phosphorylation. Data (not shown) recently obtained in our own laboratory confirmed this observation and, in addition, also showed synergy between leucine and either proline or asparagine with regard to p70S6 kinase phosphorylation. Glutamine, proline, or asparagine rapidly produce intracellular glutamate and aspartate and greatly enhance the intracellular concentrations of these amino acids. The presence of the transaminase inhibitor aminooxyacetate greatly affected the relative proportions of glutamate and aspartate but did not interfere with p70S6 kinase phosphorylation. Interestingly, intracellular generation of glutamate and aspartate from added lactate and ammonia also proved to be synergistic with leucine in stimulating p70S6 kinase phosphorylation (P.F. Dubbelhuis and A.J. Meijer; unpublished observations). Apparently leucine, combined with intracellular glutamate or aspartate, is sufficient to stimulate signaling. Because the combination of leucine with intracellular glutamate and aspartate was also highly efficient in inhibiting autophagy,[19] this is another indication in support of the view that amino acid signaling controls autophagy.

The Amino Acid Sensor

Sensing of amino acid concentrations by cells can occur intracellularly, extracellularly, or both. For all three possibilities indications may be found in the literature. Evidence in support of a plasma membrane amino acid receptor came from studies by Miotto et al.[56] They demonstrated that Leu8-MAP (but not Ile8-MAP), a nonpermeant global peptide, inhibited autophagy (compare Reference 44 and 53),[44,53] and a leucine-binding protein was isolated from the plasma membrane.[57] Additional evidence supporting the existence of a plasma membrane amino acid receptor has been provided by a recent report showing that the rate of protein synthesis in skeletal muscle in man *in vivo* responds to changes in the extracellular, but not intracellular, concentration of amino acids.[58] An amino acid sensor in the plasma membrane would be in line with a similar leucine-sensitive sensor molecule Ssy1p, which is part of the three-component SPS amino acid sensor complex in the yeast plasma membrane.[59]

Evidence in support of an intracellular amino acid receptor came from studies on xenopus oocytes. In these cells, leucine stimulated signaling only after the cell membrane had been made permeable to leucine by transfection with a leucine transporter.[60] This suggests that, at least in these cells, the amino acid sensor is intracellular rather than extracellular. In CHO cells, inhibition of protein synthesis increased mTOR signaling in amino acid-deprived cells.[61] It was concluded that, apparently, intracellular amino acids (obtained from proteolysis) rather than extracellular amino acids stimulate signaling and that the amino acid sensor must be intracellular rather than extracellular.[61] However, the possibility that proteolytically derived amino acids were transported to the extracellular fluid was not considered.

A combination of both intra- and extracellular sensing of amino acids would be in line with the situation in yeast. This organism senses intracellular amino acids through Gcn2, whereas the extracellular amino acid concentration is sensed by the SPS complex.[59] Clearly, more experiments need to be performed before this issue is settled.

AMINO ACID-DEPENDENT SIGNALING AND AUTOPHAGY: MECHANISMS

Although the mechanisms by which amino acid signaling affects autophagy are still largely unknown, some progress has been made in recent years. Possible mechanisms are discussed in the following text.

mTOR

mTOR-Associated Proteins

The mechanism by which amino acids can stimulate mTOR activity is still unknown. Recent evidence has indicated that mTOR is in a complex with raptor, a protein that functions as a scaffold for the mTOR-mediated phosphorylation of mTOR substrates (see Reference 62 for a review).[62] The protein GβL is also part of this complex.[63] Another important recent finding is that mTOR-dependent signaling appears to be restrained by association of mTOR with the tumor-suppressor proteins TSC1 (hamartin) and TSC2 (tuberin).[62] The function of TSC1/2 is suppressed by activation of PI 3-kinase class I. This may explain the synergy between insulin and amino acids with regard to mTOR-dependent signaling, as discussed earlier. The direct target of TSC1/2 is probably not mTOR itself but rather Rheb, a GTP-binding protein of the

Ras family, which is downstream of TSC1/2 and upstream of mTOR and which, like mTOR, is required for cell growth.[64] It is not known whether or not amino acids directly affect the activity of TSC1/2, Rheb, raptor, GβL, or mTOR itself. Direct activation of mTOR kinase activity by amino acids *in vitro* is considered unlikely.[62,65] Because changes in cell volume can affect molecular crowding,[66] an attractive possibility is that amino acid-induced cell swelling loosens the association of Rheb and TSC1/2, so that mTOR becomes activated. This would account for the synergy between cell swelling and activation of mTOR-dependent signaling by leucine, as discussed in the preceding text.

Phosphatidic Acid

The mitogenic messenger phosphatidic acid can also activate mTOR-dependent signaling, provided sufficient amino acids are also present, indicating that phosphatidic acid promotes signaling in parallel to amino acids.[67] This is similar to insulin signaling or the effect of cell swelling, as discussed in the preceding text. Whether insulin or cell swelling can affect phosphatidic acid concentrations or *vice versa*, or whether phosphatidic acid affects autophagy, is not known.

Polyphosphates

Inorganic polyphosphate (polyP), linear polymers of hundreds of Pi residues that are present in all cell types, has been proposed as another component that may be involved in amino acid-dependent and mTOR-mediated signaling.[51] At physiological concentrations, this compound greatly stimulated mTOR kinase activity *in vitro* and entirely mimicked the stimulatory effect of adding amino acids plus insulin on mTOR kinase activity in intact cells. Moreover, expression of a yeast exopolyphosphatase gene in a human breast carcinoma cell line inhibited the ability of insulin and amino acids to stimulate mTOR-mediated signaling with no effect on insulin-stimulated protein kinase B phosphorylation, and inhibited cell growth.[51] Whether autophagy was stimulated under these conditions was not investigated.

By contrast, in *E. coli*, polyP appears to inhibit protein synthesis and promote protein degradation. In these cells, the level of polyP increases after amino acid deprivation because of guanosinetetraphosphate (ppGpp)-mediated inhibition of exopolyphosphatase, and it stimulates ribosomal protein degradation by binding to the ATP-dependent Lon protease. In this way, the generation of amino acids for cell survival under these conditions is ensured.[68]

An intriguing possibility that has not been mentioned in the literature so far is that diadenosine polyphosphates (ApnA) are involved in mTOR stimulation. Indirect evidence comes from two independent studies: in one study, leucine stimulated Ap4A production by the mitochondria in pancreatic beta cells,[69] whereas in another study, leucine stimulated mTOR activity through increased mitochondrial oxidative metabolism.[70] Because mTOR may be associated, at least in part, with mitochondria,[71] it is tempting to speculate that Ap4A is a signal that connects mitochondrial metabolism to mTOR activity. In this context, it is important to note that Ap4A is a strong inhibitor of AMP-activated protein kinase.[72] As discussed in the section titled "mTOR, Energy, and Autophagy," inhibition of AMP-activated protein kinase stimulates mTOR-dependent signaling.

Phosphatases

Because autophagy in hepatocytes is inhibited by protein phosphatase inhibitors,[73] the possibility exists that amino acid signaling and autophagy are linked by a protein phosphatase, presumably PP2A.[13] However, evidence in support of the involvement of PP2A in mTOR-dependent signaling in mammalian cells is controversial.[47]

In yeast, the rapamycin-sensitive TOR proteins affect PP2A activity by modulating the association of PP2A with the Tap42 protein as follows. In the presence of nutrients, Tap42 becomes phosphorylated, associates with, and inhibits PP2A activity. Nutrient deprivation or rapamycin addition reverses these effects, and autophagy becomes activated.[65,74] There is evidence, however, that Tap42 does not play a role in the induction of autophagy.[74,75] Because in yeast the autophagy proteins Atg1 and Atg13 are phosphoproteins that are dephosphorylated in response to rapamycin or starvation,[74,76] it is likely that another, yet to be identified, TOR-dependent protein phosphatase is required for the control of autophagy.[75]

PI3-Kinase Class III/Beclin1

An important step forward in our understanding of the relationship between amino acid signaling and autophagy was recently obtained in experiments performed with C2C12 myotubes. It was found that autophagy, induced by amino acid depletion, was accompanied by an increase in Beclin1-associated PI 3-kinase class III activity.[24] This is the first demonstration that the production of PI(3)P for autophagy is controlled by amino acids and is stimulated when amino acid levels fall. Although the specificity of the amino acid effect was not investigated, it is, in analogy with the mTOR-TSC1/2 complex, tempting to speculate that amino acid-induced cell swelling is responsible for dissociation of the Beclin1–PI 3-kinase class III complex and thus contributes to the inhibition of autophagy. The observation in hepatocytes that inhibition of autophagy by amino acids cannot be completely reversed by rapamycin (see the preceding text) may now be explained by the fact that mTOR does not affect the Beclin1–PI 3-kinase class III complex. Another interesting observation is that, in C2C12 myotubes, amino acid addition does not result in S6 phosphorylation. However, rapamycin addition to these cells does increase autophagy.[24] Data obtained with myeloblastic cells[77] and drosophila[78] also seem to exclude a role for p70S6 kinase, and thus for S6, in the control of autophagy. This would refute our initial hypothesis, based on data with hepatocytes, that S6 phosphorylation itself may control autophagy, for example by changing the degree of occupancy of the endoplasmic reticulum with ribosomes, the ribosome-free endoplasmic retuculum being a putative source of membrane for autophagosome formation.[30] It cannot be excluded, however, that differences in mechanisms between cell types may exist.

eIF2α Kinase

Another, or perhaps additional, mechanism of amino acid sensing is one in which cells respond to changes in the charging of tRNAs. This mechanism is based on data in yeast showing that, upon amino acid starvation, free uncharged tRNA binds to the protein kinase Gcn2 because its active center strongly resembles that of aminoacyl-tRNA synthetases.[79] Gcn2 activation results in the phosphorylation of eIF2α, which then results in the derepression of the *GCN4* mRNA translation. Gcn4 is a transcriptional activator that promotes the transcription of many genes involved in nitrogen metabolism, including not only genes involved in the biosynthesis of amino acids but also genes required for autophagy.[80] The activity of Gcn2 itself becomes inhibited by phosphorylation on Ser 577, and this phosphorylation is TOR-dependent.[81]

In mammalian cells, the eIF2α kinase PKR (double-stranded RNA-dependent protein kinase), which is the equivalent of Gcn2, contributes to the control of autophagy, and in *GCN2*-disrupted yeast, PKR can rescue starvation-induced autophagy.[82] However, experiments with aminoalcohols, inhibitors of aminoacyl-tRNA synthetases, have yielded contradictory results in mammalian cells with regard to their effects on both amino acid signaling and autophagy, presumably because of lack of specificity of these compounds (see Reference 13, 44, and 61 for the literature).[13,44,61] Furthermore, amino acid deprivation in HEK-293 cells did not affect free-tRNA levels.[83]

Unexpectedly, rapamycin is still able to stimulate autophagy in *GCN2*-disrupted yeast.[82] This suggests that Gcn2 may not be downstream of Tor or that Gcn2 controls autophagy by a mechanism that is independent of Tor.

Erk1/2

In HT-29 cells, in addition to the PI 3-kinase–mTOR pathway, another amino acid signaling pathway can control autophagy. In these cells, autophagy is also controlled by the trimeric Gαi3 protein; when it is bound to GDP, it stimulates autophagy, but when it is bound to GTP, it inhibits autophagy.[84] The GTP hydrolysis rate, and thus the rate of autophagy, is controlled by the activity of the Gα-interacting protein GAIP.[85] Activity of GAIP is enhanced by Erk1/2-mediated phosphorylation. Phosphorylation and activity of Erk1/2, mediated by Raf-1 and Mek1/2 in HT-29 cells, is decreased by amino acid addition.[86,87] Amino acids do so by stimulation of phosphorylation of Ser 259 and inactivation of Raf-1.[87] Because Ca++ is required for autophagy,[88] it is suggested that the Gαi3 protein may recruit Ca++-binding proteins to the Golgi or endoplasmic reticulum, where autophagosome formation may begin.[13,89] Whether amino acid signaling to Raf-1 is mTOR-dependent is not yet known.

By contrast, in freshly prepared hepatocytes[43,90] and in C2C12 myotubes,[24] inhibition of autophagy by amino acids is not accompanied by changes in Erk1/2 phosphorylation. Apparently, differences in amino acid signaling mechanisms and in the control of autophagy may exist, depending on the cell type and, perhaps, depending on the degree of differentiation.[24]

p38^MAPK

According to Häussinger and colleagues, amino acid-induced cell swelling in the isolated perfused rat liver and in cultured hepatocytes inhibits autophagy because it activates p38^MAPK. Inhibition of this stress kinase prevents the inhibition of autophagy.[91] Integrins appear to be involved in the osmosensing mechanism.[92] In C2C12 myotubes, inhibition of autophagy is independent of p38^MAPK.[24] Because in freshly isolated hepatocytes, cell swelling *per se* has no effect on autophagy but potentiates the inhibition of autophagy by leucine (or phenylalanine and tyrosine) via mTOR-dependent signaling (see preceding text), it was suggested that, apparently, multiple osmosensing mechanisms in hepatocytes may exist, which, depending on the conditions, can be differentially linked to intracellular signaling pathways.[92]

mTOR, ENERGY, AND AUTOPHAGY

Although it is now generally accepted that mTOR functions as an amino acid sensor,[62] recent evidence suggests that it also functions as a sensor of the cellular energy state. Among many protein kinases, mTOR has a relatively low affinity for ATP, its K_m being in the millimolar range of intracellular ATP concentrations, so that mTOR kinase activity may respond to changes in the intracellular ATP concentration.[83] In addition, the effect of a decrease in ATP is also transmitted to mTOR by AMP-activated protein kinase because a small decline in cytosolic ATP results in a relatively large increase in AMP concentration via the adenylate kinase equilibrium. Activation of AMP kinase inhibits mTOR-dependent signaling[93–97] and inhibits protein synthesis.[93–95] This is in agreement with the function of AMP-activated protein kinase to turn off ATP-dependent metabolic pathways.[98] In this context, the association of mTOR with the mitochondrial outer membrane[71] is noteworthy because adenylate kinase is located in the mitochondrial intermembrane space, and mTOR is thus in a perfect position to sense changes in the ATP to AMP ratio.[94]

Because of the inhibition of mTOR signaling, one would expect stimulation of autophagy by AMP kinase activation. Yeast cells contain the AMP kinase homologue, snf1p, which is indeed required for autophagy.[99] By contrast, activation of AMP kinase in hepatocytes inhibits autophagy.[9] The reason for this discrepancy is not clear.

AUTOPHAGY, AMINO ACID SIGNALING, AND LONGEVITY

Amino acid signaling in liver and muscle declines with age,[100,101] perhaps because of a decrease in affinity for leucine.[101] It is thought that this decrease in signaling contributes to the net protein loss with age. Although protein synthesis declines with age, proteasome-catalyzed proteolysis[102] and autophagic proteolysis[103,104] also decrease. The two proteolytic mechanisms can be considered as antiaging repair mechanisms because they remove aberrant proteins and defective cell structures. Significantly, caloric restriction not only increases proteasome-catalyzed proteolysis[102] but also increases autophagy, a phenomenon that may contribute to increased longevity.[103] The increase in autophagy by caloric restriction may be related to decreased plasma insulin concentrations.[104] Autophagy as a life span-extending mechanism is apparently well conserved in evolution because it is also found in *C. elegans*.[105]

Interestingly, extended life span was also observed in fat-tissue-specific insulin receptor knockout mice that are defective in insulin signaling.[106] These mice have reduced fat mass at all ages but normal food intake. The increased longevity was tentatively ascribed to decreased generation of free radicals of oxygen. The possibility that autophagy might have increased was not considered. By extrapolation, it may be speculated that development of insulin resistance in elderly people is an adaptive mechanism to increase autophagy, which helps to increase the capacity to remove damaged cellular structures (e.g., mitochondria[11]), and thus helps to prolong life.

CONCLUSIONS

It is evident from this chapter that amino acids are important signaling molecules with insulin-like actions on protein synthesis and autophagy and thus on autophagic protein degradation. From a metabolic point of view, it is extremely efficient that the same signaling pathway controls protein synthesis and degradation. Central in the signaling pathway is mTOR, which functions both as an amino acid and energy sensor. Phosphatidylinositol phospholipids are essential [PI(3)P] or inhibitory [PI(3,4)P$_2$ and PI(3,4,5)P$_3$] for autophagy.

The mechanism by which amino acids increase signaling is still unknown. At least in hepatocytes, a combination of leucine and intracellular glutamate and aspartate (or an increase in cell volume) is sufficient to activate signaling and inhibit autophagy. The issue of whether the (leucine-specific) amino acid receptor is located intracellularly or whether it is located in the plasma membrane is not yet settled. Amino acid-induced cell swelling perhaps affects the association between mTOR, raptor, TSC1/2, and Rheb, and that between Beclin1 and PI 3-kinase class III. The nature of the proteins involved in the autophagic machinery that are the target for

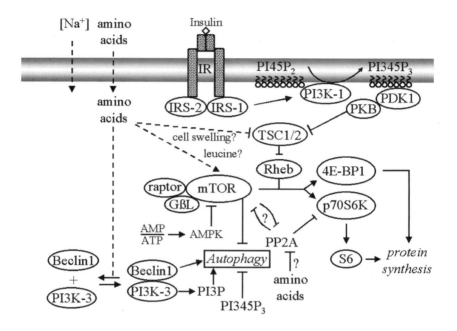

Figure 16.1   Amino acid signaling and the control of autophagy. IR, insulin receptor; IRS, insulin receptor substrate; PI3K-1, phosphatidylinositol 3-kinase class I; PI3K-3, phosphatidylinositol 3-kinase class III; PI3P, phosphatidylinositol 3-phosphate; $PI45P_2$, phosphatidylinositol 4,5-bisphosphate; $PI345P_3$, phosphatidylinositol 3,4,5-trisphosphate; PDK1, phosphoinositide-dependent kinase 1; PKB, protein kinase B; TSC, tuberous sclerosis complex; mTOR, mammalian target of rapamycin; raptor, the mammalian target of rapamycin partner; AMPK, AMP-activated protein kinase; PP2A, protein phosphatase 2A; 4E-BP1, eIF(eukaryotic protein-translation initiation factor)-4E-binding protein-1; Rheb, Ras homologue enriched in brain; S6, ribosomal protein S6; p70S6k, 70-kDa S6 kinase.

phosphorylation by mTOR (or downstream kinases) is also as yet unknown.

A summary of the possible interactions of amino acids with signaling and autophagy is given in Figure 16.1.

The importance of amino acid-dependent, mTOR-mediated signaling in cancer is growing.[65] There is also increasing evidence that suppressed autophagy contributes to cancer cell growth.[15] Compounds like rapamycin (or its analogues), which are used as anticancer agents,[107] not only inhibit protein synthesis but simultaneously accelerate autophagic protein degradation, and thus act as a double-edged sword.

ACKNOWLEDGMENTS

A.J. Meijer is grateful to the Dutch Diabetes Fund for financial support of part of his research (Grant 96.604). P. Codogno is supported by institutional funding from The Institut National de la Santé et de la Recherche Médicale (INSERM) and grants from the Association pour la Recherche sur le Cancer.

REFERENCES

1. PO Seglen, P Bohley. Autophagy and other vacuolar protein degradation mechanisms. *Experientia* 48: 158–172, 1992.

2. GE Mortimore, G Miotto, R Venerando, M Kadowaki. Autophagy. *Subcell Biochem* 27: 93–135, 1996.

3. EFC Blommaart, JJFP Luiken, AJ Meijer. Autophagic proteolysis: control and specificity. *Histochem J* 29: 365–385, 1997.

4. M Fengsrud, ES Erichsen, TO Berg, C Raiborg, PO Seglen. Ultrastructural characterization of the delimiting membranes of isolated autophagosomes and amphisomes by freeze-fracture electron microscopy. *Eur J Cell Biol* 79: 871–882, 2000.

5. T Noda, K Suzuki, Y Ohsumi. Yeast autophagosomes: de novo formation of a membrane structure. *Trends Cell Biol* 12: 231–235, 2002.

6. DJ Klionsky, SD Emr. Autophagy as a regulated pathway of cellular degradation. *Science* 290: 1717–1721, 2000.

7. DJ Klionsky, JM Cregg, WA Dunn Jr, SD Emr, Y Sakai, IV Sandoval, A Sibirny, S Subramani, M Thumm, M Veenhuis, Y Ohsumi. A unified nomenclature for yeast autophagy-related genes. *Dev Cell* 5: 539–545, 2003.

8. Y Ohsumi. Molecular dissection of autophagy: two ubiquitin-like systems. *Nat Rev Mol Cell Biol* 2: 211–216, 2001.

9. HR Samari, PO Seglen. Inhibition of hepatocytic autophagy by adenosine, aminoimidazole-4-carboxamide riboside, and N6-mercaptopurine riboside. Evidence for involvement of AMP-activated protein kinase. *J Biol Chem* 273: 23758–23763, 1998.

10. J Kopitz, GO Kisen, PB Gordon, P Bohley, PO Seglen. Nonselective autophagy of cytosolic enzymes by isolated rat hepatocytes. *J Cell Biol* 111: 941–953, 1990.

11. JJ Lemasters, AL Nieminen, T Qian, LC Trost, SP Elmore, Y Nishimura, RA Crowe, WE Cascio, CA Bradham, DA Brenner, B Herman. The mitochondrial permeability transition in cell death: a common mechanism in necrosis, apoptosis and autophagy. *Biochim Biophys Acta* 1366: 177–196, 1998.

12. JJFP Luiken, BM van den Berg, JC Heikoop, AJ Meijer. Autophagic degradation of peroxisomes in isolated rat hepatocytes. *FEBS Lett* 304: 93–97, 1992.

13. P Codogno, AJ Meijer. Signaling pathways in mammalian autophagy. In: DJ Klionsky, Ed. *Autophagy*. Georgetown, TX: Eurekah.com/Landes Bioscience, 2003, pp 26–47.

14. JJ Houri, E Ogier-Denis, D De Stefanis, C Bauvy, FM Baccino, C Isidoro, P Codogno. Differentiation-dependent autophagy controls the fate of newly synthesized N-linked glycoproteins in the colon adenocarcinoma HT-29 cell line. *Biochem J* 309: 521–527, 1995.

15. N Furuya, XH Liang, B Levine. Autophagy and cancer. In: DJ Klionsky, Ed. *Autophagy*. Georgetown, TX: Eurekah.com/Landes Bioscience, 2003, pp 244–258.

16. J Yuan, M Lipinski, A Degterev. Diversity in the mechanisms of neuronal cell death. *Neuron* 40: 401–413, 2003.

17. XM Leverve, LHP Caro, PJAM Plomp, AJ Meijer. Control of proteolysis in perifused rat hepatocytes. *FEBS Lett* 219: 455–458, 1987.

18. GE Mortimore, JJ Wert, Jr, CE Adams. Modulation of the amino acid control of hepatic protein degradation by caloric deprivation: two modes of alanine co-regulation. *J Biol Chem* 263: 19545–19551, 1988.

19. LHP Caro, PJAM Plomp, XM Leverve, AJ Meijer. A combination of intracellular leucine with either glutamate or aspartate inhibits autophagic proteolysis in isolated rat hepatocytes. *Eur J Biochem* 181: 717–720, 1989.

20. PJE Blommaart, D Zonneveld, AJ Meijer, WH Lamers. Effects of intracellular amino acid concentrations, cyclic AMP, and dexamethasone on lysosomal proteolysis in primary cultures of perinatal rat hepatocytes. *J Biol Chem* 268: 1610–1617, 1993.

21. S Mordier, C Deval, D Bechet, A Tassa, M Ferrara. Leucine limitation induces autophagy and activation of lysosome-dependent proteolysis in C2C12 myotubes through a mammalian target of rapamycin-independent signaling pathway. *J Biol Chem* 275: 29900–29906, 2000.

22. T Nagasawa, T Kido, F Yoshizawa, Y Ito, N Nishizawa. Rapid suppression of protein degradation in skeletal muscle after oral feeding of leucine in rats. *J Nutr Biochem* 13: 121–127, 2002.

23. AJ Kee, L Combaret, T Tilignac, B Souweine, E Aurousseau, M Dalle, D Taillandier, D Attaix. Ubiquitin-proteasome-dependent muscle proteolysis responds slowly to insulin release and refeeding in starved rats. *J Physiol* 546: 765–776, 2003.

24. A Tassa, MP Roux, D Attaix, DM Bechet. Class III phosphoinositide 3-kinase-beclin1 complex mediates the amino acid-dependent regulation of autophagy in C2C12 myotubes. *Biochem J*, 376: 577–586, 2003.

25. M Kadowaki, T Kanazawa. Amino acids as regulators of proteolysis. *J Nutr* 133: 2052S–2056S, 2003.

26. T Ueno, I Tanida, E Kominami. Autophagy and neuromuscular diseases. In: DJ Klionsky, Ed. *Autophagy*. Georgetown, TX: Eurekah.com/Landes Bioscience, 2003, pp 267–289.

27. D Häussinger. The role of cellular hydration in the regulation of cell function. *Biochem J* 313: 697–710, 1996.

28. AJ Meijer, LA Gustafson, JJFP Luiken, PJE Blommaart, LHP Caro, GM Van Woerkom, C Spronk, L Boon. Cell swelling and the sensitivity of autophagic proteolysis to inhibition by amino acids in isolated rat hepatocytes. *Eur J Biochem* 215: 449–454, 1993.

29. JJ Luiken, EFC Blommaart, L Boon, GM Van Woerkom, AJ Meijer. Cell swelling and the control of autophagic proteolysis in hepatocytes: involvement of phosphorylation of ribosomal protein S6? *Biochem Soc Trans* 22: 508–511, 1994.

30. EFC Blommaart, JJ Luiken, PJE Blommaart, GM Van Woerkom, AJ Meijer. Phosphorylation of ribosomal protein S6 is inhibitory for autophagy in isolated rat hepatocytes. *J Biol Chem* 270: 2320–2326, 1995.

31. S vom Dahl, F Dombrowski, M Schmitt, F Schliess, U Pfeifer, D Häussinger. Cell hydration controls autophagosome formation in rat liver in a microtubule-dependent way downstream from p38MAPK activation. *Biochem J* 354: 31–36, 2001.

32. EL Eskelinen, AR Prescott, J Cooper, SM Brachmann, L Wang, X Tang, JM Backer, JM Lucocq. Inhibition of autophagy in mitotic animal cells. *Traffic* 3: 878–893, 2002.

33. B Moazed, M Desautels. Control of proteolysis by norepinephrine and insulin in brown adipocytes: role of ATP, phosphatidylinositol 3-kinase, and p70 S6K. *Can J Physiol Pharmacol* 80: 541–552, 2002.

34. T Noda, Y Ohsumi. Tor, a phosphatidylinositol kinase homologue, controls autophagy in yeast. *J Biol Chem* 273: 3963–3966, 1998.

35. EFC Blommaart, U Krause, JP Schellens, H Vreeling-Sindelárová, AJ Meijer. The phosphatidylinositol 3-kinase inhibitors wortmannin and LY294002 inhibit autophagy in isolated rat hepatocytes. *Eur J Biochem* 243: 240–246, 1997.

36. A Petiot, E Ogier-Denis, EFC Blommaart, AJ Meijer, P Codogno. Distinct classes of phosphatidylinositol 3-kinases are involved in signaling pathways that control macroautophagy in HT-29 cells. *J Biol Chem* 275: 992–998, 2000.

37. S Arico, A Petiot, C Bauvy, PF Dubbelhuis, AJ Meijer, P Codogno, E Ogier-Denis. The tumor suppressor PTEN positively regulates macroautophagy by inhibiting the phosphatidylinositol 3-kinase/protein kinase B pathway. *J Biol Chem* 276: 35243–35246, 2001.

38. A Kihara, T Noda, N Ishihara, Y Ohsumi. Two distinct Vps34 phosphatidylinositol 3-kinase complexes function in autophagy and carboxypeptidase Y sorting in Saccharomyces cerevisiae. *J Cell Biol* 152: 519–530, 2001.

39. K Suzuki, T Kirisako, Y Kamada, N Mizushima, T Noda, Y Ohsumi. The pre-autophagosomal structure organized by concerted functions of APG genes is essential for autophagosome formation. *EMBO J* 20: 5971–5981, 2001.

40. A Simonsen, AE Wurmser, SD Emr, H Stenmark. The role of phosphoinositides in membrane transport. *Curr Opin Cell Biol* 13: 485–492, 2001.

41. XH Liang, S Jackson, M Seaman, K Brown, B Kempkes, H Hibshoosh, B Levine. Induction of autophagy and inhibition of tumorigenesis by beclin 1. *Nature* 402: 672–676, 1999.

42. A Kihara, Y Kabeya, Y Ohsumi, T Yoshimori. Beclin-phosphatidylinositol 3-kinase complex functions at the trans-Golgi network. *EMBO Rep* 2: 330–335, 2001.

43. U Krause, MH Rider, L Hue. Protein kinase signaling pathway triggered by cell swelling and involved in the activation of glycogen synthase and acetyl-CoA carboxylase in isolated rat hepatocytes. *J Biol Chem* 271: 16668–16673, 1996.

44. DA van Sluijters, PF Dubbelhuis, EFC Blommaart, AJ Meijer. Amino-acid-dependent signal transduction. *Biochem J* 351: 545–550, 2000.

45. OJ Shah, JC Anthony, SR Kimball, LS Jefferson. 4E-BP1 and S6K1: translational integration sites for nutritional and hormonal information in muscle. *Am J Physiol Endocrinol Metab* 279: E715–E729, 2000.

46. CG Proud. Regulation of mammalian translation factors by nutrients. *Eur J Biochem* 269: 5338–5349, 2002.

47. PF Dubbelhuis, AJ Meijer. Amino acid-dependent signal transduction. In: KB Storey, JM Storey, Eds. *Sensing, Signalling and Cell Adaptation*. Amsterdam: Elsevier, 2002, pp 207–219.

48. AJ Meijer. Amino acids as regulators and components of nonproteinogenic pathways. J Nutr 133: 2057S–2062S, 2003.

49. RT Peterson, BN Desai, JS Hardwick, SL Schreiber. Protein phosphatase 2A interacts with the 70-kDa S6 kinase and is activated by inhibition of FKBP12-rapamycin associated protein. *Proc Natl Acad Sci USA* 96: 4438–4442, 1999.

50. BT Nave, M Ouwens, DJ Withers, DR Alessi, PR Shepherd. Mammalian target of rapamycin is a direct target for protein kinase B: identification of a convergence point for opposing effects of insulin and amino-acid deficiency on protein translation. *Biochem J* 344: 427–431, 1999.

51. L Wang, CD Fraley, J Faridi, A Kornberg, RA Roth. Inorganic polyphosphate stimulates mammalian TOR, a kinase involved in the proliferation of mammary cancer cells. *Proc Natl Acad Sci USA* 100: 11249–11254, 2003.

52. X Tang, L Wang, CG Proud, CP Downes. Muscarinic receptor-mediated activation of p70 S6 kinase 1 (S6K1) in 1321N1 astrocytoma cells: permissive role of phosphoinositide 3-kinase. *Biochem J* 374: 137–143, 2003.

53. CJ Lynch, HL Fox, TC Vary, LS Jefferson, SR Kimball. Regulation of amino acid-sensitive TOR signaling by leucine analogues in adipocytes. *J Cell Biochem* 77: 234–251, 2000.

54. K Shigemitsu, Y Tsujishita, H Miyake, S Hidayat, N Tanaka, K Hara, K Yonezawa. Structural requirement of leucine for activation of p70 S6 kinase. *FEBS Lett* 447: 303–306, 1999.

55. U Krause, L Bertrand, L Maisin, M Rosa, L Hue. Signalling pathways and combinatory effects of insulin and amino acids in isolated rat hepatocytes. *Eur J Biochem* 269: 3742–3750, 2002.

56. G Miotto, R Venerando, O Marin, N Siliprandi, GE Mortimore. Inhibition of macroautophagy and proteolysis in the isolated rat hepatocyte by a nontransportable derivative of the multiple antigen peptide Leu8-Lys4-Lys2-Lys-beta Ala. *J Biol Chem* 269: 25348–25353, 1994.

57. GE Mortimore, JJ Wert, Jr, G Miotto, R Venerando, M Kadowaki. Leucine-specific binding of photoreactive Leu7-MAP to a high molecular weight protein on the plasma membrane of the isolated rat hepatocyte. *Biochem Biophys Res Commun* 203: 200–208, 1994.

58. J Bohe, A Low, RR Wolfe, MJ Rennie. Human muscle protein synthesis is modulated by extracellular, not intramuscular amino acid availability: a dose-response study. *J Physiol* 552: 315–324, 2003.

59. H Forsberg, PO Ljungdahl. Genetic and biochemical analysis of the yeast plasma membrane Ssy1p-Ptr3p-Ssy5p sensor of extracellular amino acids. *Mol Cell Biol* 21: 814–826, 2001.

60. GR Christie, E Hajduch, HS Hundal, CG Proud, PM Taylor. Intracellular sensing of amino acids in Xenopus laevis oocytes stimulates p70 S6 kinase in a target of rapamycin-dependent manner. *J Biol Chem* 277: 9952–9957, 2002.

61. A Beugnet, AR Tee, PM Taylor, CG Proud. Regulation of targets of mTOR (mammalian target of rapamycin) signalling by intracellular amino acid availability. *Biochem J* 372: 555–566, 2003.

62. E Jacinto, MN Hall. Tor signalling in bugs, brain and brawn. *Nat Rev Mol Cell Biol* 4: 117–126, 2003.

63. DH Kim, DD Sarbassov, SM Ali, RR Latek, KV Guntur, H Erdjument-Bromage, P Tempst, DM Sabatini. GbetaL, a positive regulator of the rapamycin-sensitive pathway required for the nutrient-sensitive interaction between raptor and mTOR. *Mol Cell* 11: 895–904, 2003.

64. AR Tee, BD Manning, PP Roux, LC Cantley, J Blenis. Tuberous sclerosis complex gene products, Tuberin and Hamartin, control mTOR signaling by acting as a GTPase-activating protein complex toward Rheb. *Curr Biol* 13: 1259–1268, 2003.

65. AC Gingras, B Raught, N Sonenberg. Regulation of translation initiation by FRAP/mTOR. *Genes Dev* 15: 807–826, 2001.

66. MM Garner, MB Burg. Macromolecular crowding and confinement in cells exposed to hypertonicity. *Am J Physiol* 266: C877–C892, 1994.

67. Y Fang, M Vilella-Bach, R Bachmann, A Flanigan, J Chen. Phosphatidic acid-mediated mitogenic activation of mTOR signaling. *Science* 294: 1942–1945, 2001.

68. A Kuroda, K Nomura, R Ohtomo, J Kato, T Ikeda, N Takiguchi, H Ohtake, A Kornberg. Role of inorganic polyphosphate in promoting ribosomal protein degradation by the Lon protease in *E. coli*. *Science* 293: 705–708, 2001.

69. F Martin, J Pintor, JM Rovira, C Ripoll, MT Miras-Portugal, B Soria. Intracellular diadenosine polyphosphates: a novel second messenger in stimulus-secretion coupling. *FASEB J* 12: 1499–1506, 1998.

70. G Xu, G Kwon, WS Cruz, CA Marshall, ML McDaniel. Metabolic regulation by leucine of translation initiation through the mTOR-signaling pathway by pancreatic beta-cells. *Diabetes* 50: 353–360, 2001.

71. BN Desai, BR Myers, SL Schreiber. FKBP12-rapamycin-associated protein associates with mitochondria and senses osmotic stress via mitochondrial dysfunction. *Proc Natl Acad Sci USA* 99: 4319–4324, 2002.

72. J Weekes, SA Hawley, J Corton, D Shugar, DG Hardie. Activation of rat liver AMP-activated protein kinase by kinase kinase in a purified, reconstituted system. Effects of AMP and AMP analogues. *Eur J Biochem* 219: 751–757, 1994.

73. I Holen, PB Gordon, PO Seglen. Protein kinase-dependent effects of okadaic acid on hepatocytic autophagy and cytoskeletal integrity. *Biochem J* 284: 633–636, 1992.

74. Y Kamada, T Funakoshi, T Shintani, K Nagano, M Ohsumi, Y Ohsumi. Tor-mediated induction of autophagy via an Apg1 protein kinase complex. *J Cell Biol* 150: 1507–1513, 2000.

75. H Abeliovich. Regulation of autophagy by the target of rapamycin (Tor) proteins. In: DJ Klionsky, Ed. *Autophagy.* Georgetown, TX: Eurekah.com/Landes Bioscience, 2003, pp 60–69.

76. H Abeliovich, WA Dunn, Jr, J Kim, DJ Klionsky. Dissection of autophagosome biogenesis into distinct nucleation and expansion steps. *J Cell Biol* 151: 1025–1034, 2000.

77. K Saeki, Z Hong, M Nakatsu, T Yoshimori, Y Kabeya, A Yamamoto, Y Kaburagi, A You. Insulin-dependent signaling regulates azurophil granule-selective macroautophagy in human myeloblastic cells. *J Leukoc Biol*, 74: 1108–1116, 2003.

78. TP Neufeld. Role of autophagy in develomental cell growth and death: insights from Drosophila. In: DJ Klionsky, Ed. *Autophagy.* Georgetown, Texas, USA: Eurekah.com/Landes Bioscience, 2003, pp 227–235.

79. J Dong, H Qiu, M Garcia-Barrio, J Anderson, AG Hinnebusch. Uncharged tRNA activates GCN2 by displacing the protein kinase moiety from a bipartite tRNA-binding domain. *Mol Cell* 6: 269–279, 2000.

80. K Natarajan, MR Meyer, BM Jackson, D Slade, C Roberts, AG Hinnebusch, MJ Marton. Transcriptional profiling shows that Gcn4p is a master regulator of gene expression during amino acid starvation in yeast. *Mol Cell Biol* 21: 4347–4368, 2001.

81. VA Cherkasova, AG Hinnebusch. Translational control by TOR and TAP42 through dephosphorylation of eIF2alpha kinase GCN2. *Genes Dev* 17: 859–872, 2003.

82. Z Talloczy, W Jiang, HW Virgin, DA Leib, D Scheuner, RJ Kaufman, EL Eskelinen, B Levine. Regulation of starvation- and virus-induced autophagy by the eIF2alpha kinase signaling pathway. *Proc Natl Acad Sci USA* 99:190–195, 2002.

83. PB Dennis, A Jaeschke, M Saitoh, B Fowler, SC Kozma, G Thomas. Mammalian TOR: a homeostatic ATP sensor. *Science* 294: 1102–1105, 2001.

84. E Ogier-Denis, JJ Houri, C Bauvy, P Codogno. Guanine nucleotide exchange on heterotrimeric Gi3 protein controls autophagic sequestration in HT-29 cells. *J Biol Chem* 271: 28593–28600, 1996.

85. E Ogier-Denis, A Petiot, C Bauvy, P Codogno. Control of the expression and activity of the Galpha-interacting protein (GAIP) in human intestinal cells. *J Biol Chem* 272: 24599–24603, 1997.

86. E Ogier-Denis, S Pattingre, J El Benna, P Codogno. Erk1/2-dependent phosphorylation of Galpha-interacting protein stimulates its GTPase accelerating activity and autophagy in human colon cancer cells. *J Biol Chem* 275: 39090–39095, 2000.

87. S Pattingre, C Bauvy, P Codogno. Amino acids interfere with the ERK1/2-dependent control of macroautophagy by controlling the activation of Raf-1 in human colon cancer HT-29 cells. *J Biol Chem* 278: 16667–16674, 2003.

88. PB Gordon, I Holen, M Fosse, JS Rotnes, PO Seglen. Dependence of hepatocytic autophagy on intracellularly sequestered calcium. *J Biol Chem* 268: 26107–26112, 1993.

89. E Ogier-Denis, P Codogno. Autophagy: a barrier or an adaptive response to cancer. *Biochim Biophys Acta* 1603: 113–128, 2003.

90. EFC Blommaart. Regulation of hepatic autophagy by amino acid dependent signal transduction. PhD thesis, University of Amsterdam, 1997.

91. D Häussinger, F Schliess, F Dombrowski, S vom Dahl. Involvement of p38MAPK in the regulation of proteolysis by liver cell hydration. *Gastroenterology* 116: 921–935, 1999.

92. S vom Dahl, F Schliess, R Reissmann, B Gorg, O Weiergraber, M Kocalkova, F Dombrowski, D Häussinger. Involvement of integrins in osmosensing and signaling toward autophagic proteolysis in rat liver. *J Biol Chem* 278: 27088–27095, 2003.

93. U Krause, L Bertrand, L Hue. Control of p70 ribosomal protein S6 kinase and acetyl-CoA carboxylase by AMP-activated protein kinase and protein phosphatases in isolated hepatocytes. *Eur J Biochem* 269: 3751–3759, 2002.

94. PF Dubbelhuis, AJ Meijer. Hepatic amino acid-dependent signaling is under the control of AMP-dependent protein kinase. *FEBS Lett* 521: 39–42, 2002.

95. DR Bolster, SJ Crozier, SR Kimball, LS Jefferson. AMP-activated protein kinase suppresses protein synthesis in rat skeletal muscle through down-regulated mammalian target of rapamycin (mTOR) signaling. *J Biol Chem* 277: 23977–23980, 2002.

96. AK Larsen, MT Møller, H Blankson, HR Samari, L Holden, PO Seglen. Naringin-sensitive phosphorylation of plectin, a cytoskeletal cross-linking protein, in isolated rat hepatocytes. *J Biol Chem* 277: 34826–34835, 2002.

97. N Kimura, C Tokunaga, S Dalal, C Richardson, K Yoshino, K Hara, BE Kemp, LA Witters, O Mimura, K Yonezawa. A possible linkage between AMP-activated protein kinase (AMPK) and mammalian target of rapamycin (mTOR) signalling pathway. *Genes Cells* 8: 65–79, 2003.

98. WW Winder, DG Hardie. AMP-activated protein kinase, a metabolic master switch: possible roles in type 2 diabetes. *Am J Physiol* 277: E1–E10, 1999.

99. Z Wang, WA Wilson, MA Fujino, PJ Roach. Antagonistic controls of autophagy and glycogen accumulation by Snf1p, the yeast homolog of AMP-activated protein kinase, and the cyclin-dependent kinase Pho85p. *Mol Cell Biol* 21: 5742–5752, 2001.

100. Y Liu, M Gorospe, GC Kokkonen, MO Boluyt, A Younes, YD Mock, X Wang, GS Roth, NJ Holbrook. Impairments in both p70 S6 kinase and extracellular signal-regulated kinase signaling pathways contribute to the decline in proliferative capacity of aged hepatocytes. *Exp Cell Res* 240: 40–48, 1998.

101. D Dardevet, C Sornet, M Balage, J Grizard. Stimulation of *in vitro* rat muscle protein synthesis by leucine decreases with age. *J Nutr* 130: 2630–2635, 2000.

102. K Merker, A Stolzing, T Grune. Proteolysis, caloric restriction and aging. *Mech Ageing Dev* 122: 595–615, 2001.

103. G Cavallini, A Donati, Z Gori, M Pollera, E Bergamini. The protection of rat liver autophagic proteolysis from the age-related decline co-varies with the duration of anti-ageing food restriction. *Exp Gerontol* 36: 497–506, 2001.

104. A Del Roso, S Vittorini, G Cavallini, A Donati, Z Gori, M Masini, M Pollera, E Bergamini. Ageing-related changes in the *in vivo* function of rat liver macroautophagy and proteolysis. *Exp Gerontol* 38: 519–527, 2003.

105. A Melendez, Z Talloczy, M Seaman, EL Eskelinen, DH Hall, B Levine. Autophagy genes are essential for dauer development and life-span extension in *C. elegans*. *Science* 301: 1387–1391, 2003.

106. M Bluher, BB Kahn, CR Kahn. Extended longevity in mice lacking the insulin receptor in adipose tissue. *Science* 299: 572–574, 2003.

107. M Hidalgo, EK Rowinsky. The rapamycin-sensitive signal transduction pathway as a target for cancer therapy. *Oncogene* 19: 6680–6686, 2000.

17

A Nutrient-Sensing Hexosamine Signaling Pathway

DONA C. LOVE AND JOHN A. HANOVER

O-GLCNAC ADDITION AND REMOVAL: A SIGNALING MODIFICATION

Recent evidence suggests that the addition and removal of a novel glycan modification, O-linked β-N-acetylglucosamine (O-GlcNAc), may function much like protein phosphorylation in critical intracellular signaling cascades. Discovered in the early 1980s, O-GlcNAc addition has been shown to be responsive to the level of nutrients and is a highly regulated and dynamic process. The enzymes of O-GlcNAc metabolism have been reviewed recently in a number of more detailed reports.[1,2] In this chapter, we will summarize evidence for a signaling cascade including the enzymatic synthesis of the precursor UDP-GlcNAc and the enzymes associated with O-GlcNAc addition and removal. We will document the high degree of conservation of this hexosamine signaling pathway in metazoan evolution and highlight the roles it may play in cellular signaling. These functions include cell proliferation, apoptosis, and cellular responses to stress and nutrient excess. In response to nutrient pools, O-GlcNAc modification of nuclear pores, transcription complexes, proteosomes, kinases, mitochondrial proteins, and the cytoskeleton appear to mediate these diverse functions.

EVIDENCE FOR A HEXOSAMINE SIGNALING PATHWAY

The pathway through which UDP-GlcNAc is synthesized from glucose is often termed the hexosamine biosynthetic pathway. A small percentage (1 to 5%) of the glucose entering cells is routed to this pathway. However, flux through the pathway is largely regulated by the levels of glucose and the rate-limiting enzymes GFAT (glutamine:fructose-6-phosphate amidotransferase) and Emeg32 (an acetyltransferase). An intriguing aspect of hexosamine biosynthesis is its apparent link to glucose-sensing pathways, in particular, that of insulin.[3–6] Overexpression of GFAT has been found to reproduce many of the physiological features of insulin resistance in vertebrates such as glucose transporter translocation and transcriptional effects.[7–13] Figure 17.1 summarizes the proposed link between the hexosamine biosynthetic pathway and the glycosylation of substrates by β-*N*-acetylglucosaminyl-transferase OGT. In this model, glucose, glutamine, and lipid-derived acetyl CoA are routed via the hexosamine biosyn-thetic pathway into the production of UDP-GlcNAc. This elevated UDP-GlcNAc level has two effects on the pathway. First, UDP-GlcNAc feedback inhibits GFAT and Emeg32 (the acetyltransferase), thus blunting synthesis, and second, it accelerates the glycosylation of proteins by OGT. The addition of O-GlcNAc to proteins would then be countered by the O-GlcNAcase in a manner that may itself be highly regulated. The range of concentrations over which OGT acts matches the cytosolic concentrations of sugar nucleotides (in the micro-molar range) in this model. One difficult aspect of this type of regulation is that the true cytoplasmic concentration of UDP-GlcNAc has not been accurately determined. Most sugar nucleotide is transported into the endomembrane system, where it is involved in membrane and secretory glycoprotein biosynthesis.[14,15] Another potential complication is the inhibi-tion of GFAT and Emeg32 by excess nucleotide, which would

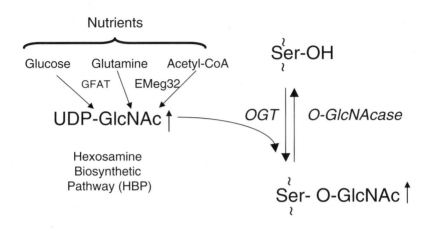

Figure 17.1 Nutrient-sensing, glycan-dependent signaling. UDP-GlcNAc is the terminal product of the hexosamine biosynthetic pathway. The three main building blocks of UDP-GlcNAc are glucose, glutamine, and Acetyl CoA. GFAT (glutamine:fructose-6-phosphate amidotransferase) and Emeg32 (an acetyltransferase) are rate-limiting enzymes in this pathway. UDP-GlcNAc is used by OGT to add O-linked GlcNAc to proteins at serine and threonine residues. The O-GlcNAcase is responsible for removal of this modification. An increase in nutrient availability leads to an increase in UDP-GlcNAc (arrow). This nutrient excess is transmitted throughout the cell as an increase in O-linked GlcNAc modified-proteins.

severely limit the range over which UDP-GlcNAc concentrations might be expected to accumulate. In spite of these caveats, a large body of evidence suggests that the hexosamine signaling pathway plays a key role in signal transduction.[7,9,11–14,16–27] To explore this glycan-dependent signaling, we will first examine the enzymes involved in the assembly and removal of O-GlcNAc.

O-LINKED GLCNAC TRANSFERASE: MOLECULAR GENETICS

Much recent work in our laboratory has focused on defining the gene encoding the O-linked GlcNAc transferase in mammalians and the genetically amenable *Caenorhabditis elegans* nematode system. We will describe the genes encoding this enzyme and attempts to decipher the multiple transcripts arising from this genetic locus.

Purification, Characterization, and Molecular Identification of O-GlcNAc Transferase (OGT)

OGT activity was originally detected in a wide range of tissues including rat liver, rabbit blood, and an enriched reticulocyte fraction.[28–30] The subcellular localization of the enzyme suggested that a substantial amount of the activity was soluble but that membrane-bound forms might also exist.[28–30] The strategies used to identify the activity initially relied heavily upon the identification of protein substrates such as the nuclear pore proteins that provided peptide substrates or could be produced in recombinant form.[28–32] Biochemical purification of the enzyme activity from these sources revealed a catalytic component of approximately 100 kDa. Microsequencing of the protein led rapidly to the molecular cloning of the transferase from a number of sources including human, rat, and *Caenorhabditis elegans*.[33,34]

Transcripts Encoding OGT

O-linked GlcNAc transferase cDNAs and the genes encoding those transcripts have been isolated from human, rat, and *C. elegans*.[33,34] We will focus on the mammalian gene, including the human, mouse, and rat OGT locus. In humans, four predominant transcripts encoding OGT, ranging from 10 Kb to 3 Kb, were found in most tissues.[34] Transcripts were particularly enriched in the pancreas.[34] Even more striking was the high concentration of OGT transcripts in the beta cells of the islets of Langerhans,[14,35] a key regulator of glucose homeostasis.[14,35] Other tissues, such as skeletal muscle, brain, fat, and cardiac muscle, also had high OGT mRNA levels. We observed fewer transcripts in the kidney.[33,34] Some confusion may have arisen from several early reports using different probes to analyze transcript distribution. We discuss this in the context of OGT isoforms derived from the OGT gene.

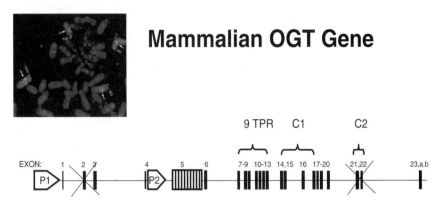

Mammalian OGT Gene

- Located on the human and murine X chromosome
- Contains 23 exons spanning over 45kb
- Alternative splicing and promoter use
- Conserved from *C. elegans* to mammals
- Knockouts are stem cell lethal
- Candidate for X-linked Parkinson-Dystonia

Figure 17.2 Genetic organization of mammalian OGT. The mammalian gene is composed of 23 exons. The two promoters are shown as P1 and P2. The longer ncOGT is generated by P1, whereas the shorter isoform, mOGT, is generated by P2 and alternative use of exon 5 (intron 4). The large Xs indicate exons that were disrupted in gene ablation studies.

The OGT Gene: Overall Structure and Isoforms

The gene encoding OGT is present as a single copy on syntenic regions of the mouse, rat, and human X chromosome (Figure 17.2). Positional cloning efforts have yet to positively identify any human diseases precisely mapping to this region, although the OGT locus remains a strong candidate gene for X-linked Parkinson's dystonia. Upon ablation of regions of the OGT gene in embryos, it has been shown to be essential for stem cell viability[36] and for embryonic development.[37] Conditional knockout experiments involving ablation of the gene in the various target tissues are now underway. Much remains to be learned about the processing and expression of the OGT gene.

Isoforms of OGT in Mammals

We have compared the mouse and human OGT genes with the partially completed rat OGT locus. Analysis of over 50 of our human cDNA clones and other expressed sequence tags (ESTs) suggest that the mammalian OGT gene encodes a number of differentially localized splice variants (Figure 17.2). The first major variant, mOGT, contains an N-terminal mitochondrial targeting sequence, a nuclear localization motif, and nine tetratricopeptide repeats (TPR). The second major variant, ncOGT, lacks the mitochondrial targeting motif but contains 3 additional TPRs (a total of 12). A third variant has a small 11-amino-acid deletion in the amino-terminal segment but is otherwise identical to ncOGT. A fourth, smaller variant containing two TPRs results from a transcript in which the amino-terminal AUGs are thrown out of frame, and an internal methionine is used for translation initiation. As shown in Figure 17.3 and Figure 17.4, we have confirmed that the two larger species represent mitochondrial and nuclear/ cytoplasmic variants of OGT. The shorter form of OGT (Figure 17.5) is also expressed in many tissues and is likely to play key regulatory roles in O-GlcNAc-dependent signaling (see Domain Structure and Posttranslational Modifications below).

DOMAIN STRUCTURE AND POSTTRANSLATIONAL MODIFICATIONS

The coding sequences of OGT isoforms suggest a tetrapartite structure consisting of an N-terminal targeting domain, a repetitive TPR domain, a linker/nuclear targeting domain, and a catalytic domain[33,34] (Figure 17.6). This overall picture of the structure arose as a result of site-directed mutagenesis and expression in *E. coli* bacculovirus, and mammalian cells.[38,39] We will summarize this evidence in the following subsections.

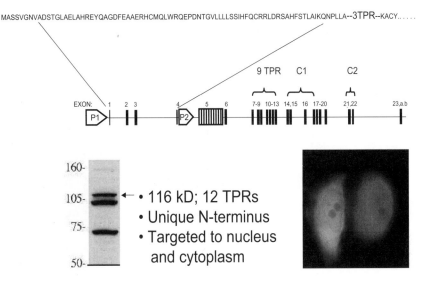

MASSVGNVADSTGLAELAHREYQAGDFEAAERHCMQLWRQEPDNTGVLLLLSSIHFQCRRLDRSAHFSTLAIKQNPLLA--3TPR--KACY......

- 116 kD; 12 TPRs
- Unique N-terminus
- Targeted to nucleus and cytoplasm

Figure 17.3 Genomic organization and intracellular localization of ncOGT. ncOGT is derived from exons 1 to 23 and utilizes promoter P1. The unique amino terminus is encoded by exons 1 to 4. Starting with the amino acid sequence KACY, ncOGT and mOGT are identical. A Western blot showing all three isoforms is shown on the left. ncOGT is 116 kDa and contains 12 TPR domains. Indirect immunofluorescence was used to demonstrate the nucleocytoplasmic localization of ncOGT.

OGT: N-Terminal Sorting Sequences

The isoforms of OGT differ dramatically at their amino terminus. For ncOGT, the amino terminus is quite long and may not contain sorting information. For mOGT, however, the shorter N-terminal has a somewhat unusual mitochondrial signal observed in some plant proteins, followed by a presumptive transmembrane helix. Our analysis suggests that proteolytic cleavage of portions of this amino terminus occurs upon import of mOGT into mitochondria.[40] The sOGT isoform has an even shorter N-terminus that almost immediately forms three TPR repeats. The subcellular localization of this isoform is unknown.

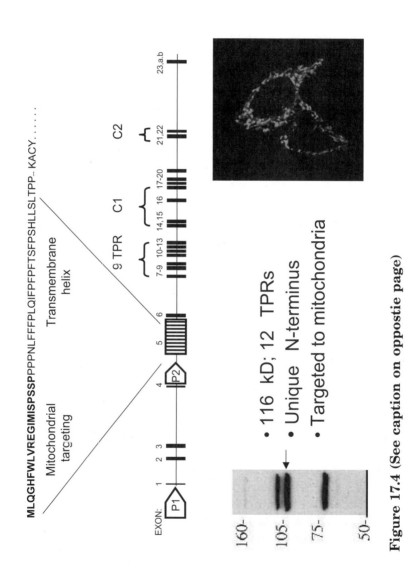

MLQGHFWLVREGIMISPSSPPPPNLFFFPLQIFPFFTSFPSHLLSLTPP- KACY. . . . .

Mitochondrial targeting

Transmembrane helix

EXON:

P1 1 2 3 4 P2 5 6 7-9 9 TPR 10-13 14,15 16 C1 17-20 21,22 C2 23,a,b

160-
105-
75-
50-

• 116 kD; 12 TPRs
• Unique N-terminus
• Targeted to mitochondria

Figure 17.4 (See caption on oppostie page)

OGT: TPR

The TPR repeats are present in 9 to 12 copies in the known O-linked GlcNAc transferases[33,34,38,41] (Figure 17.6). The TPR repeat is found in a large number of proteins and is widely thought to mediate protein–protein interactions. The TPR repeats of OGT may play a role in substrate recognition and multimerization of the enzyme.[38,39,41] In general, these repeats act in groups of three and may form a groove to accommodate peptides.[41] Recently, the x-ray structure of a TPR repeat protein has been reported, which suggests that the TPRs form a superhelix with a diameter of roughly 50 Å.[42] Other studies suggest that certain TPR proteins may serve to bridge other proteins such as those of the heat-shock proteins in the multichaperone machine.[43] We have recently obtained crystals of the TPR domain of OGT, which should prove informative about its role in enzyme assembly and target recognition.

OGT: Linker/Nuclear Targeting Domain

Following the TPR repeats in the central region of *C. elegans* OGT is a sequence that appears to be a canonic nuclear targeting signal (Figure 17.6). Overexpression of the enzyme in *C. elegans* confirmed that it could enter the nucleus when expressed in the nematode.[34] This nuclear transport motif is somewhat interrupted in human and rat OGT, and precise mutagenesis studies have not yet been performed to demonstrate whether this domain is important for the nuclear targeting of OGT. The nuclear targeting of OGT is likely to be complex because simple overexpression of the enzyme often

---

Figure 17.4 (Opposite page) Genomic organization and intracellular localization of mOGT. mOGT is derived from exons 5 to 23 and utilizes promoter P2. The unique amino terminus is encoded by alternate use of exon 5 (intron 4). Starting with the amino acid sequence KACY, ncOGT and mOGT are identical. A Western blot showing all three isoforms is shown on the left. mOGT is 103 kDa and contains 9 TPR domains. Indirect immunofluorescence was used to demonstrate the mitochondrial localization of mOGT.

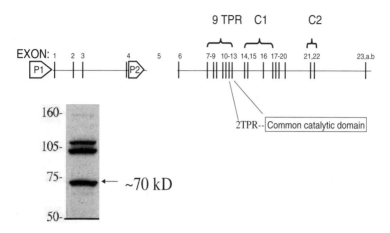

Figure 17.5 Genomic organization sOGT. sOGT is derived from an internal AUG and contains only two TPRs followed by the common catalytic domain. A Western blot showing all three isoforms is shown on the left. sOGT is 70 kDa.

leads to cytoplasmic, rather than nuclear localization. Clearly, some of the enzyme is present in the nucleus, and many substrates are nuclear. However, the nuclear trafficking of the enzyme has yet to be fully defined.

OGT: The Catalytic Domain

We have previously suggested that the catalytic domain of OGT bears some resemblance to known glycosyltransferases and certain lectins.[41] This similarity is not striking at the primary sequence level, but a number of motifs present in OGT are also present in these other enzymes (Figure 17.6). These regions of similarity are present in two subdomains designated catalytic domains 1 and 2. Recently, the crystal structure of an *N*-acetylglucosaminyltransferase I has been solved,[44] and the critical contact residues appear to be conserved between this enzyme and OGT (such as the DXD motif). Mutational analysis suggests that any substantial deletion in this region of the OGT molecule dramatically inhibits the activity of OGT.[33,34,38,39,41] Further mutagenesis and structural studies are underway.

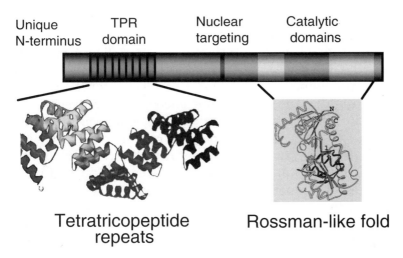

Unique N-terminus	TPR domain	Nuclear targeting	Catalytic domains

Tetratricopeptide repeats   Rossman-like fold

Figure 17.6 The domain structure of the O-linked GlcNAc transferase. Each isoform has a unique amino terminus. The number of TPRs varies between isoforms from 2 to 12. This domain is important for substrate recognition. A crystal structure of a TPR domain is shown below. The central region contains a nuclear localization sequence. The carboxy-terminal domain contains the catalytic region. Based on similarities to other glycosyltransferases, this catalytic domain is thought to contain a Rossman-like fold (crystal structure is shown below) that is important for catalysis.

Posttranslational Modifications of OGT

Although it is the product of a single gene, OGT has been shown to be modified posttranslationally in a number of intriguing ways.[33,34,38,39,41] These include autoglycosylation with O-GlcNAc and tyrosine phosphorylation. The effects of these modifications on enzyme activity or substrate recognition may provide means of regulation of the enzyme and crosstalk with other signaling pathways.

Protein Complexes Containing OGT

Using direct binding or yeast two-hybrid approaches, a number of proteins have been identified as being part of an OGT complex. The major nucleoporin p62 was shown to tightly bind to OGT in our studies, suggesting that it might be part of a larger complex.[30,32,45] In addition, other reports have suggested the interaction with other proteins.[46] These proteins included the GABA(A) receptor-associated protein GRIF-1 and its novel homologue OIP106 (for OGT-interacting protein), two highly similar proteins encoded by two separate genes. Both GRIF-1 and OIP106 contain coiled-coil domains and interact with the tetratricopeptide repeats of OGT. GRIF-1 and OIP106 were shown to be substrates for OGT. However, unlike p62, OIP106 and GRIF-1 coimmunoprecipitated with OGT, suggesting a more stable association. It was suggested that OIP106 forms an RNA polymerase II–OIP106–OGT complex that may target OGT to transcriptional complexes for glycosylation of transcriptional proteins. GRIF-1 was suggested to serve a function in GABA signaling cascades. Whether these molecules play regulatory roles or merely represent high-affinity interactions with substrates is yet to be determined. It is likely that OGT will be found in larger complexes, perhaps with other enzymes such as O-GlcNAcase, phosphatases, or kinases.

Another OGT complex has important implications for the role of O-GlcNAc in regulating transcription.[47] Transcription factors and RNA polymerase II are modified by O-GlcNAc (see the following sections). Interestingly, OGT interacts with a histone deacetylase complex by binding to the corepressor mSin3A. Functionally, OGT and mSin3A cooperatively repress transcription in parallel with histone deacetylation. The authors proposed that mSin3A targets OGT to promoters to inactivate transcription factors and RNA polymerase II by O-GlcNAc modification, which acts with histone deacetylation to promote gene silencing. The large number of interactions that are made possible by the TPR repeats of OGT may allow recruitment of other complexes, which would render them responsive to nutrient levels via the hexosamine signaling pathway.

O-GlcNAcase: MOLECULAR GENETICS

The genes encoding the enzyme responsible for O-GlcNAc removal (O-GlcNAcases) have recently been identified in nearly every metazoan organism. The structure of the gene and encoded protein suggests multiple sites for potential regulation as well as an important role in the glycan signaling pathway.

O-GlcNAcase: Purification and Identification

A number of recent studies have been directed at purifying and identifying the enzyme (or enzymes) normally responsible for O-GlcNAc removal.[14,48,49] This identification was complicated by the fact that hexosaminidases exist abundantly in the lysosomes of cells and are released upon cell lysis. One candidate enzyme, which removed O-GlcNAc, was purified from the cytosolic fraction of rat spleen. This preparation was shown to consist of a 51- and 54-kDa subunit.[48] More recently, a neutral hexosaminidase was identified in a number of tissues.[14,49,50] This enzyme was thought to be a hyaluronidase associated with meningioma[1,49,51] but has been subsequently shown to be the authentic O-GlcNAcase.[1,49,51] The gene encoding O-GlcNAcase has been defined, and is depicted in Figure 17.7.

The O-GlcNAcase Gene and Alternatively-Spliced Isoforms

The gene encoding O-GlcNAcase spans some 34 Kb on human chromosome 10 (Figure 17.7). It is termed the MGEA5 gene because it encodes a menigioma-expressed antigen. A previous analysis of the tissue distribution of this transcript suggests that it is highly evolutionarily conserved and enriched in brain, skeletal muscle, and pancreas.[51] This is similar to the known distribution of OGT transcripts.[33,34] O-GlcNAcase (MGEA5) was detected on a number of menigioma-derived cell lines and was overexpressed in these lines. In man, the gene is a candidate for an Alzheimer's susceptiblity locus and for noninsulin-dependent diabetes mellitus (NIDDM). A recent genetic examination of a Pima Indian population with a high incidence of NIDDM suggested that MGEA5 (O-GlcNAcase) was not responsible for NIDDM,[52] but this conclusion has been subsequently challenged.[53] The MGEA5 gene encodes at least two isoforms arising from alternative splicing, which were designated MGEA5 and MGEA5s. In this chapter, we will

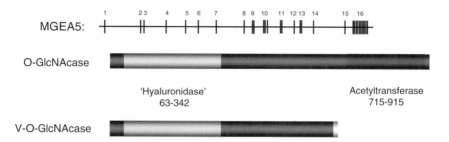

- Formerly known as Meningioma Expressed Antigen
- 34 kb on chromosome 10
- Candidate gene for NIDDM and Alzheimer disease susceptibility
- Two isoforms produced by alternative splicing

Figure 17.7 Genetic and protein structure of the O-GlcNAcase. The O-GlcNAcase is encoded by 16 exons, covering 34 kb on chromosome 10. A region at the amino terminus contains homology to other hyaluronidases (yellow), whereas the carboxy-terminus has limited homology to a Pfam acetyltransferase domain (blue). This domain is missing in the variant and is replaced by a unique sequence generated by ignoring a 5 splice site in exon 10, resulting in utilization of exon and intron 10.

refer to these as the O-GlcNAcase (OGA) and the variant O-GlcNAcase (vOGA) (see Figure 17.7).

O-GlcNAcase: DOMAIN STRUCTURE

The domain structures of the O-GlcNAcases in man are outlined in Figure 17.7. The 103-kDa OGA protein has an acidic isoelectric point.[49,51] Analysis of the protein suggests that O-GlcNAcase contains a region with similarity to the BglG antiterminator motif from *E. coli*. This RNA-binding protein prevents transcription termination of the bgl operon that is regulated by β-glucosides. The biological significance of this similarity is not yet known. The O-GlcNAcase also shows limited homology (near its amino terminus) to *Clostridium perfringens* mu toxin and to a lacto-*N*-biosidase precursor from streptomyces. These sequence similarities probably reflect the position of catalytically important residues in the molecule. Both the spliced isoforms OGA and vOGA have the same amino terminus. This observation, coupled with the sequence similarities mentioned above, allows us to tentatively assign a region corresponding to 63–283 in the amino terminus as the catalytic domain of O-GlcNAcase[1] (Figure 17.7). OGA has a partial pfam:acetyltransferase domain similar to known acetyltransferases near its C-terminus. The two isoforms OGA and vOGA differ substantially at this carboxyl terminal. The vOGA is approximately 80 kDa in molecular mass and has a shorter divergent C-terminus lacking the acetyltransferase domain. This unique C-terminus of vOGA may be involved in altering targeting of molecular and substrate recognition. Another intriguing finding is that the OGA variants contain conserved executioner caspase-3 sites. Cleavage at these sites does not lead to inactivation of the enzymes and may suggest a means of disregulating O-GlcNAcase during the induction of programmed cell death (see The Substrates Modified by O-GlcNAc and O-GlcNAcase below). Our recent analysis suggests that both the OGA and vOGA isoforms may be catalytically active under certain conditions (Lee, Love, Hanover, submitted). As suggested above for OGT, O-GlcNAcase may be part of a much larger complex containing the enzymes of O-GlcNAc addition and removal and perhaps also containing phosphatases and kinases.

THE SUBSTRATES MODIFIED BY O-GlcNAc AND O-GlcNAcase

Figure 17.8 and Figure 17.9 summarize our current efforts to identify the O-GlcNAc proteome. As shown in Figure 17.8, we are using a chemical method to define the products of hexosamine signaling, which will be described in "Efforts to Decipher the O-GlcNAc Proteome." Figure 17.9 shows the varieties of substrates modified by the enzymes of O-GlcNAc metabolism defined by this and other approaches. In general, these O-GlcNAc-modified proteins are part of macromolecular

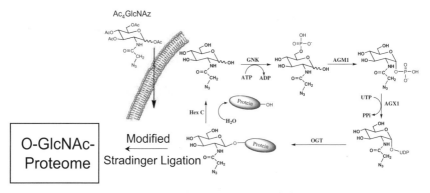

GNK, GlcNAc kinase; AGM1, Phospho-N-acetylglucosamine mutase; AGX1, UDP-GlcNAc pyrophosphorylase

Figure 17.8   A chemical approach to the O-GlcNAc Proteome. The chemical biology approach utilizes the salvage pathway for the labeling of O-GlcNAc-modified proteins. The salvage pathway bypasses the *de novo* HBP through the action of GNK and generates GlcNAc-6-phosphate. This intermediate is converted to GlcNAc-1-phosphate through the action of AGM1. *N*-actelyglucosamine-1-phosphate is converted to the end product of the HBP, UDP-GlcNAc, through the action of AGX1. OGT transfers the disaccharide moiety of UDP-GlcNAc to protein substrates within the cell. Hex C (O-GlcNAcase) acts to cleave the glycosidic linkage of posttranslationally modified proteins to liberate the protein and GlcNAc. Exogenously added Ac4GlcNAz diffuses into the cell and is deacylated through the action of intracellular esterases and then enters into the salvage pathway.

complexes; most are phosphoproteins. Among the macromolecular complexes modified are nuclear pore complexes, transcription and chromatin-remodeling complexes, and cytoskeletal complexes. We have also found that many cytoplasmic kinases and other enzymes are substrates for O-GlcNAc metabolism. We will briefly describe the glycosylation of components of these complexes and how they may function in glycan-dependent signaling in the following subsections.

Nuclear Pore Glycoproteins

Many components of the mammalian nuclear pore complex have been shown to be glycosylated[1,32,45,54–57] (see Figure 17.9). Originally, this glycosylation was imagined to be involved in the

Nuclear Pore Complex:
Nucleoporins (Nup62)
Nuclear Transport

Transcription Complexes:

Transcription factors : Sp1, steroid receptors, pol II

Cytoskeletal and Neuronal:

AP3, tau, cytokeratins, ß-catenin, E-cadherin,
ß-amyloid precursor, synuclein

Figure 17.9 The intracellular targets of OGT. O-GlcNAc-modified proteins can be divided into three main groups: proteins of the nuclear pore complex, transcriptional complexes, and cytoskeletal and neuronal proteins. Pictorial representations of each group that is modified by OGT are shown on the left.

regulation of nuclear trafficking. However, we now believe that glycosylation plays a role in assembly of the nuclear pore or activities such as transcriptional repression or silencing that may be organized by the nuclear pore. We have proposed that glycosylation of nuclear pore proteins could be a mechanism for altering the association of transcription repression complexes with the nuclear pore. In this proposal the glycosylation of nuclear pore components to which OGT-mSin3A repression complexes were bound would lead to dissociation from the pore complex leading to repression. This is similar to proposals based on boundary-trap screens in yeast, which have identified nuclear pore proteins as involved in formation of boundaries and transcriptional repression.[58]

Transcription Complexes

Nearly all transcription factors examined to date are modified by O-GlcNAc (see Figure 17.9). In addition, the CTD domain of RNA polymerase II is both phosphorylated and glycosylated under certain conditions.[59] The glycosylation of RNA polymerase II could play a role in the regulation of transcriptional elongation, or it could regulate the ubiquitination and subsequent degradation of RNA polymerase II. One protein that has been extensively studied is Sp1.[60–63] It appears likely that O-GlcNAc modulates Sp1 turnover, transactivation, and protein–protein interactions. Another protein of great interest is the hepatocyte nuclear factor 1 (HNF1). This glycosylated homeodomain-containing transcription factor has important functions in liver, kidney, intestine, and pancreas.[64] HNF1 deletion has been implicated in maturity-onset diabetes of the young, regulating sterol metabolism and noninsulin diabetes mellitus.[65–67] The transcription factor PDX-1 is posttranslationally modified by O-GlcNAc, and this modification is correlated with its DNA-binding activity and insulin secretion.[68] The steroid hormone receptors also can be modified by O-GlcNAc. Estrogen receptor β is expressed in the ovary, prostate, lung, epididymis, and hypothalamus. Some recent experiments suggest that glycosylation and phosphorylation of estrogen receptor β could play a role in its stability.[69–71] Another example of the possible interplay between phosphorylation and glycosylation is exemplified by c-myc. This transcription factor is modified on threonine 58[72–74] within the transactivation domain. A number of studies have suggested that the blocking of phosphorylation on Thr 58 results in increased protein stability and changes in transactivation. As mentioned above, the nutrient-sensitive glycosylation of components of nuclear pore and transcription complexes may suggest a role in the reversible tethering of active chromatin to the nuclear envelope. Further work will be required to fully understand the implications of these important interactions for the control of gene expression.

Cytoskeletal and Neuronal Proteins

The glycosylation of neuronal and cytoskeletal proteins has become an intriguing problem because O-GlcNAc metabolism has been implicated in neurodegenerative diseases (refer to Figure 17.9). The cytoskeletal proteins cytokeratin 8 and 18 bear O-GlcNAc.[75] In the brain, proteins such as neurofilament H, synapsin I, ankrin, tau, and MAP 1, 2, and 4 are also modified by O-GlcNAc moieties (see Reference 76 and references therein).[76] Changes in glycosylation of these proteins may occur in human disease. For example, the adapter protein 3 (AP3) is glycosylated and shows a different pattern of glycosylation in Alzheimer's disease brains. The tau protein is glycosylated and phosphorylated, and its degree of hyperphosphorylation influences its ability to form the paired helical filaments found in neurofibrillary tangles characteristic of the tauopathies (such as Alzheimer's disease) in man. From the perspective of sensing mechanisms, it should be noted that unlike other organs, the brain uses glucose almost exclusively as a source of nutrients. In Alzheimer's disease, regional reductions in glucose transport and utilization are known to occur in the affected brain regions. In fact, starvation induces hyperphosphorylation of tau in mice, suggesting that O-GlcNAc might protect tau from hyperphosphorylation in normal nutrient-rich conditions.[77] In a later section, we will highlight the close association of OGT and O-GlcNAcase with human neurodegenerative diseases such as Alzheimer's and X-linked Parkinson's dystonia.

Kinases and Other Enzymes

At least two kinases, casein kinase II and glycogen synthase kinase 3 beta, were directly shown to be substrates for OGT[34] (See Figure 17.9). The effect of the modification has not yet been determined. However, in other studies, signaling events associated with kinases are clearly altered by O-GlcNAc addition.[78] Insulin resistance of glycogen synthase is thought to be mediated by O-GlcNAc. In addition the activity of eNOS is likely to be influenced by glycosylation.[19]

Efforts to Decipher the O-GlcNAc Proteome

A number of efforts are currently underway to determine which constellation of cellular proteins is modified by O-GlcNAc. Some of these studies will undoubtedly be based on antibodies directed against O-GlcNAc. Recently we have taken a different approach to the problem of identifying O-GlcNAc-modified proteins[79] (see Figure 17.8). This method takes advantage of the promiscuity of many of the enzymes involved in hexosamine-dependent signaling, including OGT and O-GlcNAcase. By incorporating an azido derivative of O-GlcNAc (termed O-GlcNAz), we were able to metabolically label O-GlcNAc-modified proteins *en masse*. A modified Straudinger ligation was then used to concentrate and identify the proteins. Currently, this method is being applied to the problem of the O-GlcNAc proteome and shows great promise in identifying new substrates.

EVOLUTIONARY CONSERVATION OF OGT AND O-GLCNACASE

Our identification of OGT from *C. elegans* and man, and the reported sequence of rat OGT, hinted at a high degree of sequence conservation.[33,34] These early cloning and characterization papers suggested that the OGT protein was somewhat divergent between mammalian species. This initial assumption proved misleading. Our laboratory had cloned OGT from a human library, whereas another laboratory had cloned rat OGT. When comparing these two sequences, the amino-terminal ends were quite different, whereas at the carboxyl end the catalytic domains were nearly identical. The divergent amino-terminal ends were taken as species differences. Further investigation of the human genome suggested that whereas there is only a single gene for OGT, there are at least four OGT isoforms.[36] The existence of multiple, alternatively spliced OGT isoforms adequately explains the differences in the protein sequences originally reported by our laboratory and that of the Hart laboratory.

Through comparisons of the growing number of genomic databases, it is now clear that OGT is very well conserved throughout the vertebrates. Human OGT shares greater than 85% identity in the catalytic domain with bovine, murine, pig, and rat OGT. OGT has not changed significantly through evolution, as it shares 74% identity with the sea squirt (an early chordate) and *C. elegans*, a nematode.

OGT is not only well conserved and essential in animals but also appears to be essential in the plant kingdom. OGT was first described in *Arabidopsis thaliana*.[80–82] It was found as a suppressor of a deficient gibberellin (GA) pathway. GAs are plant signaling molecules that regulate many facets of plant growth and development. The genetic locus was termed *spy* for "spindly." The name given to this locus describes the phenotypic appearance of plants containing this mutation. Additional genetic experiments suggest that SPY is a repressor of GA signaling.[80,81,83] Interestingly, not all of the spy phenotypes could be attributed to deregulation of the GA pathway, suggesting that SPY may play a role in other plant pathways.[84] A database search for TPR-containing proteins yielded SEC (secret agent), another putative plant OGT.[85] Although plants containing defective SEC have no obvious phenotype, the combination of sec and spy mutations was synthetically lethal. An interesting parallel, considering that the murine knockout of OGT (rendering all isoforms catalytically inactive) is embryonic lethal (unpublished data and Reference 37).[37] The authors conducted additional searches and found that SEC-like proteins can be found throughout the plant kingdom. An amino acid comparison of the catalytic domains of both SPY and SEC with each other and with the other vertebrate OGTs gave some surprising results.[85] SEC shares more sequence similarity with other animal OGTs (53 to 59%) than with SPY (35%), whereas the sequence similarity of SPY with SEC and with other OGTs is similar (35 to 38%). The two genes for OGT (spy and sec) can be found throughout angiosperms. In contrast, vertebrates appear to have only one gene for OGT; however, multiple isoforms of this gene containing different amino termini may substitute for the two genes found in plants.

THE ENZYMES OF THE HEXOSAMINE SIGNALING PATHWAY

Of the enzymes involved in the synthesis of UDP-GlcNAc, two appear to be key to regulation of the hexosamine signaling pathway (see Figure 17.1). We will discuss (in the following subsections) the evidence implicating glutamine fructose 6-phosphate amido transferase (GFAT) and glucosamine-6-phosphate (GlcN6P) acetyltransferase (Emeg32 or GAT) in regulation of this pathway. However, it is possible that other key enzymes such as phosphoN-acetylglucosamine mutase (AGM1) and UDP-GlcNAc pyrophosphorylase (AGX1) may also play unrecognized regulatory roles in this pathway.

GFAT

The early studies of Marshall and colleagues strongly implicated hexosamine synthesis in the cellular signaling involved in mammalian insulin resistance.[3,4,6,86–88] In these studies (see Figure 17.1), amino acids and glucose were found to modulate the desensitization of the glucose transport system in 3T3 L1 preadipocytes. Such desensitization is a hallmark of mammalian insulin resistance. Subsequent studies showed that inhibition of the rate-limiting step in hexosamine synthesis prevented this desensitization.[3–7] Changes in GFAT activities and levels are have been closely associated with the development of insulin resistance.[7,16–18,89–96] Taken together, these studies strongly suggested that GFAT might play a critical role in the regulation of hexosamine synthesis and in modulating cellular signaling. A question arose as to how it does that.

Glucosamine-6-Phosphate (GlcN6P) Acetyl Transferase (Emeg32 or GAT)

The source of the acetyl group of UDP-GlcNAc is acetyl CoA derived from fatty acid oxidation in mitochondria, mitochondrial degradation of some amino acids, mitochondrial pyruvate dehydrogenase complex, and cytoplasmic acetyl CoA synthetase.

The nutrient-derived pool of acetyl CoA is then used as a precursor by the enzyme GAT or Emeg32 for the acetylation of glucosamine-6-phosphate (see Figure 17.1). When cloned and isolated, Emeg32 was found to associate with the cytoplasmic side of intracellular membrane, and it copurified with the cdc48 homologue p97/Valosin-containing protein.[97] Emeg32 is also subject to product inhibition and, perhaps, allosteric control. A demonstration of the importance of this enzyme came when it was deleted in mice.[98] Homozygous mutant embryos died at embryonic day 7.5. *In vitro* differentiated EMeg32(/) ES cells showed reduced proliferation. Mouse embryonic fibroblasts (MEFs) deficient for EMeg32 exhibited defects in proliferation and adhesiveness, which were restored by stable reexpression of EMeg32 or by restoring intracellular UDP-GlcNAc levels nutritionally. UDP-GlcNAc levels were reduced to about 10 to 15% of control, and this resulted in decreased O-GlcNAc modifications of cytosolic and nuclear proteins. There were no dramatic changes in N- or O-GalNAc glycoproteins or in the synthesis of glycosylphosphatidylinositol (GPI) anchors. Interestingly, growth-impaired EMeg32(/) MEFs withstand a number of apoptotic stimuli and express activated PKB/AKT. EMeg32-dependent UDP-GlcNAc levels were therefore shown to influence cell cycle progression and susceptibility to apoptotic stimuli (see Hexosamine-Dependent Signaling: the Involvement of OGT in Apoptosis below). Taken together, these studies have demonstrated the importance of nutrient-derived sugar nucleotide in cellular signaling. However, the question still remains about how those signals are translated by the cellular machinery. Our hypothesis has been that the enzymes of O-GlcNAc metabolism mediate these effects.

HEXOSAMINE-DEPENDENT SIGNALING:
THE INVOLVEMENT OF OGT
IN INSULIN SIGNALING

Insulin resistance and beta cell toxicity are key features of noninsulin-dependent diabetes. One leading hypothesis suggests that these abnormalities are the result of an excessive flux of nutrients through the UDP-hexosamine biosynthetic pathway described above, ultimately leading to glucose toxicity. We have shown that transgenic overexpression of OGT driven by the GLUT4 promoter to target muscle and fat cells produces the type 2 diabetic phenotype in mice.[99] Even modest overexpression of an isoform of OGT in muscle and fat produced insulin resistance and hyperleptinemia in these mice. The OGT-induced insulin resistance showed sexual dimorphism; male mice were far more susceptible than the female transgenics. Insulin resistance was confirmed in these animals by a hyperinsulinemic-euglycemic clamp. Leptin mRNA levels and circulating levels of leptin were elevated in the transgenics up to three- or fourfold, yet body weight remained relatively constant, suggesting that OGT overexpression may lead to a deregulation of the normal satiety response in mice. These data provided the first direct evidence that O-linked GlcNAc transferase participates in a hexosamine-dependent signaling pathway that is linked to insulin resistance and leptin production.[99] Because of the magnitude of the changes resulting from just subtle manipulation of OGT levels, it is likely that the addition of O-GlcNAc is an essential aspect of cellular responses to nutrients.

HEXOSAMINE-DEPENDENT SIGNALING:
THE INVOLVEMENT OF O-GlcNAcase (OGA) IN
INSULIN SIGNALING

The levels of O-GlcNAc in cells appear to be regulated both by OGT, which increases the levels, and by O-GlcNAcase, which decreased them. Recently, a link has been established between inhibition of O-GlcNAcase and mammalian insulin resistance.[78] In this study, cycling of O-GlcNAc was blocked by the pharmacological inhibition of O-GlcNAcase with *O*-(2-acetamido-2-deoxy-*d*-glucopyranosylidene) amino-*N*-phenyl-carbamate (PUGNAc). PUGNAc treatment increased levels of O-GlcNAc in 3T3-L1 adipocytes. The treatment also induced insulin resistance. Insulin receptor autophosphorylation and insulin receptor substrate 2 tyrosine phosphorylation were not affected by the PUGNAc inhibition of O-GlcNAcase. However, downstream phosphorylation of Akt at Thr 308 and glycogen synthase kinase 3 beta at Ser 9 was inhibited. The insulin resistance was associated with increased O-GlcNAc modification of insulin receptor substrate 1 and beta-catenin, both of which have been implicated in insulin signaling. These results, taken together with our previous experiments with OGT,[99] argue that elevation of O-GlcNAc levels may serve to blunt insulin signaling and contribute to the development of insulin resistance in adipocytes and muscle.

HEXOSAMINE-DEPENDENT SIGNALING: THE INVOLVEMENT OF OGT IN APOPTOSIS

The finding that an isoform of OGT is targeted to the inner mitochondrial membrane[40] suggests that hexosamine-dependent signaling may play a role in apoptosis. The abundance of OGT in the pancreatic beta cell,[14] and the action of the O-GlcNAc mimetic streptozotocin to induce beta cell apoptosis, lead to early speculation that O-GlcNAc metabolism might play a role in beta cell destruction.[14,35] These proposals have proven to be controversial. Such beta cell destruction does indeed occur in the latter stages of NIDDM and has been referred to as "glucose toxicity." Moreover, cytoplasmic O-GlcNAc-modified proteins such as AKT, GSK-3, *β*-catenin, and E-cadherin are known to be involved in apoptosis. In addition, the hexosamine pathway clearly modulates apoptosis, as evidenced by the resistance to apoptosis exhibited by cells lacking Emeg32 with lowered UDP-GlcNAc pools.[98] More direct evidence for the involvement of OGT in apoptosis came with the finding that a short form of OGT, along with bcl-x, and ornitine decarboxylase emerged in a leukemia antiapoptotic screen. It is highly likely that OGT plays a critical role in apoptosis. Our ongoing studies are directed at understanding the role of the various OGT isoforms in the regulation of apoptosis in response to nutrients. This pathway may be particularly important for such human diseases as NIDDM and neurodegenerative diseases (see later subsection titled "The Hexosamine Signaling Pathway and Human Disease").

HEXOSAMINE-DEPENDENT SIGNALING: TRANSCRIPTION, TRANSLATION, STRESS RESPONSES, AND PROTEIN DEGRADATION

O-GlcNAc addition and hexosamine-dependent signaling may regulate a large number of other cellular processes. Several recent studies suggest the hexosamine signaling pathway may modulate the transcription of key regulatory genes and alter gene expression either directly or by altering the stability of the encoded proteins. The genes encoding TGF-α and leptin are regulated by hexosamines.[100,101] In addition, a number of heat-shock proteins that mediate cellular response to stress are modified by O-GlcNAc. The role of O-GlcNAc in modulating the function of eukaryotic peptide chain initiation factor 2 (eIF-2)-associated 67-kDa polypeptide (p67) in translation is well documented.[102] Exciting recent evidence also points to modulation of the proteasome by O-GlcNAc.[103–107] As the number of O-GlcNAc proteins expands, so, too, is the number of potential functions of O-GlcNAc likely to expand.

THE HEXOSAMINE SIGNALING PATHWAY AND HUMAN DISEASE

Both genetic and biochemical evidence exists linking hexosamine signaling through O-GlcNAc to human diseases. The evidence suggests that O-GlcNAc metabolism may contribute to diabetes mellitus and neurodegenerative diseases. We have discussed this topic elsewhere[1] but will summarize the findings here. The bulk of the evidence rests on the finding that increased flux through the hexosamine biosynthetic pathway is highly correlated with the development of insulin resistance (see Figure 17.1).[4,7,11–13,17,86,99,108–112] The association is further strengthened by the finding that the O-GlcNAcase maps very near a site on chromosome 10 that has been suggested to play a role in NIDDM.[52] The significance of this association is controversial, but may indicate that O-GlcNAcase is one component of a multifactorial disease.[52,53]

The large number of glycosylated proteins in the brain and the association of OGT with a known neurodegenerative disease (X-linked Parkinson's dystonia) are also intriguing. The O-GlcNAcase also maps near a site that has been suggested to play a role in Alzheimer's disease susceptibility on 10q24 (AD6). As mentioned above, glucose utilization in the brain is altered in Alzheimer patients, suggesting that hexosamine metabolism might also be perturbed. This exciting possibility is being tested in our laboratory using transgenic mouse technology and tissue-specific knockouts.

The altered glucose metabolism observed in tumor cells suggests a role for O-GlcNAc metabolism in cancer. Evidence suggests that the oncogene c-myc and the tumor suppressor p53 are modified by O-GlcNAc.[72,73,113] Tumor cells have dramatically altered glucose metabolism, switching from oxidative to glycolytic metabolism. This "Crabtree effect" has puzzled researchers for decades.[1] It is possible that, given the role of hexosamine-dependent signaling in apoptosis and proliferation, O-GlcNAc may be involved in the deregulation of glucose metabolism in cancer.

SUMMARY AND CONCLUSIONS

Figure 17.10 summarizes our current knowledge of the hexosamine signaling pathway described in this chapter. The hexosamine signaling pathway is a nutrient-sensing signaling pathway involving the generation of UDP-GlcNAc from nutrients derived by intracellular metabolism in the mitochondrion, lipid stores, and cytosol. The enzyme of O-GlcNAc addition uses UDP-GlcNAc. Similar to phosphorylation, O-GlcNAc addition is a dynamic modification occurring in the nucleus and cytosol. Differentially targeted isoforms of OGT then utilize this sugar nucleotide to modify intracellular

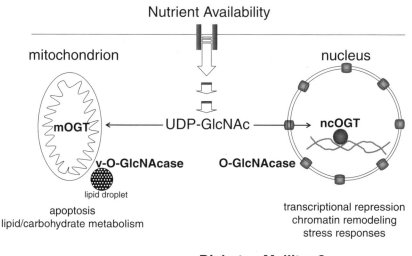

Figure 17.10 A Glycan-dependent signaling pathway. UDP-GlcNAc is generated from incoming nutrients by intracellular metabolism in the mitochondrion, lipid stores, and cytosol. Differentially targeted OGT isoforms (mOGT and ncOGT) then utilize this sugar nucleotide to modify intracellular targets. Differentially targeted forms of membrane-associated (v-O-GlcNAcase) and nucleocytoplasmic O-GlcNAcase render this a dynamic modification by removing the disaccharide moiety. O-GlcNAc metabolism appears to be intimately associated with apoptosis, and lipid or carbohydrate metabolism in the mitochondrion. In the nucleus, it is associated with transcriptional repression, chromatin remodeling, and stress responses. Deregulation of this nutrient-sensing pathway may play a role in the etiology of NIDDM and neurodegeneration.

targets. Differentially targeted isoforms of O-GlcNAcase remove O-GlcNAc, creating a dynamic means of modulating protein function. The OGT and O-GlcNAcases act in concert with kinases and phosphatases generating various isoforms of physiological substrates. O-GlcNAc metabolism appears to be intimately associated with apoptosis and lipid and carbohydrate metabolism in the mitochondrion and transcriptional repression complexes in the nucleus. In the future, we hope to determine if this signaling pathway plays a role in the pathologies associated with human cancer, diabetes mellitus, and neurodegeneration.

References

1. Hanover, J. A. 2001. Glycan-dependent signaling: O-linked N-acetylglucosamine. *FASEB J.* 15: 1865–1876.

2. Wells, L., and G. W. Hart. 2003. O-GlcNAc turns twenty: functional implications for posttranslational modification of nuclear and cytosolic proteins with a sugar. *FEBS Lett.* 546: 154–158.

3. Marshall, S., V. Bacote, and R. R. Traxinger. 1991. Complete inhibition of glucose-induced desensitization of the glucose transport system by inhibitors of mRNA synthesis. Evidence for rapid turnover of glutamine:fructose-6-phosphate amidotransferase. *J Biol Chem.* 266: 10155–10161.

4. Marshall, S., V. Bacote, and R. R. Traxinger. 1991. Discovery of a metabolic pathway mediating glucose-induced desensitization of the glucose transport system. Role of hexosamine biosynthesis in the induction of insulin resistance. *J Biol Chem.* 266: 4706–412.

5. Traxinger, R. R., and S. Marshall. 1992. Insulin regulation of pyruvate kinase activity in isolated adipocytes. Crucial role of glucose and the hexosamine biosynthesis pathway in the expression of insulin action. *J Biol Chem.* 267: 9718–923.

6. Traxinger, R. R., and S. Marshall. 1991. Coordinated regulation of glutamine:fructose-6-phosphate amidotransferase activity by insulin, glucose, and glutamine. Role of hexosamine biosynthesis in enzyme regulation. *J Biol Chem.* 266: 10148–1054.

7. Chen, H., B. L. Ing, K. A. Robinson, A. C. Feagin, M. G. Buse, and M. J. Quon. 1997. Effects of overexpression of glutamine:fructose-6-phosphate amidotransferase (GFAT) and glucosamine treatment on translocation of GLUT4 in rat adipose cells. *Mol Cell Endocrinol.* 135: 67–77.

8. Kolm-Litty, V., U. Sauer, A. Nerlich, R. Lehmann, and E. D. Schleicher. 1998. High glucose-induced transforming growth factor beta1 production is mediated by the hexosamine pathway in porcine glomerular mesangial cells. *J Clin Invest*. 101: 160–19.

9. Nerlich, A. G., U. Sauer, V. Kolm-Litty, E. Wagner, M. Koch, and E. D. Schleicher. 1998. Expression of glutamine:fructose-6-phosphate amidotransferase in human tissues: evidence for high variability and distinct regulation in diabetes. *Diabetes*. 47: 170–18.

10. Weigert, C., K. Brodbeck, R. Lehmann, H. U. Haring, and E. D. Schleicher. 2001. Overexpression of glutamine:fructose-6-phosphate-amidotransferase induces transforming growth factor-beta1 synthesis in NIH-3T3 fibroblasts. *FEBS Lett*. 488: 95–9.

11. Hawkins, M., N. Barzilai, W. Chen, I. Angelov, M. Hu, P. Cohen, and L. Rossetti. 1996. Increased hexosamine availability similarly impairs the action of insulin and IGF-1 on glucose disposal. *Diabetes*. 45: 1734–1743.

12. Hawkins, M., N. Barzilai, R. Liu, M. Hu, W. Chen, and L. Rossetti. 1997. Role of the glucosamine pathway in fat-induced insulin resistance. *J Clin Invest*. 99: 2173–2282.

13. Rossetti, L., M. Hawkins, W. Chen, J. Gindi, and N. Barzilai. 1995. *In vivo* glucosamine infusion induces insulin resistance in normoglycemic but not in hyperglycemic conscious rats. *J Clin Invest*. 96: 132–140.

14. Hanover, J. A., Z. Lai, G. Lee, W. A. Lubas, and S. M. Sato. 1999. Elevated O-linked N-acetylglucosamine metabolism in pancreatic beta-cells. *Arch Biochem Biophys*. 362: 38–45.

15. Hanover, J. A., and W. J. Lennarz. 1982. Transmembrane assembly of N-linked glycoproteins. Studies on the topology of saccharide synthesis. *J Biol Chem*. 257: 2787–2794.

16. Daniels, M. C., P. Kansal, T. M. Smith, A. J. Paterson, J. E. Kudlow, and D. A. McClain. 1993. Glucose regulation of transforming growth factor-alpha expression is mediated by products of the hexosamine biosynthesis pathway. *Mol Endocrinol*. 7: 1041–1048.

17. Buse, M. G., K. A. Robinson, T. W. Gettys,e.g.,McMahon, and E. A. Gulve. 1997. Increased activity of the hexosamine synthesis pathway in muscles of insulin-resistant ob/ob mice. *Am J Physiol.* 272: E1080–E1088.

18. Chang, Q., K. Su, J. R. Baker, X. Yang, A. J. Paterson, and J. E. Kudlow. 2000. Phosphorylation of human glutamine:fructose-6-phosphate amidotransferase by cAMP-dependent protein kinase at serine 205 blocks the enzyme activity. *J Biol Chem.* 275: 21981–21987.

19. Du, X. L., D. Edelstein, S. Dimmeler, Q. Ju, C. Sui, and M. Brownlee. 2001. Hyperglycemia inhibits endothelial nitric oxide synthase activity by posttranslational modification at the Akt site. *J Clin Invest.* 108: 1341–1348.

20. Giaccari, A., L. Morviducci, D. Zorretta, P. Sbraccia, F. Leonetti, S. Caiola, A. Buongiorno, R. C. Bonadonna, and G. Tamburrano. 1995. *In vivo* effects of glucosamine on insulin secretion and insulin sensitivity in the rat: possible relevance to the maladaptive responses to chronic hyperglycaemia. *Diabetologia.* 38: 518–524.

21. Paterson, A. J., and J. E. Kudlow. 1995. Regulation of glutamine:fructose-6-phosphate amidotransferase gene transcription by epidermal growth factor and glucose. *Endocrinology.* 136: 2809–2816.

22. Patti, M. E., A. Virkamaki, E. J. Landaker, C. R. Kahn, and H. Yki-Jarvinen. 1999. Activation of the hexosamine pathway by glucosamine *in vivo* induces insulin resistance of early postreceptor insulin signaling events in skeletal muscle. *Diabetes.* 48: 1562–1571.

23. Ross, S. A., X. Chen, H. R. Hope, S. Sun,e.g.,McMahon, K. Broschat, and E. A. Gulve. 2000. Development and comparison of two 3T3-L1 adipocyte models of insulin resistance: increased glucose flux vs. glucosamine treatment. *Biochem Biophys Res Commun.* 273: 1033–1041.

24. Sayeski, P. P., A. J. Paterson, and J. E. Kudlow. 1994. The murine glutamine:fructose-6-phosphate amidotransferase-encoding cDNA sequence. *Gene.* 140: 289–290.

25. Schleicher, E. D., and C. Weigert. 2000. Role of the hexosamine biosynthetic pathway in diabetic nephropathy. *Kidney Int.* 58: Suppl 77(2): S13–8.

26. Shankar, R. R., J. S. Zhu, and A. D. Baron. 1998. Glucosamine infusion in rats mimics the beta cell dysfunction of noninsulin-dependent diabetes mellitus. *Metabolism.* 47: 573–577.

27. Yki-Jarvinen, H., C. Vogt, P. Iozzo, R. Pipek, M. C. Daniels, A. Virkamaki, S. Makimattila, L. Mandarino, R. A. DeFronzo, D. McClain, and W. K. Gottschalk. 1997. UDP-N-acetylglucosamine transferase and glutamine: fructose 6-phosphate amidotransferase activities in insulin-sensitive tissues. *Diabetologia.* 40: 76–81.

28. Haltiwanger, R. S., G. D. Holt, and G. W. Hart. 1990. Enzymatic addition of O-GlcNAc to nuclear and cytoplasmic proteins. Identification of a uridine diphospho-N-acetylglucosamine:peptide beta-N-acetylglucosaminyltransferase. *J Biol Chem.* 265: 2563–2568.

29. Starr, C. M., and J. A. Hanover. 1990. Glycosylation of nuclear pore protein p62. Reticulocyte lysate catalyzes O-linked N-acetylglucosamine addition *in vitro. J Biol Chem.* 265: 6868–6873.

30. Starr, C. M., M. D'Onofrio, M. K. Park, and J. A. Hanover. 1990. Primary sequence and heterologous expression of nuclear pore glycoprotein p62. *J Cell Biol.* 110: 1861–1871.

31. D'Onofrio, M., C. M. Starr, M. K. Park, G. D. Holt, R. S. Haltiwanger, G. W. Hart, and J. A. Hanover. 1988. Partial cDNA sequence encoding a nuclear pore protein modified by O-linked N-acetylglucosamine. *Proc Natl Acad Sci USA.* 85: 9595–9599.

32. Lubas, W. A., M. Smith, C. M. Starr, and J. A. Hanover. 1995. Analysis of nuclear pore protein p62 glycosylation. *Biochemistry.* 34: 1686–1694.

33. Kreppel, L. K., M. A. Blomberg, and G. W. Hart. 1997. Dynamic glycosylation of nuclear and cytosolic proteins. Cloning and characterization of a unique O-GlcNAc transferase with multiple tetratricopeptide repeats. *J Biol Chem.* 272: 9308–9315.

34. Lubas, W. A., D. W. Frank, M. Krause, and J. A. Hanover. 1997. O-Linked GlcNAc transferase is a conserved nucleocytoplasmic protein containing tetratricopeptide repeats. *J Biol Chem.* 272: 9316–9324.

35. Roos, M. D., W. Xie, K. Su, J. A. Clark, X. Yang, E. Chin, A. J. Paterson, and J. E. Kudlow. 1998. Streptozotocin, an analog of N-acetylglucosamine, blocks the removal of O-GlcNAc from intracellular proteins. *Proc Assoc Am Physicians.* 110: 422–432.

36. Hanover, J. A., S. Yu, W. B. Lubas, S. H. Shin, M. Ragano-Caracciola, J. Kochran, and D. C. Love. 2003. Mitochondrial and nucleocytoplasmic isoforms of O-linked GlcNAc transferase encoded by a single mammalian gene. *Arch Biochem Biophys.* 409: 287–297.

37. Shafi, R., S. P. Iyer, L. G. Ellies, N. O'Donnell, K. W. Marek, D. Chui, G. W. Hart, and J. D. Marth. 2000. The O-GlcNAc transferase gene resides on the X chromosome and is essential for embryonic stem cell viability and mouse ontogeny. *Proc Natl Acad Sci USA.* 97: 5735–5739.

38. Lubas, W. A., and J. A. Hanover. 2000. Functional expression of O-linked GlcNAc transferase. Domain structure and substrate specificity. *J Biol Chem.* 275: 10983–10988.

39. Kreppel, L. K., and G. W. Hart. 1999. Regulation of a cytosolic and nuclear O-GlcNAc transferase. Role of the tetratricopeptide repeats. *J Biol Chem.* 274: 32015–32022.

40. Love, D. C., J. Kochan, R. L. Cathey, S. H. Shin, J. A. Hanover, and J. Kochran. 2003. Mitochondrial and nucleocytoplasmic targeting of O-linked GlcNAc transferase. *J Cell Sci.* 116: 647–654.

41. Roos, M. D., and J. A. Hanover. 2000. Structure of O-linked GlcNAc transferase: mediator of glycan-dependent signaling. *Biochem Biophys Res Commun.* 271: 275–280.

42. Das, A. K., P. W. Cohen, and D. Barford. 1998. The stucture of the tetratricopeptide repeats of protein phosphatase 5: implications for TPR-mediated protein–protein interactions. *EMBO J.* 17: 1192–1199.

43. Scheufler, C., A. Brinker, G. Bourenkov, S. Pegoraro, L. Moroder, H. Bartunik, F. U. Hartl, and I. Moarefi. 2000. Structure of TPR domain-peptide complexes: critical elements in the assembly of the Hsp70-Hsp90 multichaperone Machine. *Cell.* 101: 199–210.

44. Unligil, U. M., S. Zhou, S. Yuwaraj, M. Sarkar, H. Schachter, and J. M. Rini. 2000. X-ray crystal structure of rabbit N-acetylglucosaminyltransferase I: catalytic mechanism and a new protein superfamily. *EMBO J.* 19: 5269–5280.

45. Bailer, S. M., W. K. Berlin, C. M. Starr, and J. A. Hanover. 1995. Characterization of nuclear pore protein p62 produced using baculovirus. *Protein Expr Purif.* 6: 546–554.

46. Iyer, S. P., Y. Akimoto, and G. W. Hart. 2003. Identification and cloning of a novel family of coiled-coil domain proteins that interact with O-GlcNAc transferase. *J Biol Chem.* 278: 5399–5409.

47. Yang, X., F. Zhang, and J. E. Kudlow. 2002. Recruitment of O-GlcNAc transferase to promoters by corepressor mSin3A: coupling protein O-GlcNAcylation to transcriptional repression. *Cell.* 110: 69–80.

48. Dong, D. L., and G. W. Hart. 1994. Purification and characterization of an O-GlcNAc selective N-acetyl-beta-D-glucosaminidase from rat spleen cytosol. *J Biol Chem.* 269: 19321–19330.

49. Gao, Y., L. Wells, F. I. Comer, G. J. Parker, and G. W. Hart. 2001. Dynamic O-glycosylation of nuclear and cytosolic proteins: cloning and characterization of a neutral, cytosolic beta-N-acetylglucosaminidase from human brain. *J Biol Chem.* 276: 9838–9845.

50. Gao, Y., G. J. Parker, and G. W. Hart. 2000. Streptozotocin-induced beta cell death is independent of its inhibition of O-GlcNAcase in pancreatic Min6 cells. *Arch Biochem Biophys.* 383: 296–302.

51. Heckel, D., N. Comtesse, N. Brass, N. Blin, K. D. Zang, and E. Meese. 1998. Novel immunogenic antigen homologous to hyaluronidase in meningioma. *Hum Mol Genet.* 7: 1859–1872.

52. Farook, V. S., C. Bogardus, and M. Prochazka. 2002. Analysis of MGEA5 on 10q24.1-q24.3 encoding the Beta-O-linked N-acetylglucosaminidase as a candidate gene for type 2 diabetes mellitus in Pima Indians. *Mol Genet Metab.* 189–193.

53. Kudlow, J. E. 2002. The O-GlcNAcase theory of diabetes: commentary on a candidate gene for diabetes. *Mol Genet Metab.* 77: 1–2.

54. Davis, L. I., and G. Blobel. 1987. Nuclear pore complex contains a family of glycoproteins that includes p62: glycosylation through a previously unidentified cellular pathway. *Proc Natl Acad Sci USA.* 84: 7552–7556.

55. Hanover, J. A., C. K. Cohen, M. C. Willingham, and M. K. Park. 1987. O-linked *N*-acetylglucosamine is attached to proteins of the nuclear pore. Evidence for cytoplasmic and nucleoplasmic glycoproteins. *J Biol Chem.* 262: 9887–9894.

56. Hanover, J. A. 1992. The nuclear pore: at the crossroads. *FASEB J.* 6: 2288–2295.

57. Miller, M. W., M. R. Caracciolo, W. K. Berlin, and J. A. Hanover. 1999. Phosphorylation and glycosylation of nucleoporins. *Arch Biochem Biophys.* 367: 51–60.

58. Ishii, K., G. Arib, C. Lin, H. G. Van, and U. K. Laemmli. 2002. Chromatin boundaries in budding yeast: the nuclear pore connection. *Cell.* 109: 551–562.

59. Comer, F. I., and G. W. Hart. 2001. Reciprocity between O-GlcNAc and O-phosphate on the carboxyl-terminal domain of RNA polymerase II. *Biochemistry.* 40: 7845–7852.

60. Jackson, S. P., and R. Tjian. 1988. O-glycosylation of eukaryotic transcription factors: implications for mechanisms of transcriptional regulation. *Cell.* 55: 125–133.

61. Han, I., and J. E. Kudlow. 1997. Reduced O-glycosylation of Sp1 is associated with increased proteasome susceptibility. *Mol Cell Biol.* 17: 2550–2558.

62. Han, I., M. D. Roos, and J. E. Kudlow. 1998. Interaction of the transcription factor Sp1 with the nuclear pore protein p62 requires the C-terminal domain of p62. *J Cell Biochem.* 68: 50–61.

63. Roos, M. D., K. Su, J. R. Baker, and J. E. Kudlow. 1997. O-glycosylation of an Sp1-derived peptide blocks known Sp1 protein interactions. *Mol Cell Biol.* 17: 6472–6480.

64. Lichtsteiner, S., and U. Schibler. 1989. A glycosylated liver-specific transcription factor stimulates transcription of the albumin gene. *Cell.* 57: 1179–1187.

65. Ikema, T., Y. Shimajiri, I. Komiya, M. Tawata, S. Sunakawa, H. Yogi, M. Shimabukuro, and N. Takasu. 2002. Identification of three new mutations of the HNF-1 alpha gene in Japanese MODY families. *Diabetologia.* 45: 1713–1718.

66. Frayling, T. M., C. M. Lindgren, J. C. Chevre, S. Menzel, M. Wishart, Y. Benmezroua, A. Brown, J. C. Evans, P. S. Rao, C. Dina, C. Lecoeur, T. Kanninen, P. Almgren, M. P. Bulman, Y. Wang, J. Mills, R. Wright-Pascoe, M. M. Mahtani, F. Prisco, A. Costa, I. Cognet, T. Hansen, O. Pedersen, S. Ellard, T. Tuomi, L. C. Groop, P. Froguel, A. T. Hattersley, and M. Vaxillaire. 2003. A genome-wide scan in families with maturity-onset diabetes of the young: evidence for further genetic heterogeneity. *Diabetes.* 52: 872–881.

67. Gomez-Perez, F. J., and R. Mehta. 2003. Genetic defects of beta cell function: (MODY) application of molecular biology to clinical medicine. *Rev Invest Clin.* 55: 172–176.

68. Gao, Y., J. Miyazaki, and G. W. Hart. 2003. The transcription factor PDX-1 is posttranslationally modified by O-linked N-acetylglucosamine and this modification is correlated with its DNA-binding activity and insulin secretion in min6 beta cells. *Arch Biochem Biophys.* 415: 155–163.

69. Cheng, X., R. N. Cole, J. Zaia, and G. W. Hart. 2000. Alternative O-glycosylation/O-phosphorylation of the murine estrogen receptor beta. *Biochemistry.* 39: 11609–11620.

70. Cheng, X., and G. W. Hart. 2001. Alternative O-glycosylation/O-phosphorylation of serine 16 in murine estrogen receptor beta: posttranslational regulation of turnover and transactivation activity. *J Biol Chem.* 276: 10570–10575.

71. Cheng, X., and G. W. Hart. 2000. Glycosylation of the murine estrogen receptor-alpha. *J Steroid Biochem Mol Biol.* 75: 147–158.

72. Chou, T. Y., C. V. Dang, and G. W. Hart. 1995. Glycosylation of the c-myc transactivation domain. *Proc Natl Acad Sci USA.* 92: 4417–4421.

73. Chou, T. Y., G. W. Hart, and C. V. Dang. 1995. c-myc is glycosylated at threonine 58, a known phosphorylation site and a mutational hot spot in lymphomas. *J Biol Chem.* 270: 18961–18965.

74. Flamigni, F., I. Faenza, S. Marmiroli, I. Stanic, A. Giaccari, C. Muscari, C. Stefanelli, and C. Rossoni. 1997. Inhibition of the expression of ornithine decarboxylase and c-myc by cell-permeant ceramide in difluoromethylornithine-resistant leukaemia cells. *Biochem J.* 324(Pt 3): 783–789.

75. Chou, C. F., and M. B. Omary. 1993. Mitotic arrest–associated enhancement of O-linked glycosylation and phosphorylation of human keratins 8 and 18. *J Biol Chem.* 268: 4465–4472.

76. O'Donnell, N. 2002. Intracellular glycosylation and development. *Biochim Biophys Acta.* 1573: 336–345.

77. Yanagisawa, M., E. Planel, K. Ishiguro, and S. C. Fujita. 1999. Starvation induces tau hyperphosphorylation in mouse brain: implications for Alzheimer's disease. *FEBS Lett.* 461: 329–333.

78. Vosseller, K., L. Wells, M. D. Lane, and G. W. Hart. 2002. Elevated nucleocytoplasmic glycosylation by O-GlcNAc results in insulin resistance associated with defects in Akt activation in 3T3-L1 adipocytes. *Proc Natl Acad Sci USA.* 99: 5313–5318.

79. Vocadlo, D. J., H. C. Hang, E. J. Kim, J. A. Hanover, and C. R. Bertozzi. 2003. A chemical approach for identifying O-GlcNAc-modified proteins in cells. *Proc Natl Acad Sci USA.* 100: 9116–9121.

80. Jacobsen, S. E., and N. E. Olszewski. 1993. Mutations at the SPINDLY locus of Arabidopsis alter gibberellin signal transduction. *Plant Cell.* 5: 887–896.

81. Jacobsen, S. E., K. A. Binkowski, and N. E. Olszewski. 1996. SPINDLY, a tetratricopeptide repeat protein involved in gibberellin signal transduction in Arabidopsis. *Proc Natl Acad Sci USA.* 93: 9292–9296.

82. Thornton, T. M., S. M. Swain, and N. E. Olszewski. 1999. Gibberellin signal transduction presents ellipsis, the SPY who O-GlcNAc'd me. *Trends Plant Sci.* 4: 424–428.

83. Swain, S. M., and N. E. Olszewski. 1996. Genetic Analysis of Gibberellin Signal Transduction. *Plant Physiol.* 112: 11–17.

84. Swain, S. M., T. S. Tseng, and N. E. Olszewski. 2001. Altered expression of SPINDLY affects gibberellin response and plant development. *Plant Physiol.* 126: 1174–1185.

85. Hartweck, L. M., C. L. Scott, and N. E. Olszewski. 2002. Two O-linked N-acetylglucosamine transferase genes of *Arabidopsis thaliana* L. Heynh. have overlapping functions necessary for gamete and seed development. *Genetics.* 161: 1279–1291.

86. Marshall, S., W. T. Garvey, and R. R. Traxinger. 1991. New insights into the metabolic regulation of insulin action and insulin resistance: role of glucose and amino acids. *FASEB J.* 5: 3031–3036.

87. Traxinger, R. R., and S. Marshall. 1990. Glucose regulation of insulin receptor affinity in primary cultured adipocytes. *J Biol Chem.* 265: 18879–18883.

88. Traxinger, R. R., and S. Marshall. 1989. Role of amino acids in modulating glucose-induced desensitization of the glucose transport system. *J Biol Chem.* 264: 20910–20916.

89. Buse, M. G., K. A. Robinson, B. A. Marshall, and M. Mueckler. 1996. Differential effects of GLUT1 or GLUT4 overexpression on hexosamine biosynthesis by muscles of transgenic mice. *J Biol Chem.* 271: 23197–23202.

90. Cooksey, R. C., L. F. J. Hebert, J. H. Zhu, P. Wofford, W. T. Garvey, and D. A. McClain. 1999. Mechanism of hexosamine-induced insulin resistance in transgenic mice overexpressing glutamine:fructose-6-phosphate amidotransferase: decreased glucose transporter GLUT4 translocation and reversal by treatment with thiazolidinedione. *Endocrinology.* 140: 1151–1157.

91. Daniels, M. C., T. P. Ciaraldi, S. Nikoulina, R. R. Henry, and D. A. McClain. 1996. Glutamine:fructose-6-phosphate amidotransferase activity in cultured human skeletal muscle cells: relationship to glucose disposal rate in control and noninsulin-dependent diabetes mellitus subjects and regulation by glucose and insulin. *J Clin Invest.* 97: 1235–1241.

92. Davidson, M. B., K. Hunt, and C. Fernandez-Mejia. 1994. The hexosamine biosynthetic pathway and glucose-induced down-regulation of glucose transport in L6 myotubes. *Biochim Biophys Acta.* 1201: 113–117.

93. Filippis, A., S. Clark, and J. Proietto. 1997. Increased flux through the hexosamine biosynthesis pathway inhibits glucose transport acutely by activation of protein kinase C. *Biochem J.* 324: 981–985.

94. Hebert, L. F. J., M. C. Daniels, J. Zhou, E. D. Crook, R. L. Turner, S. T. Simmons, J. L. Neidigh, J. S. Zhu, A. D. Baron, and D. A. McClain. 1996. Overexpression of glutamine:fructose-6-phosphate amidotransferase in transgenic mice leads to insulin resistance. *J Clin Invest.* 98: 930–936.

95. Kawanaka, K., D. H. Han, J. Gao, L. A. Nolte, and J. O. Holloszy. 2001. Development of glucose-induced insulin resistance in muscle requires protein synthesis. *J Biol Chem.* 276: 20101–20107.

96. McKnight, G. L., S. L. Mudri, S. L. Mathewes, R. R. Traxinger, S. Marshall, P. O. Sheppard, and P. J. O'Hara. 1992. Molecular cloning, cDNA sequence, and bacterial expression of human glutamine:fructose-6-phosphate amidotransferase. *J Biol Chem.* 267: 25208–25212.

97. Boehmelt, G., I. Fialka, G. Brothers, M. D. McGinley, S. D. Patterson, R. Mo, C. C. Hui, S. Chung, L. A. Huber, T. W. Mak, and N. N. Iscove. 2000. Cloning and characterization of the murine glucosamine-6-phosphate acetyltransferase EMeg32. Differential expression and intracellular membrane association. *J Biol Chem.* 275: 12821–12832.

98. Boehmelt, G., A. Wakeham, A. Elia, T. Sasaki, S. Plyte, J. Potter, Y. Yang, E. Tsang, J. Ruland, N. N. Iscove, J. W. Dennis, and T. W. Mak. 2000. Decreased UDP-GlcNAc levels abrogate proliferation control in EMeg32-deficient cells. *EMBO J.* 19: 5092–5104.

99. McClain, D. A., W. A. Lubas, R. C. Cooksey, M. Hazel, G. J. Parker, D. C. Love, and J. A. Hanover. 2002. Altered glycan-dependent signaling induces insulin resistance and hyperleptinemia. *Proc Natl Acad Sci USA.* 99: 10695–10699.

100. Wang, J., R. Liu, L. Liu, R. Chowdhury, N. Barzilai, J. Tan, and L. Rossetti. 1999. The effect of leptin on Lep expression is tissue specific and nutritionally regulated. *Nat Med.* 5: 895–899.

101. Kudlow, J. E., and J. D. Bjorge. 1990. TGF alpha in normal physiology. *Semin Cancer Biol.* 1: 293–302.

102. Datta, B., M. K. Ray, D. Chakrabarti, D. E. Wylie, and N. K. Gupta. 1989. Glycosylation of eukaryotic peptide chain initiation factor 2 (eIF-2)-associated 67-kDa polypeptide (p67) and its possible role in the inhibition of eIF-2 kinase-catalyzed phosphorylation of the eIF-2 alpha-subunit. *J Biol Chem.* 264: 20620–20624.

103. Deroo, B. J., C. Rentsch, S. Sampath, J. Young, D. B. DeFranco, and T. K. Archer. 2002. Proteasomal inhibition enhances glucocorticoid receptor transactivation and alters its subnuclear trafficking. *Mol Cell Biol.* 22: 4113–4123.

104. Hay, R. T., L. Vuillard, J. M. Desterro, and M. S. Rodriguez. 1999. Control of NF-kappa B transcriptional activation by signal-induced proteolysis of I kappa B alpha. *Philos Trans R Soc Lond B Biol Sci.* 354: 1601–1609.

105. Su, K., M. D. Roos, X. Yang, I. Han, A. J. Paterson, and J. E. Kudlow. 1999. An N-terminal region of Sp1 targets its proteasome-dependent degradation *in vitro*. *J Biol Chem.* 274: 15194–15202.

106. Su, K., X. Yang, M. D. Roos, A. J. Paterson, and J. E. Kudlow. 2000. Human Sug1/p45 is involved in the proteasome-dependent degradation of Sp1. *Biochem J.* 348 (Pt 2): 281–289.

107. Zhang, F., K. Su, X. Yang, D. B. Bowe, A. J. Paterson, and J. E. Kudlow. 2003. O-GlcNAc modification is an endogenous inhibitor of the proteasome. *Cell.* 115: 715–725.

108. Akimoto, Y., L. K. Kreppel, H. Hirano, and G. W. Hart. 2001. Hyperglycemia and the O-GlcNAc transferase in rat aortic smooth muscle cells: elevated expression and altered patterns of O-GlcNAcylation. *Arch Biochem Biophys.* 389: 166–175.

109. Hawkins, M., I. Angelov, R. Liu, N. Barzilai, and L. Rossetti. 1997. The tissue concentration of UDP-N-acetylglucosamine modulates the stimulatory effect of insulin on skeletal muscle glucose uptake. *J Biol Chem.* 272: 4889–4895.

110. Rossetti, L. 1995. Glucose toxicity: the implications of hyperglycemia in the pathophysiology of diabetes mellitus. *Clin Invest Med.* 18: 255–260.

111. Rossetti, L. 2000. Perspective: Hexosamines and nutrient sensing. *Endocrinology.* 141: 1922–1925.

112. Rossetti, L., A. Giaccari, and R. A. DeFronzo. 1990. Glucose toxicity. *Diabetes Care.* 13: 610–630.

113. Shaw, P., J. Freeman, R. Bovey, and R. Iggo. 1996. Regulation of specific DNA binding by p53: evidence for a role for O-Glcyosylation and charged residues at the carboxy-terminus. *Oncogene.* 12: 921–930.

III

Insulin Release, Signaling, and Insulin Resistance

18

Pancreatic β-Cell, a Unique Fuel Sensor

MITSUHISA KOMATSU

INTRODUCTION

Pancreatic β-cells secrete insulin, the only hormone that lowers plasma glucose concentration in humans. As with many other hormones, insulin release is coordinately regulated by various stimulatory and inhibitory inputs. What is unique to the β-cell is that nutrients and nutrient-derived signals play a predominant role in the regulation of hormone exocytosis. Insulin lowers plasma glucose and, in turn, glucose is the most powerful secretagogue for insulin release. Therefore, a tight closed-loop regulation is operating *in vivo*. Accordingly, the mechanism of glucose-stimulated insulin secretion (GSIS) has been a main topic in the field of β-cell research.

Insulin release is regulated on the order of a minute to hours. This is quite different from neurotransmitter release, which occurs within milliseconds. Obviously, the dynamics of insulin secretion are perfectly matched to the timescale of whole-body nutrient excursion under normal physiology. There is an important quantitative difference between insulin and neurotransmitter release regarding the fraction of cellular hormone that is released. In the β-cell, even with the maximum stimulation, a relatively small amount of stored hormone (~15%) is released. In contrast, in the neuronal cells and in the mast cells, a vast majority of stored substance is released in response to the maximum stimulation.

Diabetes mellitus, one of the most prevalent metabolic disorders in modern society, with significant morbidity and mortality, is caused by insulin deficiency and insulin resistance. Insulin deficiency is due to a combination of reduced β-cell mass, inadequate insulin synthesis, or defective insulin release. On the other end of the spectrum, hypoglycemia, a potentially fatal condition, is caused by a sustained, inappropriately high rate of insulin secretion.

GLUCOSE-STIMULATED INSULIN SECRETION (GSIS)

Glucose Metabolism in the β-Cell

Glucose metabolism is absolutely required for glucose-induced insulin release, which is supported by the following facts: (1) glucose is efficiently metabolized by the β-cell, (2) inhibition of glucose metabolism impairs GSIS, (3) nonglucose fuels such as glyceraldehyde, which enters the glycolytic pathway as glyceraldehyde-3-P, and the amino acids that enter the tricarboxylic acid (TCA) cycle as metabolizable fuel, stimulate insulin release, and (4) fuels that are not metabolized in the β-cell, such as sucrose and lactose, are not capable of stimulating insulin release.

Glucose entry into the cell is mediated by glucose transporters (GLUT). A high-K_m GLUT, GLUT2, is abundantly expressed on the plasma membrane of the rodent β-cell, whereas a low-K_m GLUT, GLUT1, is the main transporter in the human β-cell.[1] Glucose is mainly phosphorylated by a low-affinity glucokinase (GK) (or hexokinase IV with the K_m for glucose, ~8 mM), yielding glucose-6-phosphate, which is further metabolized via the glycolytic pathway. Because the capacity of GLUT is much greater than glucose utilization, and GK activity is significantly lower than that of other enzymes of the glycolytic pathway, GK functions as a rate-limiter of glucose metabolism in the β-cell. Specific inhibitors of GK blunt GSIS in a concentration-dependent manner. In sum, GK is the β-cell "glucose sensor."[2] As expected, patients with mutation of the GK gene and heterozygote mice with disruption of the β-cell GK exhibit impaired β-cell glucose sensing, impaired GSIS, and glucose intolerance or diabetes.[3–5] The homozygote GK knockout mouse dies due to diabetic ketoacidosis with severe insulin deficiency.

Coupling of glycolysis to mitochondrial metabolism is important for insulin exocytosis. Electrons in glycolysis-derived NADH are transferred to the mitochondria through the NADH shuttle system. Metabolic flux through the NADH shuttle does not have a regulatory role but is a required step for GSIS.[6] About one half of pyruvate (a final product of glycolysis) enters the TCA cycle as oxaloacetate via the reaction catalyzed by pyruvate carboxylase, and the other half enters as acetyl CoA after decarboxylation in the reaction catalyzed by the pyruvate dehydrogenase complex. Due to the high rate of pyruvate carboxylation, glucose metabolism accompanies significant anaplerotic input into the TCA cycle in the β-cell. This is associated with export of cataplerotic TCA cycle intermediates such as α-ketoglutarate and aspartate.[7] Unlike hepatocytes in which the cataplerosis is coupled to gluconeogenesis, cholesterol, and fatty acid synthesis, the enzymes for these pathways are inactive in the β-cell. It is very likely that the key molecule or key molecules regulating insulin exocytosis are yielded by anaplerosis. Investigations aimed at its identification are ongoing (see the following text). Hydrogen ion shuttled into the mitochondria and trapped in the TCA cycle is used as energy for generation of ATP in the respiratory chain, and ATP is translocated to the cytosol. Glycogen synthesis, metabolic flux into the pentose phophate shunt, lactate formation, gluconeogenesis, and fatty acid synthesis are negligible in the β-cell.

Biphasic Insulin Release

A salient feature of GSIS is the biphasicity with the square-wave application of a stimulatory concentration of glucose.[8] Insulin release from isolated pancreatic islets increases

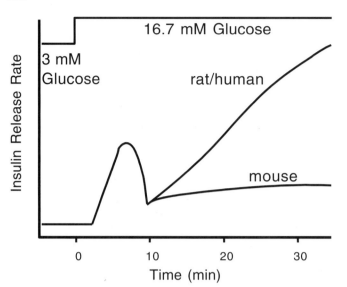

Figure 18.1 Time course of glucose-stimulated insulin secretion from rat/human and mouse pancreatic islets. A square-wave application of stimulatory concentration of glucose induces a biphasic insulin release. Note that the second phase is gradually increasing in the rat/human islets, whereas it is sustained in the mouse islets.

within 0.5 to 2 min of the step-up of extracellular glucose concentration from substimulatory (~3 mM) to stimulatory (~8 mM or higher) levels. As the single-cell-based insulin release from β-cells increases 100 to 150 sec after a step-up of extracellular concentration of glucose,[9] the lag is a reflection of the time for a series of events from glucose entry to elevation of cytosolic-free Ca^{2+} concentration ($[Ca^{2+}]i$) (see the following text). The rate of insulin release culminates in a sharp peak within 3 to 4 min, which is called the first phase. Subsequently, the rate of insulin release decreases to baseline by 10 min; followed by a gradually increasing (in the rat or human) or a sustained (in the mouse) second phase. The typical temporal profiles of GSIS from rat or human and mouse pancreatic islets are illustrated in Figure 18.1. None of the tumoral β-cell line exhibits the typical rat or human type rising second phase, although some β-cell lines show the sustained response typical for mouse β-cell.[10]

The biphasic insulin release could be observed experimentally with the perifusion of isolated pancreatic islets or perfusion of pancreas, and clinically with the hyperglycemic clamp *in vivo*. Diminution of the first phase of insulin release is a well-established marker of β-cell dysfunction. The first phase response is usually absent in patients with established diabetes and is clearly diminished in the glucose-tolerant relatives of patients with type 2 diabetes, as well as subjects with minimally impaired glucose tolerance.[11] Moreover, decreased first-phase insulin secretion is a predictor for future development of type 1 diabetes among those with positive family history of the disorder.[12] Under the regular feeding condition in normal subjects, when plasma glucose concentration (PG) gradually rises from 5 to 8 mM over a 30-min period, a large, brisk first phase-like insulin release is followed by a sustained second phase-like plateau. For the postmeal insulin response, incretins (the gut-derived peptides that activate adenylyl cyclase in the β-cell) and possibly other nonglucose stimuli play important roles (see the following text).

ATP-Sensitive K⁺ Channel-Dependent Pathway

Stimulation of the pancreatic β-cell with high glucose is associated with characteristic electrical activity.[13,14] Figure 18.2 depicts changes in membrane potential of mouse pancreatic β-cell in response to the square-wave application of 10 mM glucose. First, the membrane potential gradually increases, and when it reaches the threshold, an oscillatory electrical activity is elicited, on which action potentials are superimposed. Glucose-induced electrical activity is concentration-dependent, and 20 mM glucose evokes persistent depolarization and bursts of action potential.

ATP-sensitive K⁺ (KATP) channels play a pivotal role in the glucose-induced electrical activities. The KATP channel consists of Kir6.2, an inward rectifying K⁺ channel, and the sulfonylurea receptor 1 (SUR1), a sulfonylurea binding protein;[15]

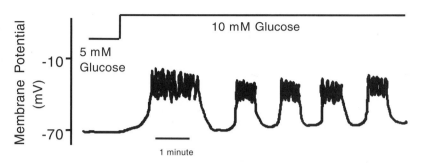

Figure 18.2 Illustration of changes in membrane potential of pancreatic beta cell in response to a square wave application of 10 m*M* glucose.

four molecules each of Kir6.2 and SUR1 form a heterooctamer channel.[16] ATP binds to the Kir6.2 and sulfonylurea binds to the SUR1, both of which result in closure of the KATP channel. On the other hand, ADP binds to the SUR1 and opens the channel. The KATP channels are spontaneously active (open) when the β-cell is exposed to a substimulatory concentration of glucose, i.e., with low intracellular ATP and ATP to ADP ratio, and the efflux of K⁺ ions through the channel generates a resting membrane potential (60 to 70 mV). The β-cell is also equipped with glucose-responsive ion channels other than the KATP channel, but the functional importance of these channels for insulin release is uncertain.

As a result of glucose metabolism, there is an increase in cellular ATP with a concomitant decrease in ADP. The rise in the ATP to ADP ratio in the cytoplasm closes the KATP channel and causes membrane depolarization and opening of the voltage-dependent Ca²⁺ channels (VDCC); L-type VDCC is the predominant Ca²⁺ channel in the β-cell. Consequently, Ca²⁺-influx through the VDCC and elevation of [Ca²⁺]i takes place, which triggers insulin exocytosis (Figure 18.3). Involvement of SNARE complex in Ca²⁺-triggered insulin release is strongly suggested.[17] Soluble *N*-ethylmaleimide-sensitive factor (NSF) attachment protein (SNAP) receptors (SNAREs) associate to form a complex machinery for membrane fusion. The vesiclar membrane proteins (v-SNAREs) synaptobrevin

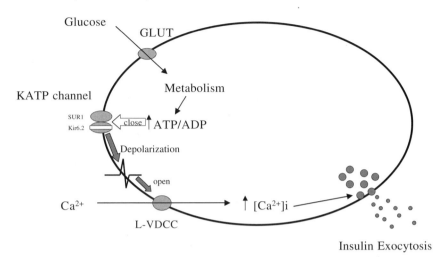

Figure 18.3 KATP channel-dependent insulinotropic action of glucose. This branch of glucose signaling is mainly responsible for triggering through induction of exocytosis from the readily releasable pool. GLUT, glucose transporter; KATP channel, ATP-sensitive K$^+$ channel; L-VDCC, L-type voltage-dependent calcium channel; [Ca^{2+}]i, cytosolic free calcium concentration.

and synaptotagmin, and the plasma membrane target proteins (t-SNAREs) syntaxin and SNAP-25, constitute the SNARE complex in which the SNAPs and NSF are assembled. Among the SNARE proteins, synaptotagmin is considered as a Ca^{2+}-sensor,[18] and a role of small-molecular-weight GTP-binding protein Rab3A in Ca^{2+}-evoked insulin exocytosis is also suggested.[19]

The KATP channel-dependent ionic pathway is essential for GSIS under regular experimental conditions. Namely, activation of the KATP channel by diazoxide, inhibition of the VDCC by Ca^{2+} channel blockers, and omission of extracellular Ca^{2+}, all effectively obliterate glucose-induced elevation of [Ca^{2+}]i and insulin release *in vitro*. Sulfonylurea (a closer of the KATP channel), forced membrane depolarization by high concentrations of K$^+$, and application of Ca^{2+} channel openers and Ca^{2+} ionophores trigger insulin release.

The KATP Channel-Independent Pathways

Glucose stimulates the β-cell in a KATP channel-independent manner as well. Unlike glucose, simple depolarization by sulfonylurea or a high concentration of K$^+$ evokes only monophasic insulin release by the β-cell. Thus, the second phase of GSIS is driven by the KATP-independent action of glucose. Experimentally, this glucose action can be seen as a strong potentiation of high K$^+$-induced insulin release (Ca^{2+}-triggered insulin release) by glucose under full activation of the KATP channel by diazoxide.[20,21] Also, glucose potently stimulates insulin release even if the KATP channel is fully closed by sulfonylurea.[22,23] In *in vitro* experiments, the KATP channel-independent glucose action can also be seen in the stringently Ca^{2+}-free media in the absence of any rise in [Ca^{2+}]i.[24,25] Time-dependent potentiation of insulin exocytosis by glucose is a well-established feature of the β-cell, which is also accomplished by the KATP-independent glucose action in the absence of rise in [Ca^{2+}]i.[26,27] Thus, the commonly held view that "KATP-independent glucose signaling is just an amplification of the Ca^{2+} sensitivity of the exocytotic machinery" is not entirely correct. The KATP channel-independent branch of glucose action may be less dependent upon glucose metabolism compared with the KATP-dependent branch.[28,29] Currently available data indicates the following scenario. Glucose triggers exocytosis from the readily releasable pool of insulin granules through the KATP channel-dependent ionic pathway, which is the first phase. Subsequently, glucose time-dependently expands the size of the readily releasable pool through the KATP channel-independent pathway, generating the second phase of exocytosis (Figure 18.4). Nevertheless, in individuals lacking functional KATP channels due to a mutation of the gene-encoding SUR1, intravenous injection of glucose elicits a brisk insulin secretion.[30] Thus, glucose triggers insulin release *in vivo* independently from its action on the KATP channel, at least under certain conditions.

It is most likely that the KATP channel-independent glucose action is mediated by several molecules originating in glucose metabolism.[31] In permeabilized β-cells, Ca^{2+} triggers

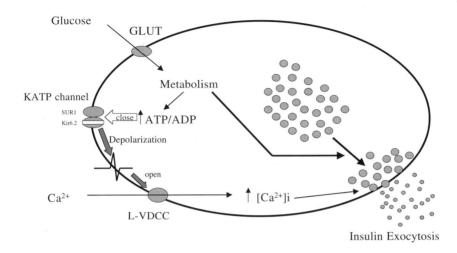

Figure 18.4 KATP channel-dependent and independent insulinotropic action of glucose. The KATP channel-independent glucose action is responsible for the second phase of insulin release through expansion of the readily releasable pool. GLUT, glucose transporter; KATP channel, ATP-sensitive K$^+$ channel; L-VDCC, L-type voltage-dependent calcium channel; [Ca^{2+}]i, cytosolic free calcium concentration.

insulin release only when millimolar concentration of ATP are present;[32] an abrupt rise in cytosolic ATP enhances subsequent insulin release triggered by Ca^{2+},[33] suggesting a role for ATP. Formycin, an ATP analogue, partially mimics the KATP channel-independent glucose action.[34] Mycophenolic acid (an inhibitor of *de novo* synthesis of GTP) strongly inhibits glucose-induced insulin release,[35] especially the KATP channel-independent portion.[36] Participation of the small molecular GTP-binding protein Cdc42 in glucose-induced (KATP channel-independent) disruption of the actin–t-SNARE complex was reported.[37] Glucose may increase GTP or the GTP to GDP ratio in the β-cell.[38] Thus, GTP and the GTP-binding protein may also be mediators of the KATP channel-independent glucose action.

Glucose increases intracellular malonyl CoA in β-cell[39] as a consequence of anaplerosis.[7] Anaplerosis may accompany increased cataplerotic output of citrate;[39] citrate is converted to acetyl CoA and then carboxylated to yield malonyl CoA. Alternatively, acetyl CoA carboxylase, a rate-limiting enzyme for malonyl CoA production, may be activated by anaplerosis.[7] Due to very low level, or absence, of fatty acid synthesis in the β-cell, malonyl CoA is not used as a substrate for it. Instead, malonyl CoA functions as a link between anaplerosis and lipid signal for insulin release. Namely, it inhibits carnitine palmitoyl transferase-1 (CPT-1), an enzyme located in the outer membrane of mitochondria responsible for incorporation of LC CoA from the cytosol to the mitochondria for oxidation. Then, cytosolic accumulation of LC CoA takes place;[40] the link between glucose metabolism and the accumulation of LC CoA is illustrated in Figure 18.5. We consider

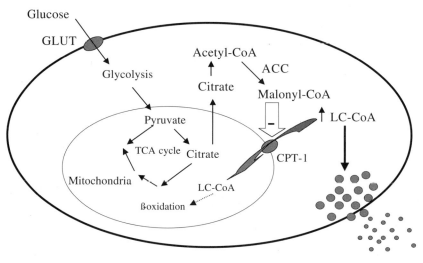

Figure 18.5 A proposed link between glucose metabolism and the accumulation of long-chain acyl CoA (LC CoA). GLUT, glucose transporter; ACC, acetyl CoA carboxylase; CPT-1, carnitine palmitoyl transferase-1; FFA, free fatty acid.

that KATP channel-independent glucose action is at least in part mediated by the accumulation of LC CoA based on the following evidence: stimulation of insulin release by an extracellular application of LC CoA in permeabilized β-cell[41] and mimicry of the KATP channel-independent glucose action by an acute exposure of β-cell to FFA, which builds up the intracellular LC CoA pool.[42,43] Modification of protein (known as acylation) occurs as a result of accumulation of cytosolic LC CoA, and cerulenin, an inhibitor of protein acylation, selectively inhibits FFA- and glucose-induced KATP channel-independent stimulation of insulin release.[44] More recently, we have identified acylated proteins in the islets with the expected characteristics as a regulator of exocytosis.[45]

Phorbol ester-insensitive PKCs, which are referred to as atypical PKCs, may be involved in the KATP channel-independent glucose action.[46,47] Potential roles of cyclic ADP-ribose and cytosolic glutamate in GSIS are controversial.[48–54]

AMINO ACID-INDUCED INSULIN RELEASE

Amino acids stimulate insulin release. In particular, arginine, leucine, and glutamine are well-established insulin secretagogues.

Arginine has been used as a nonglucose insulin secretagogue in the clinical field. Arginine induces insulin release mostly via the ionic mechanism because it is barely metabolized in the β-cell. There is a specific transporter for arginine in the β-cell.[55] Thus, elevation of extracellular concentration of arginine is associated with its intracellular accumulation. Because arginine is positively charged, its intracellular accumulation causes membrane depolarization, opening of L-VDCC, influx of Ca^{2+}, and elevation of $[Ca^{2+}]i$, leading to enhanced insulin release.

In contrast, leucine and glutamine increase the rate of insulin release via metabolism.[56] Leucine at a low concentration (~2 mM) allosterically activates glutamate dehydrogenase (GDH) in the mitochondria, an enzyme that interconverts glutamate and 2-ketoglutarate (2KG). Because the β-cell contains a significant amount of glutamate, activation of GDH is associated with increased metabolic flux from glutamate to 2KG, an intermediate of the TCA cycle. Thus, activation of GDH causes anaplerotic input into the TCA cycle and eventually raises the rate of insulin release as does glucose (see preceding text on glucose metabolism). At a higher concentration ($\sim$10 mM), leucine can be a fuel in the β-cell, in which it is metabolized as a ketogenic substance through acetyl CoA. α-Ketoisocaproic acid (αKIC) is one of the most commonly used nonglucose fuel secretagogues *in vitro*. Despite the fact that αKIC is the first catabolite of leucine, leucine and αKIC stimulate the β-cell via distinct mechanisms. GDH functions as the sensor for leucine (see preceding text), whereas transamination and resultant generation of 2KG as an anaplerotic input into the TCA cycle is important in αKIC stimulation of insulin release.[57] Glutamine is a cell-permeable amino acid that is deaminated intracellularly to glutamate. Thus, extracellular application of glutamine causes elevation of cytosolic glutamate concentration. A combination of leucine and glutamine causes activation of GDH (by leucine) on one hand and an increased substrate supply for the enzyme (by glutamine) on the other. Glutamate generated by deamination of glutamine is a substrate for GDH (see preceding text). Thus, a combination of the two amino acids strongly stimulates insulin secretion even at substimulatory concentrations of glucose.

Protein meal-induced insulin secretion may be related to the earlier-mentioned amino acid-induced insulin release. Synergistic action of multiple amino acids may be significant for insulin release because postprandial plasma concentration of total amino acids does not exceed 2 to 3.5 mM. Stimulation of insulin release by amino acids in the absence of glucose in the test meal may cause hypoglycemia. However, amino acids also stimulate glucagon release from pancreatic α-cells, which counteracts the hypoglycemic effect of insulin.

FREE FATTY ACIDS

Free fatty acid (FFA) is a nutrient that is actively metabolized in the β-cell as in many tissues, including the skeletal muscles and the liver. Nevertheless, insulinotropic effects of FFA are not considered as a result of β-oxidation in the mitochondria (see the following text).[58] Consequences of β-cell exposure to FFA are either stimulatory or inhibitory, depending on the length of exposure and the ambient glucose concentration.

During fasting, the plasma concentration of FFA is elevated as a result of lowered circulating insulin. Elevation of FFA during fasting, which does not trigger insulin release, primes the β-cell to keep it glucose competent. Thus, for normal insulin response to glucose after prolonged fasting, increased FFA during fasting is essential.[59]

Acute exposure of β-cells to FFA for several hours results in an enhancement of insulin release. The precise mechanism of enhancement of insulin release by FFA remains unknown.[60] A likely explanation for FFA stimulation of insulin release is that elevation of cytosolic LC CoA, which is produced as a result of intracellular conversion of FFA, activates the exocytotic machinery. Indeed, an extracellular application of a micromolar concentration of FFA mimics the KATP channel-independent glucose action.[42,43] Expression of FFA-responsive GTP-binding protein-coupled receptor was discovered in the β-cell, suggesting a transmembrane signal transduction for FFA.[61]

A chronic elevation of extracellular FFA, especially with high glucose, is deleterious to the β-cell, a phenomenon called *lipotoxicity*.[62] Prolonged exposure of the β-cell to FFA elevates the basal rate of insulin release and blunts GSIS.[63,64] An accumulation of islet triglycerides was proposed as a cause for lipotoxicity.[65] Lipotoxicity proceeds beyond attenuation of insulin release; prolonged exposure of β-cells to high concentratons of FFA impairs insulin biosynthesis and enhances apoptosis of the β-cell.[66,67]

HORMONAL FACTORS

Incretin and cAMP Stimulation of Insulin Release

When the same PG elevation is achieved by an intravenous glucose injection or an oral glucose load, insulin secretion is greater in the latter. Insulinotropic peptide hormones secreted from the small intestine upon oral glucose load are responsible for this phenomenon; these peptides are collectively termed *incretins*. Glucagon-like peptide 1 (GLP-1) and glucose-dependent insulinotropic peptide (GIP) are the best-characterized incretins. Plasma concentration of incretins rises in response to carbohydrate ingestion or gastrointestinal motility.

GLP-1[7-36 amide], a truncated and amidated form of GLP-1, is released from the L cells in the ileum, a distal portion of the small intestine. GLP-1 antagonists or neutralization of GLP-1 with antibodies impairs glucose tolerance in primates.[68] Targeted disruption of the gene encoding GLP-1 receptor results in impaired insulin secretion in response to glucose load.[69] Quite interestingly, in GLP-1 receptor-/- mouse, impaired glucose tolerance due to insulinopenia was seen not only upon oral glucose load, but also after intraperitoneal glucose injection. With the latter procedure, PG is elevated due to absorption of the sugar from the portal vein and therefore intestinal absorption is bypassed. The finding strongly indicates that GLP-1 not only potentiates insulin secretion after the meal but also primes β-cells to stay glucose competent during fasting. Near normalization of PG by GLP-1 was documented in patients with type 2 diabetes mellitus, and clinical trials of GLP-1 analogue or GLP-1-potentiating agents are in progress.[70] GLP-1 possesses various potentially antidiabetogenic actions other than stimulation of insulin secretion;[71] GLP-1 even increases β-cell mass by promoting β-cell proliferation and islet neogenesis, and by attenuating apoptosis.[72]

GIP is a 43-amino-acid polypeptide secreted from the endocrine cells in the duodenum and jejunum (proximal portion of the small intestine). Release of GIP is triggered by intraluminal input of carbohydrate and glucose absorption across the intestinal mucosa. Fat and protein *per se* are not able to stimulate GIP release, but these substances significantly potentiate glucose-induced GIP release. Physiological relevance of GIP is established by the findings that (1) infusion of GIP at physiological levels stimulates insulin release,[73]

(2) antibodies that neutralize GIP attenuate the insulin response to oral glucose, (3) GIP antagonists suppress meal-induced insulin release,[74] and (4) mice with disruption of the gene encoding the GIP receptor have hyperglycemia due to impaired insulin secretion after oral glucose load.[75] Interestingly, involvement of GIP signaling in development of obesity is suggested by the finding that GIP-receptor knockout mice are resistant to developing obesity in response to a high-fat diet.[76]

Pituitary adenylate cyclase-activating polypeptide (PACAP) is a neuropeptide stored in the nerve terminals located in the vicinity of pancreatic islets. Functional roles of PACAP might be similar to those of incretin. Remarkably, effects of PACAP on β-cells could be observed at extremely low concentrations ($\sim 10^{14}\ M$).[77]

GLP-1, GIP and PACAP bind to specific receptors that are coupled to the heterotrimeric G-protein. Upon binding to the respective receptors, Gs-protein and then adenylyl cyclase is activated and cellular cAMP is increased. cAMP enhances insulin exocytosis via multiple mechanisms: (1) facilitation of the KATP channel closure and opening of L-VDCC, which increases Ca^{2+} influx enhancing $[Ca^{2+}]i$, (2) enhancement of Ca^{2+}-triggered insulin release, (3) augmentation of insulin granule movement in the cell interior,[78] and (4) expansion of the readily releasable granule pool.[79] The first and second actions accelerate fusion of the granule membrane and the plasma membrane, whereas the latter two actions are most likely related to translocation, docking, and priming of the secretory granules.

All of the insulinotropic effects of cAMP are critically dependent on nutrient-derived metabolic signals, one of which is ATP. cAMP enhances insulin exocytosis in protein kinase A (PKA)-dependent and -independent manners.[80] Specific protein(s) to be phosphorylated by PKA, which mediates the insulinotropic action of cAMP, has not been identified yet. A possible role of cAMP-GEFII (Epac) as a downstream signaling molecule in the PKA-independent, Ca^{2+}-dependent cAMP action in the β-cell was reported.[81] Recently, it was found that SUR1 plays an important role in the cAMP-dependent regulation of Ca^{2+}-induced exocytosis.[82]

Acetylcholine

Activation of the parasympathetic system results in liberation of acetylcholine (Ach) from the nerve ending located in the islet. The cephalic phase of insulin release that occurs at a meal before the absorption of glucose is probably caused by parasympathetic activation. Postprandially, parasympathetic input is important for the optimal output of insulin.

Ach is a parasympathetic neurotransmitter. Ach binds to the muscarinic M_3 receptor that interacts with the heterotrimeric G-protein; subsequently, phospholipase C is activated and two distinct second messengers, diacylglycerol (DAG) and inositol-1,4,5-trisphosphate (IP_3), are generated.[83] DAG, in synergy with IP_3-induced elevation of $[Ca^{2+}]i$ (see the following text), activates classical PKC, whereas novel PKC are activated by DAG alone. Specific target proteins for phosphorylation by PKC have not been identified, as far as exocytosis of insulin is concerned. IP_3 binds to the ER IP_3 receptor and promotes Ca^{2+} release into the cytoplasm. However, it is unlikely that a Ca^{2+} rise due to this mechanism directly triggers exocytosis because the domain of increased Ca^{2+} is distant from the plasma membrane in which the granule fusion takes place. In the cells, Ca^{2+} "waves" do not propagate swiftly. Ach induced-Ca^{2+} release may facilitate insulin granule movement, augmenting insulin release through Ca^{2+}- or calmodulin-dependent phosphorylation of myosin light chain.[84] Ach depolarizes the plasma membrane by increasing the membrane permeability to Na^+.[85]

Norepinephrine and Other Inhibitory Inputs

Stress-induced inhibition of insulin release occurs mainly via sympathetic nerve activation. Norepinephrine (NE), a classic postganglionic neurotransmitter, is released from the nerve ending localized in the islet. NE binds to the α_2-adrenergic receptor coupled to the heterotrimeric G-protein on the β-cell membrane and inhibits insulin release through activation of Gi or Go proteins.[86] After NE binds to the receptor, the α- and $\beta\gamma$-subunits are dissociated, which results in activation of the KATP channel and inhibition of the L-VDCC and adenylyl cyclase. In addition, NE affects unidentified targets, suppressing insulin release; an involvement of protein acylation in this process has recently been suggested.[87] NE suppression of insulin release occurs mostly through this final mechanism.

Several other humoral factors such as epinephrine, somatostatin, prostaglandin E_2, and galanin inhibit insulin release by mechanisms similar to the one described for NE.[86] Epinephrine binds to the α_2-adrenergic receptor coupled to Gi or Go proteins and to the β-adrenergic receptor coupled to Gs protein/adenylyl cyclase. Epinephrine signaling through the former and latter paths inhibits and stimulates, respectively, insulin release, with the net effect being inhibitory. Somatostatin, prostaglandin E_2, and galanin all have respective Gi- or Go-protein-coupled receptors on the β-cell membrane. Galanin is a potent inhibitor of insulin release especially in rodent and canine β-cells. Its physiological importance in humans remains to be determined.

Significant inhibition of insulin release due to sympathetic nerve activation is seen during intense exercise and in patients with pheochromocytoma. In the former situation, inhibition of insulin release is absolutely required to maintain euglycemia because exercise increases glucose uptake by the skeletal muscle (independently from insulin action). If insulin release was not inhibited upon exercise, hepatic glucose production would not increase, and hypoglycemia would ensue. In patients with pheochromocytoma, excess production of NE or epinephrine by the adrenal tumor causes suppression of GSIS on one hand and increases hepatic glucose output on the other. Thus, diabetes with relative insulinopenia is a common metabolic abnormality in this disorder.

DIABETES AND HYPOGLYCEMIA

As outlined above, insulin release by the β-cell is intricately regulated by a balance of stimulatory and inhibitory inputs. Once the balance is lost and insulin release is pathologically lowered or heightened, glucose homeostasis is disrupted and hyperglycemia or hypoglycemia ensues.

The mechanism underlying impaired GSIS in diabetes is largely unknown except in the following situations. In maturity-onset diabetes of the young 2 (MODY2), glucose phosphorylation is attenuated due to mutation in the GK gene.[3] MODY is an autosomal dominant disease because heterozygotes develop the disease. GK mutations in MODY2 are associated with reduced affinity for glucose or attenuated maximum phosphorylating capacity. At a certain level of glucose, β-cells with the GK mutation cannot maintain normal phosphorylation of glucose. Because this is the rate-limiting step for overall glucose metabolism in the β-cell (see the preceding text), the entire glucose signaling is attenuated approximately in parallel to the degree of impairment of the GK.

In the majority of patients with maternally inherited diabetes with deafness, an A to G-transition in the mitochondrial tRNA Leu (UUR) gene at base pair 3243 is the disease locus.[88] A progressive loss of GSIS was found before development of diabetes,[89] and selective impairment of GSIS was reported in a clonal β-cell line with impeded mitochondrial respiratory chain.[90]

A polymorphism of Kir6.2, a subunit of the KATP channel, appears to be related to the susceptibility to type 2 diabetes. E23K (heterozygotes) and E23E (homozygotes) polymorphisms in the Kir6.2 allele are proposed to predispose individuals to develop type 2 diabetes.[91] The KATP channel is a heterooctamer; having one Kir6.2 molecule with E23 out of four Kir6.2 molecules is sufficient for it to experience a significant lowering of ATP sensitivity. Therefore, the opening probability of the KATP channel with E23K- or E23E-type Kir6.2 is higher than the KATP channel with K23K-type Kir6.2 at a given level of intracellular ATP. This is associated with reduced GSIS *in vitro*, and enhances susceptibility for developing diabetes. Quite interestingly, in Caucasian patients with type 2 diabetes, incidence of E23K and E23E combined was significantly higher than in nondiabetic controls.

Chronic hyperglycemia *per se* adversely affects the β-cell function. This phenomenon is called glucotoxicity. Glucotoxicity, in conjunction with lipotoxicity, contributes to the progressive deterioration of glucose homeostasis characteristic of type 2 diabetes. Chronic hyperglycemia may be a prerequisite for lipotoxicity.[62] High glucose levels induce β-cell apoptosis and impair insulin biosynthesis and insulin release; protein glycation, oxidative stress, and ER stress may be causally related to these processes.

A heterogeneous group of congenital, autosomal recessive, hypoglycemic disorders originated from genetic abnormalities of the pathways involved in GSIS; these disorders are collectively called hyperinsulinism of infancy (HI). Mutations of the subunits of the KATP channel (SUR1 or Kir6.2) were found in patients with HI.[16] In this case, β-cell is not equipped with the functional KATP channel. Such state is electrophysiologically equivalent to a permanent channel closure. Thus, insulin release is permanently elevated irrespective of PG, resulting in intractable hypoglycemia. HI can also be caused by "gain of function" mutation of the gene encoding the mitochondrial GDH, which upregulates deamination of glutamate to produce the TCA-cycle intermediate, 2KG.[56,92] Thus, anaplerotic input into the TCA cycle is exaggerated, which leads to excessive insulin release. Hypoglycemia happens especially after a high-protein meal. Because ammonia is produced with deamination of glutamate, hyperammonemia also occurs in this disorder.

Gain of function mutation of the GK gene was found in two pedigrees of HI, in which glucose phosphorylation is excessive at a given extracellular glucose concentration.[93] The set point of GSIS is shifted upward and hypoglycemia occurs. An association of type 2 diabetes and one form of mutation of SUR1 (E1506K) with reduced KATP-channel activity has recently been reported.[94] The subjects with this mutation often experience mild hypoglycemia during infancy; adult individuals exhibit an increased prevalence of diabetes. Increased β-cell apoptosis, rather than impairment of GSIS, is suggested as a possible etiology of diabetes in this case. As a dominant form of congenital hyperinsulinism, a mutation in high-affinity SUR1 has been reported.[95]

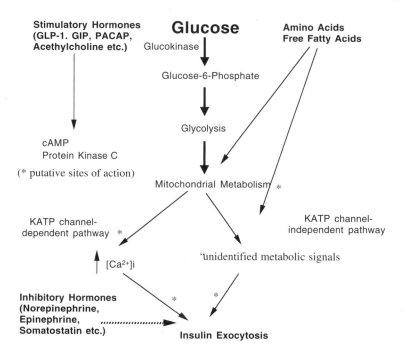

Figure 18.6 A complex signaling network in the pancreatic β-cell.

SUMMARY

For the regulation of insulin release, nutrients are playing a dominant role. Due to remarkable redundancy of the β-cell signaling network (Figure 18.6), a moderate degree of perturbation of a single step usually does not significantly affect the β-cell. The only exception for this is an up- or down-regulation of the GK, the β-cell glucose sensor. In humans, dietary intake (intestinal absorption) of glucose and glucose consumption by the skeletal muscle occur intermittently without synchronization. Moreover, there exists an uninterrupted glucose usage by tissues such as those in brain. Most importantly, brain cannot use FFA as a fuel, and its function critically depends on PG. Thus, maintenance of glucose homeostasis is of paramount importance, for which timing and quantity of insulin release are essential. A complex β-cell signaling network has evolved to meet such demand.

ACKNOWLEDGMENT

The author thanks Kiyoshi Hashizume and Toru Aizawa for careful reading of the text and constructive suggestions.

REFERENCES

1. A De Vos, H Heimberg, E Quartier, P Huypens, L Bouwens, D Pipeleers, F Schuit. Human and rat β cells differ in glucose transporter but not in glucokinase gene expression. *J Clin Invest* 96: 2489–2495, 1995.

2. FM Matschinsky. A lesson in metabolic regulation inspired by the glucokinase glucose sensor paradigm. *Diabetes* 45: 223–241, 1996.

3. N Vionnet, M Stoffel, J Takeda, K Yasuda, GI Bell, H Zouali, S Lesage, G Velho, F Iris, P Passa, P Froguel, D Cohen. Nonsense mutation in the glucokinase gene causes early-onset non-insulin-dependent diabetes mellitus. *Nature* 356: 721–722, 1992.

4. Y Terauchi, H Sakura, K Yasuda, K Iwamoto, N Takahashi, K Ito, H Kasai, H Suzuki, O Ueda, N Kamada, K Jishage, K Komeda, M Noda, Y Kanazawa, S Taniguchi, I Miwa, Y Akanuma, T Kodama, Y Yazaki, T Kadowaki. Pancreatic β-cell-specific targeted disruption of glucokinase gene. Diabetes mellitus due to defective insulin secretion to glucose. *J Biol Chem* 270: 30253–30256, 1995.

5. A Grupe, B Hultgren, A Ryan, YH Ma, M Bauer, TA Stewart. Transgenic knockouts reveal a critical requirement for pancreatic β cell glucokinase in maintaining glucose homeostasis. *Cell* 83: 69–78, 1995.

6. K Eto, Y Tsubamoto, Y Terauchi, T Sugiyama, T Kishimoto, N Takahashi, N Yamauchi, N Kubota, S Murayama, T Aizawa, Y Akanuma, S Aizawa, H Kasai, Y Yazaki, T Kadowaki. Role of NADH shuttle system in glucose-induced activation of mitochondrial metabolism and insulin secretion. *Science* 283: 981–985, 1999.

7. MJ MacDonald. The export of metabolites from mitochondria and anaplerosis in insulin secretion. *Biochim Biophys Acta* 1619: 77–88, 2003.

8. GM Grodsky. A threshold distribution hypothesis for packet storage of insulin and its mathematical modeling. *J Clin Invest* 51: 2047–2059, 1972.

9. N Takahashi, T Kishimoto, T Nemoto, T Kadowaki, H Kasai. Fusion pore dynamics and insulin granule exocytosis in the pancreatic islet. *Science* 297: 1349–1352, 2002.

10. M Noda, M Komatsu, GWG Sharp. The βHC9 pancreatic β-cell line preserves the characteristics of the progenitor mouse islets. *Diabetes* 45: 1766–1773, 1996.

11. SP O'Rahilly, Z Nugent, AS Rudenski, JP Hosker, MA Burnett, P Darling, RC Turner. β-cell dysfunction, rather than insulin insensitivity, is the primary defect in familial type 2 diabetes. *Lancet* 2: 360–364, 1986.

12. P Keskinen, S Korhonen, A Kupila, R Veijola, S Erkkila, H Savolainen, P Arvilommi, T Simell, J Ilonen, M Knip, O Simell. First-phase insulin response in young healthy children at genetic and immunological risk for Type I diabetes. *Diabetologia* 45: 1639–1648, 2002.

13. JC Henquin, HP Meissner. Significance of ionic fluxes and changes in membrane potential for stimulus-secretion coupling in pancreatic β-cells. *Experientia* 40: 1043–1052, 1984.

14. FM Ashcroft, P Rorsman. Electrophysiology of the pancreatic β-cell. *Prog Biophys Mol Biol* 54: 87–143, 1989.

15. N Inagaki, T Gonoi, IJP Clement, N Namba, J Inazawa, G Gonzalez, L Aguilar-Bryan, S Seino, J Bryan. Reconstitution of I_{KATP}, an inward rectifier subunit plus the sulfonylurea receptor. *Science* 270: 1166–1170, 1995.

16. L Aguilar-Bryan, J Bryan. Molecular biology of adenosine triphosphate-sensitive potassium channels. *Endocrine Rev* 20: 101–135, 1999.

17. TC Sudhof. The synaptic vesicle cycle: a cascade of protein–protein interactions. *Nature* 375: 645–653, 1995.

18. Z Gao, J Reavey-Cantwell, RA Young, P Jegier, BA Wolf. Synaptotagmin III/VII isoforms mediate Ca^{2+}-induced insulin secretion in pancreatic islet beta-cells. *J Biol Chem* 275: 36079–36085, 2000.

19. K Yaekura, R Julyan, BL Wicksteed, LB Hays, C Alarcon, S Sommers, V Poitout, DG Baskin, Y Wang, LH Philipson, CJ Rhodes. Insulin secretory deficiency and glucose intolerance in Rab3A null mice. *J Biol Chem* 278: 9715–9721, 2003.

20. Y Sato, T Aizawa, M Komatsu, N Okada, T Yamada. Dual functional role of membrane depolarization/Ca^{2+} influx in rat pancreatic β-cell. *Diabetes* 41: 438–443, 1992.

21. M Gembal, P Gilon, JC Henquin. Evidence that glucose can control insulin release independently from its action on ATP-sensitive K^+ channels in mouse B cells. *J Clin Invest* 89: 1288–1295, 1992.

22. U Panten, M Schwanstecher, A Wallasch, S Lenzen. Glucose both inhibits and stimulates insulin secretion from isolated pancreatic islets exposed to maximally effective concentrations of sulfonylureas. *Naunyn Schmiedebergs Arch Pharmacol* 338: 459–462, 1988.

23. L Best, AP Yates, S Tomlinson. Stimulation of insulin secretion by glucose in the absence of diminished potassium ($^{86}Rb^+$) permeability. *Biochem Pharmacol* 43: 2483–2485, 1992.

24. M Komatsu, T Schermerhorn, T Aizawa, GWG Sharp. Glucose stimulation of insulin release in the absence of extracellular Ca^{2+} and in the absence of any rise in intracellular Ca^{2+} in rat pancreatic islets. *Proc Natl Acad Sci USA* 92: 10728–10732, 1995.

25. M Komatsu, T Schermerhorn, M Noda, SG Straub, T Aizawa, GWG Sharp. Augmentation of insulin release by glucose in the absence of extracellular Ca^{2+}: new insights into stimulus-secretion coupling. *Diabetes* 46: 1928–1938, 1997.

26. N Taguchi, T Aizawa, Y Sato, F Ishihara, K Hashizume. Mechanism of glucose-induced biphasic insulin release: physiological role of adenosine triphosphate-sensitive K^+ channel-independent glucose action. *Endocrinology* 136: 3942–3948, 1995.

27. S Yamada, M Komatsu, T Aizawa, Y Sato, H Yajima, T Yada, S Hashiguchi, K Yamauchi, K Hashizume. Time-dependent potentiation of the beta-cell is a Ca^{2+}-independent phenomenon. *J Endocrinol* 172: 345–354, 2002.

28. F Ishihara, T Aizawa, N Taguchi, Y Sato, K Hashizume. Differential metabolic requirement for initiation and augmentation of insulin release by glucose: a study with rat pancreatic islets. *J Endocrinol* 143: 497–503, 1994.

29. P Detimary, P Gilon, M Nenquin, JC Henquin. Two sites of glucose control of insulin release with distinct dependence on the energy state in pancreatic B-cells. *Biochem J* 297: 455–461, 1994.

30. A Grimberg, RJ Ferry, Jr., A Kelly, S Koo-McCoy, K Polonsky, B Glaser, MA Permutt, L Aguilar-Bryan, D Stafford, PS Thornton, L Baker, CA Stanley. Dysregulation of insulin secretion in children with congenital hyperinsulinism due to sulfonylurea receptor mutations. *Diabetes* 50: 322–328, 2001.

31. M Komatsu, Y Sato, T Aizawa, K Hashizume. KATP channel-independent glucose action: an elusive pathway in stimulus-secretion coupling of pancreatic β-cell. *Endocrine J* 48: 275–288, 2001.

32. PM Jones, SL Howell. Insulin secretion studied in islets permeabilised by high voltage discharge. *Adv Exp Med Biol* 211: 279–291, 1986.

33. L Eliasson, E Renstrom, WG Ding, P Proks, P Rorsman. Rapid ATP-dependent priming of secretory granules precedes Ca^{2+}-induced exocytosis in mouse pancreatic β-cells. *J Physiol.* 503: 399–412, 1997.

34. Y Sato, JC Henquin. The K^+-ATP channel-independent pathway of regulation of insulin secretion by glucose: in search of the underlying mechanism. *Diabetes* 47: 1713–1721, 1998.

35. SA Metz, ME Rabaglia, TJ Pinter. Selective inhibitors of GTP synthesis impede exocytotic insulin release from intact rat islets. *J Biol Chem* 267: 12517–12527, 1992.

36. M Komatsu, M Noda, GWG Sharp. Nutrient augmentation of Ca^{2+}-dependent and Ca^{2+}-independent pathways in stimulus-coupling to insulin secretion can be distinguished by their guanosine triphosphate requirements: studies on rat pancreatic islets. *Endocrinology* 139: 1172–1183, 1998.

37. AK Nevins, DC Thurmond. Glucose regulates the cortical actin network through modulation of Cdc42 cycling to stimulate insulin secretion. *Am J Physiol Cell Physiol* 285: C698–C710, 2003.

38. P Detimary, GV den Berghe, J-C Henquin. Concentration dependence and time course of the effects of glucose on adenine and guanine nucleotides in mouse pancreatic islets. *J Biol Chem* 271: 20559–20565, 1996.

39. S Farfari, V Schulz, B Corkey, M Prentki. Glucose-regulated anaplerosis and cataplerosis in pancreatic β-cells. Possible implication of a pyruvate/citrate shuttle in insulin secretion. *Diabetes* 49: 718–726, 2000.

40. Y Liang, FM Matschinsky. Content of CoA-esters in perifused rat islets stimulated by glucose and other fuels. *Diabetes* 40: 327–333, 1991.

41. JT Deeney, J Gromada, M Høy, HL Olsen, CJ Rhodes, M Pretki, P-O Berggren, BE Corkey. Acute stimulation with long chain acyl-CoA enhances exocytosis in insulin-secreting cells (HIT T-15 and NMRI β-cells). *J Biol Chem* 275: 9363–9368, 2000.

42. M Komatsu, G Sharp. Palmitate and myristate selectively mimic the effects of glucose in augmenting insulin release in the absence of extracellular Ca^{2+}. *Diabetes* 47: 352–357, 1998.

43. M Komatsu, H Yajima, S Yamada, T Kaneko, Y Sato, K Yamauchi, K Hashizume, T Aizawa. Augmentation of Ca^{2+}-stimulated insulin release by glucose and long-chain fatty acids in rat pancreatic islets. Free fatty acids mimic ATP-sensitive K^+ channel-independent insulinotropic action of glucose. *Diabetes* 48: 1543–1549, 1999.

44. H Yajima, M Komatsu, S Yamada, SG Straub, T Kaneko, Y Sato, K Yamauchi, K Hashizume, GWG Sharp, T Aizawa. Cerulenin, an inhibitor of protein acylation, selectively attenuates nutrient stimulation of insulin release. A study in rat pancreatic islets. *Diabetes* 49: 712–717, 2000.

45. S Yamada, M Komatsu, Y Sato, K Yamauchi, T Aizawa, I Kojima. Nutrient modulation of palmitoylated 24-kDa protein in rat pancreatic islets. *Endocrinology* 144: 5232–5241, 2003.

46. LA Selbie, C Schmitz-Peiffer, Y Sheng, TJ Biden. Molecular cloning and characterization of PKCι, an atypical isoform of protein kinase C derived from insulin-secreting cells. *J Biol Chem* 268: 24296–24302, 1993.

47. TE Harris, SJ Persaud, PM Jones. Atypical isoforms of PKC and insulin secretion from pancreatic β-cells: evidence using Gö 6976 and Ro 31-8220 as PKC inhibitors. *Biochem Biophys Res Commun* 227: 672–676, 1996.

48. S Takasawa, K Nata, H Yonekura, H Okamoto. Cyclic ADP-ribose in insulin secretion from pancreatic β-cells. *Science* 259: 370–373, 1993.

49. GA Rutter, JM Theler, G Li, CB Wollheim. Ca²⁺ stores in insulin-secreting cells: lack of effect of cADP ribose. *Cell Calcium* 16: 71–80, 1994.

50. DL Webb, MS Islam, AM Efanov, G Brown, M Kohler, O Larsson, PO Berggren. Insulin exocytosis and glucose-mediated increase in cytoplasmic free Ca²⁺ concentration in the pancreatic β-cell are independent of cyclic ADP-ribose. *J Biol Chem* 271: 19074–19079, 1996.

51. WJ Malaisse, Y Kanda, K Inageda, O Scruel, A Sener, T Katada. Cyclic ADP-ribose measurements in rat pancreatic islets. *Biochem Biophys Res Commun* 231: 546–548, 1997.

52. P Maechler, CB Wollheim. Mitochondrial glutamate acts as a messenger in glucose-induced insulin exocytosis. *Nature* 402: 685–689, 1999.

53. S Yamada, T Aizawa, M Komatsu, Y Sato, K Hashizume. Glutamate is not the direct conveyer of KATP channel-independent insulinotropic action of glucose in rat islet β cell. *Endocr J* 48: 391–395, 2001.

54. MJ MacDonald, LA Fahien. Glutamate is not a messenger in insulin secretion. *J Biol Chem* 275: 34025–34027, 2000.

55. PA Smith, H Sakura, B Coles, N Gummerson, P Proks, FM Ashcroft. Electrogenic arginine transport mediates stimulus-secretion coupling in mouse pancreatic beta-cells. *J Physiol* 499 (Pt 3): 625–635, 1997.

56. A Kelly, C Li, Z Gao, CA Stanley, FM Matschinsky. Glutaminolysis and insulin secretion: from bedside to bench and back. *Diabetes* 51 Suppl 3: S421–426, 2002.

57. Z Gao, RA Young, G Li, H Najafi, C Buettger, SS Sukumvanich, RK Wong, BA Wolf, FM Matschinsky. Distinguishing features of leucine and alpha-ketoisocaproate sensing in pancreatic beta-cells. *Endocrinology* 144: 1949–1957, 2003.

58. GC Yaney, BE Corkey. Fatty acid metabolism and insulin secretion in pancreatic beta cells. *Diabetologia* 46: 1297–1312, 2003.

59. RL Dobbins, MW Chester, MB Daniels, JD McGarry, DT Stein. Circulating fatty acids are essential for efficient glucose-stimulated insulin secretion after prolonged fasting in humans. *Diabetes* 47: 1613–1618, 1998.

60. C Warnotte, P Gilon, M Nenquin, JC Henquin. Mechanisms of the stimulaiton of insulin release by saturated fatty acids. A study of palmitate effects in mouse β-cells. *Diabetes* 43: 703–711, 1994.

61. Y Itoh, Y Kawamata, M Harada, M Kobayashi, R Fujii, S Fukusumi, K Ogi, M Hosoya, Y Tanaka, H Uejima, H Tanaka, M Maruyama, R Satoh, S Okubo, H Kizawa, H Komatsu, F Matsumura, Y Noguchi, T Shinohara, S Hinuma, Y Fujisawa, M Fujino. Free fatty acids regulate insulin secretion from pancreatic beta cells through GPR40. *Nature* 422: 173–176, 2003.

62. V Poitout, RP Robertson. Secondary β-cell failure in type 2 diabetes-a convergence of glucotoxicity and lipotoxicity. *Endocrinology* 143: 339–342, 2002.

63. YP Zhou, VE Grill. Long-term exposure of rat pancreatic islets to fatty acids inhibits glucose-induced insulin secretion and biosynthesis through a glucose fatty acid cycle. *J Clin Invest* 93: 870–876, 1994.

64. G Boden, X Chen, J Rosner, M Barton. Effects of a 48-h fat infusion on insulin secretion and glucose utilization. *Diabetes* 44: 1239–1242, 1995.

65. Y Lee, H Hirose, M Ohneda, JH Johnson, JD McGarry, RH Unger. Beta-cell lipotoxicity in the pathogenesis of non-insulin-dependent diabetes mellitus of obese rats: impairment in adipocyte-beta-cell relationships. *Proc Natl Acad Sci USA* 91: 10878–10882, 1994.

66. RH Unger. Lipotoxicity in the pathogenesis of obesity-dependent NIDDM. Genetic and clinical implications. *Diabetes* 44: 863–870, 1995.

67. M Shimabukuro, YT Zhou, M Levi, RH Unger. Fatty acid-induced beta cell apoptosis: a link between obesity and diabetes. *Proc Natl Acad Sci USA* 95: 2498–2502, 1998.

68. DA D'Alessio, R Vogel, R Prigeon, E Laschansky, D Koerker, J Eng, JW Ensinck. Elimination of the action of glucagon-like peptide 1 causes an impairment of glucose tolerance after nutrient ingestion by healthy baboons. *J Clin Invest* 97: 133–138, 1996.

69. LA Scrocchi, TJ Brown, N MaClusky, PL Brubaker, AB Auerbach, AL Joyner, DJ Drucker. Glucose intolerance but normal satiety in mice with a null mutation in the glucagon-like peptide 1 receptor gene. *Nat Med* 2: 1254–1258, 1996.

70. JJ Holst. Therapy of type 2 diabetes mellitus based on the actions of glucagon-like peptide-1. *Diabetes Metab Res Rev* 18: 430–441, 2002.

71. DJ Drucker. Glucagon-like peptides. *Diabetes* 47: 159–169, 1998.

72. Y Li, T Hansotia, B Yusta, F Ris, PA Halban, DJ Drucker. Glucagon-like peptide-1 receptor signaling modulates beta cell apoptosis. *J Biol Chem* 278: 471–478, 2003.

73. X Jia, JC Brown, P Ma, RA Pederson, CH McIntosh. Effects of glucose-dependent insulinotropic polypeptide and glucagon-like peptide-I-(7-36) on insulin secretion. *Am J Physiol* 268: E645–651, 1995.

74. CC Tseng, TJ Kieffer, LA Jarboe, TB Usdin, MM Wolfe. Postprandial stimulation of insulin release by glucose-dependent insulinotropic polypeptide (GIP). Effect of a specific glucose-dependent insulinotropic polypeptide receptor antagonist in the rat. *J Clin Invest* 98: 2440–2445, 1996.

75. K Miyawaki, Y Yamada, H Yano, H Niwa, N Ban, Y Ihara, A Kubota, S Fujimoto, M Kajikawa, A Kuroe, K Tsuda, H Hashimoto, T Yamashita, T Jomori, F Tashiro, J Miyazaki, Y Seino. Glucose intolerance caused by a defect in the entero-insular axis: a study in gastric inhibitory polypeptide receptor knockout mice. *Proc Natl Acad Sci USA* 96: 14843–14847, 1999.

76. K Miyawaki, Y Yamada, N Ban, Y Ihara, K Tsukiyama, H Zhou, S Fujimoto, A Oku, K Tsuda, S Toyokuni, H Hiai, W Mizunoya, T Fushiki, JJ Holst, M Makino, A Tashita, Y Kobara, Y Tsubamoto, T Jinnouchi, T Jomori, Y Seino. Inhibition of gastric inhibitory polypeptide signaling prevents obesity. *Nat Med* 8: 738–742, 2002.

77. T Yada, M Sakurada, K Ishida, M Nakata, F Murata, A Arimura, M Kikuchi. Pituitary adenylate cyclase activating polypeptide is an extraordinarily potent intra-pancreatic regulator of insulin secretion from islet beta-cells. *J Biol Chem* 269: 1290–1293, 1994.

78. M Hisatomi, H Hidaka, I Niki. Ca^{2+}/calmodulin and cyclic 3,5′ adenosine monophosphate control movement of secretory granules through protein phosphorylation/dephosphorylation in the pancreatic β-cell. *Endocrinology* 137: 4644–4649, 1996.

79. S Yamada, M Komatsu, Y Sato, K Yamauchi, I Kojima, T Aizawa, K Hashizume. Time-dependent stimulation of insulin exocytosis by 3′,5′-cyclic adenosine monophosphate in the rat islet beta-cell. *Endocrinology* 143: 4203–4209, 2002.

80. E Renström, L Eliasson, P Rorsman. Protein kinase A-dependent and -independent stimulation of exocytosis by cAMP in mouse pancreatic B-cells. *J Physiol* 502: 105–118, 1997.

81. N Ozaki, T Shibasaki, Y Kashima, T Miki, K Takahashi, H Ueno, Y Sunaga, H Yano, M Y., T Iwanaga, Y Takai, S Seino. cAMP-GEF II is a direct target of cAMP in regulated exocytosis. *Nat Cell Biol* 2: 805–811, 2000.

82. L Eliasson, X Ma, E Renstrom, S Barg, PO Berggren, J Galvanovskis, J Gromada, X Jing, I Lundquist, A Salehi, S Sewing, P Rorsman. SUR1 regulates PKA-independent cAMP-induced granule priming in mouse pancreatic β-cells. *J Gen Physiol* 121: 181–197, 2003.

83. P Gilon, JC Henquin. Mechanisms and physiological significance of the cholinergic control of pancreatic β-cell function. *Endocr Rev* 22: 565–604, 2001.

84. W Yu, T Niwa, T Fukasawa, H Hidaka, T Senda, Y Sasaki, I Niki. Synergism of protein kinase A, protein kinase C, and myosin light-chain kinase in the secretory cascade of the pancreatic β-cell. *Diabetes* 49: 945–952, 2000.

85. P Gilon, JC Henquin. Activation of muscarinic receptors increases the concentration of free Na$^+$ in mouse pancreatic β-cells. *FEBS Lett* 315: 353–356, 1993.

86. GW Sharp. Mechanisms of inhibition of insulin release. *Am J Physiol* 271: C1781–1799, 1996.

87. H Cheng, SG Straub, GW Sharp. Protein acylation in the inhibition of insulin secretion by norepinephrine, somatostatin, galanin, and PGE2. *Am J Physiol Endocrinol Metab* 285: E287–294, 2003.

88. SW Ballinger, JM Shoffner, EV Hedaya, I Trounce, MA Polak, DA Koontz, DC Wallace. Maternally transmitted diabetes and deafness associated with a 10.4 kb mitochondrial DNA deletion. *Nat Genet* 1: 11–15, 1992.

89. G Velho, MM Byrne, K Clement, J Sturis, ME Pueyo, H Blanche, N Vionnet, J Fiet, P Passa, JJ Robert, KS Polonsky, P Froguel. Clinical phenotypes, insulin secretion, and insulin sensitivity in kindreds with maternally inherited diabetes and deafness due to mitochondrial tRNALeu(UUR) gene mutation. *Diabetes* 45: 478–487, 1996.

90. M Noda, S Yamashita, N Takahashi, K Eto, LM Shen, K Izumi, S Daniel, Y Tsubamoto, T Nemoto, M Iino, H Kasai, GW Sharp, T Kadowaki. Switch to anaerobic glucose metabolism with NADH accumulation in the beta-cell model of mitochondrial diabetes. Characteristics of betaHC9 cells deficient in mitochondrial DNA transcription. *J Biol Chem* 277: 41817–41826, 2002.

91. EM Nielsen, L Hansen, B Carstensen, SM Echwald, T Drivsholm, C Glumer, B Thorsteinsson, K Borch-Johnsen, T Hansen, O Pedersen. The E23K variant of Kir6.2 associates with impaired post-OGTT serum insulin response and increased risk of type 2 diabetes. *Diabetes* 52: 573–577, 2003.

92. ZY Gao, G Li, H Najafi, BA Wolf, FM Matschinsky. Glucose regulation of glutaminolysis and its role in insulin secretion. *Diabetes* 48: 1535–1542, 1999.

93. B Glaser, P Kesavan, M Heyman, E Davis, A Cuesta, A Buchs, CA Stanley, PS Thornton, MA Permutt, FM Matschinsky, KC Herold. Familial hyperinsulinism caused by an activating glucokinase mutation. *N Engl J Med* 338: 226–230, 1998.

94. H Huopio, T Otonkoski, I Vauhkonen, F Reimann, FM Ashcroft, M Laakso. A new subtype of autosomal dominant diabetes attributable to a mutation in the gene for sulfonylurea receptor 1. *Lancet* 361: 301–307, 2003.

95. PS Thornton, C MacMullen, A Ganguly, E Ruchelli, L Steinkrauss, A Crane, L Aguilar-Bryan, CA Stanley. Clinical and molecular characterization of a dominant form of congenital hyperinsulinism caused by a mutation in the high-affinity sulfonylurea receptor. *Diabetes* 52: 2403–2410, 2003.

19

Metabolic Actions of Glucagon-Like Peptides

RHONDA D. WIDEMAN AND
TIMOTHY J. KIEFFER

INTRODUCTION

Metabolism can be defined as the set of processes necessary for the maintenance of life. This includes breakdown of substances into energy and synthesis of the substances necessary for life from their individual components. These processes must be matched to food intake in order to meet, but not exceed, the metabolic needs of the body. Gastric activity must be balanced with the ability of the intestine to neutralize, break down, and absorb the nutrients arriving from the stomach. Furthermore, nutrient absorption must be coupled to the proportional liberation of hormones responsible for correct deposition and storage of the nutrients. Finally, this entire system must be adaptable to changes in nutrient availability or diet. Gastrointestinal (GI) hormones play a central role in orchestrating all of these metabolic events.

Two gastrointestinal hormones with well-described metabolic actions are glucagon-like peptide-1 (GLP-1) and glucagon-like peptide-2 (GLP-2), both products of the same prohormone from which glucagon is liberated in the pancreas.

Both are secreted from mucosal L cells of the intestine during a meal; these hormones can regulate food and water intake, gastric acid secretion and gastric emptying, intestinal permeability and nutrient transport, and pancreatic hormone secretion, as well as promote growth of pancreatic islets and the absorptive surface of the small intestine. These remarkable complementary actions are reviewed in this chapter. For further information on this topic, the reader is directed to other reviews.[1–3]

DISCOVERY OF GLUCAGON-LIKE PEPTIDES (GLPS)

Glucagon-like activity was first identified in bovine pancreatic extracts[4] and later established as glucagon, a product of islet α cells. However, following the development of assays for glucagon-like immunoreactivity, it became evident that GLPs are also present in specific areas of the intestine and the brain. The cloning of proglucagon cDNA and the isolation and subsequent sequencing of the intestinally derived GLPs led to the identification and localization of the various progluca-gon-derived peptides (PGDPs). Among the PGDPs identified, the most extensively studied are the GLPs.

SYNTHESIS AND EXPRESSION OF GLPS

The Glucagon Superfamily

GLP-1 and GLP-2 belong to the larger glucagon superfamily of peptide hormones, which also includes pituitary adenylate cyclase-activating polypeptide (PACAP), glucagon, GH-releasing factor (GRF), vasoactive intestinal polypeptide (VIP), secretin, and glucose-dependent insulinotropic polypeptide (GIP).[1] Despite diverse and sometimes opposing regulatory functions, the glucagon superfamily members bear significant similarity in terms of gene organization and peptide sequence. The proglucagon gene encodes GLP-1, GLP-2, glucagon, oxyntomodulin, glicentin, glicentin-related polypeptide, and several intervening peptides (see Figure 19.1). Of these, three bioactive peptides — glucagon, GLP-1, and GLP-2 — have

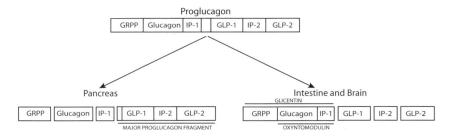

Figure 19.1 Expression and posttranslational processing of proglucagon in the pancreas, intestine, and brain. Cleavage of the proglucagon precursor at specific basic amino acids results in the production of different bioactive peptides in different tissues. In pancreatic α cells, the major bioactive peptide produced is glucagon, whereas in the intestine and brain, GLP-1 and GLP-2 are produced. GRPP, glicentin-related pancreatic polypeptide; IP-1 and IP-2, intervening peptides.

significant effects on nutrient homeostasis, whereas the function of the others remains largely unknown.

In addition to significant homology amongst the PGDPs, the GLP-1 and GLP-2 sequences are themselves highly conserved among different species; for example, GLP-1's amino acid sequence is identical among all mammalian species examined so far.[1] This high degree of conservation is indicative of the importance of the role of these peptides as key mediators of nutrient homeostasis and metabolic function.

Proglucagon Expression and Processing

The proglucagon gene is expressed in the pancreatic α cells, the intestinal L cells, and in the neurons of the caudal nucleus tractus solitarius and the hypothalamus. Gene expression is controlled by the proglucagon gene promoter in which several key transcriptional elements have been identified, largely via studies using immortalized glucagon-expressing cell lines.[1] The G1 element directs pancreatic α cell-specific expression, whereas the G2 and G3 elements are enhancers for α cell-specific expression. The presence of a cAMP response element

(CRE) is consistent with the observation that GLP-1 and GLP-2 production is cAMP-dependent.[5] The intestinal-specific element (ISE) directs intestinal L cell-specific proglucagon expression. A number of transcription factors, including Pax-6, Pax-4, Cdx-2/3, Brn-4, and Isl-1, have been shown to interact with the various gene regulatory elements in the proglucagon promoter.[1]

Whereas in other vertebrates alternative exon splicing is responsible for generating the various PGDPs, in humans it is differential posttranslational processing that results in the production of PGDPs from the 160 amino acid proglucagon precursor.[1] This processing is carried out by prohormone convertases (PCs), which are involved in the production of a wide variety of hormones. GLPs, along with oxyntomodulin and glicentin, are liberated from the proglucagon precursor via PC1-mediated processing in the brain and intestine.[6] In the intestine, the production site for GLPs is the enteroendocrine L cells, located in high concentration in the distal gut (especially the ileum and the colon). As a result of cleavage by PC2, the major products of proglucagon gene expression in pancreatic α cells are glucagon and the major proglucagon fragment. A small amount of GLP-1 appears to be cosecreted with glucagon by the pancreatic α cells. The function of this small quantity of GLP-1 remains unclear, but it is possible that it may have significant autocrine or paracrine effects in the pancreatic islets.[1] Whereas few α cells express PC1 normally, experimental induction of diabetes in rodents results in a remarkable upregulation of both PC1 and PC2 expression in α cells, with very little change in PC expression in β cells.[7] PC upregulation correlates with increased plasma and pancreatic GLP-1 levels, suggesting that GLP-1 production may be upregulated to counteract hyperglycemia and to enhance β cell performance under conditions in which β cell mass or function is insufficient to handle the nutrient loads that the body must process. This adaptability of GLP expression levels is clearly advantageous in matching peptide levels to long-term changes in metabolism.

SECRETION AND METABOLISM OF GLPS

In addition to regulation at the level of expression, GLPs are also subject to regulation at the secretory level. The intestinal L cells are ideally located for nutrient-regulated secretion of peptides because they are present in the mucosa, with microvilli reaching into the intestinal lumen. These microvilli can directly sense the presence of nutrients in the lumen and initiate appropriate levels of gut hormone secretion at the endocrine granule-rich basal lamina.[1] The diffuse distribution of L cells throughout the mucosa has complicated the study of mechanisms controlling GLP release. However, development of primary intestinal cell cultures and PGDP-secreting tumor-derived cell lines has allowed for investigation into the intracellular signaling mechanisms driving release of GLP-1 and GLP-2. It is clear that regulation is complex and integrated, involving both neural and hormonal inputs in addition to nutritional ones.

Because GLP-1 and GLP-2 are cosecreted from the L cells, it follows that their secretion is probably controlled by the same factors, though GLP-2 secretion is less well studied. The primary stimulus for secretion of GLPs into the bloodstream is ingestion of carbohydrates or fats, whereas systemic glucose administration does not enhance plasma levels of these peptides. Secretion may involve the sodium–glucose cotransporter because sugars that make use of this transporter (e.g., glucose, galactose) are potent stimulators of secretion, whereas nontransportable sugars are not.[8] It appears that secretion of GLPs is stimulated to a greater degree by long-chain, monounsaturated fatty acids than by short-chain, polyunsaturated or saturated fatty acids.[9] Ingestion of protein-containing meals also enhances secretion; however, proteins or amino acids alone do not seem to stimulate GLP-1 release.[10]

The precise mechanism by which the presence of ingested nutrients in the upper intestinal tract is coupled to secretion of GLP-1 and GLP-2 from the distal gut is not completely understood. It is unlikely that direct nutrient contact with L

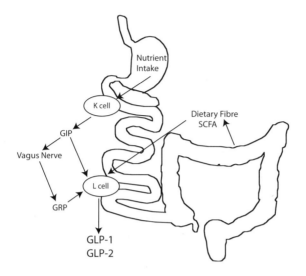

Figure 19.2 Model of the indirect mechanism driving secretion of GLPs. Entry of nutrients into the upper intestine induces GIP release from K cells. GIP stimulates the vagus nerve, which sends a signal to the L cells of the lower intestine, perhaps via release of gastrin-releasing peptide (GRP). Short-chain fatty acid (SCFA) release from the lower intestine resulting from the fermentation of dietary fibre also stimulates the L cells, resulting in secretion of GLP-1 and GLP-2.

cells is a requirement for secretion of GLPs because their levels increase within 10 min of nutrient ingestion,[8] and the majority of nutrients are absorbed prior to reaching L cells of the distal GI tract. Rather, it is likely that regulation of release occurs via hormonal and neural mediators. GIP, released from the upper intestinal tract K cells, potentiates GLP-1 and GLP-2 release from the more distal L cells via direct and indirect means (see Figure 19.2).[8] However, there is a lack of clear evidence for these pathways in humans. An alternate possibility is that L cells and K cells may be colocalized in the mid-small intestine, resulting in the simultaneous secretion of GIP and GLP-1.[11]

Several circulating forms of each of the GLPs have been identified. GLP-2_{1-33}, GLP-$1_{7-36 \text{ amide}}$, and GLP-1_{7-37} are generally understood to be the most highly bioactive forms of GLPs, whereas GLP-2_{3-33}, GLP-$1_{9-36 \text{ amide}}$, and GLP-1_{9-37} are less active early breakdown products.[1] This breakdown is catalyzed by the ubiquitous serine protease dipeptidyl peptidase IV (DPP IV). The role of DPP IV in GLP clearance is highlighted by the enhanced circulating bioactive GLP levels seen both in DPP IV-deficient animals and in animals administered DPP IV-resistant GLP analogs.[12] It is clear that the kidney also plays a role in clearance of both GLP-1 and GLP-2, based on increased circulating immunoreactive peptide levels in patients with renal failure and in nephrectomized animals.[1] Thus glucagon-like peptide activity is controlled not only at the level of secretion but also via clearance mechanisms that function as a fast metabolic "brake" to allow for minute-to-minute control over circulating GLP levels.

ACTIONS OF GLPS

The unique character of GLPs lies in their remarkable capacity to influence nutrient metabolism at virtually all levels, from food intake to gastric activity to nutrient breakdown, absorption, and storage (see Figure 19.3). The clear advantage of GLP-1 and GLP-2 having overlapping and complementary functions lies in the fact that, should the function of one of these peptides be disrupted, the other one could perhaps compensate for the loss. It is becoming increasingly evident that GLPs have characteristic general properties that are evident at all their sites of action, including an ability to regulate growth and apoptosis and an ability to adapt to changing dietary and metabolic conditions.

Effects on Food Intake

GLPs have been identified in hypothalamic areas involved in food and water intake. Intracerebroventricular (ICV) administration of GLP-1 inhibits food intake in rats, and a recent

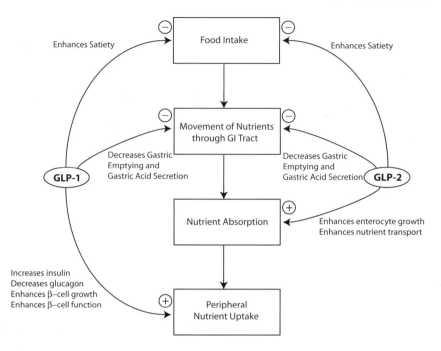

Figure 19.3 Actions of GLPs. Actions of GLP-1 and GLP-2 on each stage of nutrient metabolism. Both GLP-1 and GLP-2 enhance satiety, resulting in an inhibitory effect on food intake. Both peptides slow the movement of nutrients from the stomach to the intestine by decreasing gastric emptying and gastric acid secretion. GLP-2 enhances nutrient absorption by stimulating the growth of the absorptive surface of the small intestine and also by augmenting nutrient transport to the blood. Finally, GLP-1 has acute effects on peripheral nutrient uptake via increasing insulin and decreasing glucagon secretion, as well as longer-term effects via stimulating the growth and function of pancreatic β cells.

study reported that gastric distension activates medullary GLP-1- and GLP-2-containing neurons in rats.[13] ICV administration of GLP-2 also has short-term suppressive effects on food intake in rodents, and it is intriguing that disruption of GLP-1R signaling by genetic or pharmacological means results in enhanced inhibition of food intake by GLP-2.[14] This suggests that GLP-1 and GLP-2 may work in concert to regulate food intake. The role of GLPs in satiety has been reinforced by recent clinical studies showing that treatment with a GLP-1 analog not only enhanced glucose tolerance in patients with Type 2 diabetes but also correlated with significant weight loss.[15] Further evidence supporting the physiological relevance of GLPs in coupling nutrient handling to food intake comes from a recent study in which a protein preload of varying composition was administered to human subjects who were subsequently allowed *ad libitum* access to a buffet. Decreased food intake and lower reported hunger correlated with enhanced circulating GLP-1 levels following a whey-based preload, providing a possible example of GLPs acting as mediators controlling nutrient-specific satiety responses.[16]

In addition to stimulating the activity of central neurons involved in satiety, GLPs may have a longer-term role in protection of and growth promotion in these areas. GLP-1 enhances survival against experimentally induced cell death in neurons.[3,17] Exogenous GLP-1 is also reported to enhance learning, memory, and neuroprotection in rodents, whereas the learning deficits evident in GLP-1R knockout mice are corrected by restoration of central GLP-1R expression.[3,17] By linking food intake with metabolic needs, capacity for nutrient handling and long-term learning, memory, and neuroprotection, GLPs may provide a key integrative link providing organisms with a competitive advantage in terms of successful acquisition of food.

Effects on Nutrient Flow in the Gastrointestinal Tract

Once food has entered the body, it moves through the GI tract. For some time it has been supposed that there exists in humans an ileal "brake" that inhibits gastric emptying. By slowing the feed of gastric chyme into the small intestine, this brake ensures that the intestine receives chyme in amounts that it is capable of neutralizing, breaking down and, ultimately, absorbing. Both of the GLPs are thought to mediate the ileal brake effect by inhibiting gastric acid secretion and gastric motility, perhaps via indirect stimulation of central satiety centers.[8,18]

Together, these actions slow the enzymatic breakdown and absorption of nutrients and thereby reduce postprandial glycemic excursion.

It is well established that circulating GLP-2 levels increase significantly following bowel resection, and that this is associated with enhanced proliferation and decreased apoptosis of enterocytes.[12] Exogenous GLP-2 administration has also been found to promote crypt cell proliferation and increase crypt depth and villous height, thereby increasing mucosal tissue mass and surface area in rodents. Additionally, GLP-2 protects crypt cells and enterocytes against experimentally induced apoptosis.[8] By promoting the growth and survival of intestinal mucosa, GLPs help ensure that the absorptive surface of the GI tract is sufficient to handle food intake, fitting with the overall role of these peptides in adapting to changing metabolic conditions.

When nutrients reach the intestine, they are transported through enterocytes and absorbed into the bloodstream at the basolateral membrane. GLP-2 promotes nutrient uptake from the intestine not only by enhancing the transport and insertion of nutrient transporter proteins into the jejunal brush border membrane[19] but also by enhancing the activity of these transporters such that the efficiency of absorption is maximized.[20] Thus, it seems that nutrient intake may be coupled to enhanced absorptive ability of the intestine via increased GLP activity facilitating the disposal of ingested nutrients.

Effects on Nutrient Breakdown and Storage

Many hormones play a key role in ensuring the efficient breakdown and storage of nutrients, and it is clear that via regulating these hormones, GLPs themselves are crucial to normal nutrient disposal. In the acute sense, perhaps GLP-1's most important effect is the potent glucose-dependent insulinotropic action it exerts at the level of the pancreatic β cell. This action is potentiated in the longer term by stimulation of proinsulin gene transcription such that β cell stores of insulin are replenished.[12,21] GLP-1 can induce glucose responsiveness in otherwise unresponsive β cells.[22] Moreover, several groups have reported that GLP-1 also enhances transcription of specific genes involved in glucose sensing and metabolism (e.g., Glut-2, glucokinase, and hexokinase-1), thereby enhancing the long-term function of β cells.[23] Furthermore, GLP-1 stimulates somatostatin, inhibits glucagon secretion, and is a key counterregulatory hormone opposing the actions of leptin, a hormone which suppresses insulin synthesis and secretion.[1] The counterregulatory nature of GLP-1 and leptin is evident when either hormone is disrupted; the GLP-1R knockout mouse has increased leptin sensitivity,[24] whereas mice with mutated leptin receptors have enhanced sensitivity to GLP-1.[25]

An area of much excitement in recent years has been the identification of GLP-1-mediated trophic effects on β cells. GLP-1 secretion is increased following pancreatic injury in humans and animals, and animals treated with GLP-1R agonists have a more robust regenerative response to partial pancreatectomy.[7,26,27] The mechanism of GLP-1's enhancement of β cell mass appears to be multifold, with evidence mounting that GLP-1R agonists induce both neogenesis[28,29] and proliferation[30] of β cells in addition to inhibiting β cell apoptosis.[31] The importance of GLP-1 signaling in the normal pancreas is highlighted by the GLP-1R knockout mouse, which exhibits abnormal islet topography and mild glucose intolerance as well as impaired protection and recovery from β-cell insult.[2] The proliferative and reparative properties of GLPs may thus provide one means of keeping β cell mass in check with nutritional demands, even under conditions of injury.

In addition to regulating β cell mass and function, GLPs may have a role in regulating insulin-independent anabolism in muscle and adipose tissue.[1,8] The physiological role of GLP-1 and GLP-2 in these and other peripheral tissues remains unclear, but such actions enhancing nutrient disposal would be consistent with the emerging picture of GLPs as global regulators of nutrient homeostasis.

GLPs in Development and Adaptation

The ability of synthesis and secretion functions of GLPs to adapt to changes in overall metabolic status is highlighted by changing gastrointestinal hormone profiles in development, aging, and disease or injury. GLP secretion increases dramatically following infant weaning, presumably due to increased dietary carbohydrate levels; this enhanced secretion may be associated with the gut proliferation normally seen postweaning.[32] GLP and GLP-1 levels are elevated in older subjects, perhaps as a compensatory adjustment aimed at enhancing insulin synthesis and secretion and β cell function to compensate for the well-established age-related impairment of insulin secretion.[33] This again illustrates the ability of these peptides to bring the nutrient-handling capacity of the body in line with changing metabolic status.

Long-term changes in nutrient intake may also affect synthesis and secretion of GLPs. High fat-fed dogs have increased circulating GLP-1 levels and pancreatic GLP-1 receptor expression levels after 12 weeks of treatment.[34] This may be a compensatory effect aimed at maintaining blood glucose homeostasis by enhancing the body's ability to metabolize additional fat. In addition, elevated levels of leptin, a fat-derived satiety signal, stimulate increased GLP-1 production in rodents.[35] It is possible that increased adiposity resulting in enhanced circulating leptin triggers upregulation of GLP-1 secretion or GLP-1 receptor expression, providing a connection between the energy-absorbing intestinal tissue and the energy-storing adipose tissue. It is significant to note that a 2-week study in humans found no high-fat diet-induced increase in circulating GLP-1 levels, though this may relate to the short-term nature of the study.[36] The authors of this study did, however, note increased gastric emptying correlated with decreased pyloric tone, an adaptive mechanism aimed at enhancing the gut's ability to digest and absorb fat. The adaptive nature of glucagon-like peptides is thus evident from the adjustments made to their secretion and synthesis in response to changes in diet, age, and health status. All these adjustments seem to relate to the maintenance of normal tissue function in many of the tissues — pancreatic, intestinal, and central — involved in the regulation of nutrient flux.

MECHANISMS OF ACTION

GLPs bind receptors that have been identified as members of the seven transmembrane-spanning G protein-coupled family of receptors, which signal largely via the activation of adenylate cyclase. Adenylate cyclase catalyzes an increase in intracellular cAMP, which can have varied effects depending on the cell type in question. cAMP-dependent effects may include changes in intracellular calcium levels and ion channel activity, activation of transcription factors and subsequent changes in gene transcription levels, and exocytosis of stored hormones. In β cells, for instance, GLP-1-induced enhancement of cAMP levels results in closure of ATP-sensitive K^+ channels, which depolarizes the cell and, in turn, activates voltage-dependent Ca^{2+} channels to allow Ca^{2+} influx.[1] GLP-1 binding may also activate Ca^{2+} release from intracellular Ca^{2+} stores. The net increase in intracellular Ca^{2+} levels is the key stimulus for insulin release, coupling GLP-1 release to enhanced insulin secretion. In addition, GLP-1 regulates insulin gene transcription (and transcription of a host of other genes critical to normal β cell function) via a PKA-dependent mechanism.[1]

The GLP-1R has a strong binding affinity for GLP-1, but binds other ligands from the glucagon superfamily with poor binding affinity or none at all.[1] Exendin-4, a peptide isolated from the lizard *Heloderma suspectum,* which bears significant homology to GLP-1, is a strong GLP-1R agonist, whereas the truncated form (exendin$_{9-39}$) is a potent receptor antagonist that blocks GLP-1 binding. The broad expression pattern of GLP-1R, which has been identified in brain, lung, pancreatic, gastrointestinal, heart, and kidney tissues, reflects GLP-1's wide-ranging effects exerted in different systems.

In contrast to the GLP-1 receptor, which has been extensively characterized, much remains unknown about the GLP-2R. Study of GLP-2R has been complicated by the lack of tumor-derived cell lines and enriched populations of enteroendocrine cells expressing GLP-2R, and by the lack of available GLP-2R agonists and antagonists. GLP-2R expression has been localized to a distinct subset of intestinal enteroendocrine cells (expressing GIP, serotonin, PYY, chromogranin, or GLP-1) and not to intestinal epithelial cells.[8] This receptor distribution suggests that whereas some of GLP-2's effects may be directly exerted on the GLP-2R-bearing enteroendocrine cells, many effects may be mediated via an autocrine, paracrine, or endocrine factor released from enteroendocrine cells that acts on intestinal epithelial cells. Furthermore, GLP-2R expression in the brain suggests that GLP-2 may have central effects that are as yet unidentified.[37,38]

Treatment with GLP-2 protects fibroblasts stably transfected with the GLP-2R against apoptotic stimuli via suppression of a mediator of apoptosis and inhibition of glycogen synthase kinase 3 activity.[39,40] It is interesting to note that GLP-2 seems to inhibit apoptosis both directly by protecting GLP-2R-expressing cells and via indirect mechanisms in cells that evidently do not express the GLP-2R. GLP-2's pro-proliferative effects appear to involve GLP-2R-expressing enteric neurons, which, through an unidentified mediator, activate crypt cell proliferation.[2]

THERAPEUTIC POTENTIAL OF GLPS

Our growing understanding of GLPs as key mediators of nutrient homeostasis and metabolism has led to significant optimism about the therapeutic potential of these peptides in treating various metabolic disorders. Clinical trials investigating the properties of both GLP-1 and GLP-2 as therapeutics are currently under way.

T2D is characterized by a progressive loss of β cell function resulting in increased basal and postprandial blood glucose levels. Secretion of GLPs may be decreased in T2D,[41] suggesting that administration of exogenous peptide may be beneficial in these patients. The glucose-dependent insulinotropic and proliferative effects of GLP-1 make it an ideal candidate to enhance the function of β cells without the risk of hypoglycemia associated with exogenous insulin therapy. Several strategies for exploiting GLP-1's attractive complement of antidiabetic effects for the treatment of T2D are under investigation. These include use of DPP IV inhibitors to enhance endogenous active GLP-1 levels and the development of DPP IV-resistant GLP-1 analogs. In one recent study, a single bedtime injection of a long-acting GLP-1 derivative ($t_{1/2} \sim 10$ h) enhanced the insulin secretory response, decreased basal and postprandial glucose levels, and inhibited gastric emptying in human subjects with T2D.[42] The long-term effects of such a therapeutic approach (e.g., on β cell mass and function) will require further study, especially because the delayed gastric emptying resulting from analog treatment seems to be correlated with adverse gastrointestinal effects (i.e., nausea, cramping) in this and in other studies.[42,43]

GLP-2 may have therapeutic utility in treating intestinal disorders including small-bowel resection, colitis, drug-induced intestinal injury, inflammatory bowel syndrome, and others.[2] A degradation-resistant analog of GLP-2 is currently in clinical trials for the treatment of short-bowel syndrome, and has thus far been shown to enhance weight gain and nutrient absorption in these subjects.[8]

REMAINING QUESTIONS

It is evident that GLPs have a host of regulatory roles in coordinating metabolic function, cell survival and growth, and energy homeostasis. As we learn more, a picture of these peptides as stimulating multiple pathways in diverse tissues, at the level of cellular energy metabolism, gastrointestinal function, food intake, and perhaps even learned behavior, is emerging. The advantage of multitasking hormones with overlapping functions versus many hormones, each controlling a specific function, is clear when one considers the complex coordination necessary for the efficient and dynamic control of nutrient metabolism. Control of food intake and satiety by GLPs helps ensure appropriate food intake; regulation of nutrient transport, absorption, and uptake helps ensure that the body can efficiently make use of that which is ingested; and promotion of the survival of key cell types involved in energy balance helps ensure a functional system even in the face of damage or disease.

In many cases, however, our knowledge of these peptides remains incomplete. The signal transduction pathways driving the effects of GLPs in many cell types remain incompletely elucidated. For example, the mediators coupling GLP-2 release from enteroendocrine L cells to enhanced enterocyte survival and proliferation remain unidentified. Major gaps persist in our understanding of the function of GLP-2 under normal physiological conditions. These questions could be addressed using immunoneutralization studies or GLP-2R antagonists, or by developing targeted GLP-2R knockout models. The physiological importance of both GLPs in the CNS in humans remains to be seen and is particularly interesting, given new data suggesting a role for GLP-1 in learning and neuroprotection.[17] The role of both GLPs during early development also requires further study.

That both GLP-1 and GLP-2 are currently in clinical trials for treatment of human diseases is encouraging. With a greater understanding of the mechanisms underlying the regulated synthesis and secretion of GLP-1 and GLP-2 and the circumstances under which such regulation is altered, it is likely that we will be able to make use of these intriguing compounds to enhance nutrient metabolism and to augment or preserve β cell or intestinal epithelial mass in the setting of tissue injury or disease.

ACKNOWLEDGMENT

Support is gratefully acknowledged by RDW to the Natural Sciences and Engineering Research Council of Canada (NSERC) and the Michael Smith Foundation for Health Research (MSFHR) and by TJK to the Juvenile Diabetes Research Foundation (JDRF) and MSFHR.

References

1. TJ Kieffer and JF Habener. The glucagon-like peptides. *Endocr Rev* 20: 876–913, 1999.

2. DJ Drucker. Glucagon-like peptides: regulators of cell proliferation, differentiation, and apoptosis. *Mol Endocrinol* 17: 161–171, 2003.

3. T Perry and NH Greig. The glucagon-like peptides: a double-edged therapeutic sword? *Trends Pharmacol Sci* 24: 377–383, 2003.

4. CP Kimbal and JR Murlin. Aqueous extracts of pancreas. III. Some precipitation reactions of insulin. *J Biol Chem* 58: 337–346, 1923.

5. DJ Drucker. Minireview: the glucagon-like peptides. *Endocrinology* 142: 521–527, 2001.

6. S Dhanvantari, NG Seidah, and PL Brubaker. Role of prohormone convertases in the tissue-specific processing of proglucagon. *Mol Endocrinol* 10: 342–355, 1996.

7. Y Nie, M Nakashima, PL Brubaker, QL Li, R Perfetti, E Jansen, Y Zambre, D Pipeleers, and TC Friedman. Regulation of pancreatic PC1 and PC2 associated with increased glucagon-like peptide 1 in diabetic rats. *J Clin Invest* 105: 955–965, 2000.

8. DG Burrin, Y Petersen, B Stoll, and P Sangild. Glucagon-like peptide 2: a nutrient-responsive gut growth factor. *J Nutr* 131: 709–712, 2001.

9. AS Rocca and PL Brubaker. Stereospecific effects of fatty acids on proglucagon-derived peptide secretion in fetal rat intestinal cultures. *Endocrinology* 136: 5593–5599, 1995.

10. C Herrmann, R Goke, G Richter, HC Fehmann, R Arnold, and B Goke. Glucagon-like peptide-1 and glucose-dependent insulin-releasing polypeptide plasma levels in response to nutrients. *Digestion* 56: 117–126, 1995.

11. K Mortensen, LL Christensen, JJ Holst, and C Orskov. GLP-1 and GIP are colocalized in a subset of endocrine cells in the small intestine. *Regul Pept* 114: 189–196, 2003.

12. DJ Drucker. Biological actions and therapeutic potential of the glucagon-like peptides. *Gastroenterology* 122: 531–544, 2002.

13. N Vrang, CB Phifer, MM Corkern, and HR Berthoud. Gastric distension induces c-Fos in medullary GLP-1/2-containing neurons. *Am J Physiol Regul Integr Comp Physiol* 285: R470–478, 2003.

14. J Lovshin, J Estall, B Yusta, TJ Brown, and DJ Drucker. Glucagon-like peptide (GLP)-2 action in the murine central nervous system is enhanced by elimination of GLP-1 receptor signaling. *J Biol Chem* 276: 21489–21499, 2001.

15. MS Fineman, TA Bicsak, LZ Shen, K Taylor, E Gaines, A Varns, D Kim, and AD Baron. Effect on glycemic control of exenatide (synthetic exendin-4) additive to existing metformin or sulfonylurea treatment in patients with type 2 diabetes. *Diabetes Care* 26: 2370–2377, 2003.

16. WL Hall, DJ Millward, SJ Long, and LM Morgan. Casein and whey exert different effects on plasma amino acid profiles, gastrointestinal hormone secretion, and appetite. *Br J Nutr* 89: 239–248, 2003.

17. MJ During, L Cao, DS Zuzga, JS Francis, HL Fitzsimons, X Jiao, RJ Bland, M Klugmann, WA Banks, DJ Drucker, and CN Haile. Glucagon-like peptide-1 receptor is involved in learning and neuroprotection. *Nat Med* 9: 1173–1179, 2003.

18. E Naslund, J Bogefors, S Skogar, P Gryback, H Jacobsson, JJ Holst, and PM Hellstrom. GLP-1 slows solid gastric emptying and inhibits insulin, glucagon, and PYY release in humans. *Am J Physiol* 277: R910–916, 1999.

19. A Au, A Gupta, P Schembri, and CI Cheeseman. Rapid insertion of GLUT2 into the rat jejunal brush-border membrane promoted by glucagon-like peptide 2. *Biochem J* 367: 247–254, 2002.

20. CI Cheeseman. Upregulation of SGLT-1 transport activity in rat jejunum induced by GLP-2 infusion *in vivo*. *Am J Physiol* 273: R1965–1971, 1997.

21. DJ Drucker, J Philippe, S Mojsov, WL Chick, and JF Habener. Glucagon-like peptide I stimulates insulin gene expression and increases cyclic AMP levels in a rat islet cell line. *Proc Natl Acad Sci USA* 84: 3434–3438, 1987.

22. GG Holz, WM Kuhtreiber, and JF Habener. Pancreatic beta-cells are rendered glucose-competent by the insulinotropic hormone glucagon-like peptide-1(7-37). *Nature* 361: 362–365, 1993.

23. J Buteau, R Roduit, S Susini, and M Prentki. Glucagon-like peptide-1 promotes DNA synthesis, activates phosphatidylinositol 3-kinase and increases transcription factor pancreatic and duodenal homeobox gene 1 (PDX-1) DNA binding activity in beta (INS-1)-cells. *Diabetologia* 42: 856–864, 1999.

24. LA Scrocchi, TJ Brown, and DJ Drucker. Leptin sensitivity in nonobese glucagon-like peptide I receptor / mice. *Diabetes* 46: 2029–2034, 1997.

25. X Jia, R Elliott, YN Kwok, RA Pederson, and CH McIntosh. Altered glucose dependence of glucagon-like peptide 1(7-36)-induced insulin secretion from the Zucker (fa/fa) rat pancreas. *Diabetes* 44: 495–500, 1995.

26. K Tanjoh, R Tomita, and N Hayashi. The peculiar processing of glucagon and glucagon-related peptides in patients after pancreatectomy. *Hepatogastroenterology* 49: 825–832, 2002.

27. DD De Leon, S Deng, R Madani, RS Ahima, DJ Drucker, and DA Stoffers. Role of endogenous glucagon-like peptide-1 in islet regeneration after partial pancreatectomy. *Diabetes* 52: 365–371, 2003.

28. C Tourrel, D Bailbe, MJ Meile, M Kergoat, and B Portha. Glucagon-like peptide-1 and exendin-4 stimulate beta-cell neogenesis in streptozotocin-treated newborn rats resulting in persistently improved glucose homeostasis at adult age. *Diabetes* 50: 1562–1570, 2001.

29. G Xu, DA Stoffers, JF Habener, and S Bonner-Weir. Exendin-4 stimulates both beta-cell replication and neogenesis, resulting in increased beta-cell mass and improved glucose tolerance in diabetic rats. *Diabetes* 48: 2270–2276, 1999.

30. J Buteau, S Foisy, E Joly, and M Prentki. Glucagon-like peptide 1 induces pancreatic beta-cell proliferation via transactivation of the epidermal growth factor receptor. *Diabetes* 52: 124–132, 2003.

31. Y Li, T Hansotia, B Yusta, F Ris, PA Halban, and DJ Drucker. Glucagon-like peptide-1 receptor signaling modulates beta cell apoptosis. *J Biol Chem* 278: 471–478, 2003.

32. DG Burrin and B Stoll. Key nutrients and growth factors for the neonatal gastrointestinal tract. *Clin Perinatol* 29: 65–96, 2002.

33. L Ranganath, I Sedgwick, L Morgan, J Wright, and V Marks. The ageing entero-insular axis. *Diabetologia* 41: 1309–1313, 1998.

34. GW van Citters, M Kabir, SP Kim, SD Mittelman, MK Dea, PL Brubaker, and RN Bergman. Elevated glucagon-like peptide-1-(7-36)-amide, but not glucose, associated with hyperinsulinemic compensation for fat feeding. *J Clin Endocrinol Metab* 87: 5191–5198, 2002.

35. Y Anini and PL Brubaker. Role of leptin in the regulation of glucagon-like peptide-1 secretion. *Diabetes* 52: 252–259, 2003.

36. KA Boyd, DG O'Donovan, S Doran, J Wishart, IM Chapman, M Horowitz, and C Feinle. High-fat diet effects on gut motility, hormone, and appetite responses to duodenal lipid in healthy men. *Am J Physiol Gastrointest Liver Physiol* 284: G188–196, 2003.

37. DG Munroe, AK Gupta, F Kooshesh, TB Vyas, G Rizkalla, H Wang, L Demchyshyn, ZJ Yang, RK Kamboj, H Chen, K McCallum, M Sumner-Smith, DJ Drucker, and A Crivici. Prototypic G protein-coupled receptor for the intestinotrophic factor glucagon-like peptide 2. *Proc Natl Acad Sci USA* 96: 1569–1573, 1999.

38. B Yusta, L Huang, D Munroe, G Wolff, R Fantaske, S Sharma, L Demchyshyn, SL Asa, and DJ Drucker. Enteroendocrine localization of GLP-2 receptor expression in humans and rodents. *Gastroenterology* 119: 744–755, 2000.

39. B Yusta, RP Boushey, and DJ Drucker. The glucagon-like peptide-2 receptor mediates direct inhibition of cellular apoptosis via a cAMP-dependent protein kinase-independent pathway. *J Biol Chem* 275: 35345–35352, 2000.

40. B Yusta, J Estall, and DJ Drucker. Glucagon-like peptide-2 receptor activation engages bad and glycogen synthase kinase-3 in a protein kinase A-dependent manner and prevents apoptosis following inhibition of phosphatidylinositol 3-kinase. *J Biol Chem* 277: 24896–24906, 2002.

41. T Vilsboll, T Krarup, J Sonne, S Madsbad, A Volund, AG Juul, and JJ Holst. Incretin secretion in relation to meal size and body weight in healthy subjects and people with type 1 and type 2 diabetes mellitus. *J Clin Endocrinol Metab* 88: 2706–2713, 2003.

42. CB Juhl, M Hollingdal, J Sturis, G Jakobsen, H Agerso, J Veldhuis, N Porksen, and O Schmitz. Bedtime administration of NN2211, a long-acting GLP-1 derivative, substantially reduces fasting and postprandial glycemia in type 2 diabetes. *Diabetes* 51: 424–429, 2002.

43. J Rachman, FM Gribble, BA Barrow, JC Levy, KD Buchanan, and RC Turner. Normalization of insulin responses to glucose by overnight infusion of glucagon-like peptide 1 (7-36) amide in patients with NIDDM. *Diabetes* 45: 1524–1530, 1996.

IV

Calcium-Dependent Signaling

20

Control of Intracellular Calcium Levels

SHYAMALA DAKSHINAMURTI,
JONATHAN GEIGER, AND
KRISHNAMURTI DAKSHINAMURTI

INTRODUCTION

Calcium is arguably the most versatile and universal second messenger. As such, calcium plays a key signaling role in cells. The coupling of membrane excitation to cellular responses through calcium has long been recognized.[1] As an intracellular second messenger, cytosolic free calcium regulates, directly or indirectly, vital cellular functions. Physiological and pathological processes such as fertilization, proliferation, metabolism, membrane excitability, neurotransmitter release, synaptic plasticity, gene transcription, muscle contraction, neuronal injury through ischemia, epilepsy, and trauma, as well as cell death by necrosis or apoptosis, all involve calcium-mediated mechanisms.[2,3] Thus, mechanisms that regulate calcium homeostasis are central to physiological and pathophysiological processes. In this chapter, cellular calcium influx and calcium release from the endo/sarcoplasmic reticulum and their control by various intracellular metabolites are discussed.

Intracellular signal transduction following plasma membrane receptor activation is the principal means by which external stimuli can modulate cellular function. Calcium is a necessary component of the major messenger systems such as the cyclic nucleotide or the inositol phosphate pathways.[4,5] Transient cellular responses are initiated by rapid rise in the concentration of cytosolic ionic free calcium $[Ca^{2+}]_i$. More sustained cellular responses occur through calcium-mediated activation of protein kinases, phosphorylation of proteins that can affect cell morphology, proliferation, permeability, and cellular calcium homeostasis.[5-7] Cytoplasmic Ca^{2+} controls many aspects of cell life and death. Stimulus-response coupling phenomena, such as excitation–contraction coupling in muscle and excitation–secretion coupling in the endocrine and nervous systems, are regulated by $[Ca^{2+}]_i$. These systems are integrated and are essential for maintaining normal structure and function in various cell types and tissues.[8]

In excess, $[Ca^{2+}]_i$ is toxic to the cell. Hence, cellular $[Ca^{2+}]_i$ is tightly regulated. Under resting conditions, the intracellular concentration of free ionized calcium $[Ca^{2+}]_i$ is about 100 nM or less. Following excitation, the $[Ca^{2+}]_i$ rises by 10- to 100-fold. Even within the cytoplasm the concentration of calcium is not uniform. This has implications to cellular calcium homeostasis. The concentration of extracellular calcium (serum levels of ionized calcium) is roughly 10,000 times that found in the cytoplasm of the cell in the resting condition.[9,10] The cell coordinates a set of processes that regulate the transport and sequestration of extracellular calcium to maintain low $[Ca^{2+}]_i$ in the context of the high inwardly directed concentration and electrochemical gradients. Thus, the precise control of $[Ca^{2+}]_i$ is the central feature of mechanisms geared to achieving the optimal performance of a variety of cellular functions. It is to be noted that drastic functional changes in $[Ca^{2+}]_i$ are most likely not compatible with life.[11]

The pathways leading to the inflow of calcium from the extracellular space and the release of intracellular stored calcium must be complemented by mechanisms of calcium outflow from the cell, calcium uptake by cellular organelles, and complexing to intracellular calcium-binding proteins to maintain the resting $[Ca^{2+}]_i$. The major mechanisms of cellular influx and efflux in a composite excitable cell are represented in Figure 20.1 and comprise the following: voltage-sensitive calcium channel (VSCC), also referred to as voltage-operated, voltage-mediated, or voltage-gated calcium channel; ligand-gated calcium channel (LGCC), also termed receptor-mediated calcium channel; extracellular ATP-mediated purinergic P2X and P2Y pathways; $Na^+–Ca^{2+}$ exchange; sarco- and endoplasmic (SERCA) calcium pump; plasma membrane calcium ATPase (PMCA); ryanodine receptor- and IP_3 receptor-mediated calcium release mechanisms; and depletion-operated calcium channel and voltage-sensitive release mechanisms.

CELLULAR CALCIUM INFLUX PATHWAYS

Gated Calcium Entry

Voltage-Sensitive Calcium Channels (VSCC)

Many excitable cells express voltage-sensitive calcium channels, which translate an action potential into an increase of $(Ca^{2+})_i$. VSCCs couple plasma membrane depolarization to a variety of cellular processes such as action potential generation, neurotransmitter or hormone release, muscle contraction, neurite outgrowth, synaptogenesis, calcium-dependent gene expressions, synaptic plasticity, and cell death.[12] At the resting potential of the cell, the calcium channels are closed, and open only upon depolarization. These channels are members of a superfamily of voltage-gated calcium channels that include low-voltage activated, rapidly inactivating transient (T-type) channels, which are opened by small depolarizations from the resting potential, and high-voltage activated channels (L, N, P, Q, and R types) determined by their sensitivity to drugs and toxins. The latter channels are characterized by a high voltage of activation, large channel conductance, and slow voltage-dependent inactivation. L-type (longlasting) calcium channels are distributed widely in the cardiovascular, nervous, and endocrine systems. In muscle cells, the L-type channels are embedded in the T-tubule membrane. The receptors serve as voltage sensors associated with asymmetric

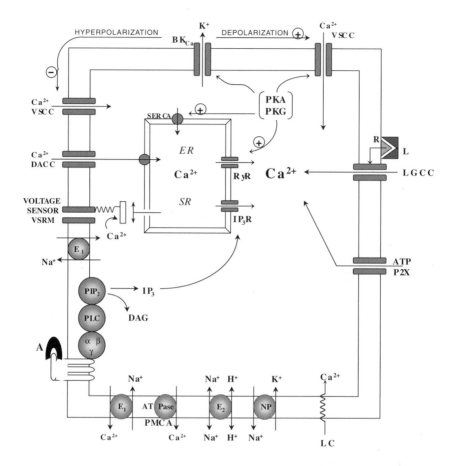

Figure 20.1 Cellular and intracellular calcium channels. VSCC, voltage-sensitive calcium channel; LGCC, ligand-gated calcium channel; A, agonist; L, ligand; P2X, P2X purinoceptor; P2Y, P2Y purinoceptor; RyR, ryanodine receptor; IP_3R, inositol trisphosphate receptor; SERCA, sarco(endo)plasmic reticulum ATPase; Bk_{ca}, calcium-dependent K^+ channel; SR, sarco(endo)plasmic reticulum; VSRM, voltage-sensitive release mechanism; DACC, depletion-operated calcium channel; IP_3, inositol trisphosphate; DAG, diacylglycerol; PIP_2, phosphatidylinositol 4, 5-bisphosphate; E_1, Na^+–Ca^{2+} exchanger; NP, sodium pump; LC, calcium leak; PMCA, plasma membrane ATPase.

charge movements consistent with depolarization to the contractile threshold.[13,14] T-type calcium channel subtypes are present in smooth muscle cells and in sinus node cells where they mediate cellular pacemaker function. Neurons contain both L-type and N-type VSCCs. L-type channels are localized in cell bodies and proximal dendrites in neurons and are present in clusters in subsynaptic membranes of some synapses.[15] In cultured cortical neurons, activation of the transcription of immediate early genes by repetitive electrical activity depends upon calcium entry through L-type channels.[16] Activation of the c-AMP response element binding protein (CREB) by cellular depolarization requires calcium influx via the L-type channel.[17] In the neuroendocrine system, activation of the L-type calcium channel is involved in persistent secretion induced by depolarization and thus in the maintenance of homeostasis of hormone secretion.

The various types of VSCCs have specific characteristics. These channels are pharmacological receptors with discrete drug-binding sites. Three chemically distinct classes of compounds have been characterized as calcium-channel antagonists: phenylalkylamines (e.g., verapamil), benzothiazepines (e.g., diltiazem), and the dihydropyridines (e.g., nifedipine). The antagonists bind at different sites on the L-type calcium channel and thereby decrease the influx of calcium into the cell through this channel.[18] Dihydropyridine analogs that have calcium agonist properties have also been synthesized. VSCCs are altered in their expression and function in many pathological conditions.[12] Calcium-channel antagonists are cardiovascular drugs of clinical significance.

Regulations of VSCC

Voltage-sensitive calcium channels are extensively modulated by a variety of cellular regulatory mechanisms including cAMP-dependent protein kinase (PKA), cGMP-dependent protein kinase (PKB), and protein kinase C. The channels are also regulated by protein tyrosine kinase, calmodulin-dependent protein kinases, and by G protein subunits.[19,20] Activation of β-adrenergic receptor pathways in cardiomyocytes causes phosphorylation and activation of L-type VSCC, thereby increasing calcium influx.[21,22] β_2-Adrenergic receptors in cardiac muscle, which can couple to both G_s and G_i, produce a localized activation of L-type VSCC through stimulation of phosphorylation of a serine residue at position 1928.[23] This stimulation of calcium channels by PKA is consistent with the well-recognized positive inotropic effect of β adrenoceptors in cardiac muscle. Even in smooth muscle there is evidence to suggest that L-type VSCCs are activated by PKA as demonstrated in cardiac muscle.[24,25]

VSCCs are also modulated by inositol phosphates and arachidonic acid.[12] PKC has been reported to increase L-type VSCC-mediated calcium-channel currents in various smooth muscle preparations. Agonists such as norepinephrine, angiotensin II, and serotonin stimulate channel activity through PKC-dependent mechanisms.[26] This effect could be masked because the agonists that activate PKC also produce inositol 1,4,5-trisphosphate-induced release of calcium from sarcoplasmic reticulum. The PKA and PKC pathways are linked to agonist-induced activation of G-protein subunits. These can regulate different ion channels not only through intracellular second messenger pathways but also by a direct gating of ion channels.[27]

Evidence that PKG directly phosphorylates calcium channels in cardiac muscle by phosphorylating serine 533 has been presented.[28] An additional mechanism by which PKG could inhibit calcium-channel activity is through dephosphorylation of the channel. In smooth muscle, PKG has been shown to inhibit calcium channels.[25] This is in addition to the role of PKG acting through the endothelium-relaxing factor, nitric oxide.

Feedback Control

The opening of single channels in plasma membrane results in increases in $[Ca^{2+}]_i$ globally by recruiting nearby calcium channels (a calcium wave) or through increases in $[Ca^{2+}]_i$ that remain highly localized. The localized increases in $[Ca^{2+}]_i$ can stimulate ryanodine receptor-regulated channels on sarco/ endoplasmic reticulum and release stored calcium leading to calcium sparks. A single calcium spark is capable of producing a very high (10 to 100 μM) local increase in about 1% of the cell volume while increasing $[Ca^{2+}]_i$ by less than 2 nM. Calcium sparks have the potential to modulate calcium-dependent processes that are not responsive to low $[Ca^{2+}]_i$. Calcium sparks could activate K_{ca} channels (also referred to as BK_{ca}). Activation of PKA or PKG increases BK_{ca} channel activity. Membrane hyperpolarization as a result of increased potassium channel activity closes the L-type calcium channel, a negative feedback effect.[29,30] Voltage-gated calcium channels are also subject to slower regulation by depolarizing stimuli. Chronic depolarization with elevation of extracellular K$^+$ downregulates calcium channel density as determined by binding of 1,4-dihydropyridine (DHP) and channel function as measured by $^{45}Ca^{2+}$ uptake. Chronic membrane depolarization reduces both low- and high-voltage-activated channels.

Capacitance Calcium Entry

Store-operated cation channels are activated by depletion of calcium store in the endo/sarcoplasmic reticulum. This concept of "capacitative calcium entry" or "depletion-activated calcium current" implies Ca^{2+} entry across plasma membrane following depletion of stored calcium. Diverse stimuli that release stored calcium such as ryanodine, thapsigargin, or IP_3 could lead to capacitative calcium entry. The signal for activating capacitative entry of Ca^{2+} may involve some diffusible signals coming from the store[31] or physical association of the channel with IP_3 receptors on the store membrane.[32] A direct activation of membrane calcium channel by IP_3, IP_4 (inositol 1,3,4,5-tetrakisphosphate) or G proteins are among the mechanisms suggested. A depletion of store calcium leads also to the production of arachidonic acid, regardless of the way in which store calcium is depleted. Arachidonic acid induces an entry of extracellular calcium and appears to act directly. A nonmetabolizable analog of arachidonic acid, 5,8,11,14-eicosatetraynoic acid, mimics its action in CHO-NTR cells, a model characterized for store depletion.[33]

Ligand-Gated Calcium Channels

Ligand-gated calcium channels represent a class of channels activated by binding of ligands to a special receptor domain on the membrane. In the nervous system, these channels are activated by excitatory neurotransmitters such as glutamate, acetylcholine, ATP, or serotonin. All ligand-gated channels are cation channels that are permeable to calcium, although to different extents.

Serotonin $5HT_3$ receptors also belong to the superfamily of ligand-gated cation channels. These receptors mediate fast synaptic transmission, are permeable to Ca^{2+}, and share many of the structural and functional aspects of other ligand-gated channels.[34]

Glutamate is the dominant excitatory neurotransmitter in the mammalian brain. The receptors for glutamate are either ionotropic — ligand-gated ion channels — or metabotropic, in which the signal is transduced to intracellular messengers. The ionotropic glutamate receptors are classified as NMDA (*N*-methyl-D-aspartate), AMPA (amino-3 hydroxy-5 methyl-4 isoxazole-propionic acid), and KA (kainic acid) receptors. The AMPA–KA receptor subtypes of glutamate receptors open and desensitize rapidly in response to agonist application and generally allow the influx of Na^+ but not that of Ca^{2+}. Calcium-permeable AMPA receptors directly participate in synaptic transmission. NMDA receptors permit the influx of both Na^+ and Ca^{2+}. The activation kinetics of NMDA receptors is slow, and channel opening requires simultaneous neurotransmitter binding and postsynaptic membrane depolarization.

NMDA receptors are essential for brain function as they are vital for the activity-dependent changes in synaptic strength and connectivity. They underlie the formation of memory and learning. The calcium transients following NMDA receptor activation *in vivo* results in the transcription of several immediate early genes controlled by, among others, the transcription factor CREB. The calcium transients are potent activators of CREB and the CREB-regulated prosurvival genes that encode the brain-derived neurotropic factor (BDNF).

Overactivation of glutamate receptors can lead to neuro-degeneration. Neurotoxicity following excessive excitation of NMDA receptors causes neuronal death following acute trauma such as stroke, mechanical injury, or epilepsy, as well as contributing to the etiology of many chronic neurodegenerative diseases.[35-37] Pathological activation of NMDAR and the consequent calcium overloading is generally believed to be responsible for glutamate excitotoxicity.[37,38] Although excess Ca^{2+} influx via NMDAR overactivation causes cell death, equivalent increase in Ca^{2+} loads generated by Ca^{2+} influx through L-type VSCC are tolerated by the cell.[39] Calcium influx through NMDAR is taken up by mitochondria through the potential-driven uniporter. This causes an initial depolarization of mitochondria, which is enhanced by the NMDAR-mediated production of nitric oxide.[40,41] Excess Ca^{2+}-loaded mitochondria contribute to the production of reactive oxygen species (ROS) which damages the mitochondrial respiratory processes and causes plasma membrane calcium extrusion, membrane polarization,[42,43] and catastrophic depletion of cytosolic ATP levels.

The NMDA receptors are linked through the C-termini of their subunits to large complexes of cytoplasmic proteins including scaffolding, adaptor, cell adhesion, and cytoskeletal proteins, as well as components of signal transduction pathways. Thus, calcium entry through NMDAR takes place in a unique cell environment that facilitates physical and functional coupling between NMDAR and Ca^{2+}-dependent neurotoxic processes.[37]

ATP is an important extracellular nucleotide that mediates its effects via plasma membrane-bound purinergic P2 receptors.[44] These receptors also respond to other nucleotides such as ADP, UTP, and UDP. The P2 purinoceptors belong to two families, P2X and P2Y, respectively, that differ in their molecular structure and transduction mechanisms. P2X receptors are agonist-gated nonselective cation channels that mediate a rapid depolarization through influx of both Ca^{2+} and Na^+. P2Y receptors act via G protein-coupled receptors that activate phospholipase C to generate inositol 1,4,5-trisphosphate, which in turn mobilizes Ca^{2+} from the endo/sarcoplasmic reticulum.[45]

P2X receptors mediate rapid and nonselective entry of cations Na^+, K^+, or Ca^{2+} into cells. Na^+, acting as the predominant charge carrier, causes membrane depolarization and effectively activates L-type VSCC. Thus, the activity of ATP-gated calcium channels might increase calcium influx through a combination of direct P2X receptor-gated Ca^{2+} influx and secondary influx of Ca^{2+} through VSCC.[46,47] Calcium-channel blockers inhibit partially the P2X purinoceptor-mediated increase in intracellular calcium.[47]

Extracellular ATP exerts significant effects on cardiovascular function and blood flow in view of its ability to modulate the contractility of vascular smooth muscle types and cardiomyocytes. P2X purinoceptors that act as ligand-gated calcium channels are expressed in many types of cardiac and smooth muscle cells but not in the adult skeletal muscle. Both glial and neuronal cell types respond functionally to P2X receptor stimulation. ATP activates rapid depolarizing currents in several types of isolated neurons, and the responses to ATP are mediated at ligand-gated ion channels.[48]

PC-12 cells, neuroendocrine cells derived from pheochromocytoma of rat adrenal medulla, express both L-type and N-type VSCC.[49] As these cells also express P2X and P2Y purinoceptors,[50] the possibility of interaction between these calcium channels was investigated. Calcium channel blockers such as DHP nifedipine inhibited the P2X purinoceptor-mediated increase in $[Ca^{2+}]_i$ and catecholamine secretion. Membrane depolarization caused by Na^+ entry through activation of P2X purinoceptor effectively activates L-type VSCC,[47] thus adding to the possibility that the VSCC- and P2X-mediated signaling in PC-12 cells might be coupled. There may be active zones on the plasma membrane where P2X and L-type VSCC might be closely located so that localized signals might interact. Similar local signals due to the influx of calcium or the release of stored calcium have been identified in a variety of tissues such as cardiac, skeletal, and smooth muscles, as well as in nonexcitable tissues.[51] P2X receptors are located on nerve terminals both post- and presynaptically, and a spatial relation between P2X receptors and N-type calcium channels has been shown.[52] Thus, a localized depolarization caused by

Na$^+$ entry via the P2X receptors might lead to activation of L-type VSCC.

Endogenous Regulation of VSCC

Several ion channels exhibit mechanosensitivity. Mechanical forces are known to stimulate many cell signaling pathways. It is possible that a mechanically advantageous association of calcium channels with cytoskeletal or membrane proteins might play a role in the gating mechanism. Such integrative mechanisms might operate through phosphorylation of channel proteins by intracellular kinases.[53] The L-type calcium channel is the major pathway for the influx of calcium into the smooth muscle cell. A functional coupling between integrins and the L-type calcium channel is indicative of the link between the extracellular matrix and vascular tone.[54] Integrin-linked soluble tyrosine kinase inhibitors block the increase in calcium current induced by α_5 integrin antibody. Tyrosine phosphate inhibition enhances the calcium current. Integrin seems to regulate tyrosine phosphorylation cascade involving a tyrosine kinase and various focal adhesion proteins that regulate the function of the L-type calcium channel.[55]

Low-density lipoprotein (LDL) and, to a greater extent, oxidized-LDL have been shown to increase cytosolic calcium concentration in smooth muscle cells.[55,56] Estrogen treatment of smooth muscle cells resulted not only in a lower resting level of $[Ca^{2+}]_i$ but also a dose-dependent inhibition of the increase in $[Ca^{2+}]_i$ caused by oxidized LDL.[57] Testosterone seems to have an opposite effect. In other work, testosterone was found to potentiate the cellular calcium response to a depolarizing stimuli in isolated vascular smooth muscle cells, while not altering the basal $[Ca^{2+}]_i$. The acute *in vitro* effect and the absence of latency suggest a direct effect of the hormones that is different from the classical steroid effects.[58] These results are also in line with suggestions of a role for androgens in the observed gender differences in blood pressure (BP) in a variety of experimental and clinical conditions. Male spontaneously hypertensive rats (SHR) have a higher mean arterial blood pressure than female SHRs. Castration or androgen-receptor blockade attenuated the gender-mediated increase in BP in these rats.[59] Women with elevated androgen levels showed a greater tendency to have elevated BP than those with lower levels of testosterone, independent of age and race.[60]

High concentrations of polyamines (putrescine, spermidine, and spermine) are found not only in tissues with a high cell turnover but also in quiescent tissues such as muscle and nerve cells, suggesting a role for polyamines in cellular processes in addition to growth and proliferation. Modulation of polyamines has been shown to alter the activities of tyrosine kinase and mitogen-activated protein kinase. A direct effect of polyamines on K^+ channels was demonstrated.[61] Polyamines modify properties of physiological regulatory systems by altering ligand binding to charged molecules and surfaces. Addition of polyamines to intestinal smooth muscle or rat portal vein causes relaxation of spontaneous and agonist-mediated contraction in association with a decrease in $[Ca^{2+}]_i$. Exogenously applied polyamines inhibit L-type VSCC. Although the inhibition of phosphatase by polyamines might be responsible for some of the physiological actions, their high affinity for negatively charged molecules and surfaces might also explain their binding to ion channels.[62]

The activities of voltage-dependent calcium channels are regulated by various protein kinases and G-protein subunits.[63] Other endogenous mechanisms controlling channel activities have not been extensively investigated. A unique modulation of neuronal N-type calcium channels by farnesol has been reported.[64] Farnesol is an isoprenoid intermediate of the mevalonate pathway that plays a key role in cell growth and differentiation. At micromolar concentrations, farnesol acts as a rapid nondiscriminatory open channel blocker of all types of high voltage-activated calcium channels. It has been shown to induce a low affinity inhibition of L-type calcium channels in vascular smooth muscle.[65] At a much lower concentration (250 nM), farnesol is a selective, high-affinity inhibitor of N-type calcium channels, suggesting that the mevalonate pathway might be implicated in neurotransmitter release in brain via regulation of presynaptic voltage-gated calcium channels.[65]

In neurosecretory anterior pituitary cells the secretion of prolactin and other hormones is tightly coupled to the activity of L-type calcium-channels. Voltage-gated calcium channels are modulated by persistent depolarizing stimuli. The neuroendocrine GH_4C_1 cells display a decrease in L-type calcium-channel density as well as in channel function following persistent exposure to elevated K^+ concentration.[66] This is analogous to hormone or other agonist-induced desensitization of pharmacologic receptors.[67]

CELLULAR CALCIUM EFFLUX

Excess $[Ca^{2+}]_i$ is toxic to the cell. In the micromolar range, Ca^{2+} would precipitate phosphate and organic acid. Chronically elevated calcium, even at much lower levels, can activate apoptosis by triggering the opening of mitochondrial permeability transition pores, destroying the energy balance, and releasing cytochrome C that activates caspases.[68] The excess calcium is removed as rapidly as possible from the cytoplasm through various mechanisms. The plasma membrane Ca^{2+}-Mg^{2+}-ATPase (PMCA, Ca^{2+} pump) produces a Ca^{2+} efflux in exchange for $2H^+$ influx, using the energy of ATP. The ATPase of smooth muscle is calmodulin-dependent. The pump has a high affinity but low capacity to pump Ca^{2+} out of the cell and is coupled to Na^+ and K^+ gradients. Another major system is the plasma membrane Na^+–Ca^{2+} exchanger, which is a high-capacity low-affinity system driven by the transmembrane Na^+ gradient and sustained by the Na^+ pump.[10]

INTRACELLULAR STORAGE

The sarcoplasmic reticulum in muscle cells and the endoplasmic reticulum in nonmuscle cells are major intracellular stores of calcium.[69] Under steady-state conditions, intracellular stores must be filled to the same extent as calcium is released from them. Sarco(endo)plasmic reticulum Ca^{2+}

ATPase (SERCA) isoforms return Ca^{2+} to the lumen of the endoplasmic reticulum (SR). The accumulation of Ca^{2+} in the lumen of the SR requires energy supplied by the hydrolysis of ATP. The SERCA calcium pump has a high affinity for Ca^{2+} and is phosphorylated by ATP. Phospholamban regulates the SERCA pump in a manner similar to that of calmodulin regulation of PMCA activity. The SERCA isoforms of cardiac muscle are regulated by Ca/CaM-dependent protein kinase and by the phosphorylation of phospholamban. Once Ca^{2+} is pumped into the lumen of the SR, it binds to proteins such as calsequestrin and calreticulin that have a large capacity for calcium binding. This decreases the luminal free Ca^{2+} concentration, thus decreasing the energy necessary to pump Ca^{2+} from the cytoplasm to this organelle.

Contrary to the earlier contention that mitochondria take up calcium only when cells are damaged, mitochondria in the proximity of plasma membrane calcium channels are exposed to local high concentrations of calcium and can store this calcium through the actions of low-affinity uniporters. The calcium is transported along electrical gradients across the mitochondrial membrane and can be released when cytosolic calcium concentrations are decreased.[70] Intranuclear calcium concentration is greater than that in the cytoplasm. A Ca^{2+} pump is reported to be responsible for pumping Ca^{2+} from the cytosol to the nucleus, and agonists increase the nuclear uptake.[10]

EFFLUX OF CALCIUM FROM SARCO/ENDOPLASMIC RETICULUM

Sarco(endo)plasmic reticulum, through its close association with plasma membrane and mitochondria, regulates precisely temporal and spatial aspects of calcium signaling. The sarco(endo)plasmic reticulum is a main storage site for intracellular calcium from which calcium can be released into cytoplasm. There are two types of voltage-independent, endoplasmic reticulum-resident calcium-release channels: ryanodine receptors (RyRs) and inositol 1,4,5 trisphosphate (IP_3) receptors (IP_3Rs).[69,71] These receptors coexist in close proximity in ER in various types of cells, but their calcium pools appear to be separate.[72]

Calcium is released from IP_3-regulated pools in response to agonists acting through phospholipase C. Agonists such as norepinephrine, acetylcholine, serotonin, and histamine interact with specific cell-surface receptors to activate phospholipase C, leading to the hydrolysis of the membrane phospholipid phosphatidyl inositol 4,5-bisphosphate (PIP_2) to produce IP_3 and diacyl glycerol (DAG). Once stimulated, IP_3 receptors mobilize the release of Ca^{2+} from ER and SR into the cytoplasm. DAG, the other product of PIP_2 hydrolysis, activates protein kinase C, which in turn phosphorylates target proteins to produce biological effects. In addition, DAG can act as an agonist of nonvoltage-dependent calcium channels. The IP_3R has binding domains for IP_3 and for cytosolic Ca^{2+} and a central aqueous channel that allows passage of Ca^{2+} from the lumen of the SR into the cytoplasm. At low concentrations of cytosolic Ca^{2+}, the IP_3-gated channel is activated, while at high concentration of cyoplasmic Ca^{2+}, the channel is inhibited. The regulation of calcium efflux by the IP_3-gated Ca^{2+} channel is similar to the efflux of Ca^{2+} through the RyR-mediated calcium channel. Calcium released from IP_3R-regulated pools is capable of eliciting the positive amplification process known as calcium-induced calcium release from ryanodine-regulated calcium pools.

Ca^{2+}-induced Ca^{2+} release is the trigger for excitation-contraction coupling in the heart. The elevation of Ca^{2+} concentration in the vicinity of SR causes a concentration-dependent release of SR Ca^{2+}, and this can be inhibited by high concentration of ryanodine.[73] A quantal release of SR Ca^{2+} (Ca^{2+} sparks) in response to opening of VSCC in myocytes provides evidence that VSCC is the primary trigger for SR-Ca^{2+} release.[74]

Intracellular Ca^{2+} release channels were identified originally by electron microscopy as part of the junctional foot protein in skeletal muscle triads.[75] Later studies showed that foot proteins bound the neutral plant alkaloid ryanodine with very high specificity and selectivity,[76] giving RyRs their alternative name. RyR constitutes a calcium channel in SR with high conductance for calcium. Ryanodine (at 10 μM) binds to the open SR membrane calcium channel and locks it in an open state. At higher concentrations, ryanodine closes the channels. RyRs share evolutionary origin, sequence homology, and structural similarities with the IP3 receptor, the other major family of intracellular calcium channels.[77]

Three subtypes of RyR have been identified. RyR type-1, the "skeletal muscle" isoform, is found in skeletal muscle and cerebellar Purkinje cells. RyR type-2, the "cardiac muscle" isoform, is heavily concentrated in heart and is distributed diffusely throughout the brain. RyR type-3, the "brain" isoform, is concentrated in hippocampus, cerebral cortex, and corpus striatum.[78] The products of these genes assemble to form very large homotetrameric structures with a molecular weight of ~2 MDa.[78,79] The three-dimensional structure of RyR has been determined, and consists of a quatrefoil-type structure comprised of four identical subunits with a central ion-conducting channel,[80] which has been observed in both open and closed states.[81–83]

Ligands that bind either to the cytosolic or the luminal face of RyR channels control RyR gating[84–87] in an isoform-specific manner.[80,88.] RyR possesses numerous modulatory binding sites including a site that binds calcium and magnesium, a site that has a high affinity for caffeine, and a site for adenosine-based compounds. Binding of caffeine and adenosine-based compounds to RyR increases ryanodine binding and sensitivity to calcium-induced calcium release.[89–91] Cyclic ADP ribose, through a calmodulin-dependent action, has been shown to modulate the release of calcium via the RyR calcium channel.[92] 3,3,5-Triiodothyronine, as well as *trans*-retinoic acid, stimulate cyclic ADP ribose production and increase the release of SR calcium.[93] These observations suggest the possibility that cyclic ADP ribose might be an endogenous modulator of SR Ca^{2+} release.

In addition to modulatory sites, a number of proteins have been shown to interact with and modulate RyR channel activity. FKBP (FK506-binding protein), a receptor for the immunosuppressant drugs FK506 and rapamycin, increases the sensitivity of RyR to caffeine and ryanodine by stabilizing the receptor in a maximal conductance state.[94,95] Sorcin, a calcium-binding protein, is expressed in neurons throughout the brain, and it, too, associates with RyR.[96] PS-1 also interacts with RyR,[97] and this interaction may play an important role in abnormal homeostasis of intracellular calcium in, and the pathogenesis of Alzheimer's disease.

All three RyR isoforms have fast activation time constants ranging from 0.5 to 1 msec,[98–100] and therefore, once activated, can liberate ER Ca^{2+} stores on a microsecond to millisecond time scale.[91,101] The Ca^{2+} dependence of channel activation and inactivations, as well as open and closed times, has been found to be isoform-specific.[87,98,102–104] Thus, the contribution of ER Ca^{2+} release appears to depend on the type and density of RyR isoforms as well as their modulations by various substances and proteins.

CALCIUM CHANNELS, PYRIDOXAL PHOSPHATE, AND CARDIOVASCULAR PROTECTION

We have characterized the moderate hypertension that develops in the pyridoxine-deficient rat.[105] As in other conditions of experimental or clinical hypertension, the pyridoxine-deficient rat was associated with significant sympathetic stimulation. The end result of centrally mediated sympathetic stimulation is an increase in peripheral resistance, a hallmark of hypertension. The increase in tone of the caudal artery segments from hypertensive rats and its decrease following the addition of calcium-channel antagonists indicated that increased peripheral resistance, resulting from increased permeability of smooth muscle plasma membrane to Ca^{2+}, might be central to the development of hypertension.[106]

The contraction of smooth muscle is dependent upon the short-term increase in cytosolic free calcium. Calcium influx occurs through the plasma membrane calcium channels, primarily the L-type VCSS. To assess the possibility that an abnormal increase in the activity of VSCC is responsible for the increased tone of VSM, we determined the intracellular calcium uptake by caudal artery segments. We found that the

influx of calcium into VSM of pyridoxine-deficient hypertensive rats was twice that of normotensive controls.[107] The intraperitoneal injection of different classes of L-type calcium
channel antagonists such as nifedipine, verapamil, or diltiazem produced a dose-dependent decrease in the systolic
blood pressure (SBP) of the hypertensive rats. The DHP-
sensitive calcium channel agonist BAY K 8644 antagonized
the hypotensive effect of nifedipine at equimolar doses. BAY
K 8644 is known to prolong the open state of the VSCC during
activation.[108]

We have investigated whether pyridoxine or, more particularly, pyridoxal-5-phosphate (PLP) could directly modulate cellular calcium uptake. BAY K 8644 stimulated the
uptake of $^{45}Ca^{2+}$ into artery segments from normotensive control animals. PLP dose-dependently reduced the BAY K 8644
stimulated calcium uptake by artery segments from normotensive rats. The uptake of calcium by artery segments from
hypertensive rats was high. Pyridoxal phosphate or nifedipine
added to the incubation medium reduced significantly the
calcium uptake of the artery segments from hypertensive rats.
The addition of BAY K 8644 to the incubation medium attenuated the *in vitro* effects of both nifedipine and pyridoxal
phosphate.[109] These *in vitro* direct antagonisms and the lack
of any latency in the action of PLP indicate the possibility
that the calcium-channel agonist BAY K 8644, the calcium-
channel antagonist nifedipine, and PLP might all act at the
same site on the voltage-sensitive calcium channel.[110] In support of this, we showed that PLP decreased the number of
[^{3}H]nitrendipine-binding sites on membrane preparations
from caudal artery of normal rats.[111.]

The positive ionotropic action of extracellular ATP is very
similar to that of norepinephrine and is associated with a
transient increase in intracellular $[Ca^{2+}]_i$. We investigated the
effect of PLP on ATP-induced positive ionotropic effect in
isolated perfused normal rat heart. The infusion of ATP
caused an immediate increase in left ventricular developed
pressure, and this was completely blocked if the hearts were
preinfused with PLP for 10 min. This antagonistic action of
PLP was dose dependent. The specificity of the effect of PLP

on the ATP-mediated effects was established, as PLP has no effect on the marked increase in contractibility induced by isoproterenol. Also, propranolol, which prevented the positive inotropic action of isoproterenol, had no effect on the positive inotropic action of ATP.[112]

Because the cardiac sarcolemmal membrane has been reported to contain both high- and low-affinity ATP-binding sites, we studied the effect of PLP on these binding sites using [^{35}S]ATP as the ligand. PLP almost completely blocked the low-affinity binding while decreasing the high-affinity binding by about 60%. The effect of PLP was comparable to those elicited by cibacron blue, suramin, and 4,4^1-disothiocyanostilbene-2,2^1-disulfonate, all known to inhibit binding of ATP at the P2X purinoceptor site. Agents such as propranolol (β adrenoreceptor blocker), prozosin (α adrenoreceptor blocker), and verapamil (L-type Ca^{2+}-channel blocker) did not have any significant effect on the high- or low-affinity ATP-binding sites, attesting to the specificity of the inhibition by PLP.[83] Sulfonic acid derivatives such as pyridoxal phosphate 6-azophenyl-2^1, 4^1-disulfonic acid (PPADS) and pyridoxal phosphate-6-azophenyl-2^1, 5^1-disulfonic acid (iso PPADS) have been shown to antagonize purinocepter-mediated responses in urinary bladder, vagus nerve, and sympathetic ganglia.[113]

CALCIUM CHANNELS, PYRIDOXAL PHOSPHATE, AND NEUROPROTECTION

The relationship between pyridoxine and the development of seizure activity was associated with decreased synthesis and secretion of the inhibitory neurotransmitter γ-aminobutyric acid (GABA).[114] Subsequently, we showed that pyridoxine, through its role in the synthesis and secretion of various neurotransmitter amines, was involved in the normal functioning of the hypothalamo-pituitary-end organ axes.[115] In addition to correcting defects associated with vitamin B_6 deficiency, high doses of pyridoxine can ameliorate degenerative diseases and protect against neurotoxicity not associated with vitamin B_6 deficiency.[110]

We reported that administration of GABA directly into rat hippocampus protected against seizures induced by the neurotoxin domoic acid.[116] In a recent study[117] we showed that mice given a single subconvulsive dose of domoic acid exhibited typical spike and wave discharges in the electroencephalographic recordings in cerebral cortex. Administration of pyridoxine simultaneously with, after, or before domoic acid administration resulted in a significant attenuation of the spike and wave activity. Administration of other antiepileptic drugs such as sodium valproate or nimodipine produced similar improvement of the domoic acid-mediated neurotoxicity. Nimodipine is an L-type VSCC calcium-channel antagonist. Sodium valproate and pyridoxine significantly attenuated domoic acid-mediated increases in the levels of glutamate and decreases in the levels of GABA, as well as increases in intracellular calcium influx. Nimodipine, sodium valproate, and pyridoxine also attenuated the increased levels of the proto-oncogenes c-*fos*, *jun*-B and *jun*-D. In hippocampal cells, domoic acid-induced increases in glutamate and calcium influx were significantly decreased by pyridoxal phosphate or nimodipine. Similarly, in neuroblastoma–glioma hybrid cells (NG 108/15), pyridoxine attenuated domoic acid-mediated increases in glutamate, an influx of extra cellular calcium, and the enhanced induction of oncoproteins.

Domoic acid, a neurotoxic contaminant of cultivated and contaminated mussels, is a rigid structural analog of glutamic acid. The autocrine mechanisms by which domoic acid causes neurotoxicity and neurodegeneration include release of the excitatory neurotransmitter glutamic acid and facilitation of NMDA receptor activation as evidenced by decreased Mg^{2+} blockade of the calcium ion channel,[118] with the resultant increase in intracellular calcium influx. The net result is an uncontrolled and sustained elevation of cytosolic Ca^{2+}, which mediates induction of early genes. Many of these early genes encode nuclear proteins that are putative transcription factors binding to activating protein 1 or CREB. Similar actions of pyridoxal phosphate in decreasing the influx of calcium through actions on cell surface calcium channels have been reported.[112] This inhibition of calcium transport is central to the neuroprotective action of pyridoxal phosphate.

Calcium channels are regulated by homologous and heterologous factors. Chronic channel activation, chronic drug exposure, hormonal influence, and specific disease states are all associated with altered expression of calcium channels, both in terms of their numbers and function. Various synthetic chemicals act as agonists and antagonists of calcium channels, and this suggests, as in the case of opiate and benzodiazepine receptors for which endogenous ligands have been characterized,[9] that calcium channels might have endogenous modulators.[119] A host of endogenous regulators have been identified as regulators of calcium-channel function, and the mechanisms of their actions need to be studied.

References

1. AK Campbell. *Intracellular Calcium: Its Universal Role as Regulator.* John Wiley & Sons, New York, 1993.

2. E Carafoli. Intracellular calcium homeostasis. *Annu Rev Biochem* 56, 395–433, 1987.

3. MJ Berridge, M.D. Bootman, P. Lipp. Calcium — a life and death signal. *Nature* 395, 645–648, 1998.

4. JH Exton. Calcium signaling in cells — molecular mechanisms. *Kidney Int* 32 (Suppl 23), S68–S76, 1987.

5. VN Smirnov, TA Voyno-Yasenetskaya, AS Antonov, ME Lukashev, VP Shirinski, VY Tertov, VA Tkachuk. Vascular signal transduction and atherosclerosis. *Ann NY Acad Sci* 598, 167–181, 1990.

6. KKW Wang, YS Du, C Diglio, W Tsang, TH Kuo. Hormone-induced phosphorylation of the plasma membrane calcium pump in cultured aortic endothelial cells. *Arch Biochem Biophys* 298, 103–108, 1991.

7. JJ Lynch, TJ Ferro, FA Blumenstock, AM Brockenuer, AB Malik. Increased endothelial albumin permeability mediated by protein kinase C activation. *J Clin Invest* 85, 1991–1998, 1990.

8. MJ Berridge. Neuronal calcium signaling. *Neuron* 21, 13–26, 1998.

9. DJ Triggle. Calcium, calcium channels and calcium channel antagonists. *Can J Physiol Pharmacol* 68, 1474–1481, 1990.

10. J Marin, A Encabo, A Briones, E-C Garcia-Cohen. MJ Alonso. Mechanisms involved in cellular calcium homeostasis in vascular smooth muscle: Calcium pumps. *Life Sciences* 64, 279–303. 1999.

11. NM Lorezo, KG Beam, Calcium channelopathies. *Kidney* I 57, 794–802, 2000.

12. L Missiaen, W Robberecht, L VanDenBosch, G Callewaert, JB Parys, F Wuytack, L Raeymaekers, B Nilius, J Eggermont, H DeSmedt. Abnormal Intracellular Ca^{2+} homeostasis and disease. *Cell Calcium* 28, 1–21, 2000.

13. SE Howlett, GR Ferrier. The 1996 Merck Frosst Award. The voltage-sensitive release mechanism: a new trigger for cardiac contraction. *Can J Physiol Pharmacol* 75, 1044–1057, 1997.

14. DA Greenberg. Neuromuscular disease and calcium channels. *Muscle Nerve* 22, 1341–1349. 1999.

15. GJO Evans, JM Pocock. Modulation of neurotransmitter release by dihydropyridine-sensitive calcium channels involves tyrosine phosphorylation. *Eur J Neurosc* 11, 279–292, 1999.

16. TH Murphy, PE Worley, JM Baraban. L-type voltage-sensitive calcium channels mediate synaptic activation of immediate early genes. *Neuron* 7, 625–635, 1991.

17. LB Rosen, DD Ginty, ME Breenberg. Calcium regulation of gene expression. *Adv. Second Messenger Protein Phosphorylation Res* 30, 225–253, 1995.

18. G Noll, TF Lüscher. Comparative pharmacological properties among calcium channel blockers: T-channel versus L-channel blockade. *Cardiology* 89 (suppl), 10–15, 1998.

19. J-B Roullet, RL Spaetgens, T Burlingame, Z-P Feng, GW Zamponi, Modulation of neuronal voltage-gated calcium channels by farnesol. *J Biol Chem* 274, 25439–25446, 1999.

20. KD Keef, JR Hume, J Zhong. Regulation of cardiac and smooth muscle Ca^{2+} channels (Ca$_v$1.2a, b) by protein kinases. *Am J Physiol Cell Physiol* 281, C1743–C1756, 2001.

21. DH MacLennan. Ca^{2+} signaling and muscle disease. *Eur J Biochem* 267, 5291–5297, 2000.

22. JN Muth, H Yamaguchi, G Mikala, IL Grupp, W Lewis, H Chen, LS Song, EG Lakatta, G Varadi, A Schwartz. Cardiac-specific overexpression of the α1-subunit of the L-type voltage-dependent Ca^{2+} channel in transgenic mice: loss of isoproterenol-induced contraction. *J Biol Chem* 274, 21503–21506, 1999.

23. Y Chen-Izu, RP Xiao, LT Izu, H Cheng, M Kuschel, H Spurgeon, EG Lakatta. G$_i$-dependent localization of β$_2$-adrenergic receptor signaling to L-type Ca^{2+} channels. Biophys J 79, 2547–2556, 2000.

24. M Kimura, T Osanai, K Okumura, S Suga, T Kanno, N Kamimura, N Horiba, M Wakui. Involvement of phosphorylation of β-subunit in cAMP-dependent activation of L-type Ca^{2+} channel in aortic smooth muscle-derived A7r5 cells. *Cell Signal* 12, 63–70, 2000.

25. V Ruiz-Velasco, J Zhong, JR Hume, KD Keef. Modulation of Ca^{2+} channels by cyclic nucleotide cross activation of opposing protein kinases in rabbit portal vein. *Circ Res* 82, 557–565, 1998.

26. M Gollacsch, MT Nelson. Voltage-dependent Ca^{2+} channels in arterial smooth muscle cells. *Kidney Blood Press Res* 20, 355–371, 1997.

27. SR Ikeda. Voltage-dependent modulation of N-type calcium channels by G-protein βγsubunits. *Nature* 380, 255–258, 1996.

28. LH Jiang, DJ Gawler, N Hodson, DJ Milligan, HA Pearson, V Porter, D Wray. Regulation of cloned cardiac L-type calcium channels by cGMP-dependent protein kinase. *J Biol Chem* 275, 6135–6143, 2000.

29. MT Nelson, H Cheng, M Rubart, LF Santana, AD Bonev, HJ Knot, WJ Lederer. Relaxation of arterial smooth muscle by calcium sparks. *Science* 270, 633–637, 1995.

30. JH Jagger, VA Porter, WJ Lederer, MT Nelson. Calcium sparks in smooth muscle. *Am J Physiol Cell Physiol* 278, C235–C256, 2000.

31. C Randriamampita, RY Tsien. Emptying of intracellular Ca^{2+} stores releases a novel small messenger that stimulates Ca^{2+} influx. *Nature* 364, 809–814, 1993.

32. K Kiselyvov, X Xu, G Mozhayeva, T Kuo, I Pessah, G Mignery, X Zhu, L Birnbaumer, S. Muallem. Functional interaction between $InsP_3$ receptors and store-operated Htrp3 channels. *Nature* 396, 478–482, 1998.

33. P Gailly. Ca^{2+} entry into CHO cells, after Ca^{2+} stores depletion, is mediated by arachidonic acid. *Cell Calcium* 24, 293–304, 1998.

34. S Sugita, K-Z Shen, RA North. 5-hydroxytryptamine is a fast excitatory transmitter at $5\text{-}HT_3$ receptors in rat amygdale. *Neuron* 8, 199–203, 1992.

35. V Dirnagel, C Iadecola, MA Moskowitz. Pathobiology of ischemic stroke: an integrated view. *Trends Neurosci* 22, 391–397, 1999.

36. E Lancelot, MF Beal. Glutamate toxicity in chronic neurodegenerative disease. *Prog Brain Res* 116, 331–374, 1998.

37. GH Hardingham, H Bading. The yin and yang of NMDA receptor signaling. *Trends Neurosci* 26, 81–89, 2003.

38. S Eimerl, M Schramm. The quantity of calcium that appears to induce neuronal death. *J Neurochem* 62, 1223–1226, 1994.

39. R Sattler, MP Charlton, M Hafner, M Tymianski. Distinct influx pathways, not calcium load, determine neuronal vulnerability to calcium neurotoxicity. *J Neurochem* 71, 2349–2364, 1998.

40. RJ White, IJ Reynolds. Mitochondria accumulate Ca^{2+} following intense glutamate stimulation of cultured rat forebrain neurons. *J Physiol* 498, 31–47, 1997.

41. J Keelan, O Vergun, MR Duchen. Excitotoxic mitochondrial depolarization requires both calcium and nitric oxide in rat hippocampal neurons. *J Physiol* 520, 797–813, 1999.

42. MW Ward, AC Rego, BG Frenguelli, DG Nicholls. Mitochondrial membrane potential and glutamate excitotoxicity in cultured cerebellar granule cells. *J Neurosci* 20, 7208–7219, 2000.

43. DG Nicholls, SL Budd. Mitochondria and neuronal survival. *Physiol Rev* 80, 315–360, 2000.

44. V Ralevic, G Burnstock. Receptors for purines and pyrimidines. *Pharmacol Rev* 50, 415–492, 1998.

45. GR Dubyak, C El-Moatassim. Signal transduction via P_2-purinergic receptors for extra cellular ATP and other nucleotides. *Am J Physiol Cell Physiol* 265, C577–C606, 1993.

46. C Fasolato, P Pizzo, T Pozzan. Receptor-mediated calcium influx in PC12 cells. ATP and bradykinin activate two independent pathways. *J Biol Chem* 265, 20351–20355, 1990.

47. E-M Hur, T-U Park, K-T Kim. Coupling of L-type voltage sensitive calcium channels to $P2X_2$ purinoceptors in PC-12 cells. *Am J Physiol Cell Physiol* 280, C1121–C1129, 2001.

48. BP Bean. Pharmacology and electrophysiology of ATP-activated ion channels. *Trends Pharmacol Sci* 13, 87–91, 1992.

49. TJ Park, KT Kim. Cyclic AMP-independent inhibition of voltage-sensitive calcium channels by forskolin in PC-12 cells. *J Neurochem* 66, 83–88, 1996.

50. LR DeSonza, H Moore, S Raha, JK Reed. Purine and pyrimidine nucleotides activate distinct signaling pathways in PC-12 cells. *J Neurosci Res* 41, 753–763, 1995.

51. MJ Berridge. Elementary and global aspects of calcium signaling. *J Physiol* 499, 291–306, 1997.

52. JA Barden, LJ Cottee, MR Bennett. Vesicle-associated proteins and P2X receptor clusters at single synaptic varicosities in mouse vas deferens. *J Neurocytol* 28, 469–480, 1999.

53. IB Levitan. Modulation of ion channels by protein phosphorylation. *Adv Second Messengers Protein Phosphorylation Res* 22, 3–22, 1999.

54. X Wu, GE Davis, GA Meininger, E Wilson, MJ Davis. Regulation of the L-type calcium channels by $\alpha_5\beta$, integrin requires signaling between focal adhesion proteins. *J Biol Chem* 276, 30285–30292, 2001.

55. B Weisser, R Locher, T Mengden, W Vetter. Oxidation of low density lipoprotein enhances its potential to increase intracellular free calcium concentrations in vascular smooth muscle cells. *Arterioscler Thromb* 12, 231–236, 1992.

56. KE Wells, R Miguel, JJ Alexander. Sex hormones affect the calcium signaling response of human arterial cells to LDL. *J Surg Res* 63, 64–72, 1996.

57. M Barbagallo, J Shan, PKT Pang, LM Resnick. DHEAS effects on VSMC cytosolic free calcium and vascular contractility. *Hypertension* 26, 1065–1069, 1995.

58. M Barbagallo, LJ Domingguez, G Licata, R Ruggero, RZ Lewanczuk, PKT Pang, LM Resnick. Effect of testosterone on intracellular Ca^{2+} in vascular smooth muscle cells. *Am J Hypertens* 14, 1273–1275, 2001.

59. JF Reckeloff, H Zhang, K Srivastava, JP Granger. Gender differences in hypertension in spontaneously hypertensive rats: role of androgen and androgen receptor. *Hypertension* 34, 920–923, 1999.

60. C Ayala, E Steinberger, A Sweeney, WH Mueller, DK Walter, R Hardy, SM Petak, LJ Rodriguez Rigau, KD Smith. The relationship of serum androgens and ovulatory status to blood pressure in reproductive-age women. *Am J Hypertens* 12, 772–777, 1999.

61. AN Lopatin, EN Makhina, CG Nichols. Potassium channel block by cytoplasmic polyamines as the mechanism for intrinsic rectification. *Nature* 372, 366–369, 1994.

62. BO Nilsson, MF Gomez, K Sward, P Hellstrand. Regulation of Ca^{2+} channel and phosphatase activities by polyamines in intestinal and vascular smooth muscle-implications for cellular growth and contractility. *Acta Physiol Scand* 176, 33–41, 2002.

63. KJ Swartz. Modulation of calcium channels by protein kinase C in rat central and peripheral neurons: disruption of G protein-mediated inhibition. *Neuron* 11, 302–320, 1993.

64. A Stea, TW Soong, TP Snutch. Determination of protein kinase C-dependent modulation of a family of neuronal calcium channels. *Neuron* 15, 929–940, 1995.

65. J-B Roullet, VC Luft, H Xue, J Chapman, R Bychkovo, CM Roullet, FC Luft, H Haller, DA McCarron. Farnesol inhibits L-type Ca^{2+} channels in vascular smooth muscle cells. *J Biol Chem* 372, 32240–32246, 1997.

66. R Peri, DJ Triggle, S Singh. Regulation of L-type calcium channels in pituitary GH4C1 cells by depolarization. *J Biol Chem* 276, 31667–31673, 2001.

67. NJ Freedman, RJ Lefkowitz. Desensitization of G protein-coupled receptors. *Recent Prog Horm Res* 51, 319–353, 1996.

68. DR Green, JC Reed. Mitochondria and apoptosis. *Science* 281, 1309–1312, 1998.

69. T Pozzan, R Rizzuto, P Volpe, J Meldosi. Molecular and cellular physiology of intracellular calcium stores. *Physiol Rev* 74, 595–636, 1994.

70. T Pozzan, R Rizzuto. High tide of calcium in mitochondria. *Nature Cell Biol* 2, E25–E27, 2000.

71. V Henzi, AB MacDermott. Characteristics and functions of calcium-and inositol 1, 4, 5-trisphosphate-releaseable stores of calcium in neurons. *Neurosci* 46, 251–273, 1992.

72. KJ Seymour-Laurent, ME Barish. Inositol 1, 4, 5-trisphosphate and ryanodine receptor distributions and patterns of acetylcholine-and caffeine-induced calcium release in cultured mouse hippocampal neurons. *J Neurosci* 15, 2592–2608, 1995.

73. A Fabiato. Effects of ryanodine in skinned cardiac cells. *Fed Proc* 44, 2970–2976, 1985.

74. LF Santana, H Cheng, AM Gomez, MB Cannell, WJ Lederer. Relation between the sarcolemmal Ca^{2+} current and Ca^{2+} sparks and local control theories for cardiac excitation-contraction coupling. *Circ Res* 78, 166–171, 1996.

75. C Franzini-Armstrong. Membrane particles and transmission at the triad. *Fed Proc* 34, 1382–1389, 1975.

76. IN Pessah, AL Waterhouse, JE Casida. The calcium-ryanodine receptor complex of skeletal and cardiac muscle. *Biochem Biophys Res Commun* 128, 449–456, 1985.

77. MJ Berridge. Inositol tri phosphate and calcium signaling. *Nature* 361, 315–325, 1993.

78. Y Ogawa. Role of ryanodine receptors. *Crit Rev Biochem Mol Biol* 29, 229–274, 1994.

79. R Coronado, J Morrissette, M Sukhareva, DM Vaughan. Structure and function of ryanodine receptors. *Am J Physiol* 266, C1485–1504, 1994.

80. V Shoshan-Barmatz, RH Ashley. The structure, function, and cellular regulation of ryanodine-sensitive Ca^{2+} release channels. *Int Rev Cytol* 183, 185–270, 1998.

81. MR Sharma, LH Jeyakumar, S Fleischer, T Wagenknecht. Three-dimensional structure of ryanodine receptor isoform three in two conformational states as visualized by cryo-electron miscroscopy. *J Biol Chem* 275, 9485–9491, 2000.

82. T Wagenknecht, M Radermacher. Ryanodine receptors: structure and macromolecular interaction. *Curr Opin Struct Biol* 7, 258–265, 1997.

83. T Wagenknecht, M Samso. Three-dimensional reconstruction of ryanodine receptors. *Front Biosci* 7, 1464–1474, 2002.

84. MA Albrecht, SL Colegrove, J Hongpaisan, NB Pivovarova, SB Andrews, DD Friel. Multiple modes of calcium-induced calcium release in sympathetic neurons. I: attenuation of endoplasmic reticulum Ca^{2+} accumulation at low [Ca^{2+}]i during weak depolarization. *J Gen Physiol* 118, 83–100, 2001.

85. V Lukyanenko, I Gyorke, S Gyorke. Regulation of calcium release by calcium inside the sarcoplasmic reticulum in ventricular myocytes. *Pfluger Arch* 432, 1047–1054, 1996.

86. R Sitsapesan, AJ Williams. Regulation of the gating of the sheep cardiac sarcoplasmic reticulum Ca^{2+}-release channel by luminal Ca^{2+}. *J Membr Biol* 137, 215–226, 1994.

87. L Xu, R Anwyl, MJ Rowan. Spatial exploration induces a persistent reversal of long-term potentiation in rat hippocampus. *Nature* 394, 891–894, 1998.

88. P Koulen, EC Thrower. Pharmacological modulation of intracellular Ca^{2+} channels at the single-channel level. *Mol Neurobiol* 24, 65–86, 2001.

89. PS McPherson, S Kim, YK Valdivia, H Knudson, CM Takekura, H Franzini-Armstrong, C Coronado, KP Campbell. The brain ryanodine receptor: a caffeine-sensitive calcium release channel. *Neuron* 7, 17–25, 1991.

90. RA Padua, JI Nagy, JD Geiger. Sub cellular localization of ryanodine receptors in rat brain. *Eur J Pharmacol* 298, 185–189, 1996.

91. CP Holden, RA Padua, JD Geiger. Regulation of ryanodine receptor calcium release channels by diadenosine polyphosphates. *J Neurochem* 67, 574–580, 1996.

92. Y Tanaka, AH Tashjian, Jr. Calmodulin is a selective mediator of Ca^{2+}-induced Ca^{2+} release via the ryanodine receptor-like Ca^{2+} channel triggered by cyclic ADP ribose. *Proc Natl Acad Sci USA* 92, 3244–3248, 1995.

93. FGS DeToledo, J Cheng, TP Dousa. Retinoic acid and triiodothyronine stimulates ADP-ribosyl cyclase activity in rat vascular smooth muscle cells. *Biochem Biophys Res Communs* 238, 847–850, 1997.

94. AM Cameron, FC Nucifora, Jr, ET Fung, DJ Livingston, RA Aldape, CA Ross, SH Snyder. FKBP12 binds the inositol 1, 4, 5-tripohsphate receptor at leucine-proline (1400–1401) and anchors calcineurin to this FK506-like domain. *J Biol Chem* 272, 27582–27588, 1997.

95. AB Brillantes, K Ondrias, A Scott, E Kobrinsky, E Ondriasova, MC Moschella, T Jayaraman, M Landers, BE Ehrlich, AR Marks. Stabilization of calcium release channel (ryanodine receptor) function by FK506-binding protein. *Cell* 77, 513–523, 1994.

96. VM Pickel, CL Clarke, MB Meyers. Ultrastructural localization of sorcin, a 22 kDa calcium binding protein, in the rat caudate-putamen nucleus: association with ryanodine receptors and intracellular calcium release. *J Comput Neurol* 386, 625–634, 1997.

97. SL Chan, M Mayne, CP Holden, JD Geiger, MP Mattson. Presenilin-1 mutations increase levels of ryanodine receptors and calcium release in PC12 cells and cortical neurons. *J Biol Chem* 275, 18195–18200, 2000.

98. T Murayama, T Oba, E Katayama, H Oyamada, K Oguchi, M Kobayaski, K Otsuke, Y Ogawa. Further characterization of the type 3 ryanodine receptor (RyR3) purified from rabbit diaphragm. *J Biol Chem* 274, 17297–17308, 1999.

99. T Akita, K Kuba. Functional triads consisting of ryanodine receptors, Ca^{2+} channels, and Ca^{2+}-activated K^+ channels in bullfrog sympathetic neurons. Plastic modulation of action potential. *J Gen Physiol* 116, 697–720, 2000.

100. NJ Emptage, CA Reid, A Fine. Calcium stores in hippocampal synaptic boutons mediate short-term plasticity, store-operated Ca^{2+} entry, and spontaneous transmitter release. *Neuron* 29, 197–208, 2001.

101. K Narita, T Akita, J Hachisuka, S Huang, K Ochi, K Kuba. Functional coupling of Ca^{2+} channels to ryanodine receptors at presynaptic terminals. Amplification of exocytosis and plasticity. *J Gen Physiol* 115, 519–532, 2000.

102. SR Chen, X Li, K Ebisamwa, L Zhang. Functional characterization of the recombinant type 3 Ca^{2+} release channel (ryanodine receptor) expressed in HEK293 cells. *J Biol Chem* 272, 24234–24246, 1997.

103. LH Jeyakumar, JA Copello, AM O'Malley, GM Wu, R Grassucci, T Wagenknecht, S Fleischer. Purification and characterization of ryanodine receptor 3 from mammalian tissue. *J Biol Chem* 273, 16011–16020, 1998.

104. AL Percival, AJ Williams, JL Kenyon, MM Grinsell, JA Airey, JI Sutko. Chicken skeletal muscle ryanodine receptor isoforms: ion channel properties. *Biophys J* 67, 1834–1850, 1994.

105. CS Paulose, K Dakshinamurti, S Packer, NL Stephens. Sympathetic stimulation and hypertension in the pyridoxine-deficient adult rat. *Hypertension* 11, 387–391, 1988.

106. M Viswanathan, R Bose, K Dakshinamurti. Increased calcium influx in caudal artery of rats made hypertensive with pyridoxine deficiency. *Am J Hypertens* 4, 252–255, 1991.

107. KJ Lal, K Dakshinamurti. Calcium channels in vitamin B_6 deficiency induced hypertension. *Hypertension* 11, 1357–1362, 1993.

108. S Kokubun, H Reuter. Dihydropyridine derivatives prolong the open state of calcium channels in cultured cardiac cells. *Proc Natl Acad Sci USA* 81, 4824–4827, 1984.

109. KJ Lal, SK Sharma, K Dakshinamurti. Regulation of calcium influx into vascular smooth muscle by vitamin B$_6$. *Clin Exp Hypertens* 15, 489–500, 1993.

110. K Dakshinamurti, S Dakshinamurti. Blood pressure regulation and micronutrients. *Nutrition Res Rev* 14, 3–43, 2001.

111. K Dakshinamurti, KJ Lal, NS Dhalla, S Musat, X Wang. Pyridoxal 5-phosphate and calcium channels. In: A. Iriarte, HM Kagan, M Martinez-Carrion, Eds. *Biochemistry and Molecular Biology of Vitamin B$_6$ and PQQ-Dependent Proteins.* Basel: Birkhauser Verlag, 2000, pp 307–314.

112. X Wang, K Dakshinamurti, S Musat, NS Dhalla. Pyridoxal 5-phosphate is an ATP-receptor antagonist in freshly isolated rat cardiomyocytes. *J Mol Cell Cardiol* 31, 1063–1072, 1999.

113. DJ Trezise, NJ Bell, BS Khakh, AD Michael, PA Humphrey. P2 purinoceptor antagonist properties of pyridoxal 5-phosphate. *Eur J Pharmacol* 259, 295–300, 1994.

114. MC Stephens, V Havlicek, K Dakshinamurti. Pyridoxine deficiency and development of the central nervous system in the rat. *J Neurochem* 18, 2407–2416, 1971.

115. K Dakshinamurti, CS Paulose, M Viswanathan, YL Siow. Neuroendocrinology of pyridoxine deficiency. *Neurosci Biobehav Rev* 12, 189–193, 1988.

116. K Dakshinamurti, SK Sharma, M Sundaram. Domoic acid induced seizure activity in rats. *Neurosci Lett* 127, 193–197, 1991.

117. K Dakshinamurti, SK Sharma, JD Geiger. Neuroprotective actions of pyridoxine. *Biochem Biophys Acta* 1647, 225–229, 2003.

118. FW Berman, KT LePage, TF Murray. Domoic acid neurotoxicity in cultured cerebellar neuron is controlled preferentially by the NMDA receptor Ca^{2+} influx pathway. *Brain Res* 924, 20–29, 2002.